AF323656

EXPERIMENTAL CHAOS

Previous Proceedings in the Series of Conferences on Experimental Chaos

	Year	Held in	Publisher	ISBN
6th	2001	Potsdam, Germany	AIP Conf. Proceedings vol. 622	0-7354-0071-7
5th	1999	Orlando, Florida	World Scientific	981-02-4561-0

Other Related Titles from AIP Conference Proceedings

665 Unsolved Problems of Noise and Fluctuations: UPoN 2002: Third International Conference on Unsolved Problems of Noise and Fluctuations in Physics, Biology, and High Technology
Edited by Sergey M. Bezrukov, May 2003, 0-7354-0127-6

661 Modeling of Complex Systems: Seventh Granada Lectures
Edited by Pedro L. Garrido and Joaquín Marro, April 2003, 0-7354-0121-7

658 Modern Challenges in Statistical Mechanics: Patterns, Noise, and the Interplay of Nonlinearity and Complexity; Pan American Studies Institute
Edited by V. M. Kenkre and K. Lindenberg, March 2003, 0-7354-0118-7

627 Computing Anticipatory Systems: CASYS 2001 - Fifth International Conference
Edited by D. M. Dubois, September 2002, 0-7354-0081-4

597 Nonequilibrium and Nonlinear Dynamics in Nuclear and Other Finite Systems: International Conference
Edited by Zhuxia Li, Ke Wu, Xizhen Wu, enguang Zhao, and Fumihiko Sakata, November 2001, 0-7354-0041-5

574 Modeling Complex Systems: Sixth Granada Lectures on Computational Physics
Edited by Pedro L. Garrido and Joaquín Marro, July 2001, 0-7354-0013-X

560 Nanoscale Linear and Nonlinear Optics: International School on Quantum Electronics
Edited by Mario Bertolotti, Charles Bowden, and Concita Sibilia, April 2001, 1-56396-993-9

553 Disordered and Complex Systems
Edited by Peter Sollich, A. C. C. Coolen, L. P. Hughston, and R. F. Streater, February 2001, 1-56396-983-1

548 Fundamental Issues of Nonlinear Laser Dynamics
Edited by Bernd Krauskopf and Daan Lenstra, December 2000, 1-56396-977-7

524 Nonlinear Acoustics at the Turn of the Millennium: ISNA 15; 15th International Symposium on Nonlinear Acoustics
Edited by Werner Lauterborn and Thomas Kurz, July 2000, 1-56396-945-9

To learn more about these titles, or the AIP Conference Proceedings Series, please visit the webpage **http://proceedings.aip.org/proceedings**

EXPERIMENTAL CHAOS

7th Experimental Chaos Conference

San Diego, California   26–29 August 2002

EDITORS

Visarath In
SPAWAR Systems Center, San Diego, CA

Ljupco Kocarev
University of California, San Diego, San Diego, CA

Thomas L. Carroll
Naval Research Laboratory, Washington, DC

Bruce J. Gluckman
George Mason University, Fairfax, VA

Stefano Boccaletti
Istituto Nazionale di Ottica Applicata, Florence, Italy

Jürgen Kurths
Universität Potsdam, Potsdam, Germany

SPONSORING ORGANIZATION
Physical Sciences Division of the USA Office
of Naval Research, Arlington, Virginia

Melville, New York, 2003
AIP CONFERENCE PROCEEDINGS ■ VOLUME 676

Editors:

Visarath In
SPAWAR Systems Center, 2363
53475 Strothe Road
San Diego, CA 92152
USA

E-mail: visarath@swapar.navy.mil

Ljupco Kocarev
Institute for Nonlinear Science
University of California, San Diego
9500 Gilman Drive
La Jolla, CA 92093
USA

E-mail: kocarev@heisenberg.ucsd.edu

Thomas L. Carroll
Code 6340
Naval Research Laboratory
Washington, DC 20375
USA

E-mail: carroll@anvil.nrl.navy.mil

Bruce J. Gluckman
Department of Physics and Astronomy·
George Mason University, Mail Stop 2A1
Fairfax, VA 22030
USA

E-mail: bgluckma@gmu.edu

Stefano Boccaletti
Istituto Nazionale di Ottica Applicata
Largo E. Fermi, 6
50125 Florence
ITALY

Jürgen Kurths
Institut für Physik
Universität Potsdam
Am Neuen Palais 10, PF 601553
14415 Potsdam
GERMANY

E-mail: jkurths@agnld.uni-potsdam.de

Cover design by Visarath In and Joseph D. Neff; graphics courtesy of Joseph D. Neff, "Chaos on a Torus from a Duffing Oscillator".

L.C. Catalog Card No. 2003108865
ISBN 0-7354-0145-4
ISSN 0094-243X
Printed in the United States of America

CONTENTS

ELECTRONICS

OPTICS

POSTER SESSION ABSTRACTS

PREFACE

Modern nonlinear dynamics began with Newton and the solution of the 1/r central potential problem. The solution of this differential equation in 3D is impressive, especially for a first stab at nonlinear dynamics theory and at differential calculus. Newton's focus was on orbits and stability, as it was with Laplace's work of developing perturbation theory to calculate short-term stability of the solar system. Poincare studied the same problem, but was inevitably drawn into the realm of instabilities, and the later developments of Kolmogorov, Arnold, and Moser delved more deeply into Hamiltonian dynamics looking for the boundaries in phase space between order and chaos. Lorenz in 1963 looked at the old problem of convection rolls (not Hamiltonian, but dissipative dynamics). He used the approach of Saltzman to derive a simplest set of interesting finite amplitude convection equations. What he found in his three coupled nonlinear equations was extreme sensitivity to initial conditions and sustained non-periodic flow. This landmark paper initially attracted little attention. A search of the citation index shows 3240 citations, but only one (a self-reference) in 1963, none in 1964, 3 in 1965, 1 in 1966, none in 1967, 1 in 1968, 2 in 1969, none in 1970, 2 in 1971 and none in 1972-1973. The citations only start to pick up in 1976, in part because of its citation in the Li and Yorke paper of 1975 on period three implies chaos. Feigenbaum's landmark 1978 paper on universal behavior in a period doubling 1D map already had 14 citations in 1980 and has 1631 citations by now. It is my belief, that part of the great interest generated by Feigenbaum was due to that quick observation by the typical theorist that first rate work could be done by studying very simple equations without investing the effort to become an expert in nonlinear differential equations. But all of the above are theoretical results. One needs experiments to make a field into a science.

On the experimental side, the Pecora and Carroll paper of 1990 on synchronized chaos in separated electrical circuits has 1191 citations, with 2 already in 1990 and 7 in 1991. The Ditto, Rauseo, and Spano experimental paper on control of chaos in a magnetoelastic ribbon, also from 1990, has 410 citations with 9 in 1991 and 19 in 1992. This is a much quicker start than Lorenz's paper, a clear sign that the field of nonlinear dynamics was growing and healthy. The synchronized chaos work has spurred development of nonlinear electronics using the chaos for a type of secure communications system. The control of chaos experiment, based on the 1990 Ott, Grebogi, and Yorke method (1477 citations) holds the promise for rapid technological developments from developing lasers that are stable at higher powers to prevention of flameout in jet engines to the design and novel operation of cardiac and epileptic pacemakers. Nonlinear dynamics has become a science.

In 1983 the Office of Naval Research (ONR) created a physics program on nonlinear dynamics with the focus on experiments. The idea was to help create and support nonlinear science and turn it into nonlinear technology. The emphasis on experiments was crucial to reach this goal. As part of this effort, the Naval Research Lab, the Naval Surface Warfare Center and ONR organized the first Experimental Chaos Conference in 1991. This seventh meeting, in 2002, was hosted by the Space and Naval Warfare Systems Center/San Diego together with the University of California at San

Diego. Speakers are chosen by the organizing committee based on who did the best experimental work over the previous two years. This, at least, leads to lively and sometimes contentious debates among the members of the organizing committee. The main idea was to have experimentalists present data and emphasize what part of their data is explained by theory, what part existing theory gets wrong, and what part theory does not yet address. Conjectures and paradoxes would be presented to challenge the community. Theorists should be drawn to this type of meeting to find out what needs to be explained. It was hoped that this mix of experimental talks with an audience with hungry theorists would be an ideal forum to create collaborations and identify fruitful directions of research. This in fact has been the case! Collaborative groups have formed and postdocs have been known to find jobs at this meeting.

The seven volumes (so far) of Experimental Chaos Conference proceedings are probably the best source available to find what has been done experimentally in the field of nonlinear dynamics. This year's proceedings follow admirably in this tradition. Electronics, optics, and fluids have been strong fields for the application of nonlinear dynamics. The trend over the years has been the growth of biophysics (especially neuroscience and cardiac dynamics), oceanography, and geophysics for the employment of the concepts and techniques of nonlinear dynamics. I predict new advances in nonlinear dynamics in computing (with chaos gates), nonlinear circuits for pattern recognition and decision making for use in autonomous robots. Enjoy this volume and come to the 2004 meeting in Firenze in Italy and let's see together what fascinating discoveries have been made and what beautiful works our colleagues have accomplished.

Michael Shlesinger
Office of Naval Research
Chief Scientist for Nonlinear Science
Arlington, Virginia
June 2003

ORGANIZERS:

Visarath In
SPAWAR Systems Center
Code 2363
San Diego, CA 92152-5001 USA
Phone: 619-553-9287
Fax: 619-553-1269
visarath@spawar.navy.mil

Ljupco Kocarev
Institute for Nonlinear Science
University of California San Diego
La Jolla, CA 92093 USA
Phone: 858-822-2011
kocarev@heisenberg.ucsd.edu
http://inls.ucsd.edu/

Stefano Boccaletti,
Istituto Nazionale di
Ottica Applicata
Largo E. Fermi, 6, I50125
Florence ITALY
Phone: +39.055.23081
Fax: +39.055.2337755
stefano@ino.it
http://www.ino.it/~stefano

Bruce Gluckman
The Krasnow Institute for
Advanced Study
Dept. of Physics and Astronomy
George Mason University
Fairfax, VA 22030 USA
Phone: 703-993-4384
bgluckma@gmu.edu

Thomas L. Carroll
Naval Research Laboratory
Code 6340
Washington, DC 20375 USA
Phone: 202-767-6242
carroll@anvil.nrl.navy.mil

Jürgen Kurths
Universitat Potsdam
Lehrstuhl Nichtlineare Dynamik
D-14415 Potsdam GERMANY
Phone: +49-331-977-1429
Fax: +49-331-977-1142
jkurths@agnld.uni-potsdam.de

SPONSOR:

Michael Shlesinger
Physical Sciences Division
Office of Naval Research 331
Arlington, VA 22217 USA
Phone: 703-696-4220
Fax: 703-696-6887
shlesim@onr.navy.mil

Experimental Chaos Conferences

ECC1   1991   Crystal City, Virginia USA
Proceedings edited by
Vohra, Spano, Shlesinger, Pecora, and Ditto

ECC2   1993   Crystal City, Virginia USA
Proceedings edited by
Ditto, Pecora, Shlesinger, Spano, and Vohra

ECC3   1995   Edinburgh, UK
Proceedings edited by
Ditto, Harrison, Lu, Pecora, Spano, and Vohra

ECC4   1997   Boca Raton, Florida USA
Proceedings edited by
Ding, Ditto, Pecora, Spano, and Vohra

ECC5   1999   Orlando, Florida USA
Proceedings edited by
Ding, Ditto, Osborne, Pecora, and Spano

ECC6   2001   Potsdam, Germany
Proceedings edited by
Boccaletti, Gluckman, Kurths, Pecora, and Spano

ECC7   2002   San Diego, California USA
Proceedings edited by
In, Kocarev, Boccaletti, Gluckman, Carroll, and Kurths

ELECTRONICS

Chaotic Circuits Based on Dependent Switched Capacitors

Toshimichi Saito* and Hidehiro Nakano*

*EECE Dept., HOSEI University, Koganei, Toyko, 184-8584 Japan (tsaito@k.hosei.ac.jp)

Abstract. This paper studies a simple chaotic circuit and a pulse-coupled networks (ab. PCN). The oscillator consists of one linear resistor, one inductor, and one dependent switched capacitor (ab. DSC). If the resistor value is negative (respectively, positive), the circuit can exhibit chaotic attractors (respectively, periodic attractors). We give the parameters condition for generation of each phenomenon. We then construct a PCN of chaotic circuits. Using a simple equivalent circuit, it is confirmed that the PCN can exhibit grouping synchronous phenomena of chaos.

INTRODUCTION

Synthesis and analysis of simple chaotic circuits have been studied with great interests. A variety of chaotic circuits and chaotic ICs are presented [1]-[11]. The chaotic circuits do not imply just realization methods of given mathematical models but important real physical systems to investigate variety of nonlinear phenomena. In fact the Van der Pol oscillator is known as the origin to study periodic attractors [12]. Also, the chaotic circuits are keys to realize interesting engineering applications including chaos-based communications systems [13] [14], improvement of power supply EMC by chaos [15] and bifurcating neural networks [16] [17].

This paper studies a simple chaotic circuit based on a dependent switched capacitor (ab. DSC) and a pulse-coupled network (ab. PCN) [18]-[22]. The DSC is a nonlinear element consisting of one capacitor and one reset switch. If the capacitor voltage reaches a threshold, the switch resets the voltage instantaneously to the base level. Using the DSC with one linear resistor and one inductor, we construct a circuit. If the resistor value is negative, the circuit can exhibit chaos characterized by one positive Lyapunov exponent [12] [23]. If the resistor value is positive, the circuit can have co-existing periodic and equilibrium attractors, and exhibits either depending on the initial states. Using plural DSC chaotic oscillators, we construct a PCN. If the PCN consists of two groups of oscillators classified by the oscillation frequencies, chaos synchronization can occur in each group but the synchronization is broken between the two groups: chaos synchronization and asynchronization can occur simultaneously. This interesting grouping phenomena are verified in the laboratory using a simple equivalent circuit. Motivations for considering such circuits are including the following. First, our DSC circuit is described by linear dynamics and nonlinear impulsive switching. It enables us to derive 1-D return map rigorously and to give the parameters condition for generation

CP676, *Experimental Chaos: 7th Experimental Chaos Conference,*
edited by V. In, L. Kocarev, T. L. Carroll, B. J. Gluckman, S. Boccaletti, and J. Kurths

of each phenomenon theoretically. Second, the impulsive switching relates deeply to integrate-and-fire type neuron models having interesting synchronization phenomena [24]-[28] but chaotic phenomena can not be observed in the existing neuron models without input. Third, our PCN is implemented easily using voltage-controlled current sources (ab. VCCSs), capacitors and controlled switches. The PCN relates deeply to engineering applications including pulse-based communications [13] [14] and image segmentations [29] [30].

THE CHAOTIC OSCILLATOR

Figure 1 shows the objective circuit where $-R$ is a linear resistor that can take both positive and negative values. If the capacitor voltage v is below the threshold V_T, the switch S is opened and the dynamics is described by Equation (1)

$$C\frac{dv}{dt} = i, \ L\frac{di}{dt} = -v + Ri, \ \text{for } v(t) < V_T. \tag{1}$$

If the voltage v increases and reaches the threshold V_T, the switch S is closed and v is reset to the base voltage E instantaneously holding i=constant.

$$(v(t^+), i(t^+)) = (E, i(t)), \ \text{for } v(t) = V_T. \tag{2}$$

Such an ideal definition of the switching is routine and is accepted widely in the switched capacitors techniques [21]. The DSC consists of the C, S and E. Let Equation (1) has complex characteristic roots $\delta\omega \pm j\omega$:

$$\delta\omega = \frac{R}{2L}, \ \omega^2 = \frac{1}{LC} - \left(\frac{R}{2L}\right)^2 > 0.$$

In this case the state vector (v, i) can vibrate below the threshold V_T. Using the dimensionless variables and parameters in (3), Equations (1) and (2) are transformed into Equation (4).

$$\tau = \omega t, \ x = \frac{v}{V_T}, \ y = \frac{1}{V_T}\left(-\delta v + \frac{1}{\omega C}i\right), \ q = \frac{E}{V_T}. \tag{3}$$

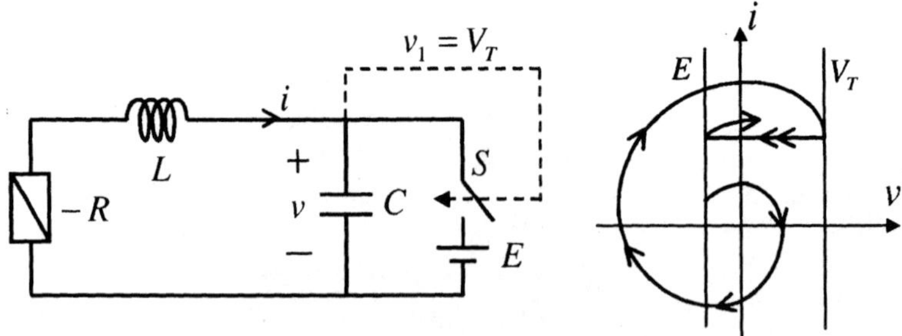

FIGURE 1.   The DSC Chaotic Oscillator

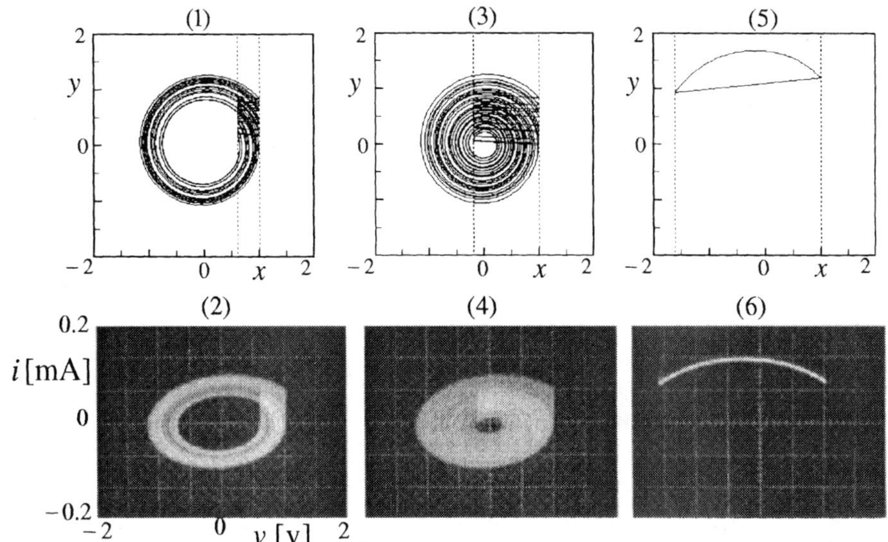

FIGURE 2. Typical attractors. $C \doteq 2.2[\text{nF}]$, $L \doteq 300[\text{mH}]$ and $V_T \doteq 1[\text{V}]$ for laboratory experiments. (1)-(2): Chaotic attractor for $(\delta, q)=(0.05, 0.6)$ and $(E \doteq 0.6[\text{V}], R \doteq 1.2[\text{k}\Omega])$. (3)-(4): Chaotic attractor for $(\delta, q)=(0.05, -0.2)$ and $(E \doteq -0.2[\text{V}], R \doteq 1.2[\text{k}\Omega])$. (5)-(6): Periodic attractor for $(\delta, q)=(-0.1, -1.6)$ and $(E \doteq -1.6[\text{V}], R \doteq 2.3[\text{k}\Omega])$. An equilibrium attractor co-exists with it.

$$\begin{bmatrix} \dot{x} \\ \dot{y} \end{bmatrix} = \begin{bmatrix} \delta & 1 \\ -1 & \delta \end{bmatrix} \begin{bmatrix} x \\ y \end{bmatrix} \text{ for } x(\tau) < 1,$$

$$(x(\tau^+), y(\tau^+)) = (q, y(\tau) + \delta(1-q)), \text{ for } x(\tau) = 1,$$

(4)

where $\dot{x} \equiv \frac{d}{d\tau}x$. This equation has two parameters: the damping δ and the base level q. The δ and q can be controlled easily by $-R$ and E, respectively. It should be noted that the generalized normal form equation can be found in Ref. [18] [19] where the $\pm R$-L subcircuit is replaced with general 1-port including one memory element. The normal form equation has three parameters and can exhibit various bifurcation phenomena [20], however, the theoretical analysis is hard. For $x < 1$, Equation (4) has the exact piecewise solution:

$$\begin{bmatrix} x(\tau) \\ y(\tau) \end{bmatrix} = e^{\delta\tau} \begin{bmatrix} \cos\tau & \sin\tau \\ -\sin\tau & \cos\tau \end{bmatrix} \begin{bmatrix} x(0) \\ y(0) \end{bmatrix} \text{ for } x(\tau) < 1,$$

(5)

where $(x(0), y(0))$ denotes an initial state vector. Figure 2 shows some typical attractors. In the numerical simulation, the attractors are calculated using this exact piecewise solution. In the laboratory experiments, the attractors are measured from a test circuit where the dependent switch is realized using a comparator, a monostable multi-vibrator and an analog switch [21]. The circuit can exhibit chaotic attractor for $\delta > 0$ and periodic attractors for $\delta < 0$. For $\delta < 0$, the origin is always the equilibrium attractor and it coexists with the periodic attractor of Figure 2 (5) and (6): the circuit exhibits either attractor depending on the initial state.

ANALYSIS OF THE CIRCUIT

The parameters condition for generation of each phenomenon is shown in Figure 3. Chaotic attractor in Figure 2 (1) is given in the subregion (I) and periodic attractor in Figure 2 (5) is given in the subregion (II). These conditions are obtained not by numerically but theoretically. In this section we explain how we obtain the conditions.

Chaotic attractors for $\delta > 0$

In this subsection, we analyze chaotic behavior for $\delta > 0$ and $1 > q > 0$ by using a 1-D return map. Analysis results for $\delta > 0$ and $0 > q$ can be found in [21]. In order to derive the 1-D return map, we define some objects for phase space of Equation (4) shown in Figure 4 left: the domain of the return map $L = \{(x,y)|x = 0, y \geq 0\}$, the threshold line $L_T = \{(x,y)|x = 1\}$, and the base line $L_q = \{(x,y)|x = q\}$. Let a point on these objects be represented by their y-coordinate. We also define a key point D on L such that a trajectory started from D touches L_T at time τ_d:

$$D = e^{-\delta \tau_d}(\sin \tau_d - \delta \cos \tau_d), \quad \tau_d = \frac{\pi}{2} + \tan^{-1} \delta. \tag{6}$$

Let us consider the trajectory started from a point y_0 on L at $\tau = 0$. If $0 < y_0 < D$, the trajectory returns to L at $\tau = 2\pi$ without reaching L_T. The return point y_1 is given by

$$y_1 = f_1(y_0) \equiv e^{2\pi\delta} y_0. \tag{7}$$

If $D \leq y_0$, the trajectory hits the threshold L_T and is reset to the base L_q. The hit point y_a and the reset point y_b are given by

$$y_a = e^{\delta \tau_a} y_0 \cos \tau_a, \quad y_b = y_a + \delta(1 - q), \tag{8}$$

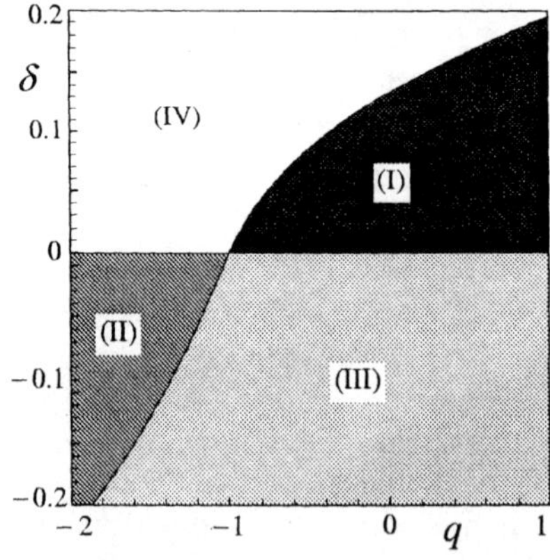

FIGURE 3.   Parameters condition for each phenomenon

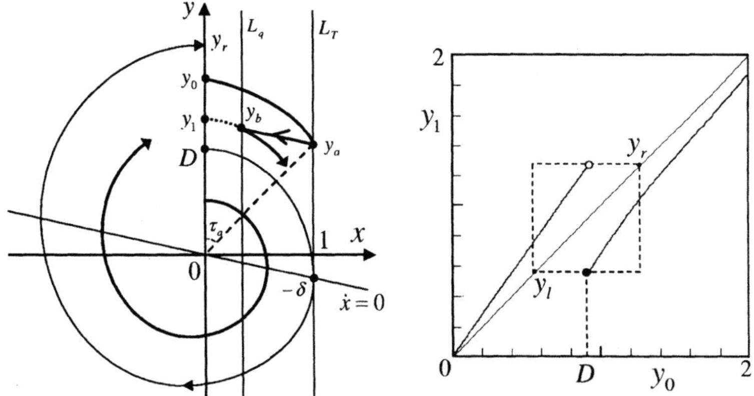

FIGURE 4. Phase space for $\delta > 0$ and $1 > q > 0$ and the 1-D return map for $(\delta, q)=(0.05, 0.6)$.

where the hit time τ_a is the root of Equation (9).

$$e^{\delta \tau_a} y_0 \sin \tau_a - 1 = 0, \quad 0 < \tau_a \le \tau_d. \tag{9}$$

Then the trajectory re-starts from y_b at time τ_a^+. Here we note existence of a point $y_1 \in L$ such that the trajectory started from it reaches y_b at time τ_b:

$$y_1 = f_2(y_0) \equiv e^{-\delta \tau_b}(-q \sin \tau_b + y_b \cos \tau_b), \quad \tau_b = \frac{\pi}{2} - \tan^{-1}\frac{y_b}{q}, \tag{10}$$

We then define the 1-D return map by Equation (11).

$$F : L \to L, \quad y_0 \mapsto y_1, \tag{11}$$

$$y_1 = F(y_0) = \begin{cases} f_1(y_0), & \text{for } 0 < y_0 < D, \\ f_2(y_0), & \text{for } D \le y_0. \end{cases}$$

Figure 4 right shows the 1-D return map corresponding to chaotic attractor in Figure 2 (1).

DEFINITION 1: If there exists some subset I in L such that $F(I) \subseteq I$, then I is said to be an invariant interval. If an invariant interval I exists and if $|DF(y)| > 1$ is satisfied for almost all $y \in I$, F on I is ergodic and has a positive Lyapunov exponent [23], where DF denotes the slope of F. In this case, F is said to generate chaos.

THEOREM 1: Let $y_l \equiv f_2(D)$, $y_r \equiv f_1(D_-)$ and let $I \equiv [y_l, y_r)$. I is an invariant interval and F generates chaos if $f_2(y_r) \le y_r$.

Proof: First we introduce explicit expressions of y_0 and y_1 for $D \le y_0$:

$$\begin{aligned} y_0 &= f(y_a) \equiv \sqrt{y_a^2 + 1}\, e^{-\delta \tau_a}, \quad \tau_a = \tfrac{\pi}{2} - \tan^{-1} y_a \\ y_1 &= g(y_a) \equiv \sqrt{y_b^2 + q^2}\, e^{-\delta \tau_b}, \quad \tau_b = \tfrac{\pi}{2} - \tan^{-1} \tfrac{y_b}{q}, \ y_b = y_a + \delta(1-q) \end{aligned} \tag{12}$$

7

Using these, we can calculate the slope of f_2

$$Df_2(y_0) = \frac{dg}{dy_a}\left(\frac{df}{dy_a}\right)^{-1} = \frac{y_b + \delta q}{y_1}e^{-2\delta\tau_b}\left(\frac{y_a + \delta}{y_0}e^{-2\delta\tau_a}\right)^{-1} = \frac{y_0}{y_1}e^{2\delta(\tau_a-\tau_b)}. \tag{13}$$

Since $\tau_a - \tau_b > 0$, we obtain

$$DF(y_0) = Df_2(y_0) > 1 \text{ if } y_1 \le y_0 \text{ and } D \le y_0. \tag{14}$$

Since $f_2(D) < D$ and f_2 is monotone, $f_2(y_r) \le y_r$ guarantees $f_2(y_0) \le y_0$ ($y_1 \le y_0$) for $y_0 \in [D, y_r)$. If there exists $y_s \in (D, y_r)$ such that $f_2(y_s) > y_s$ then there exists some fixed point such that $y_f = f_2(y_f) \in (D, y_r)$ and $0 < Df_2(y_f) < 1$. It contradicts Equation (14). Noting (14), $f_2(y_0) \le y_0$ guarantees $Df_2(y_0) > 1$. Since $Df_1(y_0) > 1$, $f_2(y_r) \le y_r$ is a sufficient condition for chaos generation. QED.

The condition $f_2(y_r) \le y_r$ is an inequality of the parameters and is satisfied in the subregion (I) for positive q in Figure 3.

Periodic and equilibrium attractors for $\delta < 0$

In this subsection, we analyze periodic behavior for $\delta < 0$ and $q < 0$. If a periodic attractor exists, it coexists with the equilibrium attractor (the origin) and the system exhibits either depending on the initial state. If $q > 0$, a trajectory started from L (positive y-axis) converges eventually to the equilibrium attractor.

In order to define the 1-D return map in a likewise manner as the previous section, we confirm some key objects in Figure 5 left: the threshold line L_T, the base line L_q and the border point D. Let us consider the trajectory started from a point y_0 on L at $\tau = 0$. If $0 < y_0 < D$, the trajectory returns to L at $\tau = 2\pi$ without reaching L_T. The return point y_1 and D are given by Equations (7) and (6). If $D \le y_0$, the trajectory hits the threshold L_T and is reset to the base L_q. Let y_a and y_b be the hit and the reset points,

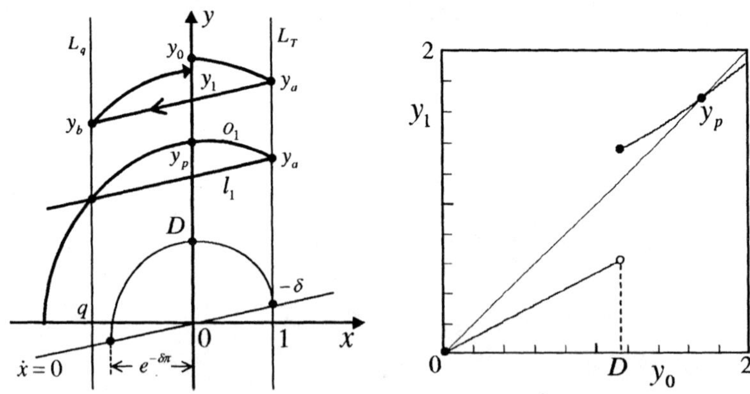

FIGURE 5. Phase space for $\delta < 0$ and the 1-D return map for $(\delta, q)=(-0.1, -1.6)$.

respectively. The points y_a and y_b are given by Equations (8) and (9). Here note that $0 < Df_1(y_0) = e^{2\pi\delta} < 1$ for $\delta < 0$. Then the trajectory re-starts from y_b at time τ_a^+ and returns to L. The return point y_1 is given by

$$y_1 = g_2(y_0) \equiv e^{\delta\tau_b}(-q\sin\tau_b + y_b\cos\tau_b), \quad \tau_b = \frac{\pi}{2} + \tan^{-1}\frac{y_b}{q}, \tag{15}$$

Since the trajectory started from L must return to L, we can define the 1-D return map

$$G : L \to L, \quad y_0 \mapsto y_1, \tag{16}$$

$$y_1 = F(y_0) = \begin{cases} f_1(y_0), & \text{for } 0 < y_0 < D, \\ g_2(y_0), & \text{for } D \le y_0. \end{cases}$$

For convenience, we introduce explicit expressions of y_0 and y_1 for $D \le y_0$:

$$\begin{aligned}
y_0 &= f(y_a) \equiv \sqrt{y_a^2 + 1}\, e^{-\delta\tau_a}, \quad \tau_a = \tfrac{\pi}{2} - \tan^{-1} y_a \\
y_1 &= h(y_a) \equiv \sqrt{y_b^2 + q^2}\, e^{\delta\tau_b}, \quad \tau_b = \tfrac{\pi}{2} + \tan^{-1}\tfrac{y_b}{q}, \quad y_b = y_a + \delta(1-q)
\end{aligned} \tag{17}$$

Figure 5 right shows a 1-D return map where two stable fixed points co-exist: The origin of G corresponds to the equilibrium attractor and y_p corresponds to the periodic attractor in Figure 2 (5).

THEOREM 2: Equation (4) has periodic attractor if $q \le -e^{-\delta\pi}$.

Proof: First we show existence of a periodic orbit. If $q = -e^{-\delta\pi}$, trajectory started from $D \in L$ becomes a periodic orbit as shown in Figure 5, where $y_a = -\delta$ and $y_b = -\delta q$. Then we assume that some $y_a > -\delta$ is given and q is unknown. For the y_a, as shown in Figure 5 left, we can define a half line l_1 from $y_a \in L_T$ with slope $-\delta$ and a trajectory o_1 started from the negative x-axis and terminated at $y_a \in L_T$ following from Equation 4. The l_1 and o_1 has unique intersection that is declared as (q, y_b) where $y_b = y_a + \delta(1-q)$. The closed curve consisting of l_1 and o_1 for $x > q$ corresponds to the periodic orbit and the q is given as a function of y_a. Here, let $y_p \in L$ be the intersection of o_1 and L. For this y_p, we obtain

$$y_p = f(y_a) = h(y_a). \tag{18}$$

Based on this relation, we can define the implicit function G of y_a and q; and can show that q is monotone decreasing for $y_a > -\delta$.

$$\ln(y_a^2 + 1) - \ln(y_b^2 + q^2) + 2\delta\left(\tan^{-1} y_a - \tan^{-1}\frac{y_b}{q}\right) - 2\delta\pi \equiv G(y_a, q),$$

$$\frac{dq}{dy_a} = -\frac{G_{y_a}}{G_q} = \frac{y_a + \delta}{q - \delta y_b}\left(\frac{y_b^2 + q^2}{y_a^2 + 1} - 1\right) < 0. \tag{19}$$

Here, $\dot{y} > 0$ at (q, y_b) guarantees $-q + \delta y_b > 0$ and $\delta < 0$ guarantees $y_b^2 + q^2 > y_a^2 + 1$. Hence the existence of a periodic orbit is guaranteed if $q \le -e^{-\delta\pi}$.

Second we show the periodic orbit is stable. Using Equation (17) in a likewise manner as Equation (13), we obtain

$$DG(y_0) = \frac{dh}{dy_a}\left(\frac{df}{dy_a}\right)^{-1} = \frac{y_0}{y_1} e^{2\delta(\tau_a + \tau_b)} \quad \text{for } y_0 \ge D \ (y_a > -\delta). \tag{20}$$

9

Substituting $y_1 = y_0 = y_p$ into Equation (20), we obtain

$$0 < DG(y_p) = e^{2\delta(\tau_a + \tau_b)} < 1. \tag{21}$$

That is, the periodic orbit is stable if $q \le -e^{-\delta\pi}$ and $y_a > -\delta$. $\hfill$ QED.

The condition $q \le -e^{-\delta\pi}$ is satisfied in the subregion (II) in Figure 3. In addition, $G(D) < D$ is satisfied for $-e^{-\delta\pi} < q < 0$. In this case, the 1-D map does not have a fixed point for $y_0 > D$. If y_f is a fixed point and there is no fixed point for $D < y_0 < y_f$, $1 < DG(y_f)$ must be satisfied. It contradicts Equation (20).

PULSE-COUPLED NETWORK

In order to construct a PCN, we introduce an equivalent circuit of the DSC chaotic oscillator. In Figure 6, the left and right VCCSs are characterized respectively by $i_2 = g_i(v_{2i} - v_{1i})$ and $i_1 = g_i v_{i2}$. Let i be a fixed integer ($i = 1$) for the independent DSC oscillator and let i be an index of oscillators for the PCN. It should be noted that advantages of such VCCS-based implementation includes 1) design simplicity based only on KCL, 2) suitability for transistor-level realization, and 3) suppression of mutual inductance for the PCNs. Let the first capacitor voltage v_{1i} be below the threshold V_T and let the two switches S_{ri} and S_{pi} be opened. In this case the circuit dynamics is described by Equation (22).

$$C_1 \frac{dv_{1i}}{dt} = g_i v_{2i}, \quad C_2 \frac{dv_{2i}}{dt} = g_i(v_{2i} - v_{1i}), \quad \text{for } v_1 < V_T. \tag{22}$$

Let the time constant $C_p R_p$ be sufficiently small and let $v_{3i} = V_L$. If the voltage v_{1i} increases and reaches the threshold V_T, the comparator closes the switch S_{pi} and the voltage v_{3i} is charged up to V_H instantaneously. The voltage v_{3i} closes the switch S_{ri} and the voltage v_{1i} is reset to E instantaneously, holding v_{2i}=constant. Then the comparator opens the switch S_{pi} and the voltage v_{3i} is reset to V_L instantaneously. We approximate these switching dynamics as the following:

$$S_{pi} = \begin{cases} \text{ON} & \text{for } v_{1i}(t) = V_T \\ \text{OFF} & \text{for } v_{1i}(t) < V_T \end{cases} \quad v_{3i}(t) = \begin{cases} V_H & \text{for } S_{pi}=\text{ON} \\ V_L & \text{for } S_{pi}=\text{OFF} \end{cases} \tag{23}$$

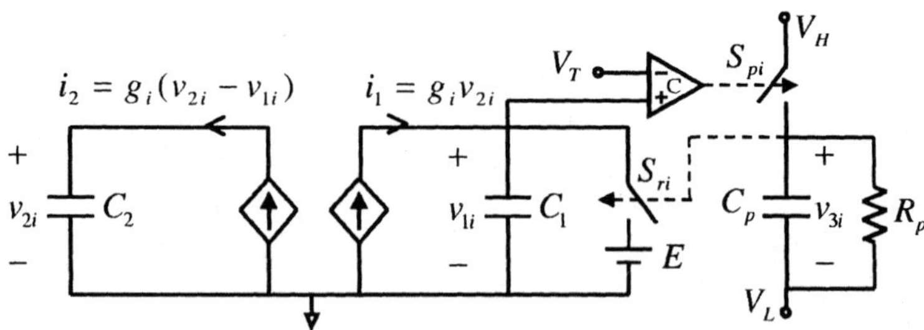

FIGURE 6. Equivalent Circuit ($C_p \doteq 120[\text{pF}]$, $R_p \doteq 5.1[\text{k}\Omega]$, $V_H \doteq -V_L \doteq 8[\text{V}]$)

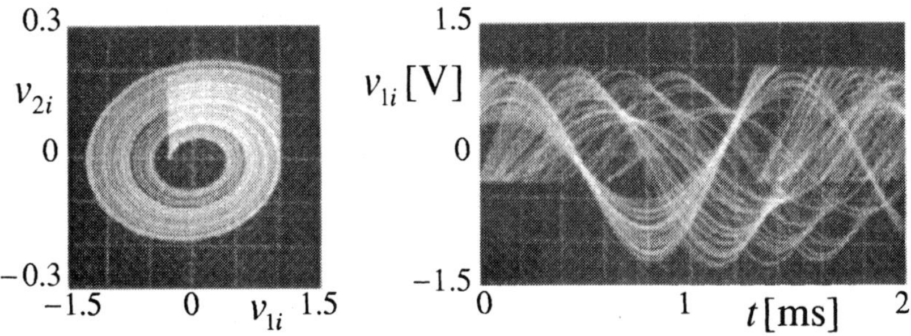

FIGURE 7. Measured chaotic attractor for $C_1 \doteq 2.2[\text{nF}]$, $C_2 \doteq 100[\text{nF}]$, $g^{-1} \doteq 15[\text{k}\Omega]$, $C_1 \doteq 2.2[\text{nF}]$, $C_2 \doteq 100[\text{nF}]$, $g^{-1} \doteq 15[\text{k}\Omega]$, $V_T \doteq 1[\text{V}]$ and $E \doteq -0.3[\text{V}]$ ($\delta \doteq 0.07$, $q \doteq -0.3$).

$$(v_{1i}(t^+), v_{2i}(t^+)) = (E, v_{2i}(t)), \quad \text{for } v_{1i}(t) = V_T. \tag{24}$$

In this approximation, v_{3i} is a pulse-train and we refer to it as an output of the DSC oscillator. The output plays an important role to the pulse-coupling in the PCN.

If Equation (22) has complex characteristic roots $\delta\omega_i \pm j\omega_i$, they are given by

$$\delta = \sqrt{\frac{C_1}{4C_2 - C_1} \frac{g_i}{|g_i|}}, \quad \omega_i^2 = \left(\frac{1}{C_1 C_2} - \frac{1}{4C_2^2}\right) g_i^2 > 0. \tag{25}$$

Using the following dimensionless variables and parameters:

$$\tau = \omega_i t, \ x = \frac{v_{1i}}{V_T}, \ y = \frac{1}{V_T}\left(-\delta v_{1i} + \frac{g_i}{\omega_i C_1} v_{2i}\right), \ q = \frac{E}{V_T}, \tag{26}$$

Equations (22) and (24) are transformed into Equations (4). This transformation guarantees that Figure 6 gives an equivalent circuit of the DSC chaotic oscillator. In order to focus on chaotic attractors, let $g_i > 0$ hereafter. Using this equivalent circuit, we have observed chaotic attractor as shown in Figure 7. There exist various implementation methods of the VCCSs and we have adopted a simple method using op-amps [22].

Using M chaotic oscillators, we construct a PCN. The dynamics is the PCN is described by Equation (27).

$$C_1 \frac{dv_{1i}}{dt} = g_i v_{2i}, \ C_2 \frac{dv_{2i}}{dt} = g_i(v_{2i} - v_{1i}), \quad \text{for } v_{3i} = V_L$$

$$(v_{1i}(t^+), v_{2i}(t^+)) = (E, v_{2i}(t)), \quad \text{for } v_{3i}(t) = V_H \tag{27}$$

$$v_{3i}(t) = \begin{cases} V_H & \text{if } (V_R < v_{1i}(t) \text{ and } v_{3N(i)}(t) = V_H) \text{ or } v_{1i}(t) = V_T \\ V_L & \text{otherwise} \end{cases}$$

where $i \in \{1, \cdots, M\}$ and $N(i)$ denotes the index of the nearest neighbor of the i-th oscillator. For example, a ladder PCN of 4 oscillators ($M = 4$) is characterized by $N(1) = 2$, $N(2) \in \{1,3\}$, $N(3) \in \{2,4\}$ and $N(4) = 3$. Note that V_R is the refractory threshold ($V_R < V_T$). As shown in Figure 8 the coupling between neighbors is realized by the pulse-train output: the voltage v_{1i} of the i-th oscillator is reset to E if the neighbor oscillator(s) outputs pulse signal ($v_{3N(i)} = V_H$) and $v_{1i} > V_R$. When either the neighbor oscillator(s) does not output pulse signal ($v_{3N(i)} = V_L$) or $v_{1i} < V_R$, the i-th oscillator behaves as an independent chaotic oscillator. As V_R approaches V_T, the probability of acceptance of the neighbor's pulse-signal is to be low and V_R may control the coupling strength. Effects of the V_R will be discussed elsewhere. In the implementation circuit, this coupling corresponding to an additional switching condition of S_p, i.e., S_p is closed if $V_R < v_{1i}$ and $v_{3N(i)} = V_H$. This switching is realized easily using comparators and analog switches [22]. Using dimensionless variables and parameters in (28), Equation (27) is transformed into Equation (29).

$$\tau = \omega_1 t, \; x_i = \frac{v_{1i}}{V_T}, \; y_i = \frac{1}{V_T}\left(-\delta v_{1i} + \frac{g_i}{\omega_i C_1} v_{2i}\right), \; z_i = \frac{v_{3i} - V_L}{V_H - V_L},$$
$$\gamma_i = \frac{\omega_i}{\omega_1} = \frac{g_i}{g_1}, \; (\gamma_1 = 1), \; q = \frac{E}{V_T}, \; a = \frac{V_R}{V_T}. \tag{28}$$

$$\begin{bmatrix} \dot{x}_i \\ \dot{y}_i \end{bmatrix} = \begin{bmatrix} \delta\gamma_i & \gamma_i \\ -\gamma_i & \delta\gamma_i \end{bmatrix} \begin{bmatrix} x_i \\ y_i \end{bmatrix} \quad \text{for } z_i = 0$$

$$(x_i(\tau^+), y_i(\tau^+)) = (q, y_i(\tau) + \delta(x_i(\tau) - q) \quad \text{for } z_i(\tau) = 1 \tag{29}$$

$$z_i(\tau) = \begin{cases} 1 & \text{if } (a < x_i(\tau) \text{ and } z_{N(i)}(\tau) = 1) \text{ or } x_i(\tau) = 1 \\ 0 & \text{otherwise,} \end{cases}$$

It should be noted that Equation (29) has $M + 2$ parameters: δ, q, a and γ_2 to γ_M. In order to execute preliminary experiments, we select γ_i as control parameters and fix other

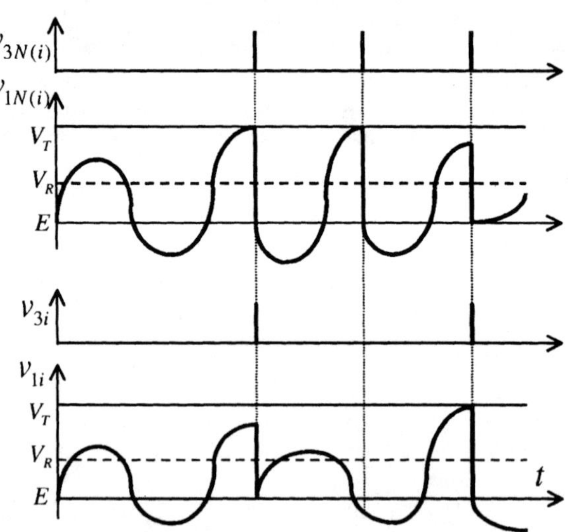

FIGURE 8.   Coupling via pulse-train

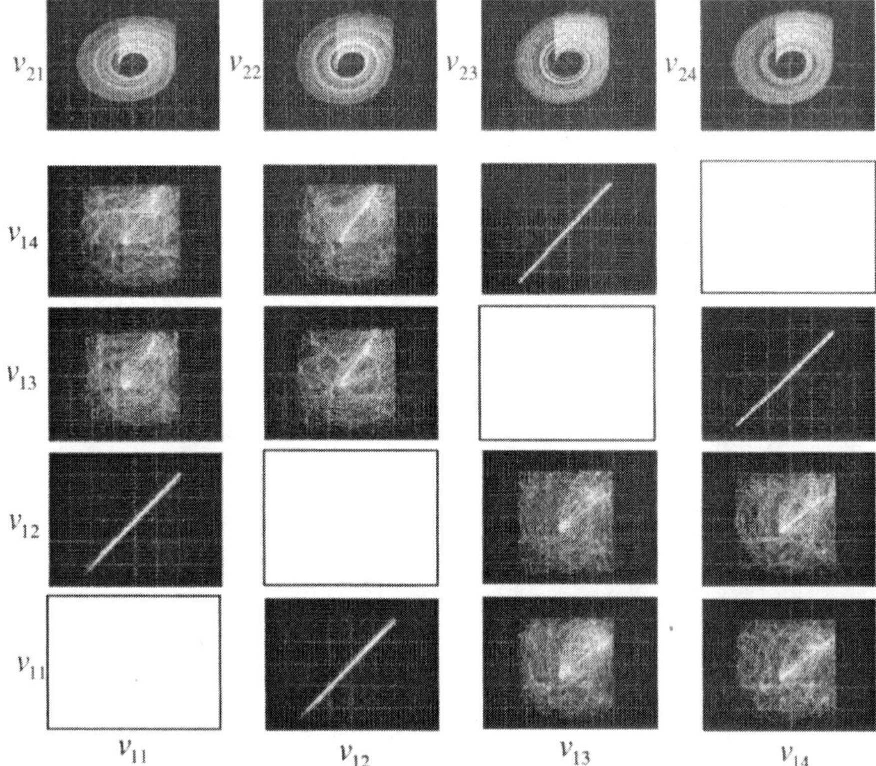

FIGURE 9. Grouping chaos synchronization ($v_{1j} : 0.5[\text{V}] / \text{div.}$ $v_{2j} : 0.1[\text{V}] / \text{div.}$) $g_1^{-1} \doteq g_2^{-1} \doteq 15[\text{k}\Omega]$, $g_3^{-1} \doteq g_4^{-1} \doteq 20[\text{k}\Omega]$, $R_p \doteq 5.1[\text{k}\Omega]$, $C_1 \doteq 2.2[\text{nF}]$, $C_2 \doteq 100[\text{nF}]$, $R_p \doteq 5.1[\text{k}\Omega]$, $V_T \doteq 1[\text{V}]$), $E \doteq -0.3[\text{V}]$), and $V_R \doteq 0.5[\text{V}]$). ($\gamma_1 \doteq \gamma_2 \doteq 1$, $\gamma_3 \doteq \gamma_4 \doteq 0.75$, $\delta \doteq 0.07$, $q \doteq -0.3$ and $a \doteq 0.5$.)

parameters by trial and error. Noting Equations (25) and (28), γ_i can be controlled independently by g_i. We have implemented a ladder PCN of 4 oscillators ($M = 4$) and have confirmed that the PCN can exhibits interesting synchronous and asynchronous phenomena. Figure 9 shows a typical example of grouping synchronization. From application viewpoints, this phenomenon relate to flexible image segmentation [22].

CONCLUSIONS

We have studied the DSC chaotic oscillator and the PCN. For the DSC chaotic oscillator, the 1D return map can be described by the exact piecewise solution and parameters condition for generation of each phenomena are elucidated theoretically. For the PCN, interesting grouping chaos synchronous phenomena are verified in the laboratory using a simple equivalent circuit. Future problems include analysis of synchronization and bifurcation phenomena in the PCN and application to information processing.

REFERENCES

1. Matsumoto,T., Chua, L. O., and Komuro, M., *IEEE Trans. Circuits and Syst.*, **32**, 798-818 (1985).
2. Newcomb, R. W., and El-Leithy, N., *Circuit Systems Signal Process*, **5**, 321-341 (1986).
3. Maggio, G. M., Feo, O. D., and Kennedy, M. P., *IEEE Trans. Circuits Syst. I*, **46**, 1118-1130 (1999).
4. Elwakil, A. S., and Kennedy, M. P., *IEEE Trans. Circuits Syst. I*, **48**, 289-307 (2001).
5. Rulkov, N. F., and Volkovskii, A. R., *IEEE Trans. Circuits Systs. I*, **47** 673-679 (2000).
6. Saito, T., *IEEE Trans. Circuits Syst.*, **32**, 320-331 (1984).
7. Saito, T., and Nakagawa, S., *Phil. Trans. R. Soc. Lond. A*, **352**, 47-57 (1995).
8. Kataoka, M., and Saito, T., *IEEE Trans. Circuits Syst. I*, **48**, 221-225 (2001).
9. Cruz, J. M., and Chua, L. O., *IEEE Trans. Circuits and Syst. II*, **40**, 614-625 (1993).
11. Restituto, M. D., Linan, M., and Vazquez, A. R., *Electron. Letters*, **32**, 795-796 (1996).
11. Arena, P., Baglio, S., Fortuna, L., and Manganaro, G., *Electron. Letters*, **31**, 250-251 (1996).
12. Guckenheimer, J., and Holmes, P., *Nonlinear oscillations, dynamical systems, and bifurcations of vector fields,* Springer-Verlag, New York, 1983, pp. 67-74.
13. Stojanovski, T., Kocarev, L., and Parlitz, U., *Phys. Review E*, **54**, 2128-2131 (1996)
14. Rulkov, N. F., Sushchik, M. M., Tsimring, L. S., and Volkovskii, A. R., *IEEE Trans. Circuits Systs. I*, **47**, 1436-1444 (2001).
15. Deane, J. H. B., and Hamill, D. C., *Electron. Letters*, **32** 1045 (1996).
16. Perez, R. and Glass, L., *Phys. Letters*, **99A**, 441-443 (1982).
17. Lee, G., and Farhat, N. H., G. Lee & N.H. Farhat, *Neural networks*, **14**, 115-131 (2001).
18. Mitsubori, K., and Saito, T., *IEEE Trans. Circuits Syst. I*, **44**, 1122-1128 (1997).
19. Mitsubori, K., and Saito, T., *IEEE Trans. Circuits Syst. I*, **47**, 105-1109 (2000).
20. Nakano, H., and Saito, T., *IEICE Trans. Fundamentals*, **E84-A**, 1293-1300 (2001).
21. Nakano, H., and Saito, T., *IEEE Trans. Neural Networks*, **13**, 92-100 (2002).
22. Nakano, H., and Saito, T., "Synchronization in a pulse-coupled network of chaotic spiking oscillators," *The Proceedings of IEEE Midwest Symposium on Circuits and Systems*, in press (2002)
23. Li, T. Y., and Yorke, J. A., *Trans. Amer. Math. Soc.*, **235**, 183-192 (1978).
24. Keener, J. P., Hoppensteadt, F. C., and Rinzel, J., *SIAM J. Appl. Math.*, **41**, 503-517 (1981).
25. Mirollo, R. E., and Strogatz, S. H., *SIAM J. Appl. Math.*, **50**, 1645-1662 (1990).
26. Catsigeras, E., and Budelli, R., *Physica D*, **56**, 235-252 (1992).
27. Izhikevich, E. M., *Neural Networks*, **14**, 883-894 (2001).
28. Izhikevich, E. M., *IEEE Trans. Neural Networks*, **10**, 508-526 (1999).
29. Hopfield, J. J., and Herz, A. V. M., *Proc. Natl. Acad. Sci.*, **92**, 6655-6662 (1995).
30. Campbell, S. R., Wang, D., and Jayaprakash, C., *Neural computation*, **11**, 1595-1619 (1999).

SYNCHRONIZATION OF OSCILLATING SYSTEMS FOR MICROWAVE ANTENNAS AND RF ELECTRONICS

Robert A. York, Paolo F. Maccarini, and James Buckwalter

University of California- Santa Barbara, Santa Barbara, CA, 93106

Abstract. During the past decade we have demonstrated that coupled nonlinear systems (oscillator arrays) can offer simple methods for phase control in microwave antenna arrays, and hence provides alternatives to conventional electronic beam scanning capability. Numerous experiments have been carried out at microwave frequencies to verify the analysis, and the experimental work proved valuable in guiding parallel theoretical efforts and demonstrating the advantages and limitations of such techniques for practical systems. During the course of this work, models have been developed and refined and used to explore new dynamical phenomena, such as "mode-locked" quasi-periodic states for pulse generation, phase noise reduction in oscillator array systems (verified experimentally at X-band), and other practical design issues. More recently, efforts have focused on the potential for coupled phase-locked-loop systems that promise more robust locking bandwidths and phase-control. We present an overview of our experimental efforts in exploiting the synchronization of oscillating systems for microwave antennas and RF electronics.

I. INTRODUCTION

Injection-locking and Phase-Locked-Loop techniques have been used to achieve synchronous operation of a number of integrated antenna oscillator elements. In addition to achieving phase coherence for power combining purposes, it has been found that such techniques also allow for the manipulation of the phase distribution without additional phase-shifting circuitry, suggesting a potential for low-cost beam scanning systems.

We review recent progress in microwave phased arrays exploiting nonlinear synchronization phenomena. Two main types of arrays are considered: coupled oscillator arrays (COAs) and coupled phase-locked loop arrays (CPLLAs). COAs have been most widely explored but are limited by a small locking bandwidth, amplitude fluctuations, and practical difficulties of creating matched oscillators. CPLLAs offer larger locking range and amplitude-independent phase relationships. Both approaches can be configured for phase-shifterless beam scanning and noise reduction. Array modeling and consequences of unit cell design and coupling schemes at microwave frequencies are discussed. In COAs, studies using simple models were able to predict observed chaotic patterns and the transient phase evolution. Recently, in CPLLAs the addition of a loop time delay enhanced the model of the single PLL far enough to predict the lower and upper boundaries for the loop gain. Additionally, the length of

CP676, *Experimental Chaos: 7th Experimental Chaos Conference*,
edited by V. In, L. Kocarev, T. L. Carroll, B. J. Gluckman, S. Boccaletti, and J. Kurths
© 2003 American Institute of Physics 0-7354-0145-4/03/$20.00

the coupling line together with the sign of the IF loop gain was proved to be an important factor in the transient and the steady-state phase distribution. Both types of arrays display interesting properties that future research and development may exploit for practical applications.

II.   COUPLED OSCILLATOR ARRAYS

The single cell used in coupled oscillator analysis is composed of an amplitude dependent negative conductance G_D, an LC resonant tank, a resistive load G_L, a noise admittance and an injection source, as in fig. 1. Usually in practice the variable capacitor controls of the oscillation frequency of the VCO. This simple parallel model is commonly used to derive the COAs' properties [1]-[4]. An equivalent series model can also be used, for which current and voltages can be interchanged to obtain analogous equations [5].   Refinements to this model may be necessary to explain certain observed behavior [6] at the expense of analytical convenience.

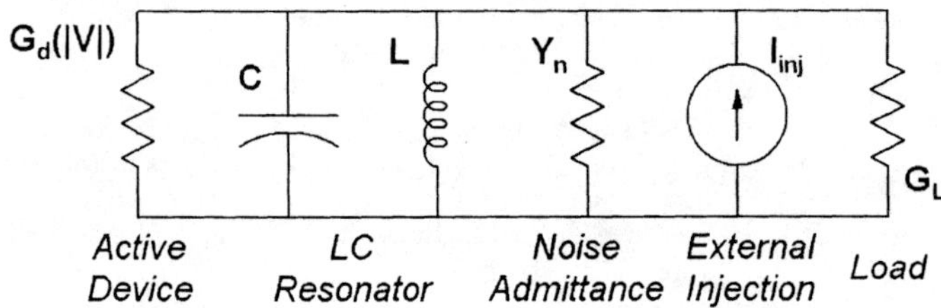

FIGURE 1. Modified VDP parallel model for COA analysis

Using basic circuit theory and with the assumptions of small phase and amplitude variations compared to the free running frequencies, the characteristic equations for the injection phenomenon of the single cell are:

$$\frac{dA}{dt} = \mu \frac{\omega_o}{2Q} A\left(1 - \frac{A^2}{\alpha^2}\right) + A_{inj}\frac{\omega_o}{2Q}\cos(\psi - \phi)$$

$$\frac{d\phi}{dt} = \omega_0 - \omega_{inj} + \frac{A_{inj}}{A}\frac{\omega_o}{2Q}\sin(\psi - \phi)$$

(1)

where: A, A_{inj}, ϕ and ψ are the amplitudes and phases of the oscillator and the injection source respectively; ω_0 and ω_{inj} are the oscillator's free-running and injection source's frequencies. α, μ and Q are the free-running amplitude, the non-linear factor (from the device admittance) and the quality factor of the oscillator.

The phase equation in (1) is referred to as Adler's equation [7] in the microwave literature, and predicts an interesting result: the phase difference between the injected signal and the oscillator is a function of the relative detunings, according to the equation:

$$\psi - \theta = \sin^{-1}\left(\frac{\omega_{inj} - \omega_0}{\Delta\omega_l}\right) \qquad \Delta\omega_l = \frac{\omega_0}{2Q}\frac{A_{inj}}{A}$$

(2)

where $\Delta\omega_l$ is defined as the "locking range". This shows that the injection-locking process can be exploited as a microwave phase-shifter, a result of significant practical interest.

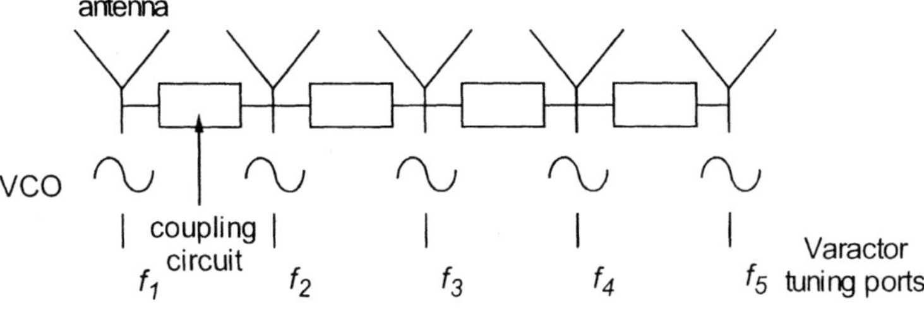

FIGURE 2. Nearest-neighbor network using transmission lines for high frequency coupling of COAs.

Several unit cells can be embedded in a coupling network to achieve the synchronization of an array of oscillators. Usually at high frequencies the nearest-neighbor coupling is most convenient to realize with the use of transmission lines that have broadband properties [4]. In this case the system is governed by the equations, for $i = 1...N$:

$$\frac{dA_i}{dt} = \frac{\mu\omega_i}{2Q} A_i (\alpha_i^2 - A_i^2) - \frac{\omega_i}{2Q} \sum_{j=1}^{N} \varepsilon_{ij} A_j \cos(\Phi_{ij} + \phi_i - \phi_j)$$

$$\frac{d\phi_i}{dt} = \omega_i - \frac{\omega_i}{2Q} \sum_{\substack{j=i-1 \\ j \neq i}}^{j=i+1} \varepsilon_{ij} \frac{A_j}{A_i} \sin(\Phi_{ij} + \phi_i - \phi_j), \quad i = 1...N \tag{3}$$

where ε_{ij} and Φ_{ij} are the coupling strengths and phases between neighbors, respectively.

Of interest is the influence of the natural frequency distribution and of the coupling strength and phase on the array operation. Several behaviors can occur as function of the coupling strength. When there is no coupling $\varepsilon_{ij} = 0$ the single cells oscillate at their free-running frequencies ω_i with their free-running amplitudes α_i. When the coupling is weak $\varepsilon_{ij} \ll 1$, we can assume [5] that only the phase dynamics will be relevant and approximately described by:

$$\frac{d\phi_i}{dt} \approx \omega_i - \frac{\omega_i}{2Q} \sum_{\substack{j=i-1 \\ j \neq i}}^{j=i+1} \varepsilon_{ij} \frac{\alpha_j}{\alpha_i} \sin(\Phi_{ij} + \phi_i - \phi_j), \quad i = 1...N \tag{4}$$

In steady state, assuming identical ε_{ij}, Φ_{ij} and α_i, from (4) it can be shown [8] that a constant stable phase distribution along the array can be easily achieved when the free-running frequencies are within the locking range $\Delta\omega_l = \varepsilon\omega_0 / 2Q$.

The coupling phase Φ affects the lock frequency and the stable phase range. For example, if the coupling phase is an even multiple of π, the free-running frequencies of the inner elements are the same and the ones of the end elements are diametrically shifted within the locking range, the phase difference ranges between $\pm 90°$, giving the possibility of a phase-shifterless broadside beam scanning. Conversely, if the coupling phase is an odd multiple of π, the phase difference will range from 0° to 180°, giving an endfire radiating array. It has been found [9] that this property depends on the model most appropriate to describe the unit cell. In the case of a series model, the relation between Φ and the beam type are reversed.

These results have being recently confirmed by a different theoretical approach, based on the transformation of the set of discrete equations into a single, approximate, continuum partial differential equation [10],[11]:

$$\frac{\partial^2 \psi}{\partial x^2} - V\,\psi - \frac{\partial \psi}{\partial \tau} = \frac{\partial \omega_{inj}}{\partial \tau} \tag{5}$$

where V describes the free-running frequency distribution, t is the time multiplied by the locking range $t\Delta\omega_l$ and $\psi = \omega - \omega_{inj}$. This equation can be solved by linear methods, and provided insight on the transient evolution of the phase (Figure 3) and thus the radiation pattern (Figure 4) for edge detuning. The array has a setting time proportional to the square of the array length inversely proportional to the locking range. Thus a larger array needs a larger locking range in order to work properly.

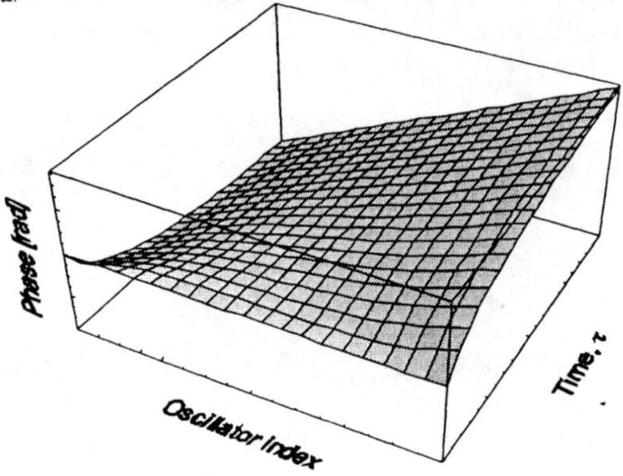

FIGURE 3. The phase evolution of a coupled oscillator array under step detuning of the end elements.

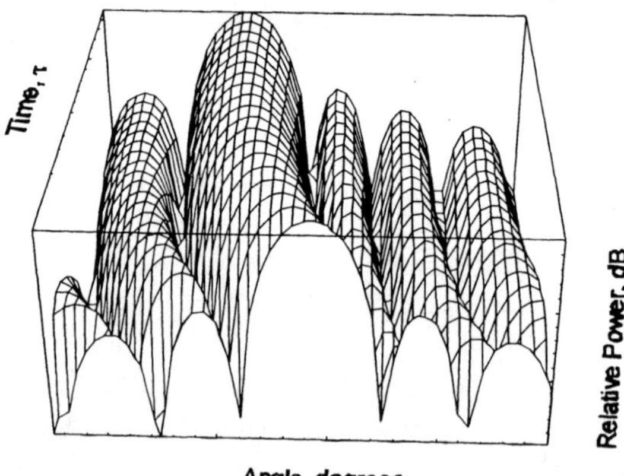

FIGURE 4. The dynamics of beam steering via step detuning of the end elements in a coupled oscillator array.

If we consider the COAs free-running distribution spread outside the limits of the locking range, more complex behavior will occur as a function of the coupling strength. As the coupling parameter reduces its amplitude, the systems goes continuously from phase-locked to mode-locked to quasi-periodic and finally to chaos below a critical coupling strength, as shown in Figure 5. A more detailed analysis of such dependence can be found in [12].

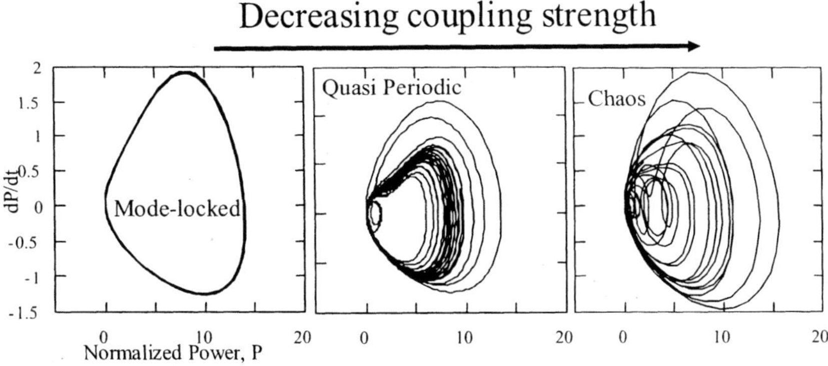

FIGURE 5. Simulated phase plane of coupled oscillator array as a function of decreasing coupling strength.

To obtain the "mode-locked" operation, the natural frequencies must be initially adjusted for approximately equal separation in frequency. The system can lock in this state, but this is usually a sensitive and difficult task [13]. Additionally, increasing the number of sources reduces the pulse width, but also makes the system more unstable and sensitive to device variations [14]. The coupling phase also plays a significant role as the different outputs adds up to form the pulse with a phase determined by the coupling [15].

The noise properties of COAs can be derived from the model presented in Figure 1. It can be shown that in all injection-locked arrays the near-carrier noise properties are governed primarily by the reference master signal, even if the oscillators themselves are quite noisy. The phase noise of free-running arrays is also shown to decrease as $1/N$, where N is the number of oscillators in the system. This has been confirmed experimentally. As shown in Figure 6 the array tracks external injection signals near carrier and returns to free-running noise far from carrier.

In practical COAs, the reduced locking range and the inevitable parasitic effects are often limiting factors in modulation performance and scan control. Moreover sidelobes increase caused by amplitude variations correlated to phase changes has been often observed in practical systems. Coupled amplitude-phase dynamics and inappropriate unit cell models are most probably the sources of such problems. To overcome these issues while trying to maintain the interesting feature of low-cost beam scanning and phase noise, recent research focused on coupled phase-locked loop arrays, first proposed by Martinez and Compton [18].

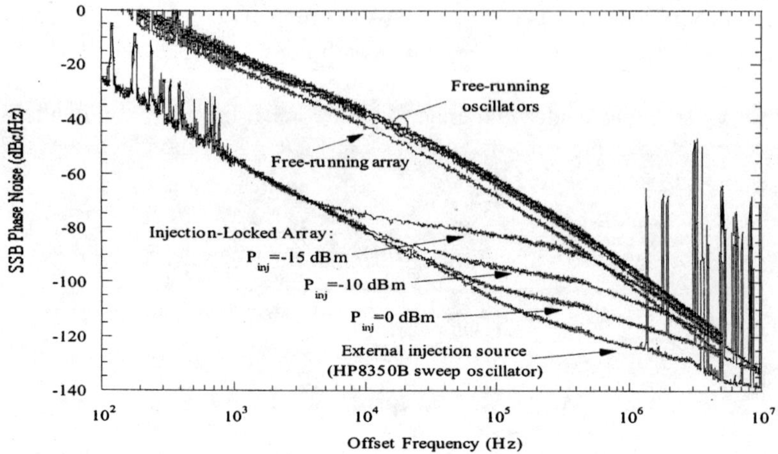

FIGURE 6. Measured phase noise for five free-running oscillators, free-running array and injection-locked array at different injection power levels.

II.   COUPLED PHASE-LOCKED LOOP ARRAYS

A simple PLL circuit is shown in fig. 7. Like injection-locked oscillators, a PLL also has the property that the phase of the oscillator can be changed relative to a reference input RF signal by adjusting a DC Offset added in the feedback loop as shown. In steady state the phase difference behaves as in the phase injection phenomenon, but shifted of 90° (Figure 7). This occurs because the the center of the stability range occurs when the two input signals to the phase-detector are in quadrature. This is solved by addition of a $\pi/2$ transmission line at the input of the phase detector.

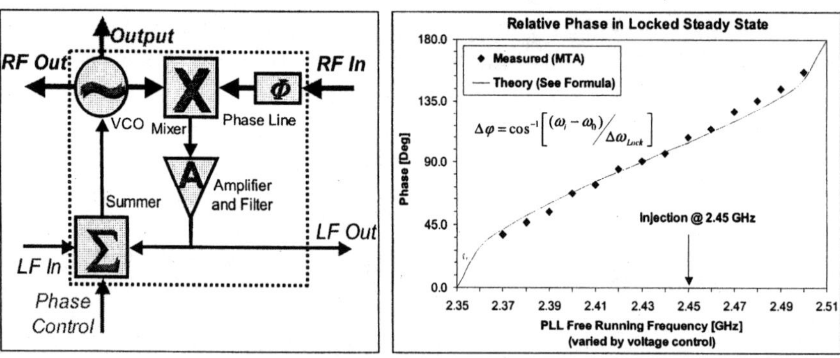

FIGURE 7. Schematics of the single phase-locked loop used in CPLLAs with free-running control, IF/RF inputs and outputs for nearest-neighbors coupling. Typical relative phase vs natural frequency measurement of a 2.45 GHz PLL.

The phase dynamics of the PLL unit cell are given by

$$\frac{d\phi}{dt} = \omega_0 - \omega_{inj} + \alpha K_v K_p \cos(\psi - \phi) \tag{6}$$

where $G = \alpha K_v K_p$ is the loop gain. Without considering filters and delays in the feedback loop, we can see from the resulting characteristic equation (6) that there is no fluctuating amplitude involved in the phase evolution. In addition the locking range is determined by the loop gain, a parameter easily controlled. Finally it is known that PLLs have lower phase noise than their open loop oscillators and the feedback loop reduces the sensibility to component tolerances.

These observations suggested that CPLLAs were attractive for beam scanning with large bandwidth modulation. The coupling scheme in Figure 8 ensures the same phase dynamics as COAs if the loops have no delay, the filters are not present and the phase detector has a sinusoidal response to phase differences.

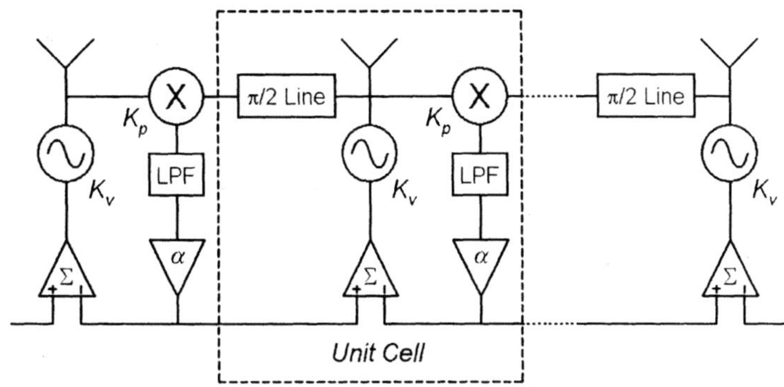

FIGURE 8. Proposed schematics for coupled PLL arrays.

When a filter loop is taken into account, higher order derivatives show up in the dynamic equation of the array. Understanding the behavior of two coupled loops helps us understand how to build larger arrays. In the case of two coupled PLLs the phase equation becomes [19]:

$$\tau_p \Delta \ddot{\phi} + (1 + \tau_z G \sin \Delta \phi) \Delta \dot{\phi} - G \cos \Delta \phi = \Delta \omega \tag{7}$$

where $\Delta \phi$ is the phase difference, $\Delta \omega$ is the frequency detuning, G is the loop gain and τ_p and τ_z are the filter zero and pole constants.

From (7) we can solve for the "hold-in" range, Ω_h, the range of frequencys within which the oscillators remain locked. This is

$$\Omega_h = 2G \tag{8}$$

This (7) predicts that the hold-in or locking range can be increased without limit by increasing the loop gain. However, in real systems increasing the gain eventually brings the system to unlock. To be able to account for this phenomenon, a delay must be introduced in the feedback loop. In this case, the solution of (7) presents a bifurcation, and a gain increase after the bifurcation causes also an increase of the acquisition time (Figure 9). Further increase of the gain eventually leads to unstable negative solutions. Thus we can now define a range for the loop gain from the optimal gain to the critical gain. This range is strongly dependent from the delay value. In a well designed and fully integrated PLL, the effect of this delay can be made negligible. In a discrete PLL, it limits the max gain loop and thus the locking range.

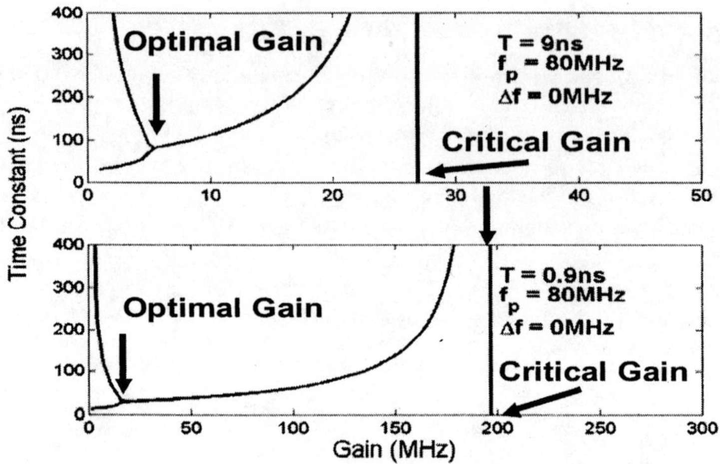

FIGURE 9. Optimal gain and critical gain limit the value of the loop gain, when introducing a delay and a filter in the PLL feedback.

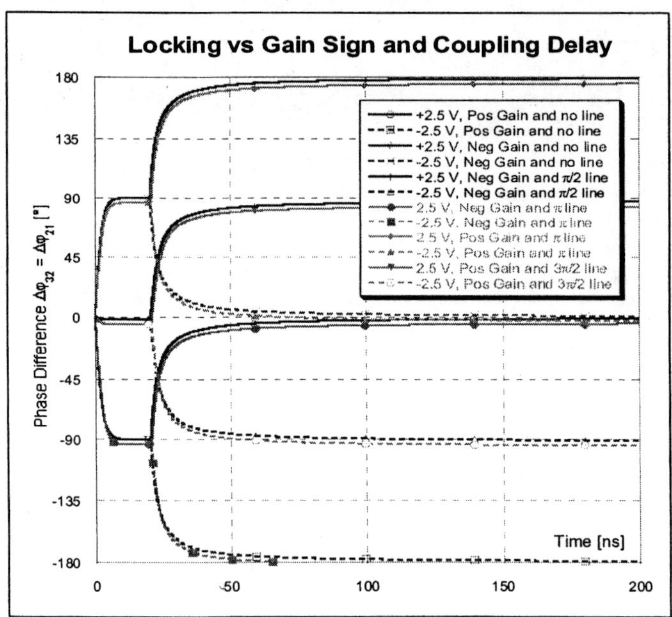

FIGURE 10. Influence of loop gain sign and coupling line length on the steady state phase difference along a 3-element CPLLA.

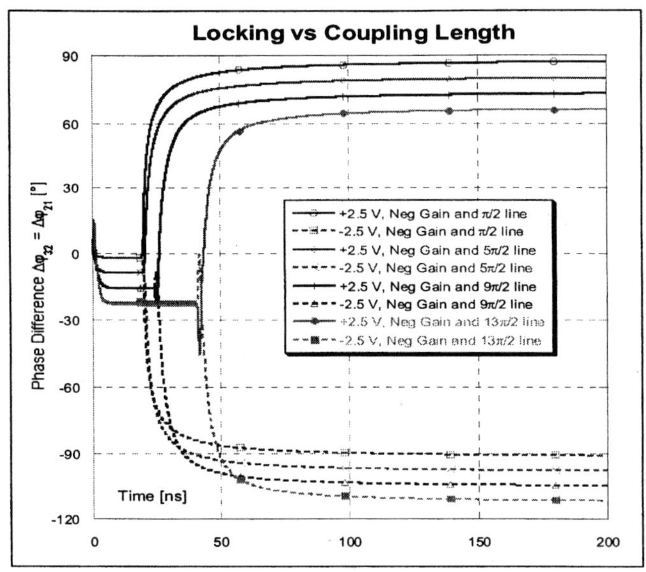

FIGURE 11. Influence of the coupling line length on the locking transient time.

The nearest-neighbor coupling scheme is experimentally convenient and allows for the implementation of COA-like phase dynamics. As with COAs, the coupling phase plays an important role, along with the sign of the loop gain, in determining where the 180° phase difference range will be centered. For example, to obtain a broadside beam, the PLLs must have negative loops and $\pi/2$ coupling lines or positive loops and $3\pi/2$ coupling lines (fig. 10). Furthermore, as intuition suggests, a longer coupling line increases the delay associated with the phase information and thus slows the phase-locking process along the array. The CPLLAs with the delay line are also intrinsically asymmetric, and this creates a phase asymmetry as shown in Figure 11.

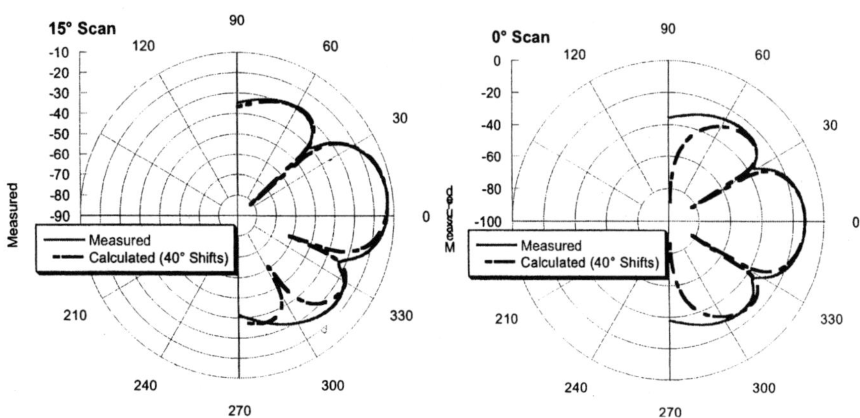

FIGURE 12. Beam scanning by edge detuning of a 5-element 2.45 GHz CPLLA

A five element CPLLA at 2.45GHz was build and the beam scanning ability by edge detuning has being experimentally verified, as shown in Figure 12

In conclusion the recent studies on CPLLAs have shown some promise and limitations. The corrections applied to the models improved our understanding of these systems, and practically they are more reliable and predictable than COAs.

ACKNOWLEDGMENTS

The authors wish to acknowledge ONR support through grant #N0014-00-1-0015. We appreciate the support and encouragement of Dr. Brian Meadows at SPAWAR, Dr. Ted Heath at Georgia Tech Research Institute, and Dr. Ronald Pogorzelski at JPL.

REFERENCES

1. B. van der Pol, "The Nonlinear Theory of Electric Oscillations", Proc. IRE, vol. 22, no. 9, pp.1051-1086, Sept. 1934.
2. J.J. Lynch, H.C. Chang, and R.A. York, "Coupled-Oscillator Arrays and Scanning Techniques" in *Active and Quasi-Optical Arrays for Solid-State Power Combining*, R.York and Z. Popović, eds., Chapter 4, Wiley: New York, 1997.
3 R.A. York, "Phase-Locking Dynamics in Active Integrated Antenna Arrays" in *Analysis and Design of Integrated-Circuit Antenna Modules*, P.S. Hall and K.C. Gupta, eds., Wiley: New York, 1999.
4. R.A. York, P. Liao and J.J. Lynch, "Oscillator Array Dynamics with Broadband N-Port Coupling Network", *IEEE Trans. Microwave Theory and Tech.*, vol. MTT-42, pp. 2040-2045, November 1994.
5. R.A. York, "Nonlinear Analysis of Phase Relationships in Quasi-Optical Oscillator Arrays", *IEEE Trans. Microwave Theory and Tech.*, vol. MTT-41, pp. 1799-1809, October 1993.
6. J. Dixon, E. Bradley and Z.B. Popović, "Nonlinear Time-Domain Analysis of Injection-Locked Microwave MESFET Oscillators", *IEEE Trans. Microwave Theory and Tech.*, vol. MTT-45, pp. 1050-1057, July 1997.
7. R. Adler, "A Study of Locking Phenomena in Oscillators", *Proc. IRE*, vol. 34, pp. 351-357, June 1946; also reprinted in *Proc. IEEE*, vol. 61, pp. 1380-1385, October 1973.
8. P. Liao, and R.A. York, "A New Phase-Shifterless Beam Scanning Technique Using Arrays of Coupled Oscillators", *IEEE Trans. Microwave Theory and Tech.*, vol. MTT-41, pp. 1810-1815, October 1993.
9. H.-C. Chang, E.S. Shapiro and R.A. York, "Influence of the oscillator equivalent circuit on the stable modes of parallel-coupled oscillators", *IEEE Trans. Microwave Theory and Tech.*, vol. MTT-45, pp. 1232-1239, August 1997.
10. R.J. Pogorzelski, P.F. Maccarini and R.A. York, "A Continuum Model of the Dynamics of Coupled Oscillator Arrays for Phase-Shifterless Beam Scanning", *IEEE Trans. Microwave Theory and Tech.*, vol. MTT-47, pp. 463-470, April 1999.
11. R.J. Pogorzelski, P.F. Maccarini and R.A. York, "Continuum Modeling of the Dynamics of Externally Injection-Locked Coupled Oscillator Arrays", *IEEE Trans. Microwave Theory and Tech.*, vol. MTT-47, pp. 471-478, April 1999.
12. R.J. Ram, R. Sporer, H.-R. Blank and R.A. York, "Chaotic Dynamics in Coupled Microwave Oscillators", *IEEE Trans. Microwave Theory and Tech.*, vol. MTT-48, pp. 1909-1917, November 2000.
13. R.A. York and R. C. Compton, "Mode-Locked Oscillator Arrays", *IEEE Microwave and Guided Wave Lett.*, vol. 1, no. 8, pp. 215-218, August 1991.
14. J.J. Lynch and R.A. York, "An Analysis of Mode Locked Arrays of Automatic Level Control Oscillators", *IEEE Trans. on Circuits and Systems-1 Fund. Theory and Appl.*, vol. 41, no. 12, pp. 859-865, December 1994.

15. R.A. York and R. C. Compton, "Automatic Beam Scanning in Mode-Locked Oscillator Arrays", *IEEE Antennas and Propagat. Int. Symp. Dig.*, pp. 629-632, Int.. Symp. 1992.

16. H.-C. Chang, X. Cao, M. J. Vaughan, U. K. Mishra, and R. A. York, "Phase Noise in Externally Injection-Locked Oscillator Arrays", *IEEE Trans. Microwave Theory and Tech.*, vol. MTT-45, pp. 2035-2042, November 1997.

17. H.-C. Chang, X. Cao, U. K. Mishra, R. A. York, "Phase Noise in Coupled Oscillators: Theory and Experiment", *IEEE Trans. Microwave Theory and Tech.*, vol. MTT-45, pp. 604-615, May 1997.

18. R.D. Martinez and R.C. Compton, "Electronic Beamsteering of Active Arrays with Phase-Locked Loops", *IEEE Microwave and Guided Lett.*, vol. 4, no. 6, pp. 166-168, June 1994.

19. James Buckwalter, Ted Heath and R. A. York, "Synchronization Design of a Coupled Phase-Locked Loop", submitted to *IEEE Trans. Microwave Theory and Tech.*

20. James Buckwalter and R. A. York, "Time Delay Considerations in High Frequency Phase-Locked Loops", submitted to *IEEE Trans. Microwave Theory and Tech.*.

Delay Line Hyper-Chaotic Oscillators in a Ring

Arūnas Tamaševičius, Gytis Mykolaitis, Antanas Čenys,
and Skaidra Bumelienė

Semiconductor Physics Institute, A. Goštauto 11, Vilnius, LT-2600, Lithuania

Abstract. We suggest to couple two or more delay line oscillators in a ring. The microwave BFG520 type transistors (f_T=9 GHz) have been used in the experiments. The power spectrum in the very high frequency range (30 to 400 MHz) is essentially smoother than that of a single oscillator. The oscillations are characterized by two-dimensional Poincaré sections.

INTRODUCTION

Electronic oscillators with delay line have attracted much attention [1-5] due to their rich dynamics, including high-dimensional and hyper-chaotic behavior [1,3,5]. For example, the analog circuit described by Namajūnas et al. [3] has been used to investigate various dynamical phenomena in the Mackey-Glass (MG) type systems, namely control [6-8] and synchronization [9] of hyper-chaos. The experimental prototypes of the MG system exhibit broadband continuous power spectra ranging from low kilohertz [1-3,6-9] to high megahertz [4-5] frequencies.

Regarding possible applications of chaotic oscillators to communications Perez and Cerdeira have shown that simple low-dimensional systems like the Lorenz system characterized by only one positive Lyapunov exponent (LE) are rather easily "breakable" because of the one-dimensional form of the Poincaré sections [10]. Their finding provoked the burst in the investigation of hyper-chaotic systems characterized by multiple positive LE as well as the development of the methods for synchronization of such systems [11-14]. The MG type oscillators having large number of positive LE [5,15] on the one hand, and admitting synchronization via a single variable [9,14] on the other hand, seem to be a good choice. However, Bünner et al. have demonstrated[1] that the time-delay systems can be identified from time series analysis [16,17]. The Bünner's method allows one to determine the delay time τ involved in the system. Moreover, the nonlinear function can be also obtained from the time series $x(t)$. Indeed, the MG type system is given by a scalar equation $dx/dt = -x + F[x(t-\tau)]$. Let us take the local maxima of the time series $x(t)$, i.e. the values $x(t_i)$ at dx/dt=0. Then it follows from the above equation, that $x(t_i)=F[x(t_i-\tau)]$. So, if one plots the Poincaré section, $[x(t_i)$ *vs.* $x(t_i-\tau)]$, where the τ is determined either empirically or by the Bünner's method, the nonlinear function F is readily recovered. In the present paper we suggest to couple simple oscillators in a ring to generate more complicated signals.

---

[1] The MG type oscillator described by Namajūnas et al. [3] has been employed in their experiments.

CP676, *Experimental Chaos: 7th Experimental Chaos Conference,*
edited by V. In, L. Kocarev, T. L. Carroll, B. J. Gluckman, S. Boccaletti, and J. Kurths
© 2003 American Institute of Physics 0-7354-0145-4/03/$20.00

CIRCUITS AND EXPERIMENTAL RESULTS

The circuit diagram of a delay line based oscillator is shown in Fig. 1. The oscillator is similar to that described previously [5]. Meanwhile several improvements have been made: (1) a buffer (Q1-stage) is inserted between the delay line DL and the nonlinear unit[2] (Q2-stage) for both, to have better matching conditions for the DL and to provide low impedance source for the nonlinear unit. In addition, the buffer ensures low influence of the load RL, (2) in the all Q1-Q4-stages large resistors R4, R9, R14, and R18 are inserted in the emitter loops to stabilize the dc currents of the transistors. Meanwhile the ac gains of the Q2-Q4-stages depend mainly on the small resistors R10, R15, and R19 (actually on R9‖R10, R14‖R15, and R18‖R19). In contrast to the circuit in [5] the Q4-stage is loaded directly with the DL, the other end of which is loaded with a matching resistor R1. This solution enables one to increase the gain of the Q4-stage by a factor of 2 or to have the same gain but with a twice larger resistance of R18‖R19. The larger values of R18‖R19 stabilize the gain of the stage.

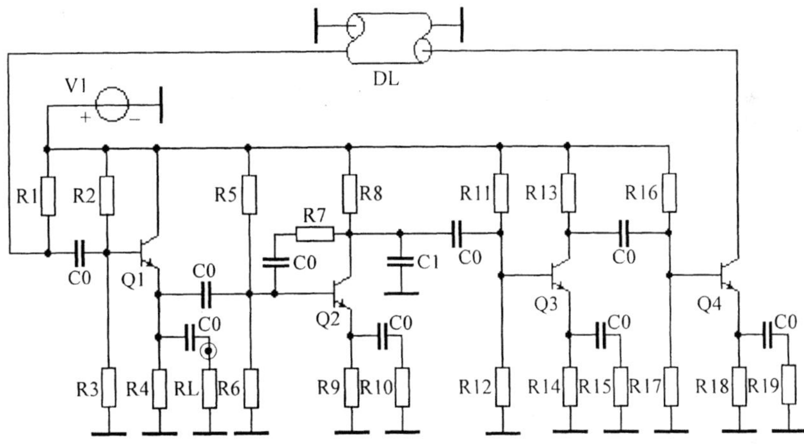

FIGURE 1. Circuit diagram of the delay line oscillator.

In a specific oscillator the circuit parameters are the following: $R_1=R_7=R_8=R_{13}=51\ \Omega$, $R_2=R_{11}=R_{16}=6.8\ \text{k}\Omega$, $R_3=R_6=R_{12}=R_{17}=2.7\ \text{k}\Omega$, $R_4=R_9=R_{14}=R_{18}=100\ \Omega$, $R_5=18\ \text{k}\Omega$, $R_{10}=11\ \Omega$, $R_{15}=R_{19}=33\ \Omega$, $C_0=47\ \text{nF}$, $C_1=22\ \text{pF}$, $T_{DL}\approx 3$ ns (RG58U type coaxial cable of length 60 cm), $V_1=7$ to 9 V. The transistors Q1 to Q4 are the BFG520 *n-p-n* type silicon devices ($f_T=9$ GHz). The RC filter is composed of R7, R8 and C1, the time constant $RC_1\approx 0.55$ ns ($R=R_7\|R_8$). The dimensionless delay $\tau=T_{DL}/RC_1\approx 5.5$.

For the further consideration it is convenient to present the oscillator in Fig. 1 and various coupled versions by the block diagrams (Fig. 2). Since a single oscillator has a ring structure (Fig. 2a) it is a natural way to couple two or more oscillators also in a ring (Fig. 2b). We note, however, that in order to have the oscillations of the coupled system in the same frequency range as the single oscillator it is necessary to ensure that the sum of the time delays of each line, T_{DL1} and T_{DL2} is equal to the time delay

---

[2] The Q2-stage operates nonlinearly due to its low dc bias (1.5 to 3 mA).

T_{DL}. Again due to the ring structure of the coupled system the two short delay lines DL1 and DL2 in Fig. 2b can be replaced with a single long line DL (Fig. 2c). Also the positions of some units can be interchanged in the ring (Fig. 2d). However in any version it is important to keep at least two RC filters. The oscillator with only one RC filter (Fig. 2e) structurally is equivalent to an oscillator in Fig. 2a. It is described by a scalar differential equation. Thus it can be identified by the Bünner's method.

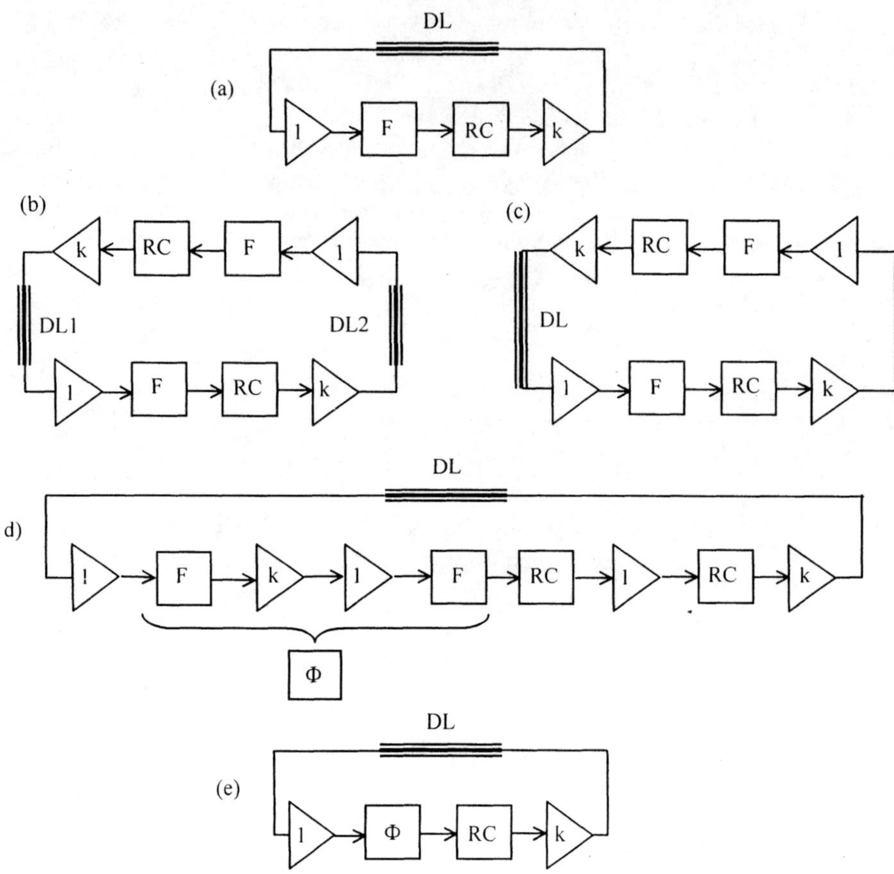

FIGURE 2. Block diagrams of the delay line oscillators. (a) single oscillator, (b), (c), (d), and (e) four different versions of the coupled oscillators. **DL** is a delay line, **1** is a buffer, **F** and Φ are the nonlinear units, **RC** is a low-pass filter, **k** is a linear amplifier. Note, that in (b) $T_{DL1}+T_{DL2}=T_{DL}$.

Experimental power spectra taken from the a- and the b-type oscillators are presented in Fig. 3. The left-top spectrum corresponds to the period-2 oscillations at a fundamental frequency $f^*=110$ MHz. The left-bottom one is for chaotic oscillations with the peaks at the f^* and its 3rd harmonic $3f^*=330$ MHz. Note, that the spectrum from the coupled oscillators (right-bottom) is essentially smother than that from the single oscillator (left-bottom). The unevenness of the spectral density is about 10 dB. Certain rises appear at about 170 MHz and 340 MHz.

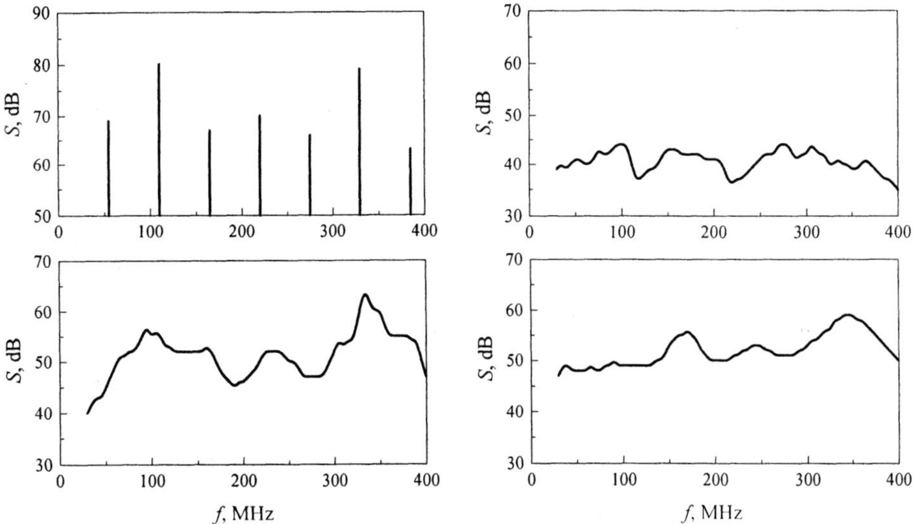

FIGURE 3. Power spectra from a single oscillator (left) and from two oscillators in a ring (right). The supply voltage V_1=7.6V (top) and V_1=8.6V (bottom).

MATHEMATICAL MODEL AND NUMERICAL RESULTS

By introducing the dimensionless variables and parameters

$$x = \frac{V_{C1}}{V^*}, \theta = \frac{t}{RC_1}, \dot{x} \equiv \frac{dx}{dt}, \tau = \frac{T_{DL}}{RC_1}, R = R_7 \parallel R_8, a = \frac{R_8}{R_7 + R_8}, b = \frac{R}{R_9 \parallel R_{10} + r}, c = \frac{V_{B0}}{V^*},$$

where the V_{C1} is the ac voltage across the C1, the V^* and the r is the breakpoint voltage in the I-V characteristic ($V^* \approx 0.7$ V for silicon transistors) and the ON-state resistance of the base-emitter junction of the Q2, respectively, the V_{B0} is the dc component of the base voltage of the Q2, $V_{B0} \approx V_1 R_6 / (R_5 + R_6)$, the following delay differential equations for the five oscillators sketched in Fig. 2 are obtained:

Case (a):
$$\dot{x} = -x + F[x(\theta - \tau)]. \tag{1}$$

Case (b):
$$\begin{cases} \dot{x}_1 = -x_1 + F[x_2(\theta - \tau_1)], \\ \dot{x}_2 = -x_2 + F[x_1(\theta - \tau_2)], \end{cases} \quad \tau_1 + \tau_2 = \tau. \tag{2}$$

Case (c):
$$\begin{cases} \dot{x}_1 = -x_1 + F[x_2(\theta - \tau)], \\ \dot{x}_2 = -x_2 + F(x_1). \end{cases} \tag{3}$$

Case (d):
$$\begin{cases} \dot{x}_1 = -x_1 + \Phi[x_2(\theta - \tau)], \\ \dot{x}_2 = -x_2 + x_1. \end{cases} \tag{4}$$

Case (e):
$$\dot{x} = -x + \Phi[x(\theta - \tau)]. \tag{5}$$

$$F(u) = \begin{cases} aku + b(c-1), & c + ku \le 1, \\ (a-b)ku, & c + ku > 1. \end{cases} \qquad \Phi = \begin{cases} akF + b(c-1), & c + kF \le 1, \\ (a-b)kF, & c + kF > 1. \end{cases}$$

Here k is the gain of the linear amplifier. The LE have been calculated from equations as the functions of the time delay τ. The consolidated results, namely the number of positive LE, n^+ and the Kolmogorov entropy H are presented in Fig. 4. The n^+ as well as the H in the coupled oscillators is noticeably larger than in the single oscillator indicating higher degree of disorder in the coupled system. This is in agreement with the smoother power spectra observed both, experimentally (Fig. 3) and by means of the fast Fourier transform of the time series $x(\theta)$ generated from the equations.

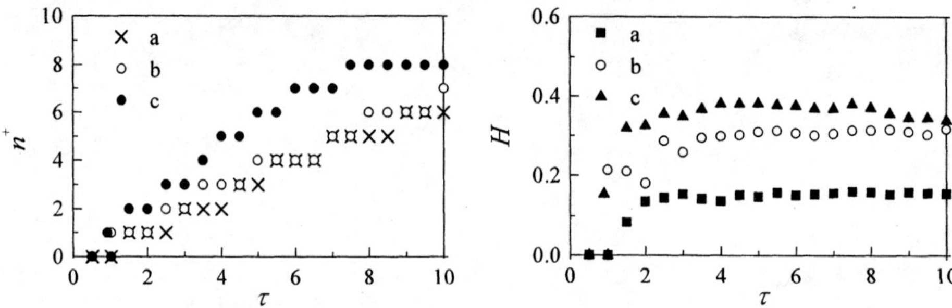

FIGURE 4. Number of positive Lyapunov exponents n^+ (left) and the Kolmogorov entropy, the sum of the positive Lyapunov exponents, $H=\Sigma\lambda^+$ (right) versus delay τ. Legends: (a) single oscillator, Eqn.(1), b=1.8, k=3, (b) two oscillators in a ring, Eqn.(2), b=1.8, k=3, $\tau=\tau_1+\tau_2$, (c) two oscillators in a ring with one RC filter omitted, Eqn.(5), b=1.3, k=3.2. Everywhere a=0.5, c=1.1.

An important result is that the Poincaré section for two oscillators coupled in a ring (Fig. 5b) appears to be a 2-dimensional[3] set in contrast to a single MG oscillator [16] that exhibits 1-dimensional set just coinciding with the nonlinear function F (Fig. 5a).

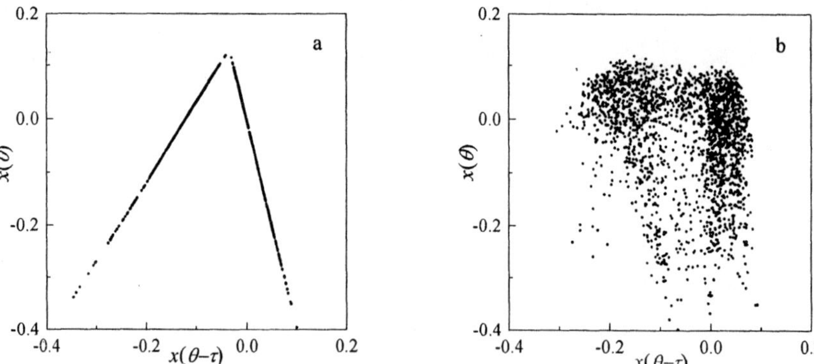

FIGURE 5. Poincaré sections (plane dx/dt=0), a=0.5, b=1.8, c=1.1, k=3. (a) single oscillator, Eqn.(1), (b) two oscillators in a ring, Eqn.(2), here $x \equiv x_1$.

---

[3] In the case of non-scalar delay differential equations [18] a single variable x is insufficient to recover the nonlinear function(s).

ACKNOWLEDGMENTS

This material is based upon work supported by the USA European Office of Aerospace Research and Development, Air Force Office of Scientific Research, Air Force Laboratory, under Contract No. F671775-01-WE065, by NATO under Contract No. PST.CLG.979003 and by the Lithuanian State Science and Studies Foundation under Contract No. T-584.

REFERENCES

1. Dmitriev, A. S., Orlov, Yu. N., and Starkov, S. O., *Radiotekhnika i elektronika* **34**, 1980-1983 (1989), in Russian.
2. Losson, J., Mackey, M. C., and Longtin, A., *Chaos* **3**, 167-176 (1993).
3. Namajūnas, A., Pyragas, K., and Tamaševičius, A., *Phys. Lett. A* **201**, 42-46 (1995).
4. Mögel, A., Schwarz, W., Lukin, K. A., and Zemlyanoy, O. V., "Chaotic Wide Band Oscillator with Delay Line" in *Proc. 3rd Int. Workshop on Nonlinear Dynamics of Electronic Systems NDES'95*, Dublin, Ireland, 1995, pp. 259-262.
5. Čenys, A., Lindberg, E., Anagnostopoulos, A. N., Mykolaitis, G., Bumelienė, S., and Tamaševičius, A., "Broadband Hyperchaotic Oscillator with Delay Line," in *Experimental Chaos: 6th Experimental Chaos Conference, Potsdam, Germany, 2001*, edited by S. Boccaletti et al., AIP Conference Proceedings 622, New York: American Institute of Physics, 2002, pp. 39-45.
6. Namajūnas, A., Pyragas, K., and Tamaševičius, A., *Phys. Lett. A* **204**, 255-262 (1995).
7. Namajūnas, A., Pyragas, K., and Tamaševičius, A., *Int. J. Bifurcation and Chaos* **7**, 957-962 (1997).
8. Namajūnas, A., Pyragas, K., and Tamaševičius, A., "Analog Techniques for Modeling and Controlling the Mackey-Glass System," in *Proc. of the Int. Workshop on Nonlinear Dynamics and Chaos, Pohang, South Korea, 1995*, edited by H. T. Moon et al., Nonlinear Science B, vol. 10, World Scientific, Singapore, 1997, pp. 264-251.
9. Tamaševičius, A., Čenys, A., Mykolaitis, G., Namajūnas, A., and Lindberg, E., *Electron. Lett.* **33**, 2025-2026 (1997).
10. Perez, G., and Cerdeira, H., *Phys. Rev. Lett.* **74**, 1970-1973 (1995).
11. Kocarev, Lj., and Parlitz, U., *Phys. Rev. Lett.* **74**, 5028-5031 (1995).
12. Peng, J.H., Ding, E. J., Ding, M., and Yang, W., *Phys. Rev. Lett.* **76**, 904-907 (1996).
13. Tamaševičius, A., Čenys, A., *Phys. Rev. E* **55**, 297-299 (1997).
14. Tamaševičius, A., Čenys, A., Namajūnas, A., Mykolaitis, G., and Lindberg, E., "Synchronizing Hyperchaotic Circuits" in *Applied Nonlinear Dynamics and Stochastic Systems near the Millenium*, edited by J. B. Kadtke and A. Bulsara, AIP Conference Proceedings 411, New York: American Institute of Physics, 1997, pp. 81-86.
15. Farmer, D., *Physica D* **4**, 366-393 (1982).
16. Bünner, M. J., Popp, M., Meyer, H., Kittel, A., Rau, U., and Parisi, J., *Phys. Lett. A* **211**, 345-349 (1996).
17. Bünner, M. J., Popp, M., Meyer, H., Kittel, A., and Parisi, J., *Phys. Rev. E* **54**, R3082-R3085 (1996).
18. Bünner, M. J., Meyer, H., Kittel, A., and Parisi, J., "The Identification of Time-Delay Systems,", in *Applied Nonlinear Dynamics and Stochastic Systems near the Millenium*, edited by J. B. Kadtke and A. Bulsara, AIP Conference Proceedings 411, New York: American Institute of Physics, 1997, pp. 75-80.

Glassy dynamics in arrays of chaotic oscillators

Normand Mousseau*, S. I. Simdyankin*, Marc-André Brière* and E. R. Hunt[†]

*Département de physique et GCM, Université de Montréal, C.P. 6128, succ. centre-ville, Montréal (Québec) Canada H3C 3J7
[†]Department of Physics and Astronomy, Ohio University, OH, USA 45701

Abstract. Arrays of coupled chaotic elements have been used as models for studying a wide range of phenomena such as synchronization and pattern formation in biological and physical systems. We present experimental and numerical results showing that these arrays can also help us understand the origin of the stretched exponential dynamics observed in glasses and other complex systems. Stretched exponential behavior has been measured over many decades in a 1D array of coupled diode-resonators, just above a crisis-induced intermittency transition. Similar results are obtained numerically in an array of identical chaotic oscillators, confirming the chaotic origin of this universal behavior. In these systems, we find that the fundamental physical quantity associated with stretched exponentials is not the auto-correlation function but, rather, the distribution of times spent in dynamical traps. Here, we review these results and discuss their relation with other systems. We will also present results obtained on higher dimensional networks.

1. INTRODUCTION

Coupled chaotic systems show a great richness of dynamical behaviors and, as such, can serve as toy-model to study many physical phenomena. This was the case for synchronization and pattern formation, for example. Numerous researchers have also found that arrays of coupled non-linear elements could behave in a way reminiscent of that of glasses [1, 2, 3], with very slow dynamics and freezing in disordered states. This is an important discovery because the problem of the glass transition is one of the oldest problems in condensed-matter physics and remains a topic of high interest.

Fragile glasses are characterized by an increase of many orders of magnitude in viscosity as the temperature drops by a few degrees. Near this region, the dynamics of relaxation of various quantities can be fitted by a stretched exponential,

$$C(t) = C_0 e^{-(t/\tau_0)^\beta},\tag{1}$$

with $0 < \beta < 1$. Glasses also undergo aging and many properties depend on the history of preparation, and on the waiting time before a measure is taken or an external strain applied. It is not clear, at this point, how these two properties are related.

In the last few years, much experimental, numerical and theoretical efforts have been targeted at trying to identify a length-scale associated with the dynamical slowing down and characterize the dynamical heterogeneities found in glasses.

We discuss here some recent results obtained in the course of a detailed study of the glassy dynamics in an array of coupled diode resonators. The experimental results can

CP676, *Experimental Chaos: 7th Experimental Chaos Conference,*
edited by V. In, L. Kocarev, T. L. Carroll, B. J. Gluckman, S. Boccaletti, and J. Kurths
© 2003 American Institute of Physics 0-7354-0145-4/03/$20.00

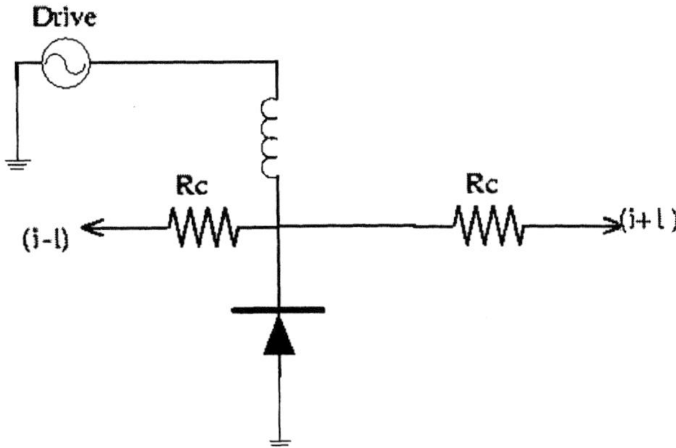

FIGURE 1. Schematic diagram of a single diode-resonator and its coupling to its nearest neighbors in a one-dimensional arrangement.

be reproduced using coupled logistic maps, suggesting that these results are universal.

In the next section, we describe the experimental set-up and the model used to reproduce some of the results numerically. We then present and discuss the results obtained both experimentally and numerically.

2. METHODOLOGY

2.1. Experimental set-up

We consider a one-dimensional chain of 256 coupled diode-resonators connected as a ring. A schematic diagram showing its basic elements is presented in Figure 1. The diode resonator circuit is simply a series combination of a 30-mHy inductor and a rectifier type pn-junction diode (1N1004) driven by a sinusoidal source at 100 kHz. This circuit has been studied extensively over the last 20 years and is known to follow the period-doubling route to chaos as the ac drive is increased [4, 5, 6]. These elements are coupled diffusively through resistors, as shown in the diagram, and period-boundary conditions are used. In spite of a careful selection, the drive voltage for a given feature, such as the period-3 fixed points, varies by up to 20% between basic elements. This one-dimensional setup has been previously studied with unidirectional coupling [7] and under the conditions of stochastic resonance [8, 9, 10].

The overall phase diagram of this system, as the drive voltage is increased, represents a kink-forming route to spatio-temporal chaos [11]. At low drive voltage, all elements follow the period-doubling route to chaos in a nearly perfect synchronous manner, with no or little current in the coupling resistors. As the drive is increased, the lattice reaches a two-band chaotic orbit with all elements, again, being in the same band at the same time. Further increasing the voltage leads to the formation of spatial structures as coarse-grained period-2 phase kinks (two-band structures with elements alternating bands at a given time) start to appear [7]. It is interesting to note that each time a new kink is formed, the orbit stabilizes itself. At it turns out, there is appreciable power dissipated in the coupling resistors at the kink position, reducing the available phase space. Increasing the driving, the number of kinks increases until all elements are on a kink, i.e., with neighbors in the opposite band. Further increasing the drive voltage, the system goes through a high-dimensional version of the well-studied attractor-merging crisis [12]. We define the critical drive voltage V_c as the point where the two attractors just touch. Just above V_c, the system oscillates between full spatio-temporal chaos and coarse-grained periodicity, being trapped for long periods in the latter state. In the case of a single chaotic element, this *crisis-induced intermittency* was shown to display exponentially distributed switching times [12]. As discussed here, we also find a form of crisis-induced intermittency in coupled systems but the trap time distribution is now dominated by stretched exponentials.

2.2. Coupled logistic maps

The experimental results can be reproduced qualitatively using an array of coupled logistic maps as introduced by Kaneko [13] and Johnson *et al.* [7] a few years ago:

$$x_i(t_{n+1}) = (1 - \alpha)f_i(t_n) + \frac{\alpha}{2}\left[f_{i-1}(t_n) + f_{i+1}(t_n)\right] \tag{2}$$

where $f_i(t_n) = rx_i(t_n)\left[1 - x_i(t_n)\right]$ and α is the coupling between oscillators. Although this map is sufficient for the problem discussed here, it does not describe the full dynamical behavior of the experimental system, which includes some memory effect which is best reproduced by a two-dimensional map [6].

Simulation are typically run on lattices with periodic boundary condition varying in size between 100 and 10000 sites. Initial conditions are set at random and the first million step discarded to avoid transient effects. Simulation length vary between 10^7 to 10^{10} time steps.

Results below are generally presented for $\alpha = 0.25$ and $r = 3.83$ or 3.8888, two sets of parameters showing a glassy behavior.

2.3. Analysis

Figure 2 shows a 4096-cycle time series for the 256 elements. The two phases of the two-band attractor are represented as black and white, and only every other time step is displayed. Note the creation, annihilation and diffusion of the boundaries of the two-band attractor. With the coupling resistors used, 150 kOhms, the coarse-grained spatial periodicity of the lattice is about 4 sites.

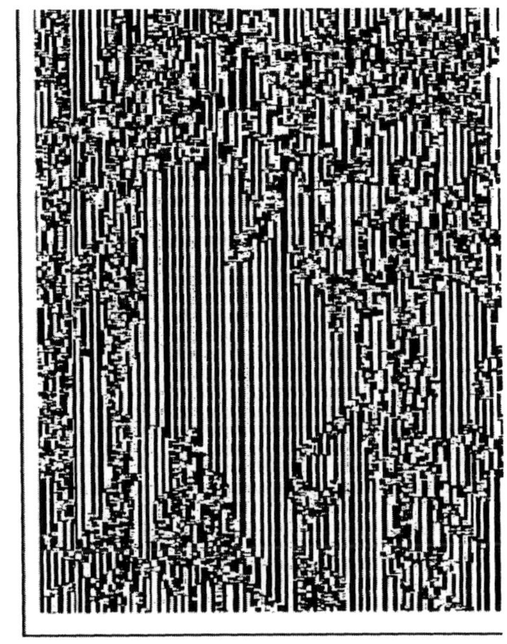

FIGURE 2. Time series of the 256 diodes in the intermittent state, just above V_c. Since the system displays a fundamental period-two oscillation in time below V_c, we show the state of the diodes at even time-steps in a binary representation.

We are interested in the statistical distribution of trap time in a period-two cycle. This quantity is formally equivalent to the distribution of time intervals between zero crossings of renewal processes such as random walks [14] and has the advantage that it can be measured experimentally and numerically to a high degree of accuracy for this system [15].

Generally, however, experimentally measurable relaxation response correspond to auto-correlation functions which are given by

$$C(t) = \langle \sigma(t')\sigma(t'+t)\rangle_{t'}. \tag{3}$$

A general framework describing the direct relation between this quantity and the distribution of trap time was recently introduced by Luck and Godrèche [14]. Applying this framework to our problem, we can show that a stretched exponential trap time distribution implies also a stretched exponential decay in the auto-correlation function, albeit with a different exponent [16].

There is therefore a one to one correspondence between the trap time distribution presented here and the more standard auto-correlation function.

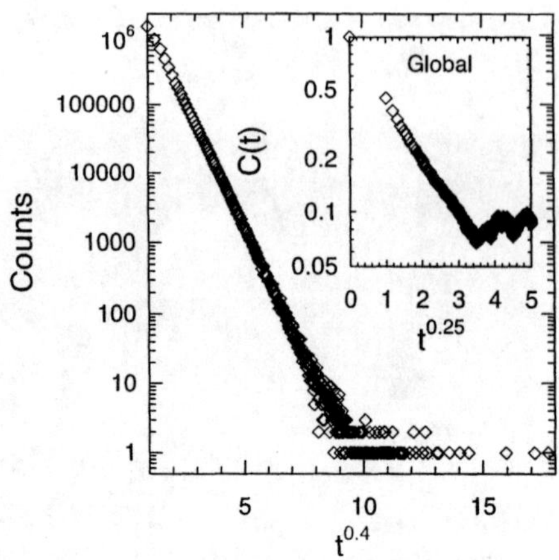

FIGURE 3. Experimental trap-time distribution for a single site measured over 10^{10} cycles. Inset: Auto-correlation function measure over 4096 cycles and averaged over 256 sites (global average).

3. RESULTS

3.1. Stretched exponential dynamics

As already mentioned, just above what would be called crisis-induced intermittancy in a single chaotic oscillator, the trap-time distribution presents an unusual stretched exponential dynamics, reminiscent of glasses. However, while for most experimental measures of stretched exponential relaxation in these systems, the results span few enough decades that they can be fitted by a number of distributions, experimental results for the coupled system presented here, taken over a full day at a rate of 10 kHz, shows a well-fitted curve over 6 decades in the distribution.

For completeness, we also plot the auto-correlation function, which also follows a stretched exponential form. Due to the way we measure the auto-correlation function, we can only get a fraction of the statistics obtained for the trap distribution. Even with less statistics, however, the auto-correlation function follows a stretched exponential over about a decade, in agreement with the theoretical analysis.

Because the set-up is disordered, with diode-resonators differing by at most 20 % in their critical voltage, after a careful selection, each site displays a different dynamics. Although all sites show a stretched exponential trap-time distribution, the stretching exponent can vary considerably. For example, Fig. 4 show the trap-time distribution for three sites on a 256-diode lattice at the same driving with β varying by more than a factor of two.

Similar effects could be produced numerically when site disorder is introduced in the variable r.

The stretching exponent β varies continuously with the external drive voltage. From about 0.35 to close to 1.0 as the system is close to fully chaotic. Closely following this behavior, the time factor, τ_0, also decreases, resulting in significantly longer traps as the

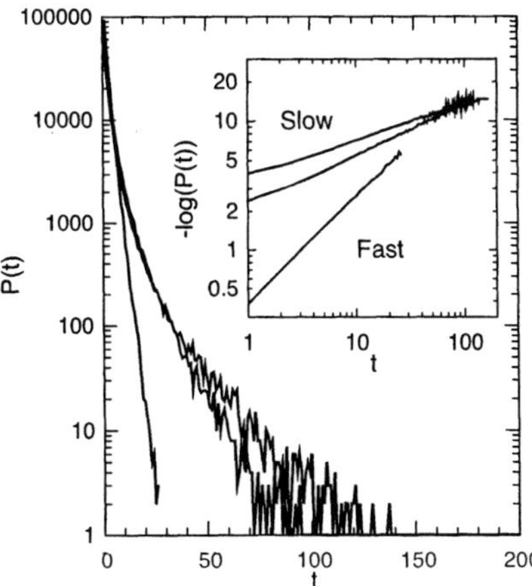

FIGURE 4. Trap-time distribution at 3 different sites on a 256-diode lattice. The main figure shows a log-normal plot of the distribution. In inset, we show a log-log vs. log plot for the three sites. The curves are translated vertically to ease viewing. The corresponding β are 0.8, 0.4 and 0.3, respectively.

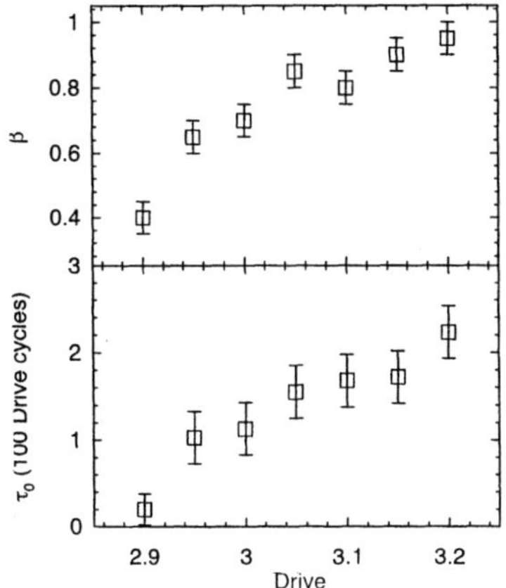

FIGURE 5. β and τ as a function of the drive voltage for the 256-diode array. These values are taken for a generic site, with an average activity.

voltage goes down. This is similar to what is seen numerically, for the model above, with $0.33 < \beta < 0.70$. In this case, the longer traps at the low β are more than 250 times longer than for $\beta = 0.70$, reaching time scale of more than 30,000 time steps.

By contrast, the increase in the length scale of the exponentially-decreasing spatial correlation function goes up from 2 to 8 sites. Moreover, as can be seen in Fig. 2, it appears that there is little correlation between the width of a trap and its duration. A parallel can be drawn between these surprising features and the absence of any diverging length scale in glasses as the relaxation times increase rapidly, near the glass-transition temperature.

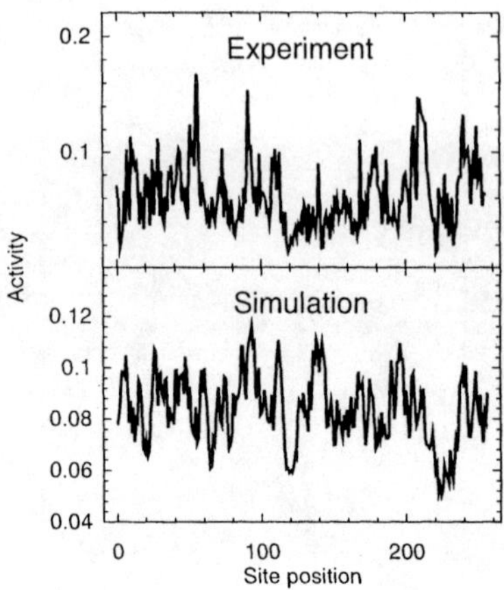

FIGURE 6.   Top: Activity as a function of site in a 256-site experimental setup. Bottom: Activity in the coupled map lattice as a function of the coupling α and the driving parameter r.

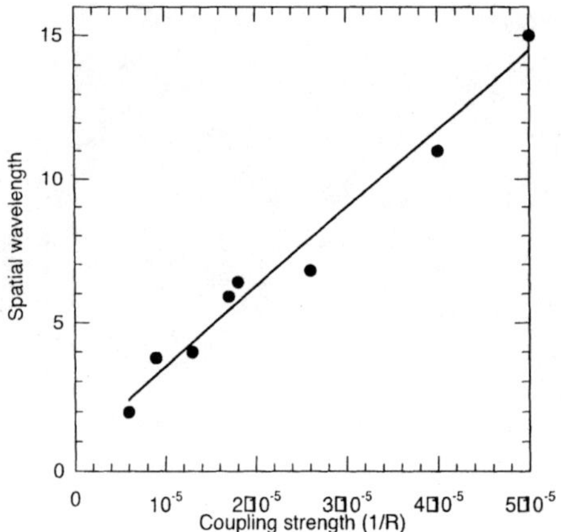

FIGURE 7.   Spatial wavelength as a function of the coupling strength for a 256-element network. Results presented in this paper are for a coupling,R_c, of 10^{-5} Ohms^{-1}

The absence of long-range spatial correlations is also revealed in Fig. 6 which plots the number of times a site has flipped from one phase to another normalized by the number of time steps in the simulation. A low activity indicates a site that remains trapped for long time. Although the experiment is taken on a relatively short period of time, 2048 time steps sampled at a frequency 10 times that of the basic clock tick, it is clear that spatial correlation is minimal.

While spatial correlation increases only slightly with decreasing driving, the spatial *periodicity* remains unchanged. As it turns out, this quantity depends only on the coupling strength and not on the driving. Figure 7 shows that the spatial wavelength follows linearly the coupling strength. The stretched exponential behavior is mostly constrained

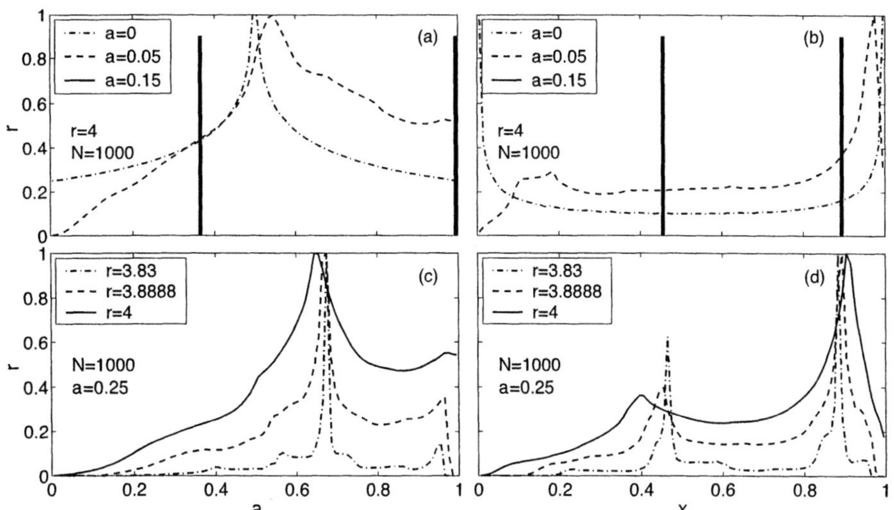

FIGURE 8. Natural invariant density for the external perturbation(left-hand panels) and an isolated site submitted to this external perturbation (right-hand panels), in a numerical simulation, as described by Eq. (4). Panels (a) and (b) show the variation of the density, $\rho(a)$ and $\rho(x)$ as a function of the coupling α while panels (c) and (d) show the variation of the density for various driving, r.

to a spatial periodicity between 4 and 5 sites.

3.2. Origin of the stretched exponential

In order to better understand the origin of the stretched exponential, we turn to the numerical model. The coupled-map model introduced above reproduces closely the features seen experimentally. This demonstrates that the stretched exponential behavior is a generic property of coupled chaotic systems and not caused by frozen disorder [15].

In order to investigate the origin of the stretched exponential behavior in this system, we rewrite Eq. (2) as an external perturbation, a^t on an isolated site:

$$x(t+1) = (1 - \alpha)f[x(t)] + \alpha a(t). \tag{4}$$

Within this picture, it is convenient to relax the coarse-grained picture and to introduce the natural invariant density, $\rho(x)$. This density is defined such that $\rho(x)dx$ gives the fraction of the time the orbit spends in the interval dx around x. A similar quantity can be defined for the external perturbation, $\rho(a)$.

For the uncoupled map in the fully chaotic, the natural invariant $\rho(x)$ has the well-known "U" shape [12]. $\rho(a)$, representing the distribution for $\left[f(x_{i-1}) + f(x_{i+1})\right]$ is already very Gaussian-like and peaked at 0.5.

These densities are modified significantly with the coupling. For the single site natural invariant, $\rho(x)$, the peaks at 0 and 1 have moved inside. The competition between the

39

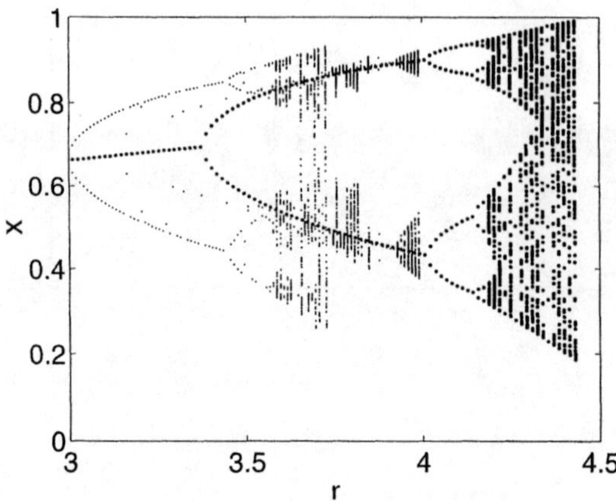

FIGURE 9. Bifurcation diagrams for a single site on the coupled map lattice with $N = 16$ (smaller dots) and for the map $x(t+1) = 0.75f[x(t)] + 0.1625$ (larger dots). The perturbation in the second map corresponds to the contribution of the neighbors in a spatial period-four regime [the peak at 0.67 in Fig. 8(d)].

period oscillations and the chaotic behavior are clearly seen there, by a non-vanishing density between the peaks at $x = 0.45$ and $x = 0.85$. The central peak for $\rho(a)$ is shifted to the right and a shoulder appears near 1. The central peak arises when the two neighbors of site i are in opposite band, i.e., when the lattice is in a spatial period-four. The effect of this perturbation on site i in Eq. (4) is, in effect, to shift the bifurcation map from a chaotic to a temporal period-two regime, setting the site into a trap (see Fig. 9).

While Eq. (4) can help understand the origin of the traps in these coupled systems, it also underlines the need for local spatial organization. For example, selecting the perturbation $a(t)$ from the distribution $\rho(a)$ presented in Fig. 8 generates only an exponential trap-time distribution. If we introduce a bias in the distribution, by favoring with a stretched exponential probability, a value of $a(t)$ corresponding to the peak in $\rho(a)$, then the map also shows stretched exponential distributions. This demonstrates that once a stretched exponential is present in the dynamics, it will force it on the rest of the lattice.

This is confirmed by constructing $a(t)$ by adding the state of two sites selected at random on a large lattice and using this perturbation on an isolated map. The trap-distribution on this map also displays a stretched-exponential trap distribution, with a stretched exponent β larger than that of the lattice.

Since there is no direct communication between the two sites selected at random (the spatial correlation is very short range — less than 10 sites), it is even sufficient that each neighbor be locked into a stretched-exponential trap-distribution independently to impact similarly the central site.

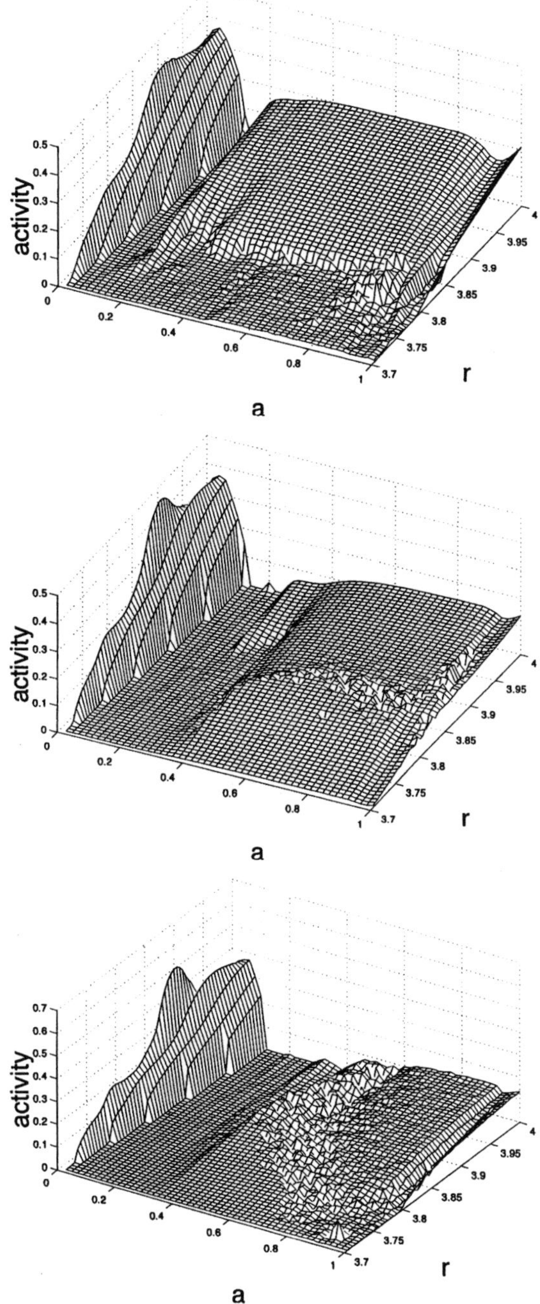

FIGURE 10. Activity in the coupled map lattice as a function of the coupling α and the driving parameter r for the 1D, 2D and 3D numerical models.

3.3. Higher-dimensional lattices

In view of the existence of 3D configurational glasses, it is interesting to assess whether the stretched-exponential trap-time distribution persists in higher dimension. In view of the discussion of the previous section, one could expect that the stretched exponential behavior disappear rapidly with increasing dimensionality. How fast exactly has to be established experimentally.

Fig. 10 shows the phase diagram of the coupled map, in terms of the level of activity as function of the driving r and the coupling parameter α. While the overall level of activity, save for very small couplings, decreases significantly with the dimensionality of the lattice, the qualitative structure of phase diagram remains similar.

In all dimension, there is a significant frozen region above $\alpha \simeq 0.1$, where the lattice locks into a spatial and temporal period two. At very high coupling, the activity increases again as the whole lattice starts behaving as a single, perfectly in step, oscillator. The interesting region of the phase space, for our purposes is therefore in the medium coupling, $0.2 < \alpha < 0.4$, where stretched exponential dynamics occurs in 1D.

Changes are significant enough, however, that we might wonder whether or not the close relation between experiment and simulation will survive higher dimensions. While a 3D experimental set-up is currently being put together, simulations indicate that the parameter window for stretched exponential behavior decreases in size. There are, however, some regions of the phase diagram where stretched exponential trap-distribution occur in both 2D and 3D but a full characterization of those has not been completed at this point.

4. CONCLUSION

Coupled arrays of chaotic elements display are remarkably rich range of dynamics. We have found that for a wide range of parameters, these coupled lattices behave in ways very reminiscent of glasses, with frustration inducing stretched exponential distributions. These results, which are seen both experimentally and numerically appear to be universal. Because the dynamical systems can be much more easy to measure and simulate, they can provide us with a new way to study the origin of the stretched exponential relaxation in glasses.

Spatial organization appears critical, here, to generate the stretched exponential trap-time distribution. The stability of traps is ensured by the presence of a spatial period-four region. Nevertheless, the short range spatial correlations indicate that there is not much stability gained by extending greatly the period-four region. As β decreases from 0.50 to 0.33 and the longer trap increase in time 250-fold, the spatial correlation length goes up from 2 to 8. This can be compared with the absence of a diverging length scale observed in glasses as the dynamics slows down to a halt.

Although we have made much progress in understanding this new dynamics in arrays of coupled chaotic elements, there are still a few questions that remain to be answered regarding the effects of dimension and the links between these systems and configurational glasses. We are hopeful that studying further these phases will help us identify the origin of stretched exponential relaxation, a question which has been with us for more than 150 years by now.

ACKNOWLEDGMENTS

This work is supported in part by the ONR (EH), NSERC and NATEQ (NM). NM is a Cottrell Scholar of the Research Corporation.

REFERENCES

1. Kaneko, K., *Physica D*, **37**, 60–82 (1989).
2. Tsang, K. T., and Ngai, K. L., *Phys. Rev. E*, **54**, R3067–R3070 (1996).
3. Tsironis, G. P., and Aubry, S., *Phys. Rev. Lett.*, **77**, 5225–5228 (1996).
4. Linsay, P. S., *Phys. Rev. Lett.*, **47**, 1349 (1981).
5. Testa, J., Pérez, J., and Jeffires, C., *Phys. Rev. Lett.*, **48**, 714 (1982).
6. Rollings, R. W., and Hunt, E. R., *Phys. Rev. Lett.*, **49**, 1295–1298 (1982).
7. Johnson, G. A., Löcher, M., and Hunt, E. R., *Physica D*, **96**, 367 (1996).
8. Löcher, M., Cigna, D., Hunt, E. R., Johnson, G. A., Marchesoni, F., Gammaitoni, L., Inchiosa, M. E., and Bulsara, A. R., *Chaos*, **8**, 604 (1998).
9. Löcher, M., Cigna, D., and Hunt, E. R., *Phys. Rev. Lett.*, **80**, 5212 (1998).
10. Löcher, M., Chatterjee, N., Marchesoni, F., Ditto, W. L., and Hunt, E., *Phys. Rev. E*, **61**, 4954–4961 (2000).
11. Gade, P. M., Chatterjee, N., Mousseau, N., and Hunt, E. R., *N/A* (2002), in preparation.
12. Ott, E., *Chaos in Dynamical Systems*, Cambridge University Press, Cambridge, 1993.
13. Kaneko, K., editor, *Theory and Applications of Coupled Map Lattices*, Wiley, Chichister, 1993.
14. Godrèche, C., and Luck, J. M., *J. Stat. Phys.*, **104**, 489–524 (2001).
15. Hunt, E., Gade, P., and Mousseau, N., cond–mat/0204179 (2002).
16. Simdyankin, S., and Mousseau, N., in preparation.

Local and Global Intermittency Observed in Phase-Locked Loops

Tetsuro Endo

Department of Electronics and Communication, Meiji University 214-8571 Higashi-Mita, Tama, Kawasaki Japan

Abstract. In this paper, local and global intermittency are introduced in phase-locked loops by theory, computer simulation and experiments. Here, the local intermittency means the Pomeau and Manneville Type I intermittency and the global intermittency means the crisis-induced intermittency due to I-D chain. The experimental results agree at least qualitatively with computer simulation results of the model equation.

INTRODUCTION

Phase-locked loops (PLLs) are versatile functional device widely used in communication systems; namely, they are used as a signal tracker from satellites, pulse wave synchronization in TV sets and radar systems, and frequency synthesis and conversion in quarts crystal oscillators. PLL devices are so useful that they are now available as a one chip integrated circuit.

Practical PLLs have been used in phase-locked condition where linear theory dominates. However, it has rich nonlinear dynamics when it is out of lock. Nonlinear analysis of PLL has not been performed due to its complexity, although it presents various interesting phenomena including chaos. In fact, it can be considered as an electrical analog of the forced swing equation. Therefore, it will be meaningful to investigate nonlinear dynamics of PLL. Since it is implemented by integrated circuit, the realization is quite easy.

In this article, we will introduce local and global intermittency observed in PLL both by computer simulation and experiments [1], [2]. Here the local intermittency means the Pomeau and Manneville Type I intermittency and the global intermittency means the crisis-induced intermittency due to I-D chain. The experimental results agree well with computer simulations of the model equation.

MODEL EQUATION

Figure 1 presents one of the experimental circuits made by using a PLL IC module CD4046B. In this module, a phase detector (PD) and a voltage- controlled oscillator (VCO) are included and a low pass filter (LPF) is added outside the module. Thus the

CP676, *Experimental Chaos: 7ᵗʰ Experimental Chaos Conference,*
edited by V. In, L. Kocarev, T. L. Carroll, B. J. Gluckman, S. Boccaletti, and J. Kurths

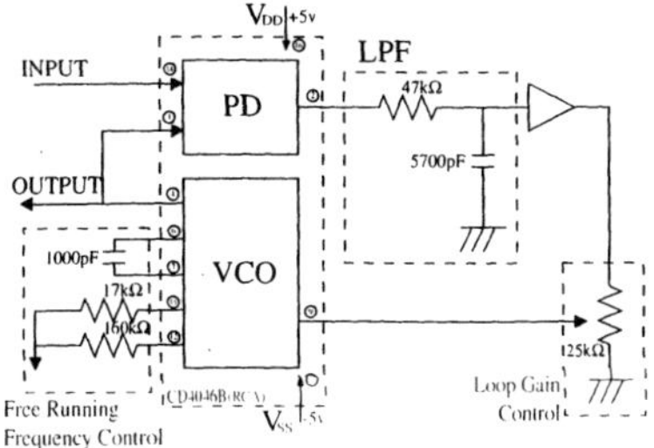

FIGURE 1. An experimental circuit of a PLL

diagram of a PLL can be written as in Fig.2. By defining

$$y_{in} = A\sin(\omega_0 t + \tilde{\theta}_i(t))$$
$$y_{out} = B\cos(\omega_0 t + \tilde{\theta}_0(t)) \tag{1}$$

and $\phi(t) = \tilde{\theta}_i(t) - \tilde{\theta}_0(t)$, the characteristic of PD can be written as

$$y_{PD} = K_{PD}h(\phi) \tag{2}$$

where $h(\phi)$ is a $2\pi-$periodic function with respect to ϕ such that $h'(0) = 1$. Namely, if PD is a mixer type, then $h(\phi)$ becomes a sinusoidal characteristic: $h(\phi) = \sin\phi$. If PD is an exclusive OR type, then $h(\phi)$ becomes a triangular characteristic. If PD is flip-flop type, then $h(\phi)$ becomes a sawtooth characteristic.

Next, LPF is a single-pole low-pass filter consisting of R and C whose transfer function F(s) can be written as

$$F(s) = \frac{1 + s\tau_2}{1 + s\tau_1}, \quad \tau_1 \equiv C(R_1 + R_2), \quad \tau_2 \equiv CR_2. \tag{3}$$

This filter is called a lag-lead filter, and it is called a lag filter in the special case of $\tau_2=0$. At last, the VCO can be expressed as integral characteristic: K_{VCO}/s. By combining

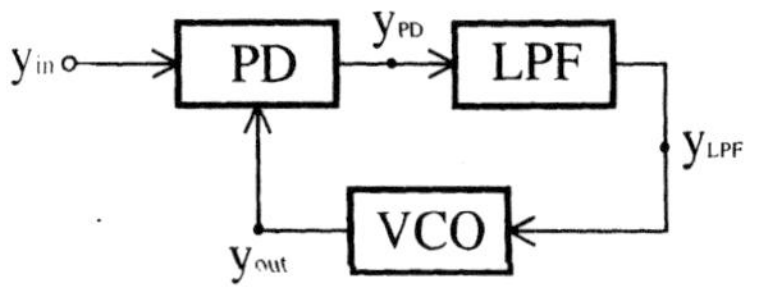

FIGURE 2. Block diagram of a PLL

these characteristics, Fig.2 can be rewritten as Fig.3 ($K_0 \equiv K_{PD} \cdot K_{VCO}$). Since this diagram is expressed by phases $\tilde{\theta}_i$, $\tilde{\theta}_0$ and ϕ, this is called the phase diagram. The phase diagram is very important to understand the PLL dynamics. Namely, the fundamental differential equation can be derived directly from this diagram as follows:

$$\tilde{\theta}_0 = \frac{1}{s} K_0 F(s) h(\phi) = \tilde{\theta}_i - \phi \tag{4}$$

Regarding $s = d/dt$, eq. (4) with a lag-lead filter can be written as

$$\frac{d^2\phi}{dt^2} + \frac{1}{\tau_1}(1 + K_0\tau_2 h'(\phi))\frac{d\phi}{dt} + \frac{K_0}{\tau_1}h(\phi) = \frac{d^2\tilde{\theta}_i}{dt^2} + \frac{1}{\tau_1}\frac{d\tilde{\theta}_i}{dt} \tag{5}$$

Here we assume that the input signal is modulated by a sinusoidal waveform with detuning $\Delta\omega$ such that

$$\frac{d\tilde{\theta}_i}{dt} = \Delta\omega + M\sin\omega_m t \tag{6}$$

The detuning $\Delta\omega$ is the difference between carrier frequency ω_c and VCO free-running frequency ω_0; namely, $\Delta\omega = \omega_c - \omega_0$. By combining eqs.(5) and (6) and by changing the time t into $t' = \omega_n t$ where $\omega_n = \sqrt{K_0/\tau_1}$, one can obtain the following normalized PLL equation(t' is returned to t again for simplicity):

$$\begin{aligned}
\frac{d^2\phi}{dt^2} + \beta[1 + (2\zeta - \beta)h'(\phi)/\beta]\frac{d\phi}{dt} + h(\phi) \\
= \beta\sigma + \beta m\sin\Omega t + m\Omega\cos\Omega t
\end{aligned} \tag{7}$$

where $\omega_n \equiv \sqrt{K_0/\tau_1}$ is called the natural frequency, and $\zeta \equiv (1 + K_0\tau_2)/2\sqrt{K_0\tau_1}$ is called the damping coefficient, and where $\beta = \omega_n/K_0 = 1/\sqrt{K_0\tau_1}$ is called the normalized natural frequency, $\sigma = \Delta\omega/\omega_n$ is called the normalized frequency detuning, $\Omega = \omega_m/\omega_n$ is called the normalized modulation frequency, $m = M/\omega_n$ is called the normalized maximum frequency deviation. Note that when $2\zeta = \beta$ is held, eq.(7) becomes a PLL with a lag-filter. Equation (7) can be rewritten as a system of first-order equations:

$$\begin{aligned}
\dot{\phi} &= y \\
\dot{y} &= -\beta[1 + (2\zeta - \beta)h'(\phi)/\beta]y - h(\phi) + \beta\sigma \\
&\quad + \beta m\sin\Omega t + m\Omega\cos\Omega t
\end{aligned} \tag{8}$$

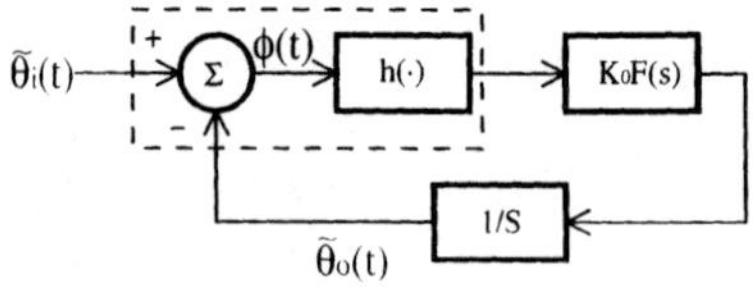

FIGURE 3. Phase diagram of a PLL

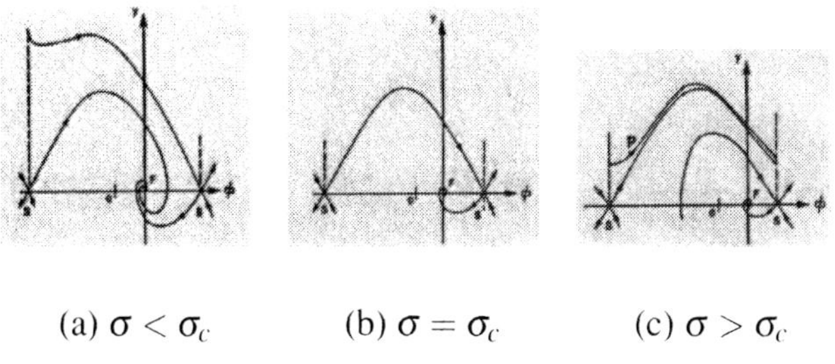

(a) $\sigma < \sigma_c$ (b) $\sigma = \sigma_c$ (c) $\sigma > \sigma_c$

FIGURE 4. Three phase diagrams for m=0.

At first, we will investigate the phase diagram of eq.(8) for m=0.There are three typical phase diagrams as shown in Figs.4(a), (b) and (c). For $\sigma < \sigma_c$, all flows converges to a stable focus F as seen in Fig.4(a). At $\sigma = \sigma_c$, a homoclinic orbit connecting a saddle to the next saddle appears as seen in Fig.4(b). For $\sigma > \sigma_c$, flows starting above the unstable manifold of the left saddle S converges to the periodic orbit labeled as P, and flows starting below it converges to the stable focus F as seen in Fig.4 (c). In this paper, we focus our attention to the case of Fig.4(c), and investigate what kind of bifurcations occur when the external force m is applied.

LOCAL BIFURCATIONS OF P

At first, we investigate the local bifurcations of P. Actually, P is a periodic orbit rotating around the cylindrical co-ordinate system. Namely, since ϕ is a phase and the equation is invariant by replacing ϕ by $\phi + 2n\pi$, one can identify the left S and the right S in Figs.4 (a), (b) and (c). Thus one can make a cylindrical co-ordinate system as seen in Fig.5, and P is a periodic orbit rotating around the cylinder. Figure 6 presents a local bifurcation diagram of P with respect to m and Ω, where solid curves denote the saddle-node bifurcation sets and where dotted curves denote the flip bifurcation sets($\beta = 2\zeta = 0.56, \sigma = 1.27$). The region surrounded by two solid curves and one dotted curve is the stable region of the *m/n-periodic solution*. When a parameter point

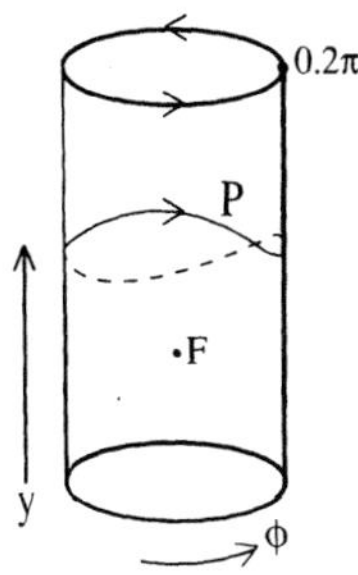

FIGURE 5.   The periodic orbit of the second type on the cylindrical phase space.

47

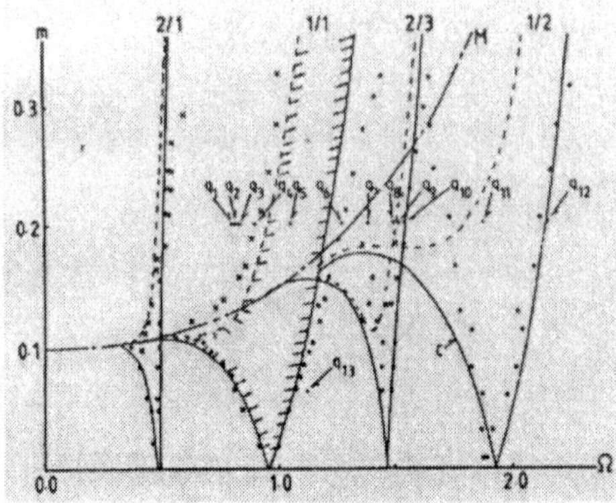

FIGURE 6. The two paramter bifurcation diagram of P for the sinusodal PD case($\beta = 2\zeta = 0.56$, $\sigma = 1.27$).

goes across the solid curve from inside to outside, the periodic solution is lost and either torus or intermittent chaos appears. Namely, if the point is above the Melnikov homoclinic curve M, intermittent chaos will occur. If it is below the curve M, torus will appear. When a parameter point goes across the dotted curve from right-hand side to left-hand side, an n-periodic point bifurcates to be a 2n-periodic point. The flip bifurcation repeats itself infinitely many times to the left direction (in the diagram only one dotted curve is drawn for simplicity). Therefore, if one move a parameter to the left going across many flip bifurcation curves, one may obtain chaos eventually. Fig.7 shows an experimental set-up of the FM demodulation PLL. The PLL is built by using a mixer type phase detector MC1594L and a sinusoidal VCO ICL 8038 and a lag-lead type low-pass filter. Therefore, the PD characteristic $h(\phi)$ becomes sinusoidal. We observe $\sin\phi$ as the X-axis and $\dot\phi$ as the Y-axis. Many circles in Fig.6 presents experimentally-confirmed saddle-node bifurcation points, and crosses presents also experimentally confirmed flip bifurcation points. They agree at least qualitatively.

Next, we observe some Lissajous of $\sin\phi$ and $\dot\phi$ and their Poincare maps in Fig.8

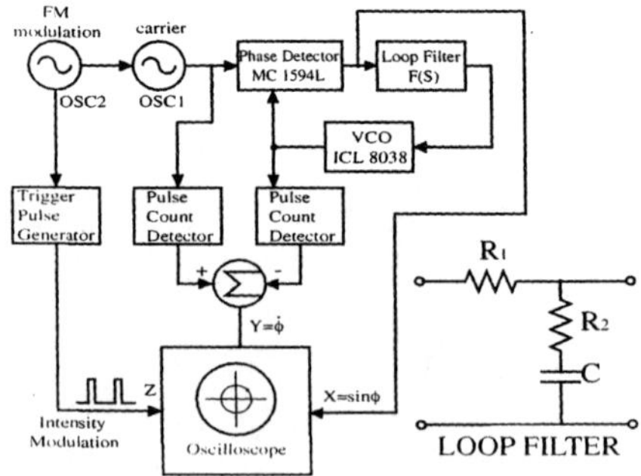

FIGURE 7. An experimental set-up for the PLL with frepuency - modulated carrier signal.

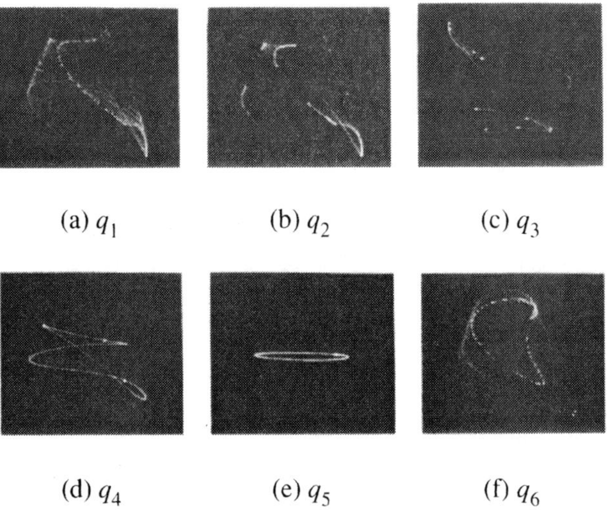

(a) q_1 (b) q_2 (c) q_3

(d) q_4 (e) q_5 (f) q_6

FIGURE 8. An experimentally - observed Lissajous of $sin\phi$ versus $\dot{\phi}$ and their Poincare maps of various attractors bifurcated from the 1/1 - periodic state (m=0.2). (a) 1-band chaos at q_1 ($\Omega = 0.81$). (b) 2-band chaos at q_2 ($\Omega = 0.825$). (c) period 4 at q_3 ($\Omega = 0.834$). (d) period 2 at q_4 ($\Omega = 0.895$). (e) period 1 at q_5 ($\Omega = 1.05$). (f) Intermittency at q_6 ($\Omega = 1.27$).

at parameter points $q_1, q_2, ..., q_6$ in Fig.6. At q_1 a one-band chaos can be seen for $(\Omega, m) = (0.81, 0.2)$ in Fig.8(a). At q_2 the one-band chaos is splitt into two-band chaos for $(\Omega, m) = (0.825, 0.2)$ in Fig.8(b). At q_3 the two-band chaos becomes a period-4 orbit for $(\Omega, m) = (0.834, 0.2)$ in Fig.8(c). At q_4 the period-4 orbit becomes a period-2 orbit for $(\Omega, m) = (0.895, 0.2)$ in Fig.8(d). At q_5 the period-2 orbit becomes a period-1 orbit for $(\Omega, m) = (1.05, 0.2)$ in Fig.8(e). At q_6 the period-1 orbit becomes an intermittent chaos for $(\Omega, m) = (1.27, 0.2)$ in Fig.8(f) . Fig.9 presents the associated power spectra for the 1-band chaos for q_1, for the period-4 orbit for q_3, for the period-2 orbit for q_4, for the period-1 orbit for q_5, and for the intermittency for q_6. The period doubling process can be seen clearly that the fundamental frequency becomes half in Fig.9 (d) and (c), and in Figs.9 (c) and (b). The evidence that the power sepctrum in (e) is really an intermittency is shown in Fig.10. In Fig.10(a) a sinusoidal time waveform continues for some time and interrupted by a short burst and this is repeated. Further, in Figs.10(b) and (c), plots of X_n versus X_{n+1} and those of Y_n versus Y_{n+1} are tangent to the 45 degree line. In addition, if parameter points are apart from the saddle-node curve, the plots leave the 45 degree line as seen in Figs. 10 (d) and (e). In the same manner, we can observe period-3 and period-2 and chaos at parameter points q_7 to q_{12}.

GLOBAL BIFURCATIONS OF P AND F

In this section we investigate a global bifurcation called "Explosion". The stable focus F for $m = 0$ becomes a periodic orbit for small m, and as m is increased, the periodic orbit repeats period-doubling bifurcations infinitely many time to become chaos. We call all attractors bifurcated from F the P_1-attractors generically. The stable periodic solution of the second type P for $m = 0$ repeats torus and locking and period-doubling to become chaos, as m is increased. We call all attractors bifurcated from P the P_2-attractors generically. At some values of m the P_1- and the P_2-attractors explode suddenly. Figure 11 shows a global bifurcation diagram of the P_1, P_2, $P_1 \& P_2$-attractors

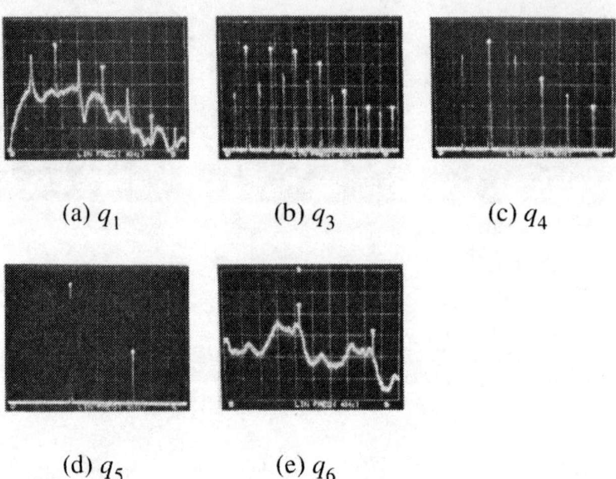

(a) q_1 (b) q_3 (c) q_4

(d) q_5 (e) q_6

FIGURE 9. The associated power spectra for the experimentally-observed attractors of Fig.8.

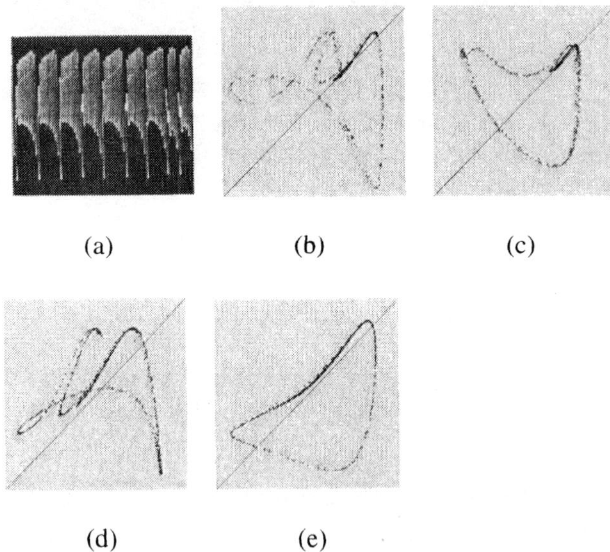

(a)   (b)   (c)

(d)   (e)

FIGURE 10. A time waveform and first-return maps showing intermittency.

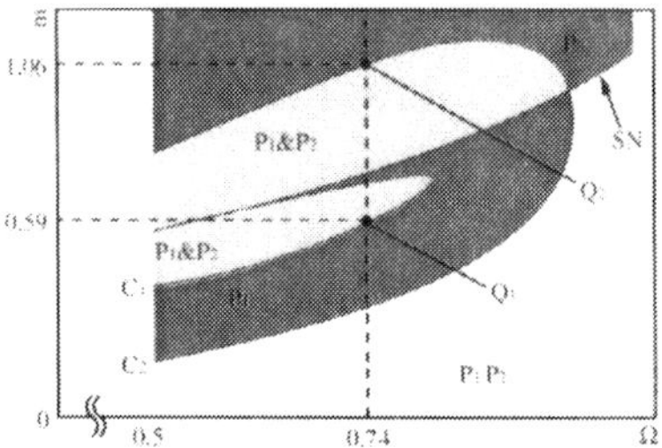

FIGURE 11. Stable region of the P_1- and P_2,and the $P_1\&P_2$-attractors and observed points Q_1 and Q_2.

$(\beta = 2\zeta = 0.566, \sigma = 1.488, h(\phi)$:triangular PD). In the region outside C_2 and below SN, the P_1 and the P_2-attractors co-exist. On curve C_2 the P_2-attractor disappears due to the I-D chain explained later. Therefore,in the region inside C_2, outside C_1 and below SN, only the P_1-attractor exists. In the same manner, above SN and outside C_2, only the P_2-attractor exists. Now we ask what kind of phenomenon will occur when one changes a paremeter across the points Q_1 and Q_2. Below Q_1 only the P_1-attractor exists and when the parameter is changed above C_1, this P_1-attractor is destroyed via the I-D chain. Similarly, when the parameter is changed across Q_2 downward, the P_2-attractor is destroyed. Therefore, in the region labelled as $P_1\&P_2$, neither P_1- nor P_2-attractor exists. We will present the attractor explosion from the P_1- to the $P_1\&P_2$-attractor around Q_1 as an example.

Figure 12 presents a transition obtained from computer simulation on the Poincare

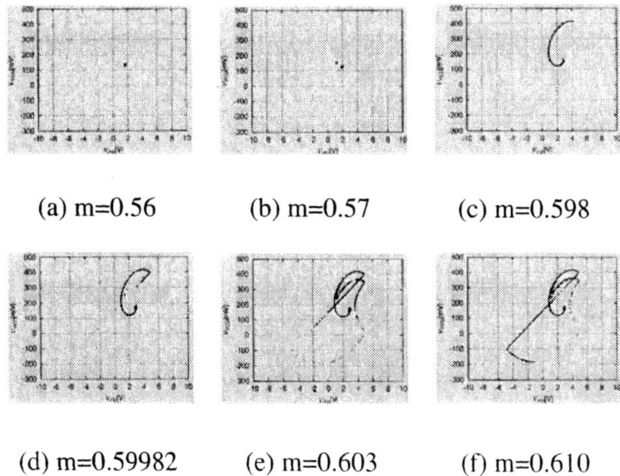

(a) m=0.56 (b) m=0.57 (c) m=0.598

(d) m=0.59982 (e) m=0.603 (f) m=0.610

FIGURE 12. Poincare maps showing the transition from the P_1-attractor to the $P_1\&P_2$-attractor in the (V_{PD}, V_{VCO})-plane.

51

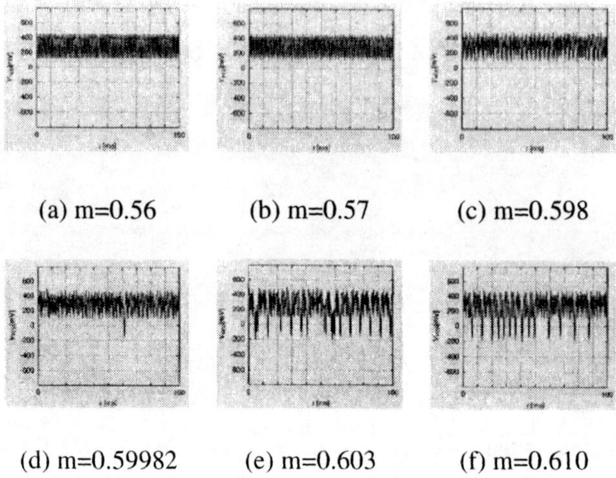

(a) m=0.56   (b) m=0.57   (c) m=0.598

(d) m=0.59982   (e) m=0.603   (f) m=0.610

FIGURE 13. The V_{vco} time waveforms in the transition of $P_1 \rightarrow P_1\&P_2$. Parameter values of m in $(a) \sim (f)$ are the same as those in Fig12

map from the P_1-attractor to the $P_1\&P_2$-attractor in the (V_{PD}, V_{VCO})- plane. In Figs.12(a), (b) and (c) a series of period-doubling bifurcations to chaos of the P_1-attractor can be observed. In Fig.12(d) the P_1-strange attractor suddenly explode to become a $P_1\&P_2$-strange attractor, but the main part is still on the locus of the P_1-strange set and scattered points are distributed on the P_2-strange set. These points appear as bursts in time domain. As m is increased, the scattered points are spread over the P_2-strange set as seen in Fig.12(e), and in Fig.12(f) mapped points are equally distributed on the P_1- and the P_2-strange sets to form a complete $P_1\&P_2$-strange attractor. Figure 13 presents a variation of the V_{VCO}-time waveforms before and after the explosion. In particular, the time waveforms are restricted to 140 to 450 mV before the explosion as seen in Figs. 13(a), (b) and (c). However, after the explosion bursts appear and increase as the parameter m is increased as seen in Figs. 13 (d), (e) and (f).

Now we investigate the mechanism of burst appearance. Figure 14 presents the behavior of unstable manifold of I (=an inversely unstable saddle fixed point which is the

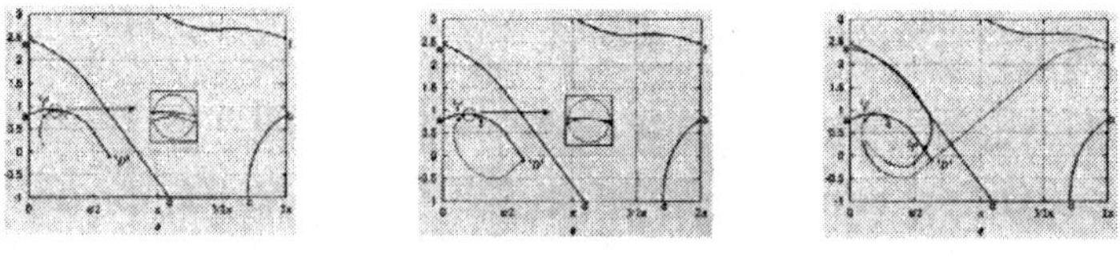

(a) Before the explosion   (b) At the explosion   (c) After the explosion

FIGURE 14. The invariant curves before at and after the explosion in the transition process of $P_1 \rightarrow P_1\&P_2$. (a)m=0.59 (b)m=0.592 (c)m=0.5998. Inverse time evolution is $a \rightarrow b \rightarrow c \rightarrow d \rightarrow e \rightarrow f$.

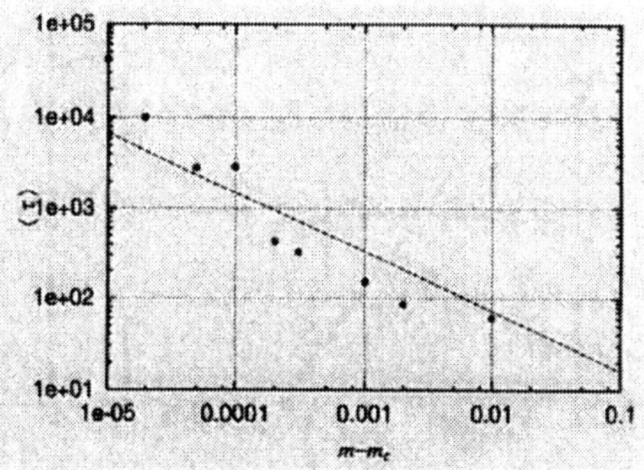

FIGURE 15. The mean burst period in terms of $m - m_c$ in the transition process of $P_1 \rightarrow P_1 \& P_2$.

core of the P_1-strange attractor) and the stable manifold of D (=a directly unstable saddle fixed point. The I and D appear in pairs). Before the explosion these two manifolds do not touch as seen in Fig.14(a). At the explosion they are tangent with each other, and after the explosion they intersect as seen in Figs.14(b) and (c). From this fact, the explosion of an attractor occurs when the unstable manifold of I is *tangent* to the stable manifold of D. Therefore, this bifurcation is called the I-D chain or heteroclinic tangency crisis [3], [4]. Figure 15 presents the mean burst period as a function of $m - m_c$ after explosion. Each point is calculated from computer simulation of the V_{VCO}-time waveform. The dotted line is drawn by the following formula:

$$\langle \tau \rangle \propto |m - m_c|^{-\gamma} \tag{9}$$

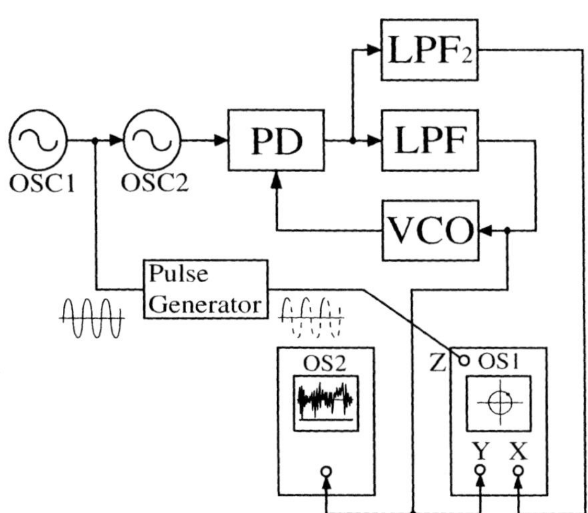

FIGURE 16. Schematic diagram in the experiments.

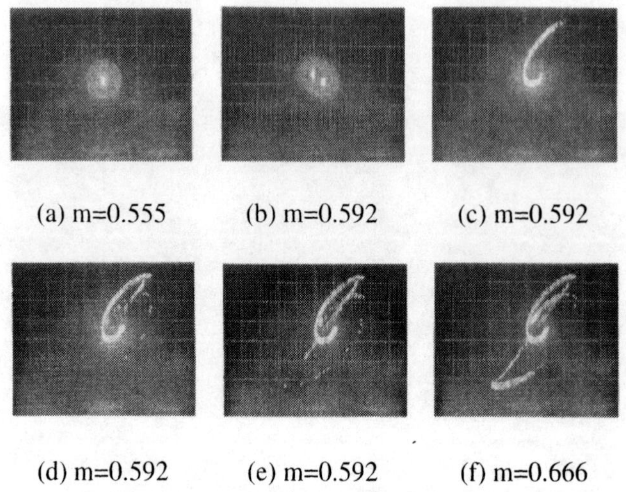

(a) m=0.555   (b) m=0.592   (c) m=0.592

(d) m=0.592   (e) m=0.592   (f) m=0.666

FIGURE 17. The experimental observation in the transition process of $P_1 \rightarrow P_1 \& P_2$ in the (V_{PD}, V_{VCO})-plane map. (a)m=0.555 (b)m=0.592 (c)m=0.592 (d)m=0.592 (e)m=0.592 (f)m=0.666.

where

$$\gamma = \frac{1}{2} + \frac{\ln|\lambda_1|}{|\ln|\lambda_2||} \tag{10}$$

and where λ_1 is an expanding eigenvlaue, and λ_2 is a contracting eigenvalue of the inversely unstable fixed point I ($|\lambda_1| > 1, |\lambda_2| < 1$). In this case, λ_1 and λ_2 are given by

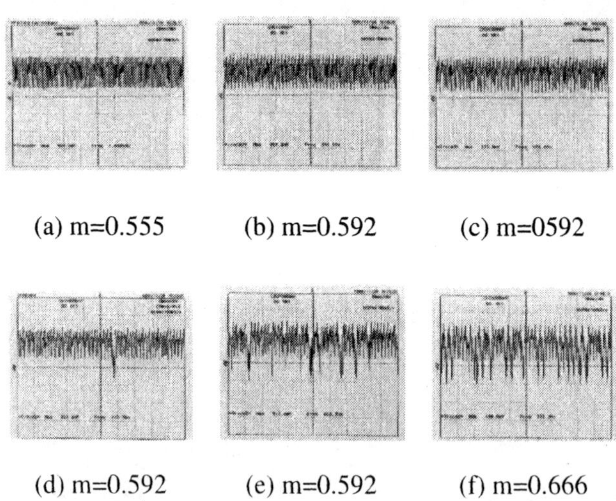

(a) m=0.555   (b) m=0.592   (c) m=0592

(d) m=0.592   (e) m=0.592   (f) m=0.666

FIGURE 18. The experimentally-observed V_{VCO} time waveforms in the transition process of $P_1 \rightarrow P_1 \& P_2$. Parameter values of m in $(a) \sim (f)$ are the same as those in Fig.17.

$\lambda_1 = -2.4727$ and $\lambda_2 = -0.003286$, and hence γ becomes 0.6583. The dotted line is drawn in this manner.

At last we will describe the experimental results of attractor explosion. The schematic diagram of our experiments is shown in Fig.16. This is almost the same as that in Fig.7 except that the input of Y-axis is the VCO input V_{VCO} which can be written as

$$V_{VCO} = K_{VCO}(\sigma + m\sin\Omega t - y) \approx -K_{VCO}y \propto -y. \tag{11}$$

Therefore, we apply $K_{PD}h(\phi)$ to X-input and $-K_{VCO}y$ to Y-input approximately. In addition, we apply short-width pulses synchronized with OSC1 to Z-axis to observe the Poincare map. Figure 17 presents the attractor variation around point Q_1 which corresponds to Fig.12. Fig.18 presents the VCO input time waveform V_{VCO} corresponding to Fig.13. Comparing these experimental results with the associated computer simulation results, one can recognize good coincidence with each other.

CONCLUSIONS

In this paper, local and global intermittency observed in an actual PLL is demonstrated. Local intermittency means the Pomeau and Manneville Type I intermittency and the global intermittency means the crisis-induced intermittency. The experimental results agree well with computer simulation results.

REFERENCES

1. T.Endo, and M.Imai, *IEICE*, **J72-A**, 1973–1981 (1989).
2. T.Endo, W.Ohno, and Y.Ueda, *Int.J.of Bifurcation and Chaos*, **10**, 891–912 (2000).
3. Y.Ueda, "Explosion of strange attractors exhibited by Duffing's equation," in *Ann.NY.Acad.Sci.357*, 505, 1980, pp. 422–434.
4. E.Ott, *Chaos in Dynamical Systems*, Cambridge University Press, 1993, pp. 280–283.

Chaos Synchronization Via The Transmission Of Symbolic Information

Shawn D. Pethel, Ned J. Corron and Quitisha Underwood

U. S. Army Aviation and Missile Command,
AMSAM-RD-WS-ST, Redstone Arsenal, AL 35898

Krishna Myneni

Science Applications International Corp., 6725 Odyssey Drive, Huntsville, AL 35806

Abstract. We report high-quality chaotic synchronization of one-way coupled electronic circuits through a communication channel that transmits only symbolic dynamical information. This is accomplished by converting the synchronization signal into a discrete symbol sequence at the transmitter and then decoding the symbols at the receiver in real time. We find that symbol information is critical in the synchronization process and that an arbitrarily low rms synchronization error can be maintained by only transmitting a single 1-bit symbol per cycle. This indicates that chaotic synchronization is robust to severe band limiting and/or noise in the coupling channel provided symbol information is preserved. The experimental setup also allows us to explore so-called *achronal* synchronization in which the receiver lags or leads the transmitter by fixed time. We discuss fundamental trade-offs between the accuracy to which the transmitter state can be known, the quality of synchronization, and the delay or anticipation created in the receiver.

Separate chaotic systems can synchronize if allowed to share information over a coupling channel [1]. The fundamental and practical implications of this phenomenon have stimulated a great amount of research recently in both scientific and engineering disciplines. Many issues remain concerning the robustness of synchronization. Central to these issues is the fact that chaotic oscillators have nonzero Shannon entropy--that is, they act like information sources. For synchronization to occur this information must be encoded at the transmitter, passed through a coupling channel of some sort, and then successfully decoded at the receiver. If there is a loss of information in the channel, then one expects synchronization to suffer. Exactly how this will affect the quality of synchronization is not always clear in the context of a typical synchronization scenario.

The tools provided by the theory of symbolic dynamics provide a good framework for addressing these issues. Recently, Stojanovski *et. al.* [2] brought symbolic dynamics to bear on the question of how much channel capacity is required for chaotic synchronization in unidirectionally coupled systems. They establish a relationship between the channel capacity and the Komolgorov-Sinai entropy such that synchronization error can be made arbitrarily small.

Here we extend this work both theoretically and experimentally. We find that the transformation into symbol sequences reveals tradeoffs between the quality of synchronization desired, the accuracy of the detection process at the transmitter, and

CP676, *Experimental Chaos: 7th Experimental Chaos Conference,*
edited by V. In, L. Kocarev, T. L. Carroll, B. J. Gluckman, S. Boccaletti, and J. Kurths
2003 American Institute of Physics 0-7354-0145-4

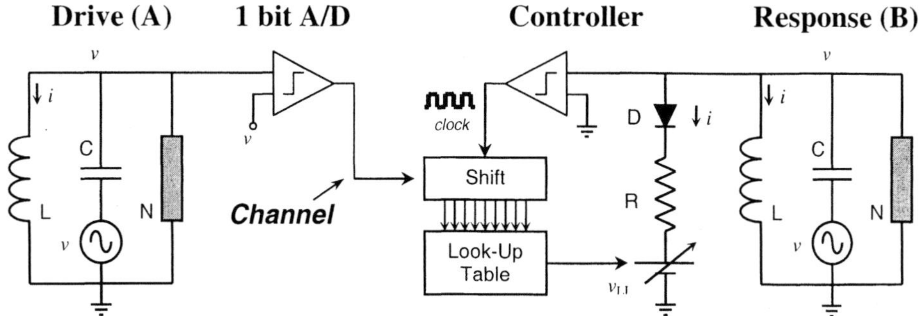

FIGURE 1. Diagram of the experimental setup showing the drive oscillator (A), 1-bit detector, channel, controller, and response oscillator (B).

the time delay or advance of the receiver with respect to the transmitter. For confirmation, we devise an experiment to synchronize two chaotic electronic circuits by transmitting only 1-bit symbols. Below we describe these findings prefaced with a short review of synchronization and symbolic dynamics.

Consider two identical chaotic oscillators, A and B that are connected through a unidirectional coupling channel. Coupling into and out of the channel is facilitated by a detector at A and a controller at B. The detector signal can be a function of some or all of the dynamic variables of A. Let us say that that these systems have the property that synchronization will occur if B is provided with a scalar signal $x(t)$ from A. We say synchronization is achieved when all dynamical variables of B follow the corresponding dynamical variables of A to within some specified fidelity. Typically, one compares the states of A and B at the same time; however, we will also consider so-called *achronal* synchronization in which the receiver can lag or lead the transmitter by a fixed amount of time.

Let us say that system A can be represented as a return map through the method of Poincare. Accordingly, the differential system is now a return map of the form $x(t_{n+1}) = f(x(t_n), x(t_{n-1}), \ldots)$. Therefore, given the sequence $x(t_n), x(t_{n-1}), \ldots$ with N_b bits of accuracy from system A, system B can predict $x(t_{n+1})$ using the return map f with an accuracy slightly less than N_b bits. Therefore, to maintain synchronization, system A need only send the bits required to make the predicted value of $x(t_{n+1})$ fully N_b bits accurate. The theory of symbolic dynamics provides a transformation that allows us to isolate the relevant N_b bits of information.

The conversion of a chaotic signal into a symbol sequence begins with a coarse partitioning of the chaotic attractor. Each region produced by the partitioning is labeled with an alphanumeric symbol such as "0","1","A","B", etc. When the system state enters a new region the corresponding symbol is generated. In this way a trajectory can be mapped into a symbol sequence. Every initial condition on a chaotic attractor produces a unique trajectory and a corresponding symbol sequence. If the chosen partition is *generating*, then every observed symbol sequence can be identified with a unique initial condition.

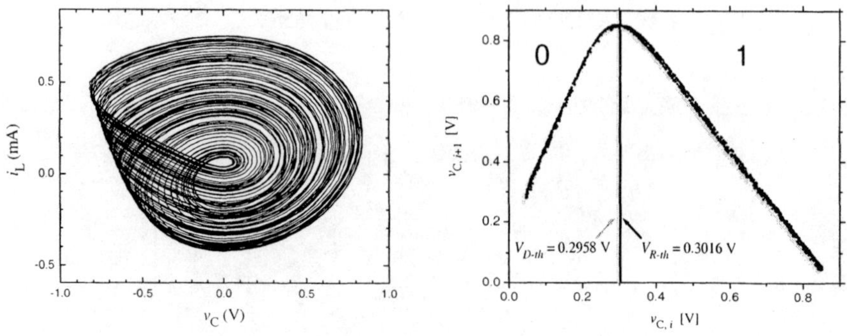

FIGURE 2. Return map taken from successive voltages peaks for the drive (black) and receiver (gray). The voltage corresponding to the generating partition is shown along with the symbols designating each region. To the right is the phase space of the drive system.

With regard to synchronization we can arrive at some immediate conclusions. The first is that *it the channel need only transmit one symbol per cycle to maintain synchronization between A and B.* This is because only one new symbol is generated per Poincare crossing; all the other symbols describing the state were generated previously and have already been transmitted. For a two symbol system this means that 1 bit per cycle is enough information to maintain high-quality synchronization. This situation suggests that the full analog chaotic waveform transmitted in most synchronization experiments is highly redundant and should therefore be resistant to significant levels of noise and distortion. This conclusion, however, is not consistent with the vast majority of reports which show chaotic synchronization as very fragile phenomena. The reason, as we will show, is not due to a fundamental physical limitation, but rather to a suboptimal controller at B.

We now present experimental data indicating that synchronization via the transmission of symbols is realizable in physical flows. The configuration of the synchronization experiment is shown in Figure 1. Systems A and B are piecewise-linear LC circuits of a type used and documented previously [3], with the modification that all capacitances have been halved. These oscillators exhibit a simply folded band attractor and generate a nearly sinusoidal time-varying voltage across the capacitor with a dominant spectral peak near 1350 Hz. The chaos manifests itself as peak-to-peak amplitude fluctuations over a -1.5 to 1.5 volt range. To eliminate phase drift between the two circuits, a function generator producing a 30 mV peak-to-peak sine wave at 1350 Hz is connected in series with the tank capacitor in each circuit. This low-level signal entrains the phase of both oscillators to a common reference but has a negligible effect on the amplitudes, which remain chaotic and uncorrelated. We note that it is not strictly necessary to use a common external oscillator; separate oscillators can be phase locked by an additional piece of information sent before the amplitude synchronization process begins.

The return map and attractor observed for the drive oscillator are shown in Figure 2. This map, which is constructed using consecutive waveform peaks, indicates

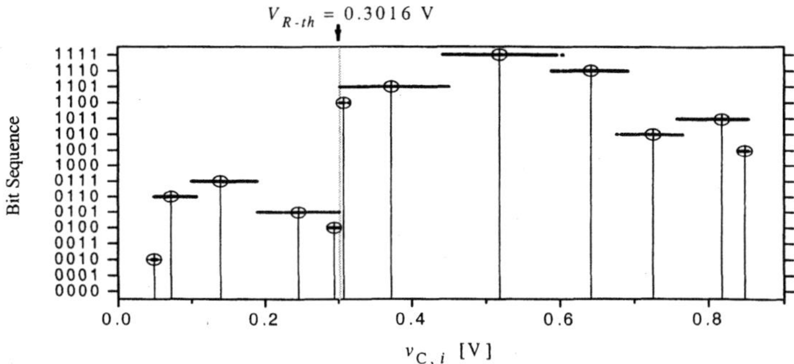

FIGURE 3. A 4-bit look-up table that maps peak voltages into symbol sequences. To invert the map one chooses the mean voltage represented by the circles.

that the system dynamics are well approximated by a 1-dimensional return map containing a single quadratic maximum. For such maps the symbolic dynamics consist of two symbols and a unique generating partition lies at the maximum, which we have estimated to be at $x_c = 0.296$ V for the drive oscillator. Following the usual convention, we say the drive system generates the symbol "0" for a peak $x < x_c$ and the symbol "1" otherwise. Also in Figure 2, we show the return map for the oscillator used in the response system. The partition derived from the maximum in the return map for the response oscillator is $x_c = 0.302$ V.

In Figure 3, we show a partitioning of the observed peaks by the 4-bit symbol sequence exhibited by the oscillator starting from that state. We observe that the partition divides the return map into several contiguous segments corresponding to different symbol sequences with some ambiguity evident at boundaries due to experimental imprecision. The drop lines in the figure shows the mean peak voltage for each segment: we use these mean voltages as control points as described below. The set of allowable sequences is termed the *grammar* of the system, and the grammar of the drive and response systems must be consistent to enable synchronization.

On the response oscillator an active limiter is attached to enable dynamic limiting as described previously [3]. Dynamic limiting controls an oscillatory system by providing feedback proportional to the peak overshoot of a system state compared to a limiter level that is set for each cycle. As such, the mean peak values shown in Figure 3 are directly applicable as limiter levels for using dynamic limiting to control the response oscillator to the corresponding symbol sequence. This particular technique was chosen out of convenience; the outcome of the experiment is not dependent on the type of controller used.

The remainder of the experimental configuration shown in Figure 1 is implemented using a DSP card hosted in a PC. A single state of the drive oscillator is sampled continuously by the DSP card at ~100kHz, from which voltage peaks are found and used to generate the corresponding symbols. At the response system the received symbols are put into a binary shift register, which is clocked at the symbol arrival rate. The state of the shift register provides an index into a look-up table

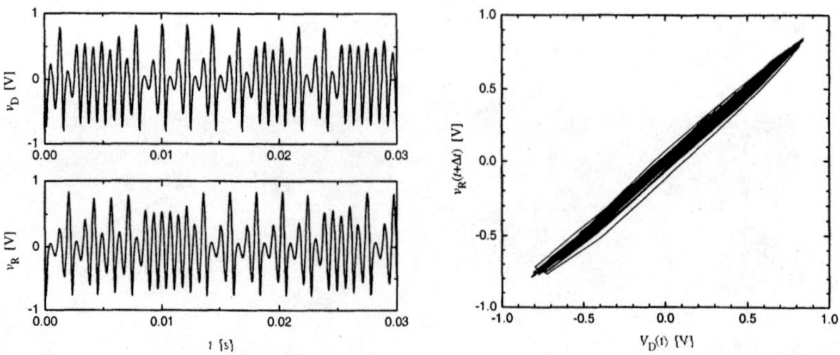

FIGURE 4. Drive (V_D) and response (V_R) voltage waveforms showing high quality synchronization with an 8-cycle lag.

containing the mean peak voltages, and the table output is converted to an analog voltage that provides the limiter levels used to control the response system.

Figure 4 shows the tank voltages of the drive and response systems observed when coupled through the symbolic channel. As shown, the controller forces the response system to accurately synchronize to the drive waveform. Importantly, the synchronization was achieved using only *small* control perturbations, i.e., comparable to that required to stabilize periodic orbits [3]. For this figure, we used a look-up table generated by partitioning the response system return map using 8-bit symbol sequences. In general, the table can be generated using any symbol sequence length M, with the choice of M having several important effects on the synchronization. First, the receiver lags the transmitter by M cycles, due to the depth of the shift register at the response system. In the figure the drive waveform is shifted by 8 cycles to account for this. Second, the synchronization error decreases·exponentially as M increases, since the look-up table doubles in precision (and size) with each additional bit and the drive state can be recreated more accurately. Third, the magnitude of the average control perturbation becomes exponentially smaller, since the corrections required to follow the drive state also become finer. In this experiment there is a practical limit to these trends at $M \approx 12$, due to hardware limitations and the noise floor.

It is significant to note a duality that emerged in this analysis: precision vs. delay. To understand this relationship, view the M-bit symbol sequence as a binary representation of an oscillator's state. The leftmost symbol is the most significant bit, while the rightmost is the least significant. At the transmitter we choose to detect the most significant bit, since we can do that simply and with the least chance for error. At the receiver we provide the correction to the least significant bit, since it requires the least control perturbation to the oscillator's natural dynamics. The delay is due to the time it takes for the information inserted at the least significant bit to percolate up to the most significant bit, i.e., M cycles. Incrementing M increases this delay one more cycle and doubles the precision of the look-up table. This, then, is why the synchronization error can be made arbitrarily small at the expense of longer delay.

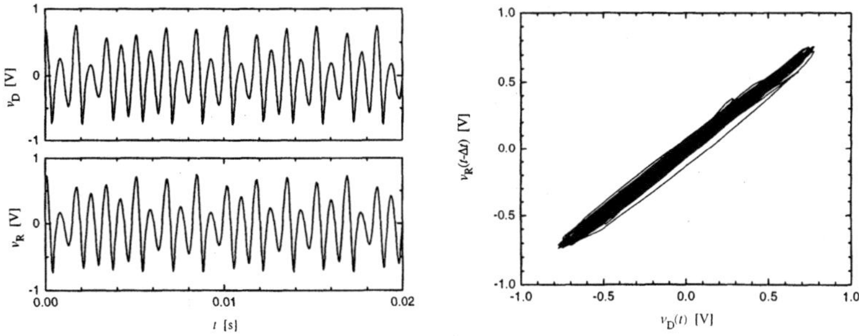

FIGURE 5. Drive (V_D) and response (V_R) voltage waveforms showing high quality synchronization with 2-cycle anticipation.

It is also possible to decrease the delay while maintaining synchronization quality. If the drive system can reliably detect and transmit a less significant bit, say the Nth bit, the time lag due to the shift register in the response system is M-N+1 cycles. Notably, synchronization without a delay requires N = M+1, implying comparable precision in the detectors of both the drive and response systems. Note this does not increase the information capacity requirements of the transmission channel: synchronization still demands transmitting only 1 symbol per cycle. The trade-off here is in the additional accuracy required by the detector at the transmitter. If one has a very precise detector that exceeds the resolution of the look-up table, it is possible that the synchronization delay could be made negative so that the response system actually leads the drive. In figure 5 we have set N = 12 and M = 9 so that the receiver now leads the transmitter by two whole cycles. This effect has been seen previously [4] and has been described as *anticipating* synchronization. In this case we can clearly see how transmitter accuracy (N) and synchronization quality (M) determine the degree of anticipation that is possible.

In summary, analog synchronization signals are highly redundant and occupy far more bandwidth than is strictly necessary for high-quality synchronization. This redundancy should be exploitable to make synchronization tolerant to noise and channel distortions, though perhaps at the cost of delay. The tradeoffs between detector accuracy and delay become apparent when considering the transmission of symbols. In the future, we hope to extend this understanding to mutually coupled systems and to large arrays.

REFERENCES

1. Pecora, L. M., and Carroll, T. L., *Phys. Rev. Lett.* **64**, 821 (1990).
2. Stojanovski, T., Kocarev, T., and Harris, R., *IEEE Transactions on Circuit and Systems* **44**, 1014 (1997).
3. Corron, N. J., and Pethel, S. D., *Chaos* **12**, 1 (2002).
4. Voss, H. U., *Phys. Rev. E* **61**, 5115 (2000).

Phase Synchronization of Shil'nikov Chaos in Coupled Chua's Oscillators

Syamal Kumar Dana[1], Prodyot Kumar Roy[2], Biswapriya Mukhopadhyay[3], Satyabrata Chakraborty[1]

[1]*Instrument Division, Indian Institute of Chemical Biology, Kolkata 700032, India*
[2]*Department of Physics, Presidency College, Kolkata 700073, India*
[3]*Jadavpur Vidyapith, Kolkata 700032, India*

Abstract. An experiment on generation of homoclinic chaos using two unidirectionally coupled Chua's oscillators is described here. Homoclinic chaos is obtained at the response oscillator for weak coupling limit in the range of phase synchronization. Stable homoclinic oscillation is obtained by forcing periodic pulses to the driver. Phase locking of homoclinic oscillation to forcing pulse has been observed with different locking ratios (m:n), when the frequency of the pulse is close to time period of the homoclinic oscillation.

INTRODUCTION

Phase synchronization (PS) of homoclinic chaos of Shil'nikov type [1] has been implemented in laser [2] with sinusoidal forcing. Its possible application in information encoding in the interspike intervals [3] of homoclinic chaos has been explored recently for secure communication. The mechanism of PS near homoclinic bifurcation has been explained earlier [4,5]. Complete synchronization (CS) in master-slave type [6] chaotic oscillators is susceptible to channel noise [7], which causes intermittent loss of synchronization. PS of homoclinic chaos uses the time sequence of spiking oscillations, whose shape may be changed by channel noise yet the time sequence of spike trains is not disturbed. Homoclinic chaos has thus advantage to other deterministic form of chaos in communication applications, particularly, in the context of chaotic pulse position modulation scheme (PPM) [8]. Moreover, the question still remains unanswered how sensory system of neuron assembly encodes [9] external information in the form of interspike intervals. Studies on PS of homoclinic chaos may help in understanding how neurons communicate information with each other before transmitting to the brain.

Homoclinic orbits are highly unstable and very difficult to obtain both numerically and experimentally [10] in a single Chua's oscillator. In this paper, an experimental procedure is described for generating homoclinic chaos using two coupled Chua's oscillators. Further, experimental evidence of phase locking of homoclinic orbits to external periodic pulses is presented. Two non-identical Chua's oscillators are coupled in drive-response unidirectional mode. The driver is kept in single scroll oscillatory mode (near period-3, period-4 and so on). The response oscillator is kept in point attractor mode, the response oscillator produces homoclinic chaos in the weaker coupling limit near PS.

This work is partially funded by BRNS/DAE, Government of India under project grant #2000/34/13/BRNS/1924

CP676, *Experimental Chaos: 7th Experimental Chaos Conference,*
edited by V. In, L. Kocarev, T. L. Carroll, B. J. Gluckman, S. Boccaletti, and J. Kurths

"

GENERATION OF HOMOCLINIC OSCILLATION

Numerical procedures are available to obtain homoclinic orbits using HOMCONT [11]. Homoclinic chaos has also been observed in electronic circuits [12], but it is always difficult to obtain stable homoclinic orbits due to inherent instabilities of saddle focus equilibrium. Homoclinic chaos [13] has been identified in single Chua's circuit, as a rigorous proof of chaos in such system. An experimental setup is proposed here to generate homoclinic chaos (henceforth called as homoclinic oscillation) using two non-identical Chua's oscillators coupled in unidirectional mode. The parameters of the drive circuit are selected close to period-3, period-4 single scroll oscillations while the response circuit is kept in point attractor mode (uncoupled) with an appropriate choice of circuit parameters. The response circuit produces homoclinic orbit with intermittent instability when the coupling strength is in the weaker limit of PS. The circuit diagram of two non-identical Chua's oscillators coupled in drive response mode is shown in Figure 1. Each Chua's oscillator consists of one inductor with leakage resistance, two capacitors and a nonlinear resistance.

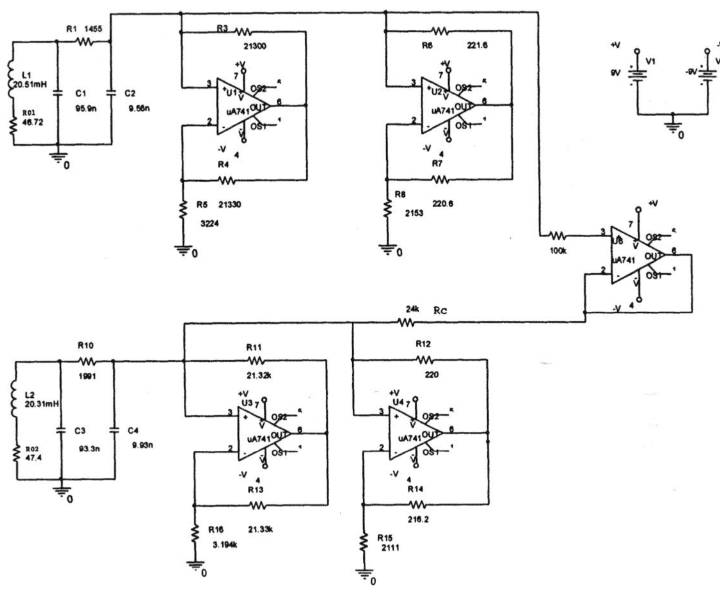

FIGURE 1. Circuit schematic of coupled Chua's oscillators

The circuit components of the driver are chosen as R_1= 1455ohm, inductance L_1=20.51mH with series resistance R_{01}=46.72, and capacitors C_1=95.9nF, C_2=9.56nF, and the nonlinear resistance is simulated by using two op-amps (uA741) with resistances R_3=21.3kohm, R_4=21.33kohm, R_5=3224ohm, R_6=221.6ohm, R_7=220.6ohm, R_8=2.153kohm. The components of the response circuit are selected as R_{10}=1991ohm, L_2=20.31mH with series resistance R_{02}=47.4, C_3=93.3nF, C_4=9.93nF, and the nonlinear resistance with R_{11}=21.32kohm, R_{12}=220ohm, R_{13}=21.33kohm, R_{14}=216.2ohm, R_{15}=2111ohm, R_{16}=3194ohm. The coupling resistance R_c is chosen as

24kohm in the PS limit. For the selected circuit components, the driving Chua's oscillator is near single scroll period-3 and period-4 oscillation and even chaotic when R_1 is decreased from 1455ohm to 1437ohm. The response oscillator is a point attractor. The governing equations of the coupled circuit are given as

$$\frac{dV_{C_2}}{dt} = \frac{G_1}{C_2}(V_{C_1} - V_{C_2}) - \frac{1}{C_2} f(V_{C_2}) \tag{1a}$$

$$\frac{dV_{C_1}}{dt} = \frac{1}{C_1}[I_{L1} - G_1(V_{C_1} - V_{C_2})] \tag{1b}$$

$$\frac{dI_{L1}}{dt} = -\frac{1}{L_1}(V_{C_1} + r_{01}I_{L1}) \tag{1c}$$

$$\frac{dV_{C_4}}{dt} = \frac{G_2}{C_4}(V_{C_3} - V_{C_4}) - \frac{1}{C_4} g(V_{C_4}) + \frac{1}{C_4 R_c}(V_{C_2} - V_{C_4}) \tag{1d}$$

$$\frac{dV_{C_3}}{dt} = \frac{1}{C_3}[I_{L2} - G_2(V_{C_3} - V_{C_4})] \tag{1e}$$

$$\frac{dI_{L2}}{dt} = -\frac{1}{L_2}(V_{C_3} + r_{02}I_{L2}) \tag{1f}$$

where $f(\bullet)$ and $g(\bullet)$ represents the nonlinear resistances of driver and response respectively and are given, in general form, by

$$f(V) = G_b V + 0.5(G_a - G_b)\left[\,|V+1| - |V-1|\,\right] \tag{1g}$$

and $G_1 = 1/R_1$, $G_2 = 1/R_{10}$ and f(V_{C2}), g(V_{C4}) are the piecewise-linear function representing the DP characteristics of the nonlinear resistances of the driver and the response respectively. $G_a(G_{a'})$ and $G_b(G_{b'})$ are the inner and outer slopes of the characteristic of nonlinear resistance of the driver (response).

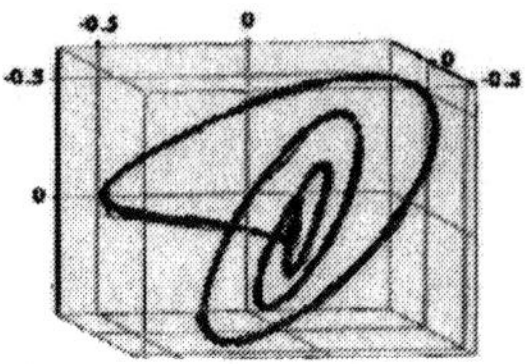

ABC

FIGURE 2. Experimental Homoclinic orbits (A) time series of V_{C3}, V_{C4} of response oscillator (B) phase portrait of V_{C3} and V_{C4} (C) 3D homoclinic orbit

Experimental results are shown in Figure 2 (from display of TEKTRONIX TDS 220 digital oscilloscope). The time series of V_{C3} (upper trace), V_{C4} (lower trace) in the response circuit are shown in Figure 2A, and phase portrait of V_{C3} and V_{C4} is shown in Figure 2B. The 3D diagram of homoclinic chaos in Figure 2C is reconstructed using delay coordinate from experimental data of V_{C4}.

MECHANISM OF HOMOCLINIC CHAOS GENERATION

A Chua's oscillator has three saddle focus equilibrium, one in the inner region D_0 at zero and other two in the odd symmetric outer regions D_+, D_- [10]. For the selected circuit parameters of the coupled circuit, the eigen values of the driver and response oscillators for both inner and outer regions are given in Table 1.

Table 1 Eigen values of Coupled Chua's Oscillators

Circuit	Inner Region	Outer Region
Driver	5.568, $-1.378 \pm 1.996i$	-5.531, $0.12 \pm 2.596i$
Response	2.65, $-1.205 \pm 3.805i$	-3.232, $-0.173 \pm 3.315i$

Two Chua's oscillators have been coupled here in unidirectional mode for generation of homoclinic chaos. The driver circuit parameters are selected for single scroll chaotic oscillation near period-3, period-4 orbits, when the odd symmetric equilibrium points are saddle focus with one negative real eigen value and a pair of complex conjugate eigen values with positive parts. The trajectory near period-3 orbit in driver oscillator spirals away from the outer region into the inner region D_0 and even down to zero (for period-3 orbit) at (0,0,0) equilibrium, when they are folded [10] back to the outer region (D_+ or D_-) by the positive real eigenvector of inner region D_0. For single scroll attractor near period-4 and higher periods, the trajectory expands spirally from the outer region and even crosses the zero equilibrium point in the inner region. Subsequently, the trajectory twisted around the positive real eigenvector of zero region and folded back to its original outer region. The response oscillator parameters are selected in the point attractor mode. The equilibrium points at the outer regions of the response oscillator are stable focus with one negative real eigen value and a pair of complex conjugate eigen values with negative real parts. But the equilibrium at (0,0,0) in the inner region of the response oscillator is still a saddle focus with a positive real and a pair of complex conjugate eigen values with negative real parts.

When the response oscillator is driven by near period-3, period-4 or even higher periodic oscillation of the driver, the response oscillator follows the driver with a time lag for weaker coupling limit until the driver trajectory reaches zero. At the instant the driver trajectory reaches zero, the response oscillator trajectory fold back fast to its stable focus in the outer region along the negative real eigenvector. The driver grows beyond zero again to repeat its periodic trajectory. Then the trajectory of the response oscillator moves away spirally from its stable focus again following the trajectory of the driver and repeats. The trajectory of the response oscillator is repeatedly forced away from its stable focus by the driver and comes back along its negative real eigenvector at an instant when the driver trajectory reaches zero. At this instant, the driver and response are phase locked. Although the homoclinic oscillation appears very regular as periodic, the return time varies. The phase locking between driver and response has been elaborated in Figure 3, where X(t) is defined as

$$X(t) = V_{C2}(t) - A.V_{C4}(t) \qquad (2)$$

X(t) is the difference between voltages $V_{C2}(t)$ at driver capacitor C_2 and $V_{C4}(t)$ at response capacitor C_4. A is constant arbitrarily chosen to adjust the difference of amplitude between $V_{C2}(t)$ and $V_{C4}(t)$. The difference signal X(t) gives a measure of phase difference between drive and response signal. It is evident from Figure 3, that the phase difference is zero at the instant $V_{C2}(t)$ folds back while $V_{C4}(t)$ folds back to its point attractor equilibrium.

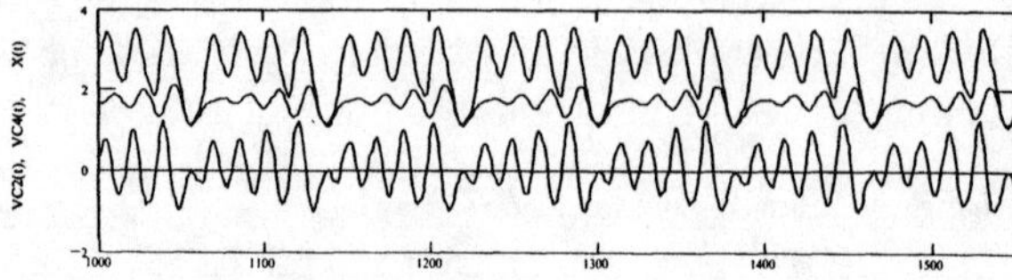

Figure 3. Phase relation between driver and response: upper trace shows $V_{C2}(t)$, middle trace shows $V_{C4}(t)$ and lower trace shows the difference signal X(t)

The coupling strength plays an important role in homoclinic oscillation, which must be in the weaker limit of PS. CS [14] of chaotic oscillators is possible for strong coupling, while LS has been observed for weaker coupling and PS is possible after further weakening of coupling strength. Another route to PS from CS through intermittency has been reported [15] recently.

PHASE LOCKING WITH FORCING PULSE

The coupled Chua's oscillators generates unstable homoclinic oscillations at the response oscillator, which has been stabilized by applying periodic pulse at the driver capacitor C_2. The unstable homoclinic oscillation readjusts to different stable homoclinic oscillation. The amplitude of external forcing pulse is selected small enough while the pulse period is chosen as close to the period of homoclinic oscillation. The external pulse is phase locked to the homoclinic oscillations with different locking ratios. For the selected parameters, the natural period of the Chua's oscillator (driver) is about $T_n = 1/f_n = 390\mu s$. When the pulse period is close to $T_p \cong T_n \pm T_s$ ($T_s \cong 14\mu s$, a small limit of time period), 1:1 phase locking has been observed (Figure 4A). Homoclinic oscillation is a period-5 oscillation, hence the period of a homoclinic cycle is $T_h = 5T_n = 1950us$ while the forcing pulse has an amplitude of 248mV and time period $T_p = 1936us$ (hence $T_s = 14\mu s$). The pulse amplitudes are measured by the oscilloscope at the source. Actual amplitude of the pulse forced at the capacitor C_2 are much less since the pulse source is connected to the capacitor through a 25kohm series resistance.

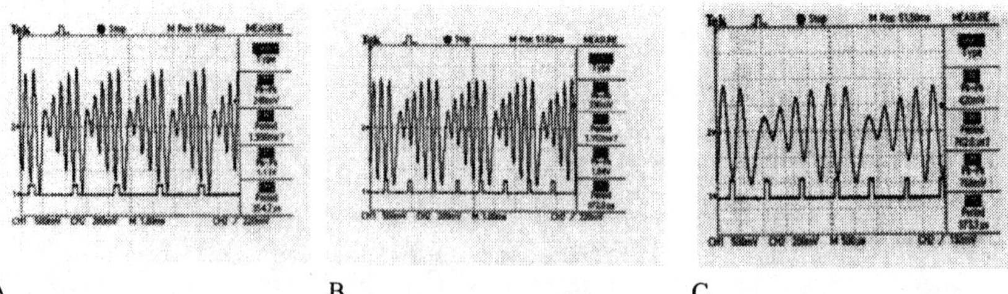

A   B   C

Figure 4. Phase locking of homoclinic orbits with external pulse (A) 1:1 phase locking with pulse amplitude=248mV, T_p=1936us (B) 1:2 phase locking with pulse amplitude=336mV, T_p=1152us (C) 1:3 phase locking with pulse amplitude=420mV, T_p=762us

For pulse amplitude 336mV and time period T_p=1152us, the homoclinic oscillation is a stable perioid-6 orbit (Figure 4B), whose time period is given by T_h=6T_n= 2340us. The pulse is 1:2 phase locked to the period-6 homoclinic orbit; T_h=2T_p+T_s =2340us (Ts=36us). For pulse amplitude of 420mV and T_p=762us, the homoclinic oscillation is stable at perioid-6 orbit (Figure 4C), whose time period is given by T_h=6T_n= 2340us. The pulse is 1:3 phase locked to the period-6 homoclinic orbit; T_h=3T_p+T_s =2340us (T_s=54us). If the difference in time period between forced pulse and homoclinic orbit is much larger beyond a limit (say, ~54us or more), phase slips occurs and finally loses phase. Details are not shown here due space limitations.

CONCLUSION

Generating homoclinic chaos in a single Chua's oscillator is challenging due to inherent instabilities of saddle focus equilibrium of the circuit. Experimental evidences are given here that it can be generated in a set of two Chua's oscillators coupled unidirectionally and when the coupling strength is in the weaker limit of PS. When forced external periodic pulses, phase locking with different locking ratios have been observed if the forcing period is close to the natural period of the oscillator and its sub-harmonics.

REFERENCES

[1] C.P.Silva, Shil'nikov's theorem-a tutorial, *IEEE Trans.Cir.Syts.*-I, **40** (10) 675-682 (1993)

[2] E.Allaria, F.T.Arecchi, A.Di Garbo, M.Meucci, Synchronization of homoclinic chaos, *Phys.Rev.Lett.*, **86** (5) 791-794 (2001)

[3] I.P.Mariño, E.Allaria, M.Meucci, S.Boccaletti, F.T.Arecchi, Information encoding in homoclinic chaotic systems, *preprint arXiv.nlin.CD/0109o26 v1*, 20 Sept, 2001

[4] D.Postnov, S.K.Han, H.Kook, Synchronization of diffusively coupled oscillators near the homoclinic bifurcation, *Phys.Rev. E*, **60**, no.3, pp.2799-2807 (1999)

[5] D.E.Postnov, A.G.Balanov, N.B.janson, E.Moslekilde, Homoclinic bifurcation as a means of chaotic phase synchronization, *Phys.Rev. Lett.*, **83** (10), 1942-1945 (1999)

[6] T.L.Carroll,L.M.Pecora, Synchronizingchaotic circuits, *IEEE Trans Cir.Systs.*,**38**(4),453-456 (1991)

[7] M.Sushchik, N.Rulkov, H.Abarbanel, Robustness and stability of synchronized chaos:An illustrative model, *IEEE Trans. Cir.Systs-I*, **44** (10), 867-873 (1997)

[8] M.Sushchik, N.Rulkov, L.Larson, L.Tsimring, H.Abarbanel, K.Yao, A.Volkovskii, Chaotic pulse position modulation: A robust method of communicating with chaos, *IEEE Commun. Lett*, **4** (4), 128-130 (2000)

[9] E.M.Izhikevich, Neural excitability, spiking and bursting, *Int. J.Bifur.Chaos*, **10**, 1171-1266, 2000

[10] M.P.Kennedy, Three steps to chaos-Part II: A Chua's circuit primer, *IEEE Trans. Cir.Systs-I*, **40** (10), 657-674 (1993)

[11] O. de Feo, G.M.Maggio, M.P.Kennedy, The Colpitts oscillator:Families of periodic solutions and their bifurcations, *Int.J.Bifur.Chaos*, **10** (5), 935-958 (2000)

[12] A.R.Champneys, Y.A.Kuznetsov, B.Sandstede, A numerical tools for homoclinic bifurcation analysis, *Int.J.Bifur. Chaos*, **6**, 867-887 (1996)

[13] L.O.Chua, M.Komuro, T.Matsumoto, The double scroll family, *IEEE Trans.Cir.Systs.*, **33**, 1072-1118 (1986)

[14] M.G.Rosenblum, A.S.Pikovsky, J.Kurths, From phase to lag synchronization in coupled chaotic oscillators, *Phys.Rev.Lett.*, **78**, 4193-4196 (1997)

[15] S.Rim, I. Kim, Y-J. Park, C-H. Kim, Routes to complete synchronization in coupled nonidentical chaotic oscillators, *Phy.Rev.E*, **66**, 015205 (R) (2002)

OPTICS

Evidence of Noise Induced Synchronization and Coherence Resonance in Homoclinic Chaos

R. Meucci[*], C.S. Zhou[†], E. Allaria[*], F.T. Arecchi[***], S. Boccaletti[*] and J. Kurths[†]

[*]*Istituto Nazionale di Ottica Applicata, Largo E. Fermi, 6 150125 Florence, Italy*
[†]*Institute of Physics, University of Potsdam PF 601553, 14415 Potsdam, Germany*
[**]*Dept. of Physics, University of Florence, Florence, Italy*

Abstract. We present experimental and numerical investigation of noise-induced synchronization (NIS) and coherence resonance (CR) in a system displaying homoclinic chaos. These two effects have been observed in one experimental laser system by using weak noise intensities that do not change the main geometrical features of the unperturbed dynamics. These nontrivial effects originating from the intrinsically strong nonuniform dynamics of homoclinic chaos are of grat relevance to neuroscience.

Effects of noise in nonlinear systems have been a subject of great interest in the context of stochastic resonance (SR) [1] and coherence resonance (CR) [2]. With SR, noise can optimize a system's response to an external signal, while with CR, pure noise can generate the most coherent motion in the system.

On the other hand, two identical systems which are not coupled, but subjected to a common noise may synchronize, as has been reported both in the periodic [3] and in the chaotic [4] cases. For noise-induced synchronization (NIS) to occur, the largest Lyapunov exponent (LLE) ($\lambda_1 > 0$ in chaotic system) has to become negative [4]. However, whether noise can induce synchronization of chaotic systems has been a subject of intense controversy [5, 6, 7]. The debate mainly focuses on the effect of the mean value of noise [6, 7]. It has been shown that the nonzero mean of the applied noise plays a decisive role by shifting the dynamics toward a stable regime (fixed point or periodic orbit smeared by noise) [6, 7]. However, a general conclusion [7] that an unbiased noise cannot lead to synchronization has been disproved by recent examples where NIS has been observed for unbiased noise[8]. Recently, we have clarified this long-standing controversy by showing that the key mechanism of NIS is the existence of a large *contraction region* in the phase space where all the eigenvalues of the Jacobian matrix have a negative real part and nearby trajectories converge to each other [9]. Noise changes the competition between contraction and expansion, and synchronization ($\lambda_1 < 0$) occurs if the contraction becomes dominant. However, in the systems of Ref.[8, 9], the dynamical structure has been significantly deformed when NIS occurs at large enough noise intensity [8, 9].

Common noise is of great relevance in several disciplines, e.g. neuroscience. Neurons connected to another group of neurons, receive a common input signal which often approaches a Gaussian distribution as a result of integration of many independent synaptic

CP676, *Experimental Chaos: 7th Experimental Chaos Conference,*
edited by V. In, L. Kocarev, T. L. Carroll, B. J. Gluckman, S. Boccaletti, and J. Kurths
© 2003 American Institute of Physics 0-7354-0145-4/03/$20.00

currents. The shape of neuronal spike is preserved in the presence of a noisy input, which is important for biological information processing. This Letter addresses the following question: in what type of systems a significant contraction region can exist and NIS can be observed while the basic dynamical structure is preserved in spite of the presence of noise?

We study a general class of chaotic systems possessing the structure of a saddle point S embedded in the chaotic attractor, i.e. homoclinic chaos [10]. There chaotic trajectories starting from a neighborhood of the saddle point will have very close recurrence to S. There exists generically an expansion region close to the unstable manifold of S, which is the origin of the chaoticity ($\lambda_1 > 0$), and a large contraction region close to the stable manifold of S so that expanded trajectories converge for the close recurrence to the neighborhood of S. Such a structure underlies spiking behavior in many neuron [11], chemical [12] and laser [13] systems. The dynamics is characterized by rather regular orbits in the phase space and widely fluctuating time intervals T between successive returns, because the trajectory slows down considerably and T depends on how close the orbit approaches S. It is important to note that this type of systems has intrinsically highly nonuniform dynamics and the sensitivity to small perturbations is only high in the vicinity of S. Hence, the basic geometrical structure of the orbits is preserved, while T may be changed significantly. Especially, noise may reduce the fluctuation of T and enhance the coherence of spike sequences.

In the following, we demonstrate both NIS and CR of homoclinic chaos in a single mode CO_2 laser with a feedback proportional to the output intensity, numerically as well as experimentally.

We first present numerical simulation of the laser model recently introduced [14]:

$$
\begin{aligned}
\dot{x}_1 &= k_0 x_1 (x_2 - 1 - k_1 \sin^2 x_6), & (1)\\
\dot{x}_2 &= -\gamma_1 x_2 - 2k_0 x_1 x_2 + g x_3 + x_4 + p_0, & (2)\\
\dot{x}_3 &= -\gamma_1 x_3 + g x_2 + x_5 + p_0, & (3)\\
\dot{x}_4 &= -\gamma_2 x_4 + z x_2 + g x_5 + z p_0, & (4)\\
\dot{x}_5 &= -\gamma_2 x_5 + z x_3 + g x_4 + z p_0, & (5)\\
\dot{x}_6 &= -\beta \left(x_6 - b_0 + \frac{r x_1}{1 + \alpha x_1} \right) + D\xi(t). & (6)
\end{aligned}
$$

Here, x_1 represents the laser output intensity, x_2 the population inversion between the two resonant levels, x_6 the feedback voltage signal which controls the cavity losses, while x_3, x_4 and x_5 account for molecular exchanges between the two levels resonant with the radiation field and the other rotational levels of the same vibrational band. Furthermore, k_0 is the unperturbed cavity loss parameter, k_1 determines the modulation strength, g is a coupling constant, γ_1, γ_2 are population relaxation rates, p_0 is the pump parameter, z accounts for the number of rotational levels, and β, b_0, r, α are respectively the bandwidth, the bias, the amplification and the saturation factors of the feedback loop. A Gaussian noise term $\xi(t)$ with zero mean, δ-correlation in time and intensity D is added to the feedback loop. In order to compare our results with the experimental trials, in the following we set $k_0 = 28.5714$, $k_1 = 4.5556$, $\gamma_1 = 10.0643$, $\gamma_2 = 1.0643$,

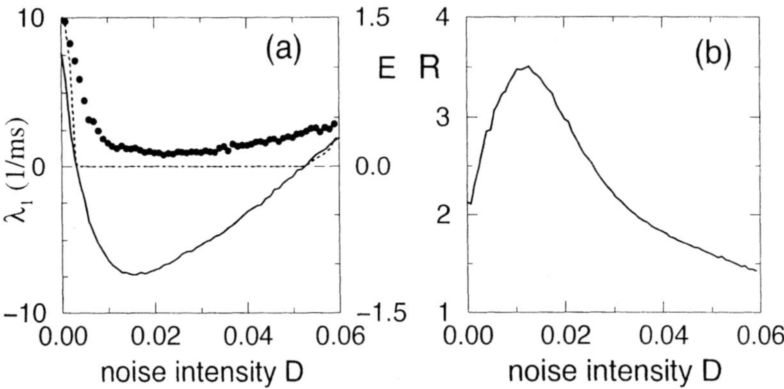

FIGURE 1. Noise-induced synchronization (NIS) and coherence resonance (CR) in the the noisy laser model. (a) Solid line: the LLE λ_1, dotted line: normalized synchronization error E between two fully identical lasers **x** and **y** and filled cycles: E between two lasers with a small amount of independent noise (intensity $D_1 = 0.0001$) vs. intensity D of the common noise. (b) Coherence factor R vs. D.

$g = 0.05, p_0 = 0.016, z = 10, \beta = 0.4286, \alpha = 32.8767, r = 160$ and $b_0 = 0.1031$, which describe very accurately the regime of homoclinic chaos observed experimentally [14].

We calculate the LLE λ_1 as a function of the noise intensity D (Fig. 1a, solid line). λ_1 undergoes a transition from a positive to a negative value at $D_c \approx 0.0031$. Beyond D_c, two identical lasers **x** and **y** with different initial conditions but the same noisy driving force achieve complete synchronization after a transient, as shown by the vanishing normalized synchronization error $E = \frac{\langle |x_1 - y_1| \rangle}{\langle |x_1 - \langle x_1 \rangle| \rangle}$ (Fig. 1a, dotted line). Shortly before D_c, the behavior is characterized by on-off intermittency where synchronization is interrupted by short bursts of desynchronized spikes.

The synchronization behavior in this system is determined by the competition between contraction and expansion in the phase space. Without noise, the laser intensity displays large spikes, followed by a fast damped train of a few oscillations and a successive longer train of chaotic bursts which on average appears as a growing oscillation, as seen by the time series of the laser output (Fig. 3a,b). The damping oscillation manifests the contraction in the phase space, while the growing one manifests the expansion, which can be described approximately as

$$X(t) \sim X_0 \exp[\lambda_u(t - t_0)] \cos \omega(t - t_0), \tag{7}$$

where $\lambda_u \pm i\omega$ are the eigenvalues of the unstable manifold of S and X_0 is the distance from S at time t_0. Thus, the closer the trajectory comes to S, the longer is the time taken to spiral out. In the presence of noise, the trajectory on average cannot come closer to S than the noise level. With a larger X_0, the system spends a shorter time following the guidance of the unstable manifold so that the degree of expansion is reduced. When contraction becomes dominant, the LLE takes a negative value, and hence identical systems synchronize. At larger noise intensities, expansion becomes again significant, and the LLE increases and synchronization is lost when λ_1 becomes positive for $D >$ 0.052. Notice that even when $\lambda_1 < 0$, the trajectories still have access to the expansion region where the largest local Lyapunov exponent is positive so that small distances between them grow temporally. As a result, when the systems are subjected to additional perturbations due to parameter mismatch or discrepancies between the noisy driving forces, which are unavoidable in real systems, synchronization is lost intermittently, especially for D close to the critical values. The transition to synchronization should be therefore smeared out in real systems. To take into account this unavoidable nonidentity in experimental systems, we introduce into the equations x_6 and y_6 a small amount of independent noise (with intensity $D_1 = 0.0001$) in addition to the common one $D\xi(t)$, and calculate the synchronization error E again. Parameter mismatches have similar effects. By comparison, it is evident that the sharp transition to a synchronized regime in fully identical model systems (Fig. 1a, dotted line) is smeared out by this nonidentity (Fig. 1a, filled cycles), as observed in the experimental data that will be presented below.

Furthermore, we remark that NIS is obtained already for a very low noise intensity D. If compared with the peak amplitude of the spikes in x_1 (multiplied by r and β), the critical intensity $D_c \approx 0.0031$ represents only about 1% perturbation to the feedback loop. In the presence of nonidentity, the onset of synchronization occurs at about 3% perturbation to the feedback loop. This tiny amount of noise only affects the system's behavior close to the saddle S, while it does not change the main geometrical feature of the orbit. This is at variance with other systems [8, 9] where the dynamical structure has been distorted significantly when NIS occurs at large noise intensities, because there noise makes the systems explore regions in the phase space usually not accessible to the trajectory in the absence of noise.

Based on the above considerations, we expect also noise enhanced coherence of spike sequences. The interspike interval T becomes shorter on average due to larger X_0 for increasing noise intensity. It is also noted from Eq. (7) that, when X_0 is larger, a variation ΔX_0 of it may result in a smaller difference in ΔT, because they are related approximately by $\Delta T \sim \Delta X_0 / X_0$, so that the fluctuation of T becomes smaller. When the noise is rather large, it affects the dynamics not only close to S but also during the spiking, so that the spike sequence becomes fairly noisy. We observe the most coherent spike sequences at a certain intermediate noise intensity. We quantify the coherence by R,

$$R = \langle T \rangle / \sqrt{\operatorname{var}(T)}, \tag{8}$$

as a function of noise intensity D (Fig. 1b). When D increases, R reaches a maximal value and decreases again, exhibiting coherence resonance (CR), similar to that of excitable systems [2], but with a different mechanism. At variance with excitable systems [2], where noise induces spiking by kicking the system over an energy barrier, here in the homoclinic chaos, the spike sequence is generated by chaotic close recurrence to the saddle S, and CR occurs as a consequence of a small noise that changes the time spent in the neighborhood of S. This mechanism is different from noise-induced coherent jumping among coexisting attractors [15].

In our model NIS and CR persist over a large region of the parameters space.

These effects predicted in the model have been tested in a real laser experiment. The numerical parameters considered above refer to a regime of homoclinic chaos in a CO_2

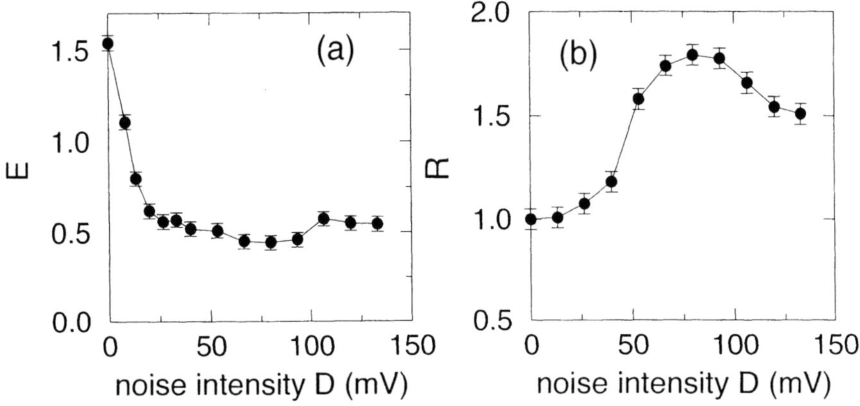

FIGURE 2. (a) Experimental observation of NIS and CR. The normalized synchronization error E (a) and coherence factor R (b) vs. noise intensity D.

laser with electro-optic feedback as recently reported in Ref. [16]. We use the same experimental setup, consisting of a CO_2 laser with an intracavity loss modulator, driven by a feedback signal which is a function of the laser output intensity. Here, in order to investigate the role of external noise and its capability to induce NIS and CR, we use a random noise generator with a 50 kHz Gaussian distribution. Precisely, a noise signal, recorded by means of a real time Input-Output PC board with a sampling time of 2 μs, is added to the feedback loop for a finite duration of 400 ms. This long time is well beyond the correlation time of the unperturbed homoclinic dynamics [16].

To study NIS experimentally for each noise intensity D, we repeat the experiment twice with the same noise. The normalized synchronization error E between the laser output intensities x_1 and y_1 of the two experimental runs is then calculated and shown in Fig. 2a as a function of the noise intensity D. Because in repeated experiments the systems are subjected to a small inevitable hardware noise in addition to the common external one, and thus are not fully identical, E does not reach zero exactly, and it increases slightly at large D, as is consistent with the numerical results (Fig. 1a, filled cycles).

The experimental coherence factor R (Fig. 2b) again resembles numerical results. It is important to emphasize that NIS and CR are obtained for rather small noise intensities. In particular, the onset of NIS occurs at an experimental noise ($D \approx 20$ mV) of about 3% of the feedback driving signal, whereas the maximal coherence is obtained for a noise ($D \approx 66$ mV) of about 10% of the feedback signal.

To present a more descriptive view of NIS and CR, we report in Fig. 3 some portions of the numerical (a,c,e) and experimental (b,d,f) laser intensity data for three noise intensities. Precisely, Fig. 3a,b are obtained without a common noise, Fig. 3c,d for the noise value D_{max} at which the coherence factor R is maximal, and Fig. 3e,f are obtained for a noise intensity $D \approx 2D_{max}$. In the absence of a common noise, the two signals are unsynchronized, the spike intervals T are large on average and fluctuate strongly.

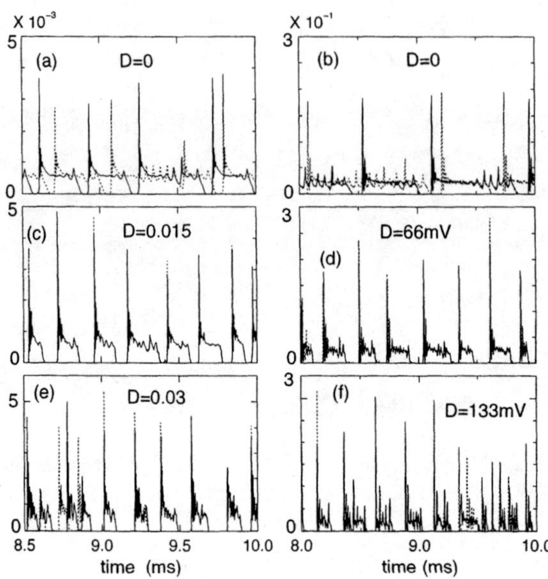

FIGURE 3. Time series of the laser output intensities (arbitrary units) x_1 (solid lines) and y_1 (dotted lines) with a common noise at difference intensities. The left panel (a, c, e) is for the models including independent noise components (with intensity $D_1 = 0.0001$). The right panel (b, d, f) is for the experimental systems.

For $D \approx D_{max}$, both experimental and numerical results demonstrate the existence of almost complete synchronization induced by the external noise, and in parallel the spike sequences are most coherent with a smaller average T, because the escaping time from the vicinity of the saddle S has been reduced; yet, the main geometry of spikes is preserved. Finally, at larger noise, synchronization is intermingled with short epochs of desynchronization. All features of NIS and CR in the model and in the experimental system are in good agreement.

Let us summarize the main results presented in this Letter, and discuss their relevance. We have given evidence for the first time that NIS and CR can be both found in the same experiment. Unbiased noise is sufficient to induce NIS. Moreover, NIS is not necessarily associated with large noise intensities, but in the case of homoclinic chaos can be induced by weak noise which does not significantly alter the geometry of the dynamics. This finding should be of great relevance to neuroscience. In this regard, there are observations [17] of a remarkable reliability of repetitive spike sequences in neucortical neurons in the response to repeated fluctuating stimuli (colored Gaussian noise) that resemble real synaptic currents, a feature which is not observed in the response to constant input currents. This reliability is of significance for information encoding by spiking timing precision. We have also investigated a chaotic Hodgkin-Huxley model of thermally sensitive neurons, which mimics various types of spike train patterns in electroreceptors from dogfish and catfish, and from facial cold receptors and hypothalamic neurons of the rat [18]. This model has been demonstrated to display homoclinic bifurcation when the control parameter temperature is varied [19]. Here we also find NIS at rather small noise intensity in a wide parameter range. By synchronization, an ensemble of neurons can establish a collective response to a common fluctuating stimulus. Therefore, the mechanism presented here can interpret the experimental observations of repetitive spiking in Ref. [17] and bring new understanding of neuron encoding and biological information processing.

ACKNOWLEDGMENTS

This work is supported by MPG (Germany), Humboldt-Foundation (Germany), and EC Network n. HPRN CT 2000 00158.

REFERENCES

1. R. Benzi, A. Sutera, and A. Vulpiani, J. Phys. A **14**, L453 (1981); K. Wiesenfeld and F. Moss, Nature **373**, 33 (1995); L. Gammaitoni, P. Hänggi, P. Jung, and F. Marchesoni, Rev. Mod. Phys. **70**, 223 (1998).
2. G. Hu, T. Ditzinger, C.Z. Ning, and H. Haken, Phys. Rev. Lett. **71**, 807 (1993); A.S. Pikovsky and J. Kurths, *ibid.* **78**, 775 (1997)). B. Hu and C. Zhou, Phys. Rev. E **61**, R1001 (2000); C. Zhou, J. Kurths and B. Hu, Phys. Rev. Lett. **87**, 98101 (2001); G. Giacomelli, M. Giudici, S. Balle, and J. R. Tredicce, *ibid.*, **84**, 3298 (2000); J. L. A. Dubbeldam, B. Krauskopf, and D. Lenstra, Phys. Rev. E, **60**, 6580 (1999).
3. A.S. Pikovsky, Radiophys. Quantum Electron. **27**, 576 (1984); R.V. Jensen, Phys. Rev. E **58**, R6907 (1998).
4. K. Matsumoto and I. Tsuda, J. Stat. Phys. **31**, 87 (1983); A.S. Pikovsky, Phys. Lett. A **165**, 33 (1992); L. Yu, E. Ott and Q. Chen, Phys. Rev. Lett. **65**, 2935 (1990).
5. A. Mritan and J.R. Banavar, Phys. Rev. Lett. **72**, 1451 (1994).
6. H. Herzel, and J. Freund, Phys. Rev. E **52**, 3238 (1995); G. Malescio, *ibidi* **53**, 6551 (1996); P.M. Gade, and C. Basu, Phys. Lett. A **217**, 21 (1996).
7. E. Sanchez, M.A. Matias and V. Perez-Munuzuri, Phys. Rev. E **56**, 4068 (1997).
8. C.H. Lai, and C. Zhou, Europhys. Lett. **43**, 376 (1998); R. Toral, C.R. Mirasso, E. Hernandez-Garcia, and O. Piro, Chaos 11, 665 (2001); L. Baroni, R. Livi and A. Torcini, Phys. Rev. E, **63**, 036226 (2001).
9. C. S. Zhou and J. Kurths, Phys. Rev. Lett. (submitted).
10. L. P. Shilnikov, Math. USSR Sb. **10**, 91 (1970).
11. A. L. Hodgkin and A. F. Huxley, J. Physiol. **117**, 500 (1952).
12. F. Argoul, A. Arnéodo, and P. Richetti, J. Chem. Phys. **84**, 1367 (1987).
13. F. T. Arecchi, R. Meucci, and W. Gadomski, Phys. Rev. Lett. **58**, 2205 (1987).
14. A.N. Pisarchik, R. Meucci and F.T. Arecchi, Eur. Phys. J. D **13**, 385 (2001).
15. C. Palenzuela *et al*, Europhys. Lett. **56**, 347 (2001); C. Masoller, Phys. Rev. Lett. **88**, 034102 (2002).
16. E. Allaria, F.T. Arecchi, A. di Garbo and R. Meucci, Phys. Rev. Lett. **86**, 791 (2001).
17. Z.F. Mainen and T.J. Sejnowski, Science **268**, 1503 (1995); A.C. Tang, A.M. Bartels, and T. Sejnowski, Cerebral Cortex **7**, 502 (1997).
18. H.A. Braun, M.T. Huber, M. Dewald, K. Schäfer and K. Voigt, Int. J. Bifurcation Chaos, **8**, 881 (1998).
19. U. Feudel *et al*, Chaos **10**, 231 (2000).

Noise Induced Burst Synchronization in Fiber Ring Lasers

David J. DeShazer, Brian P. Tighe, Jürgen Kurths[*] and Rajarshi Roy

*Department of Physics and Institute for Research in Electronics and Applied Physics
University of Maryland College Park, College Park, Maryland 20742*
Institute of Physics
University of Potsdam PF 601553, 14415 Potsdam, Germany

Abstract. Can bursts in dynamical systems be synchronized by a weak, common, noise background? We observe large, uncorrelated bursts of intensity fluctuations in two nearly identical erbium-doped fiber ring lasers, initiated by common injection of a weak, constant intensity optical signal. The magnitude and frequency of the chaotic bursting depends upon the injected wavelength and intensity. Significant synchronization of the bursts is obtained for noise and sinusoidal modulation of the injected light intensity. The degree of synchronization is quantified using an envelope constructed from the amplitude of the analytic signal, and is studied as a function of the noise variance. The noise induced burst synchronization can be viewed as a form of generalized synchronization of the fiber ring lasers to the common modulation signal, though no obvious relationship between the drive and response dynamics is directly observed.

INTRODUCTION

The synchronization of bursting systems is highly relevant to understanding aspects of communication. A question of great general interest is whether two dynamical systems subject to a common noise background will synchronize their bursts; and if so, is there a noise threshold at which the synchronization occurs? Here, we report experiments carefully designed to study these questions quantitatively in a way that is difficult, if not impossible, at present with biological systems. We note that the synchronization of uncoupled chaotic dynamical systems driven by a common noise source is a topic that has raised such controversy over the years [1-4] Although the debate has persisted, and many numerical studies have been carried out, there have been only very limited quantitative experiments performed to examine this issue [3].

The study of synchronization of nonlinear, chaotic systems began more than a decade ago and focused at that time on what is now often called identical synchronization [5]. It was soon realized that other, weaker types of synchronization, such as generalized and phase synchronization are widespread in nature [6-8]. This is particularly true for systems that are not identical in parameter values or even their geometrical aspects.

Burst synchronization is a further example of weak synchronization - the individual oscillations within an envelope need not be synchronized in amplitude or phase - only the burst envelopes are correlated between the systems. Here we study synchronization of burst dynamics in erbium doped fiber ring lasers (EDFRL)[9]. The intensity bursts are initiated by injection of a common, weak, optical signal into two almost identical EDFRLs [10]. When the injected monochromatic signal is constant in

CP676, *Experimental Chaos: 7th Experimental Chaos Conference*,
edited by V. In, L. Kocarev, T. L. Carroll, B. J. Gluckman, S. Boccaletti, and J. Kurths
© 2003 American Institute of Physics 0-7354-0145-4/03/$20.00

power, the intensity bursts in the two lasers are quite unsynchronized with a broad distribution of inter-burst times, as would be expected. In these experiments, we show that a small level of broadband noise or sinusoidal modulation of the common injected optical signal is sufficient to synchronize the bursts that occur in the two lasers to a remarkable extent.

Experimental Setup

The experimental set-up is displayed in Fig. 1. The active element of the two EDFRLs is an erbium-doped fiber amplifier (EDFA. Both ring-lasers are operated far above threshold. The uni-directionality of each ring-laser is enforced by Faraday optical isolators internal to the EDFAs. The total fiber ring lengths are approximately 41.5 m and are matched to within 1% of each other. We select a mode of operation where both EDFRLs display a single-peaked spectrum and are tuned to within 5 GHz of each other and centered at ($\lambda = 1557.7$ nm), with a full width at half maximum of approximately 0.6 nm.

A common, constant, optical signal, also at $\lambda_{inj} = 1557.7$ nm, is injected into the rings using 70/30 fiber-optic evanescent field couplers, stimulating large intensity bursts in the ring-lasers. The constant signal source is a tunable external cavity semi-conductor diode laser (ECSL). This signal may be amplitude modulated with a lithium-niobate electro-optic modulator (EOM).

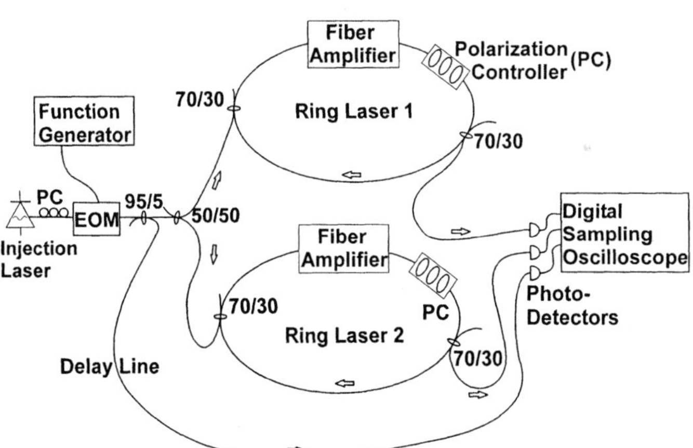

Figure 1. Experimental Setup: Two nearly identical Erbium-Doped Fiber Ring-Lasers (EDFRLs) of cavity length ≈ 41.5 m are injected with a common, weak optical signal. The injection laser output may be amplitude modulated with the Electro-Optic Modulator (EOM) driven by the Function Generator.

Noise-Induced Burst Synchronization

The bursting dynamics of the EDFRLs, with an injection power $P_{inj} = 32\,\mu W$ and a circulation power of $P_{RL} \approx 9$ mW circulating in the ring cavity, are plotted in Fig. 2. Figure 2(a) displays the case when the injected signal is unperturbed, resulting in the expected uncorrelated bursting. When a small amplitude, unbiased noise-modulation is applied to the injection signal a dramatic shift to burst synchronization is observed, as seen in Fig. 2(b). The strength of the noise modulation is measured by the ratio of the noise standard deviation to the mean injected power; $M_{noise} \equiv \sigma(P_{inj}/\langle P_{inj}\rangle = 2\times 10^{-2}$. Contrasting Figs. 2(a) and (b), the shift from uncorrelated to synchronized bursting is qualitatively evident.

The EDFRL intensities display bursts, consisting of large amplitude relaxation oscillations at about $v_{rel} = 45$ kHz. The relative amplitudes and phases of these fluctuations are not necessarily matched.

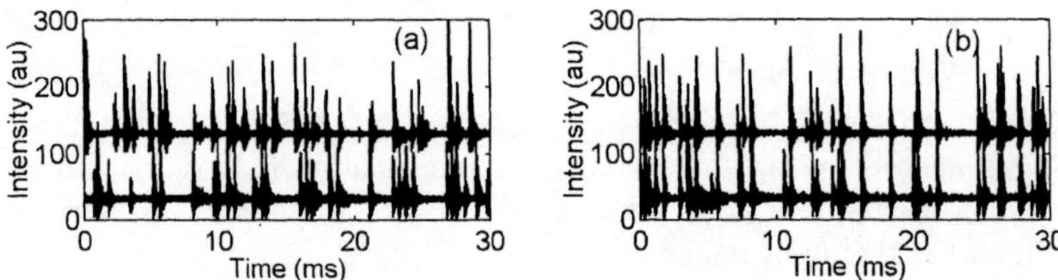

Figure 2. The bursting dynamics of the EDFRLs are uncorrelated with constant injection (a), but shift to synchronized behavior with unbiased noise modulated modulation of the optical injection signal (b).

While the relative behavior of the ring dynamics shifts dramatically towards synchronization the statistics of the bursting dynamics of an individual ring remain unaffected.

DEFINITION AND APPLICATION OF A BURST ENVELOPE

While the presence of burst synchronization is qualitatively evident in Fig. 2(b), the fluctuations of the EDFRL dynamics at $v_{rel} = 45$ kHz obscure direct attempts to determine the degree of synchronization. However, it is relatively straightforward to define a burst envelope for the dynamics using the low pass filtered analytic signal [10,11]

$$V(t) = \frac{1}{\pi}\int_0^{v_c} e^{ivt} u(v)dv. \tag{1}$$

$V(t)$ is the low-pass filtered complex analytic signal and $u(v)$ is the Fourier transform of the intensity time-series. We are not concerned with the nanosecond time-scale fluctuations intrinsic to the EDFRL dynamics. We only wish to maintain the integrity of the bursting dynamics observed in the injected ring laser. Therefore, we select a cut-off frequency, $v_c = 50$ kHz, slightly larger than the relaxation oscillations observed internal to the individual bursts. Burst synchronization is studied using the real amplitude $A(t)$ of the centered complex analytic signal $V(t) - \langle V(t) \rangle = A(t)e^{i\phi(t)}$. Hence $A(t)$ defines an envelope over the bursting intensity dynamics of a ring-laser.

Re-examining burst synchronization we again observe the shift from unsynchronized bursting with a constant injection signal, Fig. 2 (a), to burst synchronization with noise modulated injection, Fig. 2(b). In Figs. 4(a) and (b) we have shown the xy-synchronization plots of the burst envelopes, $A_2(t)$ vs. $A_1(t)$. It is clear that burst synchronization is not present in the constant injection case, Fig. 4(a). The application of noise modulation centers the plot about the diagonal, demonstrating the shift to burst synchronization, Fig. 4(b). Viewed from this perspective, burst synchronization of the two systems driven by common noise may be regarded as an example of generalized synchronization [8].

Phase synchronization of the intra-burst fluctuations may be studied using the phase $\phi(t)$ of the analytic signal $V_A(t)$) [6]. Here we have chosen to shift our cut-off frequency to infinity, hence returning to the unfiltered complex analytic signal. In Figs. 4(c) and (d) the relative phase difference $\Delta\phi(t) = (\phi_1(t) - \phi_2(t))/2\pi$ is plotted for the initial 5 ms of the unmodulated injection and noise modulated injection cases

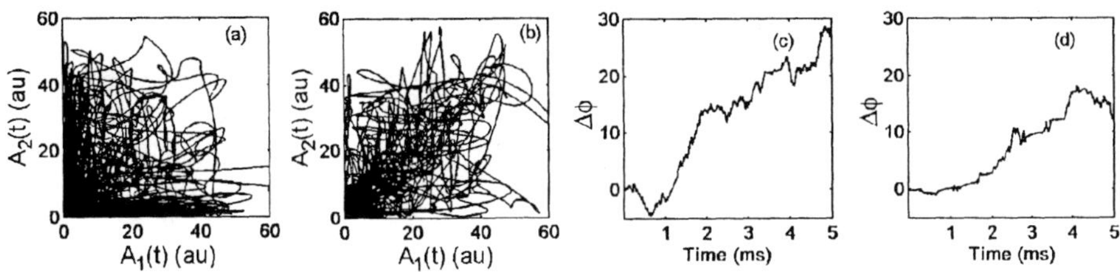

FIGURE 4. Burst and intermittent phase synchronization. (a) and (b) are xy-synchronization plots for the burst envelopes for Figs 2(a) and (b), respectively, and illustrate burst synchronization with noise modulation. Examining the analytic phases of the initial 5 ms of the bursting time series, the relative phase differences of the two EDFRLs are shown in (c) and (d). Nearly continuous phase slipping with varying slopes is observed in (c), indicating an absense of phase synchronization. The flat sections in (d) indicate phase synchronization.

respectively. The nearly continuous phase slipping with varying slopes observed in Fig. 4(c) demonstrates a lack of phase synchronization in the constant injection case. In Fig. 4(d) we note that $\Delta\phi(t)$ is roughly constant within bursts. Closer inspection shows that these phase synchronized time segments may have non-zero slopes, indicating imperfect phase-synchronization.

QUANTIFYING BURST SYNCHRONIZATION

In determining a level of burst synchronization we are not necessarily concerned with the amplitude or phase of the bursting dynamics, merely whether or not the bursts occur at the same time. Therefore we reduce the time series to symbolic dynamics indicating if the ring lasers are bursting or not at a given point in time. We define an appropriate threshold A_{th} to identify bursts, and generate a new time series $B(t)$, where

$$B(t) \equiv \begin{cases} 1, A(t) > A_{th} \\ -1, A(t) \le A_{th} \end{cases}. \tag{2}$$

Burst synchronization is easily identified and quantified with this simplified burst time-series, which has the character of a random telegraph signal [x]. To determine the level of burst synchronization we calculate the zero-delay, normalized cross-correlation

$$C \equiv \frac{1}{N} \frac{\sum_{n-1}^{N} (B_1(n) - \langle B_1 \rangle)(B_2(n) - \langle B_2 \rangle)}{\sigma(B_1)\sigma(B_2)}. \tag{3}$$

Here B_1 and B_2 are the burst time series for EDFRL 1 and EDFRL 2, respectively, and N is the total number of sampled values in the time series. With these definitions for $B(t)$ and C we have a particularly transparent measure of synchronization; $C = 1$ represents perfect synchronization, $C = -1$ indicates complete anti-synchronization, and for $C = 0$ there is no relation on the average between the bursts of EDFRL 1 and EDFRL 2.

Synchronization with Modulation Strength

Burst synchronization is studied as a function of the modulation strength of the injected power in Fig. 6. The effect of an increasing noisy modulation is displayed in Fig. 6(a), and increasing sinusoidal modulation amplitude in Fig. 6(b). The noisy modulation strength is measured by the ratio $M_{noise} \equiv \sigma(P_{inj})/\langle P_{inj} \rangle$ and the size of the sinusoidal modulation is given by $M_{\sin e} \equiv 0.5((\max(P_{inj}) - \min(P_{inj}))/\langle P_{inj} \rangle$. The minimal modulation point corresponds to no modulation in both cases. The mean injected power $\langle P_{inj} \rangle$ and the pumping of the EDFRLs are held constant, such that $\langle P_{RL} \rangle = 9 \, \text{mW}$, and $\langle P_{inj} \rangle / \langle P_{RL} \rangle = 3.6 \times 10^{-3}$. A transition from uncorrelated bursts $(C \sim 0)$ to significant synchronization $(C \sim 0.4)$ is found to occur at modulation strengths of about 4×10^{-3}.

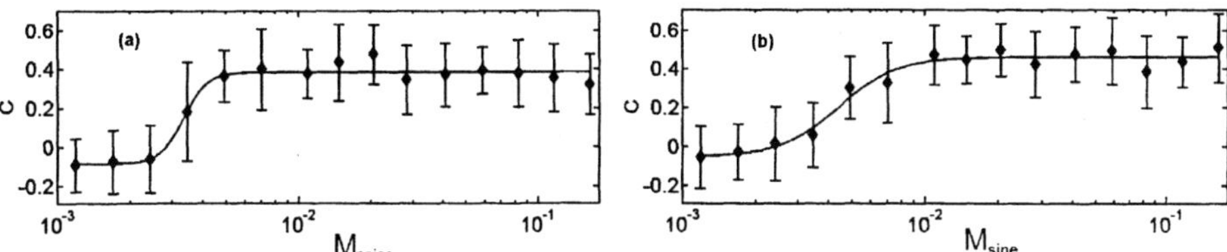

FIGURE 6. Burst synchronization is studied as a function of modulation strength. Both white-noise modulation (a) and sine modulation at 45 kHz (b) show a significant transition to synchronization with increase of the modulation strength. The solid lines are empirical fits with a sigmoidal function. M_{noise} is the ratio of the standard deviation of the noise fluctuations to the average injected power, and M_{sine} is the ratio of the amplitude of the sinusoidal modulation to the average power.

DISCUSSION

We have demonstrated that burst synchronization, as well as intermittent phase synchronization may occur in an experimental system of two nearly identical, uncoupled lasers in the presence of a common, weak modulation background. A clear transition to synchronization was demonstrated for both broad-band noise and sinusoidal modulation. Surprisingly, the noise modulation was as effective as the sinusoidal modulation. While the transition to burst synchronization is clear, the mechanism of synchronization is not. The burst synchronization suggests generalized synchronization of the EDFRLs to the modulation signal, as indicated by the replicant method [8].

ACKNOWLEDGMENTS

We gratefully acknowledge support from the Office of Naval Research (Physics) and discussions with Henry Abarbanel and Ed Ott.

REFERENCES

1. Maritan, A. and Banavar, J.R. *Phys. Rev. Lett.* **72**, 1451-1454 (1994).
2. Pikovsky, A.S. *Phys. Rev. Lett.* **73**, 2931 (1994).
3. Sanchez, E., Mathias, M.A., and Perez-Munuzuri, V. *Phys. Rev. E* **56**, 4068-4071 (1997).
4. Lai, C.-H. and Zhou, C.S. *Europhys. Lett.* **43**, 376-380 (1998).
5. Pecora, L.M. and Carroll, T.L. *Phys. Rev. Lett.* **64**, 821-824 (1990).
6. Rosenblum, M.G., Pikovsky, A.S., and Kurths, J. *Phys. Rev. Lett.* **76**, 1804-1807 (1996).
7. Rosenblum, M.G., Pikovsky, A.S., and Kurths, J. *Phys. Rev. Lett.* **78**, 4193-4196 (1997).
8. Abarbanel, H.D.I., Rulkov, N.F., and Sushchik, M.M. *Phys. Rev. E,* **53**, 4528-4535, (1996).
9. DeShazer, D.J., Tighe, B.P., Kurths, J., and Roy, R. *submitted Phys. Rev. E,* (2002).
10. DeShazer, D.J., Garcia-Ojalvo, J., and Roy, R. *to be submitted Phys. Rev. E,* (2002).
11. Gabor, D., *J. IEEE London* **93**, 429 (1946).
12. Born, M. and Wolf, E., *Principles of Optics*, 7th ed., Cambridge University Press, New York, 1999.
13. Papoulis, A., *Probability, Random Variables and Stochastic Processes*, 3rd edition, McGraw-Hill, New York, 1991.

Sub-critical Hopf bifurcation in a time-delayed differential dynamic.

Laurent Larger*, Thomas Erneux[†] and Jean-Pierre Goedgebuer*

*Lab. d'Optique P.M. Duffieux, UMR CNRS 6603, Univ. de Franche-Comté, 16 route de Gray,
F–25030 Besançon Cedex - France -
[†]Groupe d'Optique Théorique Non Linéaire, Univ. Libre de Bruxelles, Campus Plaine CP231,
B–1050 Bruxelles - Belgium -

Abstract. Experimental investigations are reported, which are intended to carry out a scalar nonlinear time delay dynamical system exhibiting a sub–critical Hopf bifurcation. These investigations are guided by analytical results from a perturbative approach near the Hopf point, in the limit of a large delay. An experimental bifurcation diagram is reported, which exhibits a characteristic hysteresis with bistability between a cycle and a fixed point.

INTRODUCTION

In the recent past years, nonlinear delayed differential equations (NLDDE) have attracted more and more attention in the field of nonlinear dynamics. On the one hand, they are indeed good candidates to build simple models for the description of high complexity dynamics met in the natural world, like hematologic disorder [1], predator-pray eco–system [2], neuronal network simulation [3]. On the other hand, this type of equations is also able to describe and explain the high complexity dynamics of engineered physical systems, like the Ikeda optical ring cavity [4], external cavity semiconductor lasers [5], or optoelectronic oscillators [6, 7, 8], even when using dynamical models as simple as Eq.(1):

$$\varepsilon \dot{y}(t) = -y(t) + f[\lambda, y(t-1)], \tag{1}$$

where y is the dynamical variable, $f[.]$ is the nonmlinear process, and ε is the characteristic time τ normalized to a time delay T, i.e. $\varepsilon = \tau/T$.

From the experimental and engineering point of view, these dynamics have been particularly studied in the field of Optics. One reason for this is that the necessary large time delay for these high–dimensional dynamics can be easily performed using for example a fiber optic length, or any sufficiently long propagation medium for an optical wave carrying the dynamical fluctuations. Due to fast optical transient time, the condition of a large delay compared to the typical response time (the time scale of the fluctuations) is met easily. Another reason is that a strong nonlinear transformation is possible in Optics through electrooptic, acoustooptic, and/or interference effects. Thus, high complexity dynamical behaviour can be generated in optical systems, what is currently intensively exploited for secure communications using chaos. More generally, various dynamical behaviours found numerically were nicely confirmed by experiments, such as for ex-

CP676, *Experimental Chaos: 7th Experimental Chaos Conference,*
edited by V. In, L. Kocarev, T. L. Carroll, B. J. Gluckman, S. Boccaletti, and J. Kurths
© 2003 American Institute of Physics 0-7354-0145-4/03/$20.00

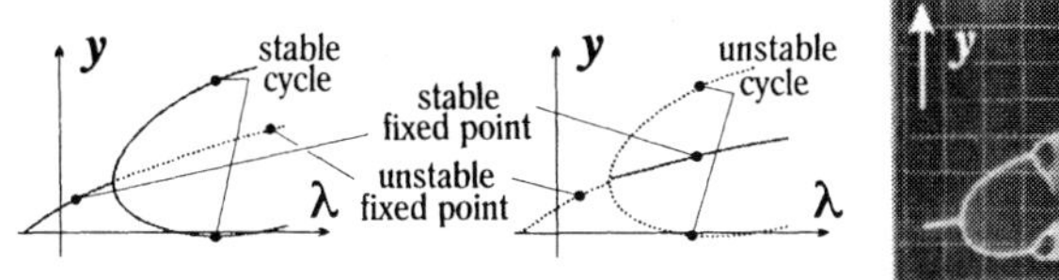

FIGURE 1. Super and sub–critical forward Hopf bifurcation, and a usual bifurcation diagram.

ample the well known period–doubling route to chaos. When looking in the litterature at the bifurcation diagrams obtained from NLDDE dynamics, the first bifurcation of the cascade is usually of the super–critical type, and as far as we know, no sub–critical Hopf bifurcation was reported so far (see fig.1). Independently of our principal field of research (high–dimensional chaos for secure communications [9, 10]), we were also interested in understanding if the sub–critical situation was possible, and under which conditions.

The paper is organized as follow. In the first section we will detail the principal of operation of an experimental setup modeled by an NLDDE. Section II will summerize the perturbative approach of the Hopf bifurcation, which allowed us to extract the conditions for a super– or sub–critical situation. Section III will describe the experimental results obtained when applying the analytical results to the design of an NLDDE system.

EXPERIMENTAL SETUP MODELED BY AN NLDDE

From an experimentalist viewpoint, a physical system modeled by Eq.(1) with a large delay, may be interpreted as a trial to carry out experimentally a discrete dynamic (the mapping $y_{n+1} = f[\lambda, y_n]$. Such an approach is illustrated in the block diagrams of Fig.2. The left bloc diagram (mapping) only needs a nonlinear process and a sampling and holding time. The first bloc enables complex dynamical behaviours (such as chaos), and the second is representative of the computation time needed for the nonlinear transformation, before its output is fed back to its input (thus performing the iteration process). In the right bloc diagram, the continuous time dynamic might exhibit similar behaviours, if the transient term $\dot{y}$ can be neglected. Such an asumption might be fulfilled in some cases, if we introduce a time delay in the feedback loop, which value is much larger than the transient time of the dynamical process (response time τ). The continuous time dynamic also involves a linear converter, which is required in the feedback scheme since the input and output physical variables are generally not of the same nature.

The experimental realizations of NLDDEs in Optics might be separated into two types, the ones involving intrinsic optical and dynamical phenomena, and the ones building separately each bloc, and organizing them according to the feedback scheme in Fig.2. The Ikeda ring cavity, and the external cavity semiconductor laser are of the first type. They involve (very fast) response times caused respectively by the dynamics of the χ^2–optical nonlinear medium (atom level life time), and the electrical pumping dynamic in the semiconductor laser cavity (carrier and photon lifetimes; the external cavity semi-

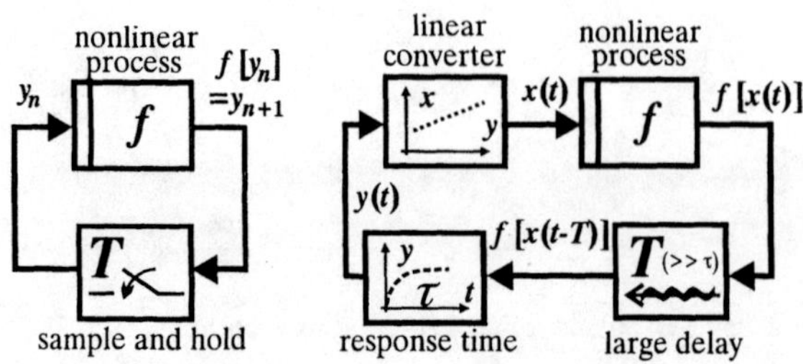

FIGURE 2.   Bloc diagrams of discrete time dynamic and the corresponding continuous time NLDDE

conductor laser case is actually not scalar, but vectorial). Due to the fast time scales, the large delay is obtained with quite short cavities (less than a meter), in which the optical waves travels without distorsion, before it interacts in a nonlinear manner through a feedback. The nonlinear process is here typically generated by an interference effect of the coherent feedback optical wave.

Several optical extrinsic dynamics (the second type of NLDDE experimental setup) are found in the literature. They are performed using an optoelectronic feedback loop, in which transients are originated by electronic, electrooptic, or optoelectronic response times. The delays are obtained usually with optic, or electronic, delay lines. The nonlinear process is also usually produced by an interference effect, and it involves practically different couples of input–output variables, e.g. output interference intensity vs. input voltage in the case of electrooptic modulators [6, 11, 12]. The linear converter consists in a photodiode and a transimpedance amplifier. In the case of the wavelength–chaos generator [8] with which the bifurcation diagram in Fig.1 were recorded, the input variable is the wavelength of a laser beam entering a birefringent filter, and the filter output intensity is the nonlinearly transformed physical variable. The linear converter is performed by a photodiode driving a tunable laser diode. The main advantage of the last setup is the potential flexibility in choosing the nonlinear process, which is defined by the filtering function of any optical filter. Following this idea, we developed the electronic analog of the wavelength–NLDDE generator, i.e. an electronic frequency–NLDDE generator [9](see Fig.3). Hence, the nonlinear process is determined by the (nonlinear) amplitude–profile of any electronic filter, which input variable is the frequency of a sine waveform. The linear converter consists of a amplitude demodulator and a Voltage Controlled Oscillator (VCO). The dynamical process (response time τ) is controlled by a first order electronic filter, operating at frequencies much lower than the ones of the VCO. With this electronic frequency dynamic, generating an NLDDE with any kind of nonlinear process consists in the design of an electronic filter. More precisely, we will see in the next section that the sub–critical Hopf bifurcation should be observable when particular nonlinear process are generated.

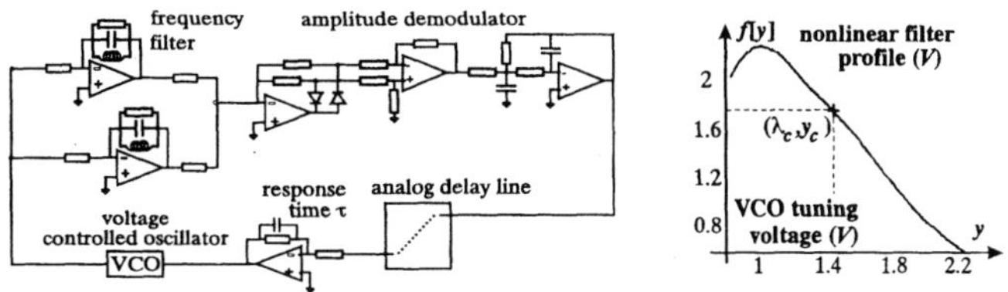

FIGURE 3.   The electronic version of the wavelength dynamic, and the corresponding nonlinearity.

PERTURBATIVE APPROACH AT THE HOPF POINT

First order results

We assume first the system is in a steady state (λ, y_s) defined by the condition $y_s = f[\lambda, y_s]$. λ is the bifurcation parameter, which variations might induce the stability gain or loss of the steady state at the critical Hopf point (λ_c, y_c). A first order stability analysis gives the eigenvalue at any point (λ, y_s), in the form $\sigma = i\omega + (\lambda - \lambda_c)(a + ib)$. ω is the angular frequency of the cycle, and (a, b) are constants that can be expressed as functions of f and its partial derivatives evaluated at (λ_c, y_c). More precisely, we get:

$$a = f_y \left[f_{yy} f_\lambda / (1 - f_y) + f_{y\lambda} \right] \cdot \left[f_y^2 + \varepsilon \right] / \left[(f_y^2 + \varepsilon)^2 + \omega^2 \varepsilon^4 \right], \tag{2}$$

which sign gives the direction for the stability change (if $a > 0$, stability is lost when increasing λ through λ_c, and conversely if $a < 0$).

Location of the limit cycle

The detailed calculation of the applied higher order perturbation method is out of the scope of the paper. We will only give some guidelines of this calculation, and we refer the reader to the literature for more details [13].

To determine if the limit cycle exists for λ greater or smaller λ_c, a third order perturbation method is used. Doing this, we assume there exists a 2π–periodic solution in the re–scaled time variable $\xi = \sigma t$. Such a solution can be written as a Taylor expansion with respect to the fundamental amplitude δ of the considered periodic solution, $y(\xi) = y_s(\lambda) + \delta y_1(\xi) + \delta^2 y_2(\xi) + \ldots$ Similar Taylor expansions are done for the angular frequency ω and the bifurcation parameter λ, for continuity reasons of those quantities in the vicinity of the Hopf point: $\sigma = \omega + \delta^2 \omega_2 + \ldots$ and $\lambda = \lambda_c + \delta^2 \lambda_2 + \ldots$ Introducing these expansions into Eq.(1), and assuming the time dependence $y_1(\xi) = e^{i\xi} + c.c.$ and $y_2 = (p_2 e^{2i\xi} + c.c.) + p_0$, we are able to solve the equality up to the third order in δ, and we finally obtain equations in λ_2. The latter coefficient gives practically the location of the cycle around λ_c on the bifurcation axis. When comparing this location with the stability analysis results (Eq.(2), the super– or sub–critical nature of the Hopf bifurcation is obtained. In the limit of the large delay ($\varepsilon \to 0$), the Hopf bifurcation nature corresponds to the conditions summerized in table 1.

TABLE 1. Nature of Hopf bifurcation in an NLDDE with larger delay

	$-\frac{1}{2}f_\lambda f_{yy} - f_{y\lambda} > 0$	$-\frac{1}{2}f_\lambda f_{yy} - f_{y\lambda} < 0$
$\frac{3}{4}f_{yy}^2 + \frac{1}{2}f_{yyy} > 0$	super–critical, forward	super–critical, reverse
$\frac{3}{4}f_{yy}^2 + \frac{1}{2}f_{yyy} < 0$	sub–critical, forward	sub–critical, reverse

Since the conditions involve partial derivatives with respect to y up to the third order, a local behaviour of the nonlinear function verifying the proper conditions should be at least the one of a third degree polynomial. Moreover, when replacing $f[.]$ by the sine model of a two waves interference function, such as the one usually involved in Optics, it is found that the sub–critical conditions cannot be met. The next section will report the realization of such a nonlinear function, on the basis of the electronic frequency dynamic of the first section.

EXPERIMENTAL RESULTS

A quite general writing of a third degree polynomial is: $f(\lambda,y) = y_s(\lambda) + [y - y_s(\lambda)][A(\lambda) + B(\lambda)y + C(\lambda)y^2]$. The conditions of table 1 applied to this model lead to: $A(\lambda_c) = -1$ and $C(\lambda_c) < -2B^2(\lambda_c)$. For sake of simplicity, we use $(1,0)$ as the Hopf point. A simple nonlinear function meeting the sub–critical case then could be: $f[\lambda,y] = -\lambda y[1 + (\alpha y)^2]$, where λ acts as a multiplication factor (experimentally performed through a tunable gain in the feedback loop). The nonlinear function is unfortunately not bounded, but since only a local cubic behaviour is needed, we might effectively use an electronic filter of the bounded form:

$$f(\lambda,y) = -\lambda y[1 + (\alpha y)^2]/[1 + (\beta y)^4], \qquad (3)$$

where the parameter β has to be greater than α, so that the bounding effect of the denominator operates far away from a local interval of y centered in y_c (in which the cubic behaviour of the numerator is expected). As a consequence of the bounding of $f[.]$, a non local stable limit cycle might be generated. In that case, when observing the sub–critical Hopf bifurcation, the dynamic would jump in a crisis from the steady state onto that stable non local limit cycle. When tuning back the bifurcation parameter, a hysteresis behaviour would then be observed in the reverse bifurcation diagram; the steady state is recovered near the connection of the stable cycle with the continued unstable cycle generated at the sub–critical Hopf bifurcation.

From the model of Eq.(3), we built an electronic frequency dynamical system, as described in the first section and depicted in Fig.3. The nonlinear function is performed by two parallel RLC filters, which resonance frequencies, quality factors and relative weight are adjusted to match qualitatively the nonlinear profile of Eq.(3). The resulting nonlinear transformation is depicted in Fig.3, and the recorded bifurcation diagram is shown in Fig.4 with two periodic waveforms observed inside the hysteresis range of λ. Inside this range, the central trace of the bifurcation dagram corresponds to the steady state, it is observed while increasing λ; the upper and lower traces correspond to the non local stable cycle, they are obtained when decreasing λ from a value greater than

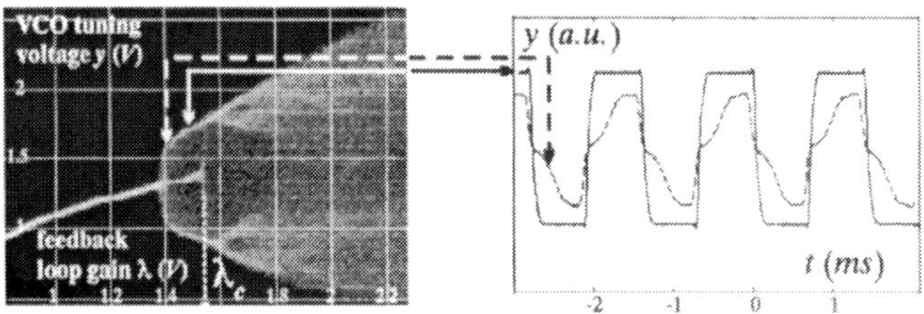

FIGURE 4. Experimental bifurcation diagram and two waveforms of the hysteresis domain.

λ_c. The non–square waveform occurs just before the bifurcation of the cycle back to the steady state, while λ is further decreased. We also checked by a continuous tuning of the nonlinear profile, that the stable cycle is indeed connected to the unstable cycle born at the sub–critical Hopf point. As a final remark, one could notice the global behaviour (out off the sub–critical Hopf situation) is very similar to the one in Fig.1), thus underlining the local character of the observed phenomena.

CONCLUSION

The general architecture of an experimental setup modeled by a scalar nonlinear delayed differential equation was presented. According to this, and following analytical results on the conditions for a sub–critical Hopf bifurcation, we built an experimental setup to demonstrate the possibility of such an unusual bifurcation. A typical hysteresis behaviour was experimentally observed in the bifurcation diagram, indicating that the sub–critical case was effectively obtained.

REFERENCES

1. Mackey, M. C., and Glass, L., *Science*, **197**, 287 (1977).
2. May, R. M., *Nature*, **261**, 459 (1976).
3. Wei, J., and Shigui, R., *Physica D*, **130**, 255–272 (1999).
4. Ikeda, K., *Opt. Commun.*, **30**, 257 (1979).
5. Lang, R., and Kobayashi, K., *IEEE J. Quantum Electron.*, **16**, 347 (1980).
6. Gibbs, H. M., Hopf, F. A., Kaplan, D. L., and Schoemacker, R. L., *Phys. Rev. Lett.*, **46**, 474 (1981).
7. Arecchi, F., Gadomski, W., and Meucci, R., *Phys. Rev. A*, **34**, 1617 (1986).
8. Larger, L., Goedgebuer, J.-P., and Mérolla, J.-M., *IEEE J. Quantum Electron.*, **34**, 594 (1998).
9. Larger, L., Udaltsov, V. S., and Goedgebuer, J.-P., *Electron. Lett.*, **36**, 199 (2000).
10. Larger, L., Lee, M. W., Goedgebuer, J.-P., Erneux, T., and Elflein, W., *JOSA B*, **18**, 1063 (2001).
11. Okada, M., and Takisawa, K., *IEEE J. Quantum Electron.*, **17**, 2135–2140 (1981).
12. Neyer, A., and Edgar, V., *IEEE J. Quantum Electron.*, **18**, 2009–2015 (1982).
13. Giannakopoulos, F., and Zapp, A., *J. Math. Anal. Appl.*, **237**, 425–450 (1999).

Chaotic Synchronization on Optical Phases in Bandwidth- Broadened Semiconductor Lasers Subject to Optical Injection

How-foo Chen and Jia-ming Liu*

*Department of Electrical Engineering, University of California, Los Angeles
Los Angeles, California 90095-159410

Abstract. Chaotic optical communication at high bit rates has attracted great attention because of its potential application in high-speed optical spread spectrum communications. Among several proposed setups, all-optical systems, such as the optical injection system and the optical feedback system, can achieve chaotic synchronization on the amplitude, the slowly-varying phase of the chaotic envelope, and the fast-varying phase of the optical field. However, all of the experimental work reported so far demonstrated only the synchronization of the intensity waveform or, equivalently, the amplitude waveform of the laser field. This report experimentally demonstrates chaotic synchronization on the fast-varying and the slowly-varying optical phases of the bandwidth-broadened semiconductor lasers subject to the optical injection. The result shows a good synchronization quality.

INTRODUCTION

Chaotic synchronization has attracted much attention for its potential application in private communications and spread spectrum communications due to the characteristics of a chaotic system: sensitivity to initial conditions, noise-like waveform as a message carrier and spectrum broadening. Among several proposed optical systems for chaotic communications [1, 2, 3, 4, 5, 6], the setups based on semiconductor lasers have their special roles in this development of using chaotic synchronization on communications due to their compact sizes, high speed capability and integrability to present optical communication systems. The chaotic synchronization in all-optical systems, such as the optical injection system and the optical feedback system, is completed through the synchronization on the amplitude, the slowly-varying phase of the chaotic envelope, and the fast-varying phase of the optical field. However, all of the experimental work reported so far for full optical systems demonstrated only the synchronization of the intensity waveform or, equivalently, the amplitude waveform of the laser field. Utilizing the synchronizations on both slowly-varying and fast-varying phases, as well as the synchronization on the amplitude, message can be extracted through optical destructive interference between the transmission signal and the output of the receiver. When the system is used to transmit high bit rate message, the complexity and expense of high-speed electronics for retrieving the high bit rate message can be avoided. The optical injection system also has the characteristic of bandwidth broaden [7]. This report experimentally demonstrates chaotic synchronization on the fast-varying, the slowly-varying optical phases

CP676, *Experimental Chaos: 7th Experimental Chaos Conference*,
edited by V. In, L. Kocarev, T. L. Carroll, B. J. Gluckman, S. Boccaletti, and J. Kurths
© 2003 American Institute of Physics 0-7354-0145-4/03/\$20.00

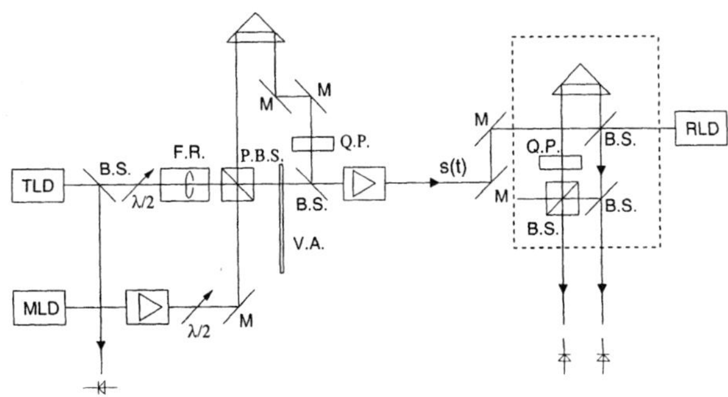

FIGURE 1. Schematic experimental setup. MLD: Master Laser, TLD: Transmitter Laser, RLD: Receiver Laser, M: Mirror, BS: Beam Splitter, PBS: Polarized beam splitter, VA: Variable Attenuator, PD: Photodetectors, QP: quartz plate, FR: Faraday Rotator, $\lambda/2$: Half-wave Plate, dashed-line block: interferometer

of the semiconductor lasers subject to the optical injection. The result shows a good synchronization quality.

Since the synchronization quality is very sensitive to optical phase, the synchronization quality subject to both phase match and phase mismatch is also measured in both time and frequency domains to prove the achieved synchronization is not chaos driven. The synchronization quality of each case is quantitatively expressed by the correlation coefficient.

EXPERIMENTAL SETUP

The chaotic waveform is generated throught the optical injection scheme. When the transmitter is subject to the optical injection field, it can be driven into chaos through period-doubling path [8, 9]. Based on the synchronization concept proposed by Kocarev et. al.[10], the two lasers have to be identical except that $\gamma_c^R = \gamma_c^T + 2\eta\alpha$, where γ_c is the cavity decay rate, α is coupling strength between the transmitter and the receiver [3].

The experimental setup to accomplish the chaotic synchronization based on the proposed numerical model [3] is shown in Fig. 1. The output of the master laser, denoted as MLD, is separated into two beams through a variable beam splitter. One beam goes through a Faraday rotator, a half-wave plate, a beam splitter, and then injects into the transmitter to drive the transmitter into chaotic state; the other beam goes through the prism, two mirrors, a quartz plate, and combines with the output of the transmitter with proper relative phase. The quartz plate is used to control this relative phase since the chaotic synchronization of this system is very sensitive to this relative phase. The combined optical field is then injected into the receiver to achieve chaotic synchronization.

The phase synchronization is detected through the interferometer built near the receiver, which is indicated by the dashed-line block. The quartz plate in this dashed-line block is used for controlling the relative phase of two beams. When the chaotic synchronization on the slowly-varying phase and the fast-varying phase is achieved, the detector at the end of this interferometer will detect the constructive interference and the destructive interference.

The semiconductor lasers used in this setup are InGaAsP/InP single-mode DFB semi-conductor lasers with the wavelength about 1.3 *um* from the same wafer with careful pick up to match the laser parameters within the mismatch tolerance by try-and-error method.

ANALYSIS AND DISCUSSION

In the experiment, the semiconductor laser served as the transmitter is biased at 50 mA and the surrounding temperature is controlled at 21° C. The master laser is operated at the condition that the frequency detuning to the transmitter is about 2.73 GHz. The injection strength from MLD to TLD is adjusted so that TLD is driven at the chaotic state with the power spectrum shown in Fig. 2(a). From the power spectrum we can see that the bandwidth of the chaotic waveform is more than 10 GHz in comparison of the resonance frequency of the free-running transmitter laser, which is about 6 GHz at this operation condition.

In order to achieve optima synchronization quality, beside considering the mismatches of the intrinsic parameters of the semiconductor lasers, the mismatches of operation conditions have to be considered also, such as the biased current and the cold-cavity frequency. Because of the existence of the intrinsic parameter mismatches, the match conditions on the bias current and the cold-cavity frequency will be offset from those for the perfect synchronization when all the parameters are exactly matched. The effect of the bias current mismatch and the cold-cavity frequency mismatch combined with the optical phase mismatch on the synchronization quality will be discussed in both numerical analysis and experiment in the future publication.

The coupling between the transmitter and the receiver is adjusted as well as the injection strength from the master laser to the receiver to obtain the optima synchronization quality. The receiver is operated at the surrounding temperature 16.38^oC. The experimental results shown in this paper are obtained when the receiver is biased at 41.00 mA. The receiver biased at 41.00 mA performs the optima synchronization observed.

Under this operation condition, we adjust the quartz plate to match the relative optical phase between the output of the transmitter and that of the master laser to the receiver

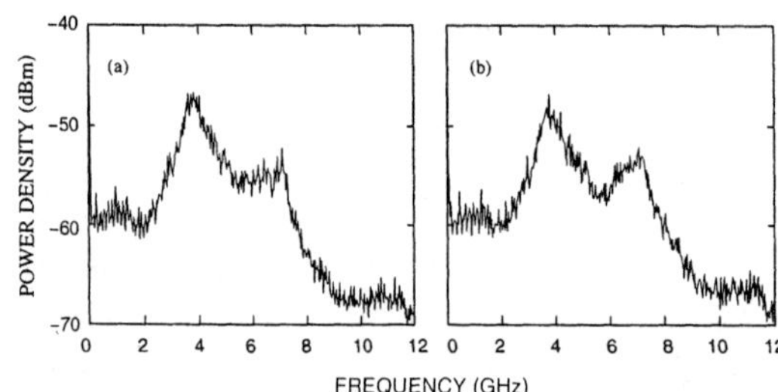

FIGURE 2. Power spectra when phase matches: (a) power spectrum of the transmitter chaotic output, (b) power spectrum of the receiver output.

when two laser beams are combined together as the transmission signal. Under this careful match on the relative optical phase, we obtained the output of the receiver laser with power spectrum very similar to that of the transmitter output. The power spectra are shown in Fig. 2b and Fig. 2a in respective. The time series of the transmitter output and that of the receiver output are shown in Fig. 3(a) and (b), respectively. In comparison of the chaotic output of the transmitter and that of the receiver in time domain and frequency domain, we can see that good chaotic synchronization on the amplitude of the output beams has been observed. The quality of this synchronization can also be realized through the correlation plot, which is shown in Fig. 3(c). The synchronization quality can also be measured quantitatively in term of correlation coefficient defined as the following:

$$\rho = \frac{< |X(t) - < X(t) > ||Y(t) - < Y(t) > | >}{\sqrt{< |X(t) - < X(t) > |^2 >}\sqrt{< |Y(t) - < Y(t) > |^2 >}}$$

The analysis shows that $\rho = 0.85$ when the relative optical phase is 0^0.

Since the relative optical phase has vital effect on the synchronization quality, the effect of the relative optical phase on the synchronization quality serves as a criteria to judge if the achieved synchronization is the synchronization predicted by the theory of chaotic synchronization. Therefore it is necessary to check the synchronization quality with 180^0 phase mismatch. The power spectra of the transmitter and the receiver when this phase mismatch occurred are shown in Fig. 4(a) and (b) in respective. The intensity of the transmitter output and that of the receiver output in time domain are shown in Fig. 5(a) and (b) in respective, and the correlation diagram is shown in Fig. 5(c). In comparison of this results with the results when optical phase is matched, we can see that phase mismatch significantly degrade the synchronization quality, which proves what has been observed a chaotic synchronization. The synchronization quality is measured $\rho = 0.32$ when the relative optical phase is 180^0. We can see that the relative optical phase does affect the synchronization quality.

After verifying the achieved synchronization as the chaotic synchronization, we expect to see the slowly-varying phase and the fast-varying phase of the receiver output are also synchronized to those of the transmitter output. The synchronization on both phases can be observed from the interferometer shown in Fig. 1. Figure 6 shows the op-

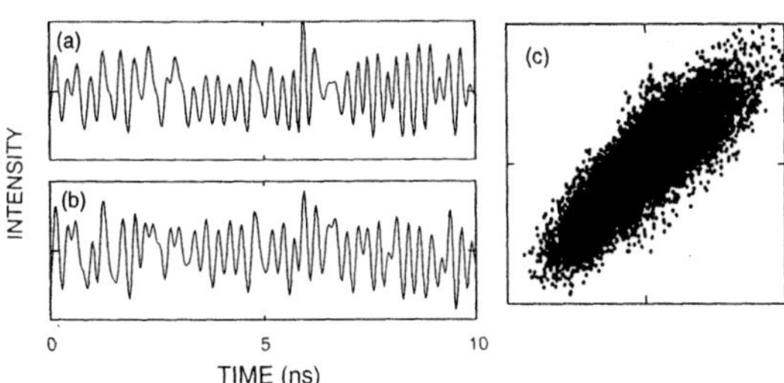

FIGURE 3. Time series when phase matches: (a) intensity of the transmitter output, (b) intensity of the receiver output, (c) correlation diagram

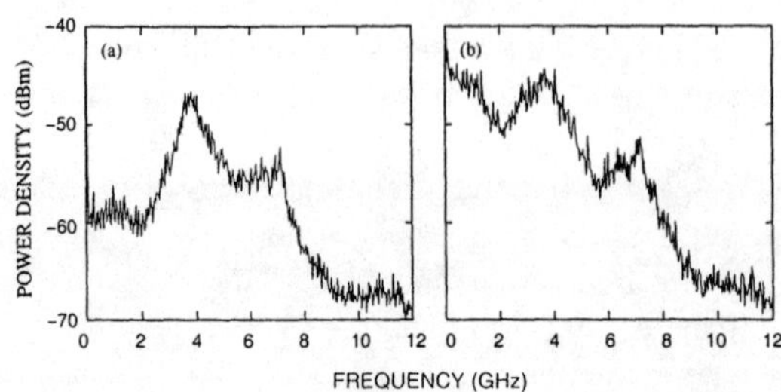

FIGURE 4. Power spectra when phase mismatch: (a) power spectrum of the transmitter output, (b) power spectrum of the receiver output

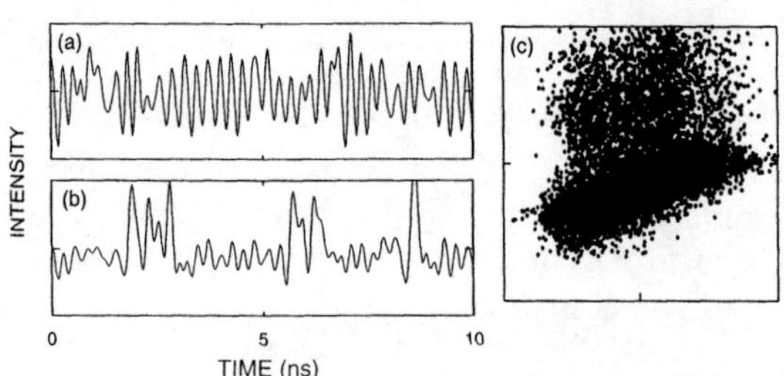

FIGURE 5. Time series when phase mismatch: (a) intensity of transmitter output, (b) intensity of receiver output, (c) correlation diagram

optical interference of the transmission signal and the receiver output in time domain. The constructive interference is shown in Fig. 6(a), and the destructive interference is shown in Fig. 6(b). By comparing the constructive interference and the destructive interference, we measure the extinguish ratio to be larger than 5, which shows the chaotic

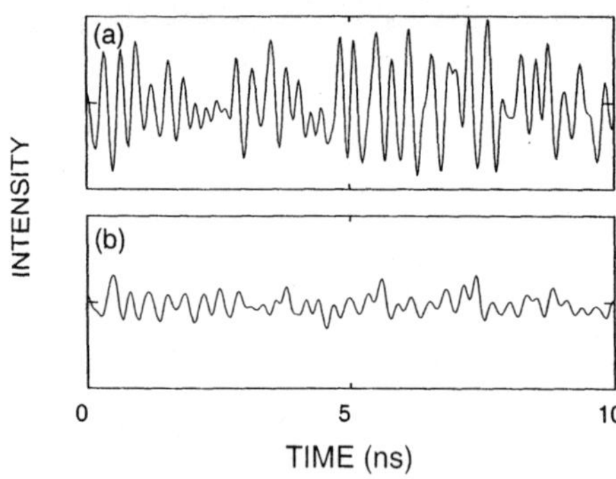

FIGURE 6. Interference in time domain: (a) constructive interference, (b) destructive interference

on the slowly-varying phase, the fast-varying phase, and the amplitude as well.

CONCLUSION

In conclusion, utilizing optical injection on the semiconductor laser to generate chaotic waveform has guaranteed the broadening of the chaotic bandwidth larger than the laser intrinsic bandwidth. For the optical injection system, the synchronization is achieved on the slowly-varying phase, the fast-varying phase, and the amplitude as well, which proves what the chaotic synchronization theory has predicted. This phase synchronization and the bandwidth enhancement characteristics of this optical injection system make a broadband and high-speed chaotic communications in optics possible without the need of expensive and complex electronic device to retrieve the message. Under proper control of optical phase and the stabilization of the optical paths, the chaotic synchronization can be achieved with good synchronization quality. The effect of phase mismatch observed in experiment distinguishes the result as chaotic synchronization from chaotic driven.

ACKNOWLEDGMENTS

This work is supported by the U. S. Army Research Office under contract No. DAAG55-98-1-0269.

REFERENCES

1. G. D. VanWiggeren, and R. Roy, "Optical Communication with Chaotic Waveform,"*Phys. Rev. Lett.* vol. 81, no. 16, pp. 3547-3550, Oct. 1998.
2. L. G. Luo, P. L. Chu, and H. F. Liu," 1-GHz Optical Communication System Using Chaos in Erbium-Doped Fiber Lasers,"*IEEE Photon. Technol. Lett.,* vol. 12, no. 3, pp. 269-271. Mar. 2000.
3. H. F. Chen, and J. M. Liu, "Open-loop Chaotic Synchronization of Injection-Locked Semiconductor Lasers with Gigahertz Range Modulation,"*IEEE J. of Quantum Electronics,* vol. 36, no. 1, pp. 27-34, Jan. 2000.
4. S. Tang and J. M. Liu , "Synchronization of High-frequency Chaotic Pulses,"*Optics Lett,* vol. 26, no. 9, pp. 596-598, May 2001.
5. Liu, Y., and Davis, P., "Synchronization of Chaotic Mode Hopping,"*Optics Lett.,* vol. 25, pp.475-477, 2000.
6. J. P. Goedgebuer, L. Larger, and H. Porte,"Optical Cryptosystem Based on Hyperchaos Generated by a Delayed Feedback Tunable Laser Diode," *Phys. Rev. Lett.,* vol. 80, pp.2249-2252, 1998.
7. T. B. Simpson and J. M. Liu,"Enhanced Modulation Bandwidth in Injection-Locked Semiconductor Lasers,"*Photon. Technol. Lett.,* vol. 9, no. 10, pp. 1322-1324, Oct. 1997.
8. V. Kovanis, A. Gavrielides, T. B. Simpson, and J. M. Liu,"Instability and Chaos in Optically Injected Semiconductor Lasers," *Appl. Phys. Lett.,* vol. 67, pp.2780-2782, 1995.
9. S.K. Hwang and J.M. Liu, "Dynamical Characteristics of an Optically Injected Semiconductor laser," *Opt. Commun.,* vol. 183, pp. 195-205, 2000.
10. L. Kocarev, and U. Parlitz, "General Approach for Chaotic Synchronization with Applications to Communication," *Phys. Rev. Lett.,* vol. 74, pp. 5028-5031, 1995.

MODAL INTERFERENCE AND DYNAMICAL INSTABILITY IN A SOLID-STATE SLICE LASER WITH ASYMMETRIC END-PUMPING

KENJU OTSUKA AND JING-YUAN KO

Department of Human and Information Science, School of Information Technology and Electronics, Tokai University, 1117 Kitakaname, Hiratsuka, Kanagawa 259-1292, Japan
E-mail: ootsuka@keyaki.cc.u-tokai.ac.jp

We observed complicated emission patterns consisting of many transverse modes and associated intensity pulsations at beat frequencies between some pairs of transverse eigenmodes in a solid-state thin slice laser with laser-diode asymmetric end-pumping. The dependence of transverse patterns on pump power and crystal rotation has been demonstrated. The pump-power dependent pulsation frequency and the resonant excitation of chaotic oscillation have been observed. The interference among non-orthogonal transverse eigenmode fields, which are formed in a thin-slice Fabry-Perot cavity possessing an asymmetric gradient potential for optical waves, is shown to result in intensity modulations

1   Introduction

Formations of microcavity modes such as whispering gallery modes [1] and scarred modes [2], and related chaotic waves [3,4] in microdisk and deformed microcylinder lasers without feedback mirrors have attracted much attention in recent years for understanding wave formations in microstructure resonators. These microcavity modes can be understood to be formed by light reflections at surrounded hard walls in terms of ray optics [3,4]. Recently, thin-slice Fabry-Perot lasers such as vertical-cavity surface-emitting laser diodes (VCSEL's) have revisited in a different context as promising candidates for investigating wave formations in microcavity lasers. Mulet and Balle studied spatio-temporal dynamics numerically in VCSEL's and showed that stable transverse pattern consisting of different linearly polarized orthogonal LP_{ml} eigenmodes ($m = 1,2,..., l = 0,\pm1,...$), which are formed in a thin-slice cavity possessing a parabolic transverse refractive-index potential due to thermal lensing through injection current, appears in the stationary state [5].

We report here on the formation of a variety of transverse lasing mode patterns, which differ from conventional orthogonal Hermite-Gaussian modes in Fabry-Perot resonators, in a solid-state thin slice laser with laser-diode asymmetric end-pumping. We show that successive structural changes take place with changing the pump power or with rotating the laser crystal. Intensity pulsations at beat frequencies among coexisting closely-spaced transverse eigenmodes are demonstrated. The modal beats between non-orthogonal transverse modes are proposed to result in intensity pulsation through the interaction with the atomic system. The

CP676, *Experimental Chaos: 7th Experimental Chaos Conference,*
edited by V. In, L. Kocarev, T. L. Carroll, B. J. Gluckman, S. Boccaletti, and J. Kurths
© 2003 American Institute of Physics 0-7354-0145-4/03/$20.00

observed pulsation has been reproduced by numerical simulations of a proposed laser equation including modal interference effect of non-orthogonal transverse lasing fields.

2 Experimental configuration

The experimental setup is shown in Fig. 1(a). A 7-mm-square c-plate Nd-direct compound $LiNdP_4O_{12}$ (LNP) laser with a 0.3-mm-thick plane-parallel Fabry-Perot resonator configuration was attached to a Cu heat sink that had a 5-mm-diameter hole in the center. The present thin-slice laser cavity possesses an extremely large Fresnel number of 1.6×10^5 which ensures plane-wave approximation for light propagation along the short laser cavity. In short, lasing eigenmodes can be expressed by the product of transverse eigenmodes and longitudinal standing-wave patterns which satisfy the lasing boundary condition. An end-surface was coated to be transmissive at the LD pump wavelength of 808 nm (85% transmission) and highly reflective (99.9%) in the lasing wavelength range of 1000-1100 nm. The other surface was coated to be 1% transmissive in the lasing wavelength range. The thickness of the LD active layer along the vertical direction y was 1 μm and the emitting width along the horizontal direction x was 100 μm, where the LD pump light was linearly polarized along the x-axis. We collimated the LD pump beam and the resultant collimated beam was directly focused on the laser crystal surface by using a microscope objective lens of 10xM magnification. The mode profile of the focused LD beam on the input mirror of the crystal was rectangular with 20 μm x 2 μm size, yielding the aspect ratio of 10. The absorption length for the LD light was about 100 μm and the pump light was strongly absorbed near the input mirror. However, due to thermal diffusion and heat dissipation into the air, a strongly asymmetric temperature distribution is thought to be created in the x-y plane along the laser axis z in the equilibrium. An example of two-dimensional temperature distributions calculated by the boundary-element method is shown in Fig. 1(b). The resultant refractive-index distributions due to the thermally induced refractive index change are depicted in Fig. 1(c) by dashed curves. In short, a strongly "deformed" Fabry-Perot cavity possessing an asymmetric graded refractive-index distribution in the transverse direction is formed in the stationary state by sheet-like LD pumping. The thermally-induced potential for optical waves possessing a strongly asymmetric transverse optical confinement effect [6] loses its lateral symmetry which ensures usual Gaussian cavity modes. The deformed microcavity structure and corresponding mode pattern are expected to change depending on the pump power, i.e., temperature rise.

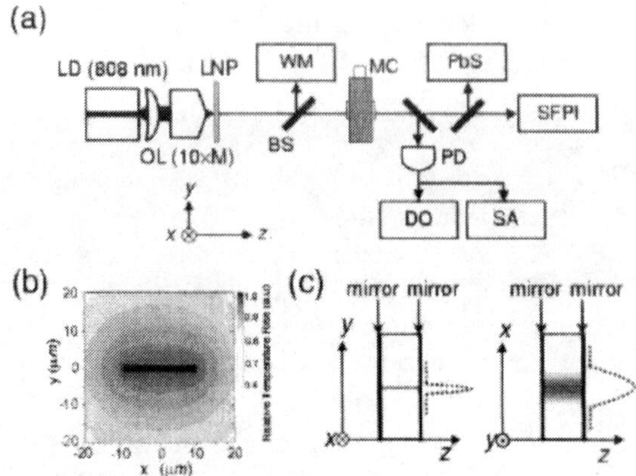

Figure 1. (a) Experimental configuration of an LNP slice laser with sheet-like pumping. LD: laser diode, OL: microscope objective lens, LNP: LiNdP$_4$O$_{12}$ laser, BS: beam splitter, WM: multi-wavelength meter, MC: monochrometer, PD: photo-detector, DO: digital oscilloscope, SA: rf spectrum analyzer, SFPI: scanning Fabry-Perot inteferometer, PbS: PbS photo-image tube. (b) Calculated relative temperature distribution in the two-dimension. Thermal conductivity of LNP, K=0.032 J/sec•cm•°K, was assumed. (c) Thermally- induced refractive index distribution within the crystal

3   Structural change of emission patterns [7]

In LNP lasers possessing anisotropic spectroscopic properties, the stimulated emission cross section along the b-axis is the largest and the linearly-polarized emission along the b-axis occurs independently of the pump condition, e.g., polarization direction of the pump beam. When the pump power was increased above the threshold, a first lasing occurred on the $^4F_{3/2}(1) \rightarrow {}^4I_{11/2}(1)$ transition at 1048 nm. As the pump power was increased, a second lasing associated with the $^4F_{3/2}(1) \rightarrow {}^4I_{11/2}(2)$ transition appeared at 1055 nm, and finally a third lasing on the $^4F_{3/2}(1) \rightarrow {}^4I_{11/2}(3)$ transition joined at 1060 nm. A typical input-output characteristic and an example of global oscillation spectra measured by a multi-wavelength meter are shown in Figs. 2(a)-(b). In all experiments, each transition exhibited a single longitudinal mode family and no adjacent longitudinal mode families separated by the free spectral range of $\lambda^2/2nL$ = 1.18 nm were observed on each transition (λ: wavelength, n: refractive index, L: laser thickness) by a multi-wavelength meter with 0.2-nm resolution. This is an example of a multi-transition-oscillation (MTO) resulting from resonant reabsorption in highly-doped lasers [8,9]. It should be noted that each transition emission exhibited an optical spectrum consisted of similar closely-spaced eigenmodes, as shown in Fig. 2(c). Detailed optical spectra of individual transition emissions were measured by a scanning Fabry-Perot interferometer (free-spectral range: 2 GHz; resolution: 6 MHz) after passing the output beam through a monochrometer. From this figure, it is found that several common transverse patterns are embedded in different transition emissions. However, the MTO itself is not essential for transverse pattern formations discussed in this paper. Let us first show examples of lasing patterns emitted from the LNP slice laser with sheetlike end-pumping. Near-field and far-field lasing patterns were measured by an infrared PbS photo-image tube followed by a monitor. Far-field patterns for each transition are shown in Fig. 3 for increasing pump power where LNP *b*-axis was set parallel along the *x*-axis. The transverse pattern is found to make successive structural changes with increasing pump power. The structural difference between near-field and far-field patterns was not observed. According to the symmetry of the thermally-induced deformed cavity, the global structure of observed transverse patterns was symmetric with respect to the *x(b)* and *y(a) axes.* In the low pump-power region, thermally-induced optical confinement effect is weak and usual Hermite-Gaussian modes were observed. However, as the pump power was increased these modes became unstable due to the increased strongly asymmetric transverse optical confinement effect and complicated transverse patterns appeared. It should be pointed out that the dynamical instability which will be shown later was not observed when the LNP laser thickness was increased over 1 mm, in which the two-dimensional optical confinement effect along the whole laser cavity was absent because of much longer cavity length than the absorption length and the larger diffraction effect.

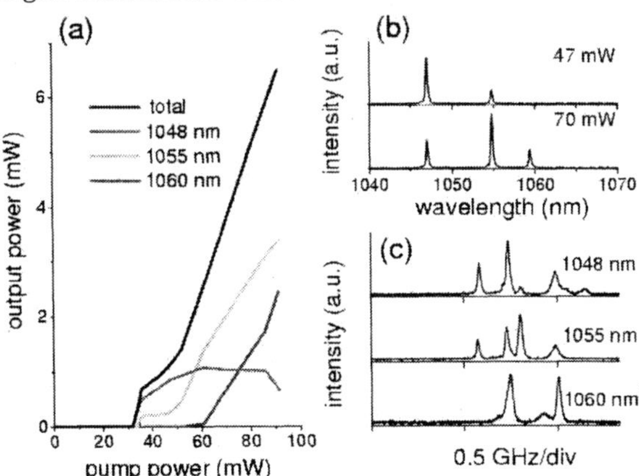

Figure 2. (a) Input-output characteristics, (b) global oscillation spectra and (c) detailed optical spectra of the sheet-pumped LNP laser.

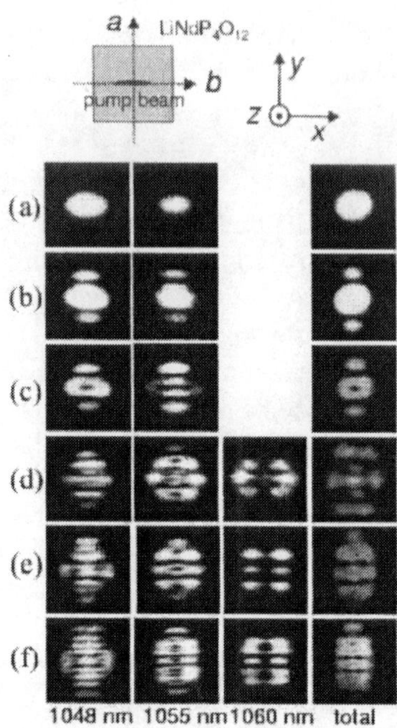

Figure 3. Far-field pattern change with increasing pump power when b-axis is parallel to x-axis. Pump power: (a) 42 mW (b) 47 mW (c) 52 mW (d) 61 mW (e) 70 mW (f) 91 mW.

4 Modal interference and dynamical instability

Next, let us focus on dynamical behaviors of the present laser. Temporal evolutions were measured by an InGaAs photoreceiver with 1GHz bandwidth followed by a digital oscilloscope. In the experiment, the entire beam was focused on the detector. Typical optical spectra, intensity waveforms and corresponding power spectra are shown Fig. 4 for different pump powers. Here, results for 1048-nm emission are shown, but similar results were obtained for other transitions. From this figure, a significant feature should be addressed. Some eigenmode pairs in many eigenmodes are coupled with each other and produce intensity pulsations at beat frequencies corresponding to the energy (i.e., eigenfrequency) differences between eigenmode pairs. Mode pairs, which are indicated by arrows, induce intensity pulsations at different frequencies. In the case of Fig. 4(a), "breathing" oscillation appears at the difference frequency of two pulsations, i.e.,(f_1 - f_2)<< f_1 . As for Fig. 4(b), more complicated intensity pulsation involving many mode pairs takes place. Note that pulsation frequencies can be two orders of maginitude higher than the intrinsic relaxation frequency of the laser of about 1 MHz.

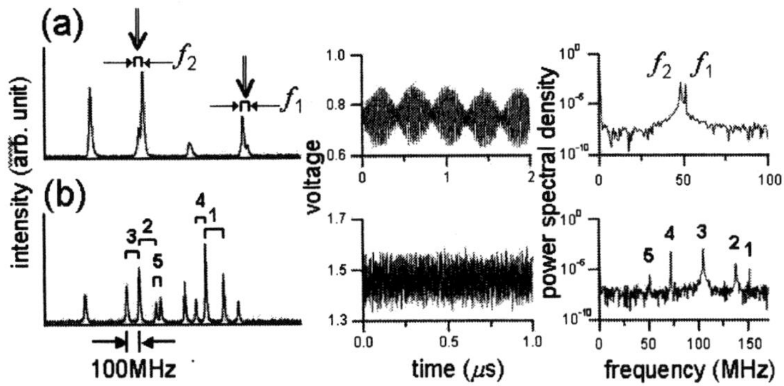

Figure 4. Fine structures of optical spectra, pulsation waveforms and corresponding rf power spectra of the LNP laser with asymmetric end pumping. Pump power: (a) 58 mW, (b) 80 mW.

These features can never be expected in conventional Hermite-Gaussian laser modes, in which perfect mode-orthogonality is established. In this case, beat notes between orthogonal sets vanish and transverse mode interaction through modal interference is absent. In the present system, however, many transverse eigenmodes are excited, in which some of them with the same parity violate mode-orthogonality and may result in modal beats through the transverse-field interference effect.

Intensity modulations were not observed in the circular (symmetric) LD end-pumping scheme, although complicated spectra featuring fine structures similar to Fig. 4 appeared in the high pump-power region where the symmetric optical confinement resulting in LP_{ml} modes similar to VCSEL's was established [5]. A typical far-field pattern and the corresponding optical spectrum are shown in Fig. 5. The circular pumping was achieved by inserting anamorphic prism pairs before the objective lens to transform the elliptical beam into the circular one. This result clearly implies that the optical confinement by thermally-induced asymmetric refractive index potential resulting in non-orthogonal transverse eigenmodes is essential for the appearance of instabilities.

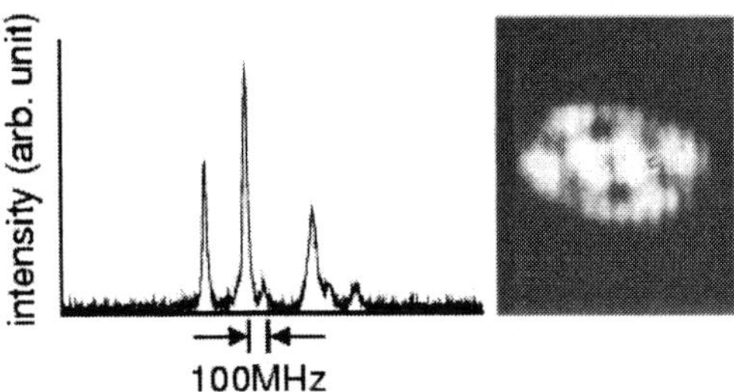

Figure 5. Far-field patterns and oscillation spectra of the LNP laser with circular end pumping. Pump power: 140 mW.

Oscillation spectra and frequency spacing between eigenmodes were found to change depending on the pump power and the aspect ratio of the pump beam, which critically changes on the crystal position along the

z-axis (i.e., defocusing of the pump beam). This is because deformed microcavity structure changes with these parameters, yielding different transverse eigenmodes. The modal-beat (i.e., pulsation) frequency is plotted as a function of pump power in Fig. 6(a). It should be pointed out that mode spacing (i.e., energy difference) decreases first and then increases with increasing pump power featuring rich bifurcation structures in the high pump-power regime. In the low pump-power region, usual Hermite-Gaussian mode appeared without exhibiting beat-notes. As the pump power was increased, the optical confinement became effective and Hermite-Gaussian modes were deformed, resulting in modal-beat mediate intensity modulation at a decreased beat frequency (i.e., mode spacing) appeared above a critical pump power. In the high pump-power region, on the other hand, the average pulsation frequency is considered to have increased with deformation of the gradient refractive-index profile with increasing the pump power (i.e., refractive index gradient). This tendency was also identified by changing the aspect ratio of the focused pump beam. These observations may have some connection with the theoretical prediction of quantum billiard (chaos) systems that the increased deformation results in larger energy differences between eigenmodes, implying stronger chaos with increased energy-level repulsion [10]. In addition, in the present system, when the beat frequency coincided with the intrinsic relaxation oscillation frequency of the laser, chaotic relaxation oscillations were found to be resonantly excited as shown in Fig. 6(b).

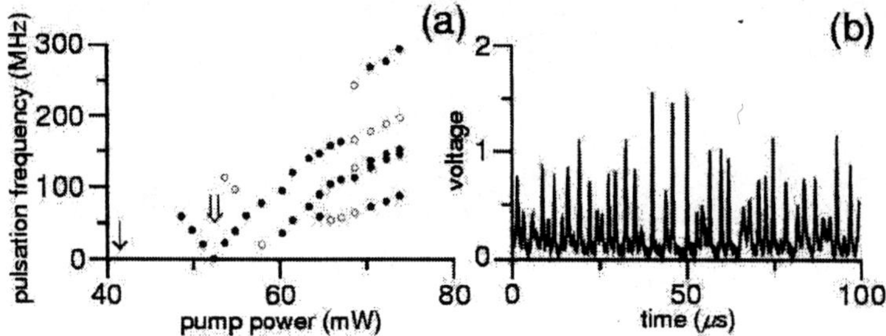

Figure 6. (a) Pulsation frequencies as a function of pump power. Solid (open) circle corresponds to stable (unstable) pulsation frequency, where "unstable" implies that the pulsation-waveforms are unstable in time. The lasing threshold is 41 mW ($\downarrow$). (b) Chaotic relaxation oscillation observed when the beat frequency f_{QB} coincided with the relaxation frequency $f_R = 1$ MHz ($\Downarrow$).

5 Numerical results

Let us show a numerical result indicating modal-beat mediate intensity modulation. The interference of non-orthogonal complex transverse fields $\hat{E}m(x,y)$ and $\hat{E}n(x,y)$ introduces gain (stimulated emission) modulation to the laser in the form of $N_0 B \int_x \int_y \hat{E}m(x,y)\hat{E}n(x,y)^* dxdy + $ c.c. , where N_0 is the population inversion density and B is the Einstein's stimulated emission coefficient. Then, the model equations are given:

$$\frac{dN_i}{dt} = \frac{1}{K}\left[w - 1 - N_i - (1 + 2N_i)\left(E_i^2 + \beta \sum_{i \neq j} E_j^2\right)\right], \tag{1}$$

$$\frac{dE_i}{dt} = N_i E_i + gE_i E_{i+1} \cos\phi_{i,i+1} + gE_i E_{i-1} \cos\phi_{i,i-1}, \tag{2}$$

$$\frac{d\phi_{i,i+1}}{dt} = \Delta\Omega_{i,i+1} + D_i\xi_i(t), \quad i, j = 1, 2, 3. \tag{3}$$

where N_i is the normalized excess population-inversion density, E_i is the normalized field amplitude averaged over the transverse direction, $\phi_{i,i+1}$ is the phase difference between i-th lasing mode and its adjacent mode, w is the relative pump power normalized against the threshold, β is the cross-saturation parameter, K is the lifetime ratio of fluorescence lifetime τ to photon lifetime τ_p, g is the interference parameter, $\Delta\Omega_{i,i+1} = \Delta\omega_{i,i+1}\tau_p$ is the normalized lasing frequency difference between i-th lasing mode and its adjacent mode, and t is scaled to the photon lifetime. The last term in Eq. (3) express the phase noise where D_i is the phase noise strength and $\xi_i(t)$ is the Gaussian white noise, which satisfies $\langle\xi_i(t)\rangle = 0$ and $\langle\xi_i(t)\xi_j(t')\rangle = \delta_{ij}\delta(t-t')$. Numerical intensity waveforms ($E^2 = \sum_i E_i^2$) indicating

modal-beat mediate pulsation featuring breathing mode and chaotic relaxation oscillation corresponding Figs. 4(a) and 6(b) are shown in Fig. 7.

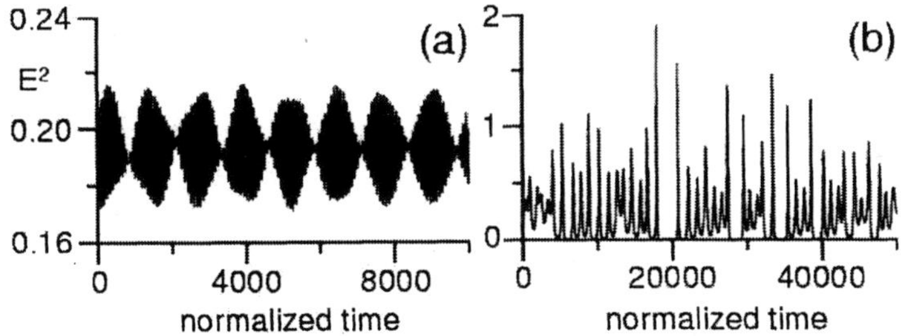

Figure 7. (a) A numerical breathing intensity modulation for two mode pairs. The parameters are w=1.05, β=0.667, K=2000, g=0.002, $\Delta\Omega_{12}$=0.08, $\Delta\Omega_{23}$=0.075, D_1=3×10⁻⁶, and D_2=3×10⁻⁶. (b) A chaotic relaxation oscillation for one mode pair as $\Delta\Omega_{12}$ approach coincided with the relaxation oscillation frequency. $\Delta\Omega_{12}$=0.008, where other parameters are the same as (a).

6 Symmetry-breaking and kaleidoscopic pattern change

Formation of complicated structures has been found to depend not only on the pump power and the asymmetry of the pump beam, but also on symmetry-breaking of thermally-induced refractive-index distribution as well as rotation of the crystal in the x-y plane. Depending on these pump conditions, a variety of structural changes were observed. Let us show patterns when the symmetry of the thermally-induced microcavity was changed. The experiment was carried out by tilting the focused beam as shown in Fig. 8, in which the polarization direction of the LD pump-beam was fixed to the $b(x)$-axis by the use of a polarizer. At this moment, the cavity symmetry was lowered and global structures of transverse patterns exhibited two-fold symmetry as expected as shown in Fig. 8. When the crystal is rotated, on the other hand, the thermally-induced deformed microcavity structure is expected to change through the anisotropy of absorption coefficients because the temperature distributions changes when an angle between the polarization direction of the pump beam and the crystal $b(a)$-axis changes. Indeed, a large number of transverse lasing patterns with different topologies appeared at fixed pump powers by successive rotations of the crystal like a kaleidoscope, namely "kaleidoscopic laser". Figure 9 shows an example of structural changes observed even for small rotation. In both cases of Figs. 8 and 9, intensity pulsations based on modal interference were also observed similar to Fig. 4.

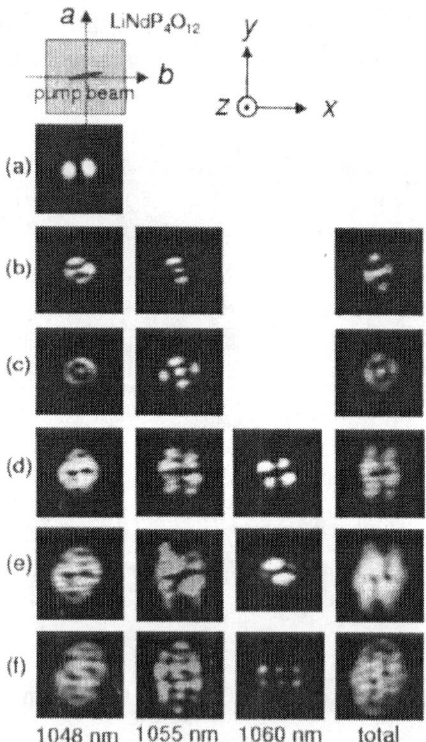

Figure 8. Pump-dependent structural change when the symmetry of thermally-induced refractive-index distribution was broken. Pump power: (a) 34 mW (b) 39 mW (c) 42 mW (d) 57 mW (e) 65 mW (f) 71 mW.

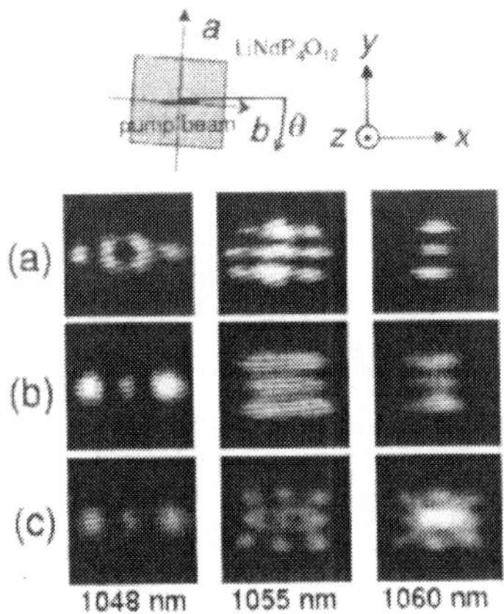

Figure 9. Kaleidoscopic structural change when the crystal was rotated. The angle between a-axis and y-axis: (a) θ=0°, (b) θ=4°, (c) θ=6°.

7 Summary

We showed the formation of non-orthogonal transverse modes and associated dynamical instabilities in a thin solid-state slice laser by using sheet-like end pumping. The observed emission patterns showed complicated structures consisting of different transverse eigenmodes, which resemble patterns formed in the quantum billiard systems. Modal-beat mediate intensity modulations for same-parity mode pairs without mode-orthogonality have been demonstrated and numerically reproduced by the simple model laser equation. The present finding opens a new research direction for wave formations in general thin-slice Fabry-Perot lasers with asymmetric end-pumping and provides nonlinear dynamic aspects into microcavity lasers. Further quantitative theoretical investigations on the difficult eigenvalue problem, that is wave formations in asymmetric gradient refractive-index potentials, coupled with atomic variables are strongly anticipated. The basic idea of sheet-like pumping of crystalline lasers for generating a large number of different transverse emission patterns from the same laser and achieving high-speed modulations, which are not restricted by the intrinsic relaxation oscillation, is generally applicable to many solid-state laser materials and VCSEL's with an asymmetric (non-circular) contact. Similar high-speed pulsations due to the interference between non-orthogonal transverse modes have been observed in LD-pumped Nd:YAG and Nd:YVO₄ lasers [11].

Acknowledgement

The authors are indebted to Drs. H. Makino and T.-S. Lim for discussions.

References

1. S. L. McCall, *et al.*, *Appl. Phys. Lett.* **60**, 289 (1992).
2. S.-B. Lee, *et al.*, *Phys. Rev. Lett.* **88**, 033903 (2002).
3. J. U. Nöckel, A. D. Stone, *Nature* **385**, 45 (1997).
4. C. Gmachl, *et al.*, *Science* **280**, 1556 (1998).
5. J. Mulet and S. Balle, IEEE J. Quantum Electron. **38**, 291 (2002).
6. P. B. Wilkinson, *et al.*, *Phys. Rev. Lett.* **86**, 5466 (2001).
7. K. Otsuka, *et al. Phys. Rev. Lett.* **87**, 083903 (2002).
8. K. Otsuka, et al., *Opt. Lett.* **23**, 201 (1998).
9. R. Kawai, Y. Asakawa, and K. Otsuka, *IEEE J. Quantum Electron.* **35**, 1542 (1999).
10. O. Bohigas, *Random matrices and chaotic dynamics*, in "Chaos and Quantum Physics" edited by M.-J. Giannoni, A. Voros, and J. Zinn-Justin, (LesHouches Session LII, North-Holland, 1989).
11. K. Otsuka, et al., *Opt. Lett.* **27**, to be published in the October 2002 issue.

Collective Chaos Synchronization among Modes in a Chaotic Three-Mode Laser

Takayuki Ohtomo, Jing-Yuan Ko, and Kenju Otsuka

Department of Human and Information Science, Tokai University,
1117 Kitakaname, Hiratsuka, Kanagawa 259-1292, Japan

Abstract. We observed three types of collective chaos synchronization featuring phase, generalized and lag synchronization among pairs of modes for different pumping conditions in a three-mode solid-state laser subjected to self-mixing modulations. The direction of information flow among modes formed in different collective synchronizations was characterized by information circulation analysis of the experimental time series. The combined effect of self-mixing modulations and coherent delayed optical feedback has been investigated experimentally and random switching between stable and collective synchronization states has been demonstrated.

1.Introduction

Cooperative behavior of coupled chaotic systems has been notified for theoretical significance and practical application. Chaos synchronization discovered by Pecora and Carroll is the one of most attractive subjects for elucidating complex behaviors occurring in coupled chaotic oscillators[1][2]. Chaos synchronization was considered initially that a system (drive system) had driven to follow another system (response system) as these dynamics X(t)=Y(t) (i.e. complete synchronization)[3][4]. In recent studies, however, other forms of chaos synchronization, such as *generalized, phase, lag,* and *anticipating synchronization,* have been found. Generalized synchronization means that there is a functional relationship between jointed systems. In phase synchronization, amplitudes of coupled systems are uncorrelated but their phases have a correlation. Lag and anticipating synchronization feature time-delays.

In optics, these types of chaos synchronization have demonstrated in different operation schemes of two-coupled lasers. In multimode solid-state lasers, much theoretical and experimental studies have led to recognition that the oscillating modes are globally coupled through cross-saturation of population inversions that results in the inherent antiphase dynamics[5]. In the free-running condition, N relaxation-oscillation components appear at $f_1 > f_2 > \cdots > f_N$ (N: number of modes) in the modal output[6]. In the total output, on the contrary, a unique relaxation oscillation component at f_1 remains while lower frequency relaxation oscillation components cancel each other out[7][8]. The present antiphase dynamics have been demonstrated to hold also in periodic and chaotic regimes[5],[7]-[9]. The cross-saturation coefficient among modes which governs the antiphase dynamics depends on the pump power, i.e. modal intensity ratios.

The first purpose of this study is to demonstrate collective behaviors of globally-coupled three-mode microchip chaotic lasers, featuring various forms of synchronizations among pairs of modes, operating in different pump regimes, i.e., modal intensity ratios, subjected to Doppler-shifted self-mixing modulations[10].

CP676, *Experimental Chaos: 7th Experimental Chaos Conference,*
edited by V. In, L. Kocarev, T. L. Carroll, B. J. Gluckman, S. Boccaletti, and J. Kurths
© 2003 American Institute of Physics 0-7354-0145-4/03/$20.00

On the other hand, phase-noise driven random switching between stable and chaotic pulsations (namely, random chaotic bursting) has been observed and reproduced numerically in a microchip two-mode laser with external feedback[11]. The second purpose is to identify collective behaviors occurring in the switching process in the case of globally-coupled three-mode operations. The effect of self-mixing modulations on random chaotic bursting has been also demonstrated.

The dynamic interplay among modes have been characterized in terms of "information circulation"[12], which was proposed based on the transinformation rate introduced by Paluš *el al*[13].

2.Experimental Setup

The experimental scheme is shown in Fig. 1. Experiments were carried out by using a laser-diode(LD)-pumped 300μm-thick LiNdP$_4$O$_{12}$(LNP) laser. As the level of pump power was increased from zero, single-frequency oscillation appeared at 1048 nm representing the ^{4}F$_{3/2}$(1). ^{4}I$_{11/2}$(1) transition. As the pump-power level was increased, a second-lasing mode appeared at 1055 nm [^{4}F$_{3/2}$(1). ^{4}I$_{11/2}$(2)] and this was joined by a third-lasing mode at 1060 nm [^{4}F$_{3/2}$(1). ^{4}I$_{11/2}$(3)]. In the first experiment, the output beam is divided to obtain two beams. One beam (the total output) is made to impinge on a rotating turntable via which self-mixing Doppler-shift feedback modulation is applied to the laser. In the second experiment, the laser was subjected to external coherent feedback with a 10-m optical fiber and/or self-mixing modulations as shown in Fig. 1. The other beam is split to produce three beams and each beam is passed through a monochrometer to allow the simultaneous observation of modal evolution over time.

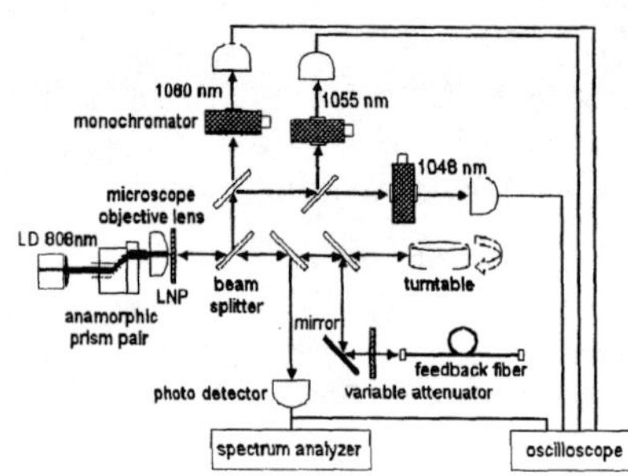

Fig.1. Experimental scheme of three-mode LNP laser subjected to external perturbations.

3. Collective Chaos Synchronization in Self-Mixing Feedback

Modal input-output characteristics in a self-mixing laser are shown in Fig. 2. We carried out at different pump-power levels that correspond to different output intensity ratios (i.e., cross-saturation coefficients)[7].The modulation frequency f_D was turned to in-phase relaxation oscillation component f_1. As the feedback ratio was increased, chaotic relaxation oscillations appeared.

Let us show results obtained for different pump-power regions I, II and III in Fig. 2. Examples of output intensity evolutions over time observed in region I, where $I_1 > I_2 > I_3$ (I_k: modal intensity), are shown in Fig. 3(a). Amplitude correlations for the three possible pairs of modes are plotted in Fig. 3(b). Here, output voltages were accurately calibrated from the intensity ratios of the average modal output as measured by multi-wavelength meter. Using the Hilbert transformation of time series which was introduced for characterizing phase synchronization[14], we calculated analytic phase correlations for pairs of modes. The analytic phase ϕ is related to the analytic signal V_A and its time average $\langle V_A \rangle$ by $V_A(t) - \langle V_A(t) \rangle = R_A(t)e^{i\phi(t)}$. Here $V_A(t) = I(t) + iI_H(t)$ where $I(t)$ is the time series of scalar intensity and $I_H(t)$ is its Hilbert transform. Results are shown in Fig. 3(c). The strong phase correlation among all modes is apparent, in which their amplitude correlations are not good enough to ensure complete synchron i-zation as shown in Fig. 3(b). Meanwhile, let us call this form of collective synchroni-zation, featuring phase synchronization among all the modes, type I .

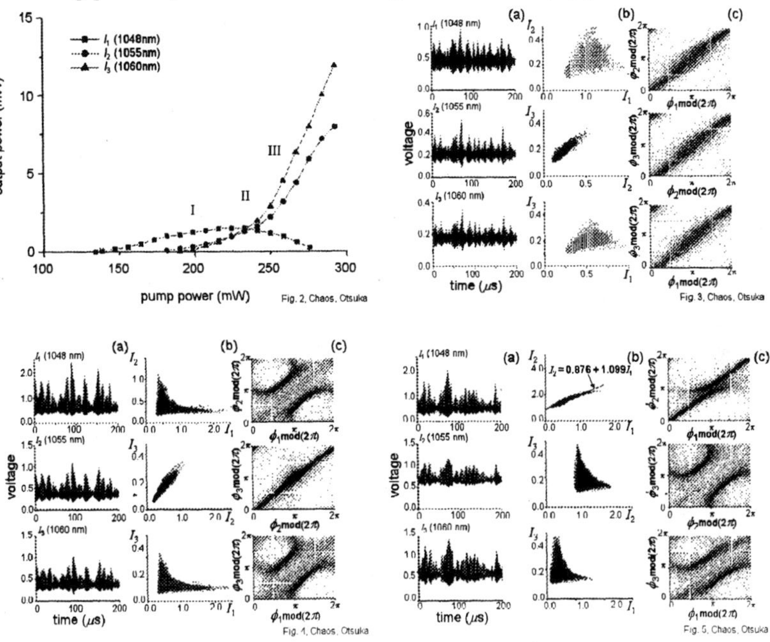

Fig. 2. Input-output characteristic of a LNP laser oprating in multi -transition -oscillation regime.
Figs. 3(upper right), 4(lower left) and 5(lower right). Fig. 3: type-I collective synchronization, Fig. 4: type-II collective synchronization, Fig. 5: type-III collective synchronization. In all figures: (a) ex-perimentally obtained modal intensity time series, (b) synchronization plots of the time series of inten-sity I_k for pairs of modes, and (c) analytic phase ϕ_k correlations for pairs of modes.

As the pump power was increased into region II where modal intensities of 1055 and 1060 nm modes approached the 1048 nm mode intensity, i.e., $I_1 \geq I_2 \approx I_3$, another type of chaos synchronization among pairs of modes was o bserved. Near the crossing point around which I_1 saturates, the highest frequency f_1 does not change with increas-ing pump power resulting in plateau. Modal intensity correlation plots are shown in Fig. 4. In this case (type II), 1048 nm mode exhibits strong . out -of-phase correlations against 1055 and 1060 nm modes as shown in Fig. 4(b) and (c), where each pair shows synchronization when the time series of the 1048 nm mode is shifted by one pulsation (i.e., relaxation oscillation) period about $1/f_1$, i.e., analytical phase shift of π. As for type II, a clustering occurred in the form of [1048,(1055,1060)].

The third form of collective chaos synchronization (type III) was observed in region III after the crossing point where 1060 nm mode intensity became lager than 1055 nm mode intensity and 1048 nm mode intensity decreased, i.e., $I_1 > I_2 > I_3$. In this region,

109

the in-phase relaxation oscillation at f_1 was dominated by the strongest 1060 nm mode. Then, a clustering occurred in the form of [(1048,1055),1060], in which 1048 and 1055 nm modes exhibited generalized synchronization while 1060 nm mode showed lag synchronization with other modes. From 30,000 data points, the intensity relation was estimated to be $I_2 = A + B \times I_1$ where $A = 0.876 \pm 0.001$ and $B = 1.099 \pm 0.002$.

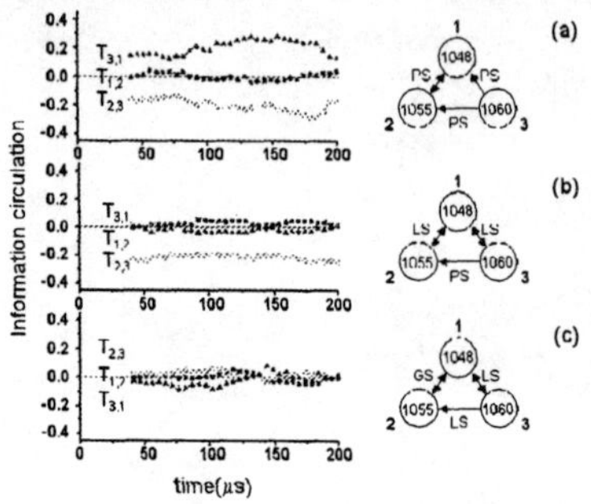

Fig. 6. Graph of information circulations for three types of synchronizations calculated from experimental time series shown in Fig 3, 4 and 5. (a) type I, (b) type II and (c) type III. The chart of the flows of information is given to the right of each.

Here, let us characterize dynamic interplay among modes by the use of "information circulation" that identifies the direction of information flows among modes which results in the observed self-organized chaos synchronization:

$$T_{X,Y} = T_{X \to Y} - T_{Y \to X},$$ (1)

where

$$T_{X \to Y} = \frac{1}{\tau^*} \sum_\tau S(Y, Y_\tau | X) - \frac{1}{\tau^*} \sum_\tau S(Y, Y_\tau)$$ (2)

is the information transfer rate from a time series X={x(t)} to Y={y(t)} [13], and $S(Y,Y.)$ is the self-mutual information for Y, i.e., a measure of the information that Y is able to provide about $Y.=\{y(t)+.\}$ (or of the predictability of $Y.=\{y(t)+.\}$) given Y). $S(Y,Y.|X)$ is the conditional self-mutual information of the time series X given the series Y. τ^* is the first local minimum of $S(Y,Y.)$. When $T_{X,Y} > 0$ (< 0), the information flow is from X(Y) to Y(X). Graphs of information circulation as calculated from the time series for three types of collective chaos synchronizations show in Fig. 3, 4 and 5 are given in Fig. 6 (a)-(c). In type I (upper trace) corresponding to Fig. 3, the 1060 nm mode simultaneously transfers information in a similar fashion to both the 1048 and 1055 nm modes among which bi-directional balanced information flow is established in this case. From the definition of information circulation based on the intensity distribution, it is reasonable that the direction of information flow is not unique for phase synchronization. However, the well-organized symmetric information flows exhibiting a mirror-image may suggest some form of $\sum T_{i,j} \approx 0$. The information flow chart is depicted on the right-hand side of Fig. 6. In type II (middle trace) corresponding to Fig. 4, balanced information flows are established between the 1048 nm mode and other modes which exhibit lag synchronizations with the 1048 nm mode, while the 1060 nm mode transfers the information to 1055 nm mode in this case. As for type III (bottom trace) corresponding to Fig. 5, balanced bi-directional information flows are established among three modes. The balanced bi-directional information flows for lag and generalized chaos synchronizations in type II and type III are reasonable from the concept of information circulation. It is interesting to point out that the number of strongly correlated mode pairs that exhibit bi-directional information flow increases with increasing pump power in the present system. As for $f_D = f_2$, f_3 modulations, in-phase dynamics of modes at these frequencies are inhibited resulting from the anti-phase dynamics and no synchronizations were observed[15].

4. Collective Behavior and Information Circulation in Random Chaotic Bursting

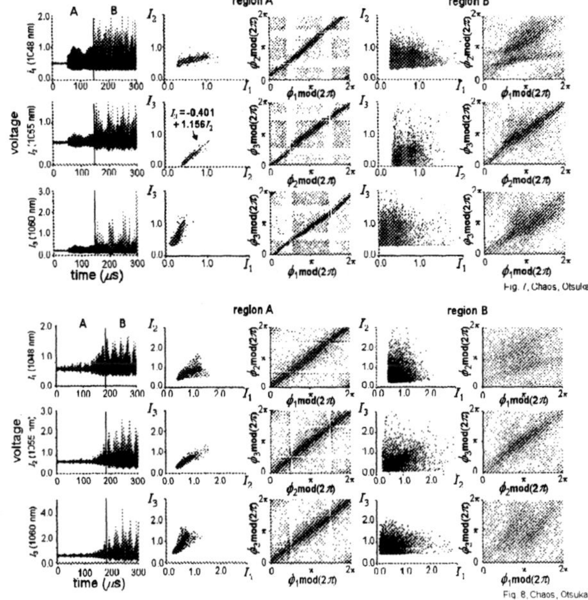

Figs. 7 and 8. Fig. 7(upper traces): Chaotic bursting observed by fiber feedback. Fig. 8(lower traces): Chaotic bursting subjected to self-mixing modulation. (a) Experimentally obtained modal intensity time series, (b) correlation plot for intensity I_k and analytic phase ϕ_k for pairs of modes in region A, (c) correlation plots in region B.

The issue of chaotic instabilities in the output of laser subjected to external feed-back was initiated by pioneering work of Lang Kobayashi in 1980[16]. In microchip solid-state lasers, the random chaotic bursting has been demonstrated in multimode regimes and reproduced numerically[11]. This phenomenon can be interpreted in terms of phase-noise-induced switching between stable and unstable states featuring degenerate Hopf bifurcations. The causal drive-response relationship exhibiting unidirectional information flow has been shown to be established in the buildup region from stable to chaotic spiking operations in the two-mode regime[12].

Here, let us study collective behavior associated with random chaotic burstings in globally-coupled three-mode regime and the effect of self-mixing modulation on chaotic bursting.

Figure 7 shows typical example of switching waveforms, amplitude-, and phase-correlations in the chaotic bursting, in which correlation plots in different regions A and B are depicted. In the switching transient A, collective synchronization featuring generalized synchronization among 1055-nm and 1060-nm modes in the form of $I_3 = A + B \times I_2$ is found to be established where $A = -0.401 \pm 0.0017$ and $B = 1.15 \pm 0.0031$, while synchronization fails in chaotic spiking region B. In this operation, modal intensity relation was $I_1 > I_2 = I_3$ and collective phase synchronization was observed for self-mixing modulation at f_1 similar to type-I. When self-mixing modulation was applied to the laser with external feedback, the transient behavior developing into chaotic spiking (region A) was found to be greatly affected. Results are shown in Fig. 8. The self-induced collective synchronization nature without modulations is replaced by that of the self-mixing laser in the absence of external in region A. However, the effect of self-mixing modulation is less effective in spiking region B, in which well-developed chaos takes place.

Information circulations and corresponding flow charts are summarized in Fig. 9 for (a) external feedback only, (b) self-mixing modulation only, and (c) combined case. A substantial change of information circulations between A and B regions is evident and parallels Fig. 8.

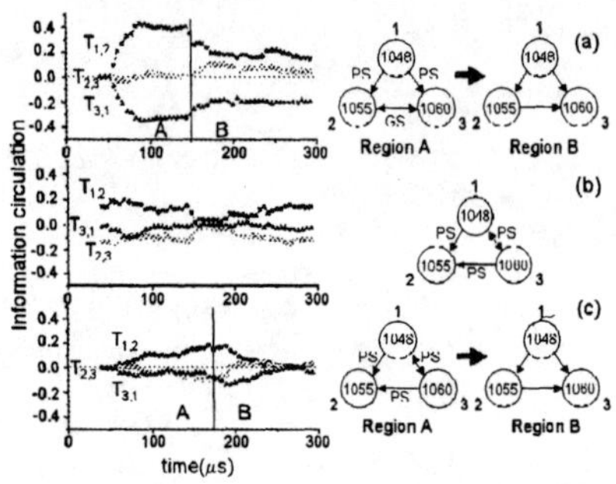

Fig. 9. Graph of information circulations calculated for (a) fiber feedback case, (b) modulated case, and (c) combined case. The chart of the flows of information is given to the right of each. As for combined case, flow charts for regions A and B are shown.

5.Conclusion

Three forms of collective synchronization behaviors have been observed in globally coupled nonidentical chaotic oscillators in the system of a modulated three mode microchip solid-state laser operating in different regimes of modal couplings. Each pair of modes has been shown to exhibit different types of chaos synchronizations. In the second experiment, different chaos synchronizations associated with chaotic bursting have been observed in delayed coherent feedback scheme, and the effect of self-mixing modulation on chaotic bursting has been identified. We have characterized such collective behaviors in different operation schemes by the information circulation, which was introduced by combining the ide as of circulation and rates of information transfer between coupled elements, as an aid in she dding light on the mechanism that lies behind such collective behaviors ruled by the inherent antiphase dynamics, in terms of the direction of info rmation flow among the modes.

REFERENCES

[1] L. M. Pecora and T. L. Carroll, Phys. Rev. Lett. **64**, 821 (1990).
[2] T. L. Carroll and L. M. Pecora, IEEE Trans. Circuits Sys. **II 38**, 453 (1991).
[3] R. Roy and K. S. Thonburg, Jr., Phys. Rev. Lett. **72**, 2009 (1994).
[4] T. Sugawara, M. Tachikawa, T. Tsukamoto, and T. Shimizu, Phys. Rev. Lett. **72**, 3502 (1994).
[5] K. Otsuka, M. Georgiou, and P. Mandel, Jpn. J. Appl. Phys., Part 2 **31**, L1250 (1992).
[6] K. Otsuka, *Nonlinear Dynamics in Optical Complex Systems* (Kluwer Academic, Dordrecht, 1999).
[7] P. Mandel, M. Georgiou, K. Otsuka, and D. Pieroux, Opt. Commun. **100**, 341 (1993).
[8] P. Mandel, K. Otsuka, J. Wang, and D. Peroux, Phys. Rev. Lett. **76**, 2694 (1996).
[9] P. Mandel, B. A. Nguyen and K. Otsuka, Quantum Semiclassic. Opt. **9**, 365 (1997).
[10] K. Otsuka, R. Kawai, S.-L. Hwong, J.-Y. Ko, and J.-L. Chern, Phys. Rev. Lett. **84**, 3049 (2000).
[11] K. Otsuka, J.-Y. Ko, J.-L. Chern, K. Ohki, and H. Utsu, Phys. Rev. A. **60**, 5 (1999).
[12] K. Otsuka, J.-Y. Ko, T. Ohtomo, and K, Ohki, Phys. Rev. E **64**, 056239 (2001).
[13] M. Paluš, V. Komárek, Z. Hrn.í., and K. Št. rbová, Phys. Rev. E 63, 04211 (2001).
[14] M. G. Rosemblum, A. S. Pikovsky, and J. Kurths, Phys. Rev. Lett. **76**, 1804 (1994).
[15] K. Otsuka, T. Ohtomo, A. Yoshioka and J.-Y. Ko, Chaos **12**, 678 (2002).
[16] R. Lang and K. Kobayashi, IEEE J. Quantum Electron. **QE-16**, 347 (1980).

GEOPHYSICS, OCEANOGRAPHY, AND METEOROLOGY

Synchronized Chaos In Climate Dynamics

Gregory S. Duane

IMA, University of Minnesota, 207 Church St. SE #400Lind, Minneapolis, MN 55101[1]

Abstract. Two parallel, quasi-two-dimensional fluid dynamical channel models, each of which vacillates chaotically between meteorologically significant dynamical regimes, synchronize when only the medium-scale eddy components of the flow are coupled. That the smallest scales need not be coupled suggests a low-dimensional dynamics, in accordance with the theory of inertial manifolds, and validates the use of low-order models in synchronization studies. Generalized synchronization of two channels with forcing terms localized in different sectors implies a relationship between large-scale circulation patterns over the Atlantic and Pacific oceans. However, the resulting correlations are small because of the large time lag in the coupling.

In a different interpretation of the two channels - one as "truth" and the other as "model" - synchronization suggests that a limited band of spatial frequencies in observed data will suffice to update weather prediction models, and provides a framework for the assessment of data assimilation strategies.

INTRODUCTION

Loosely coupled chaotic oscillators will fall into stably synchronized motion along their strange attractors, irrespective of initial conditions, in a wide variety of scenarios [1]. However, in contrast to the ubiquitous synchronization of oscillators with limit cycle attractors in Nature, the widespread occurrence of synchronized chaotic oscillators in naturally occurring systems has not been demonstrated. Rather, the phenomenon of synchronized chaos has been explored primarily in low-order or man-made systems.

Here, we show that two fluid-dynamical channel models, abstractly conceived as parallel, synchronize completely with various limited forms of coupling between corresponding points. Two applications of this synchronization tendency are explored. First, we show how the parallel channels can be used to represent the interaction between two sectors of a single channel, such as might represent the Atlantic and Pacific sectors of the Northern midlatitude climate system. These sectors are known to vacillate chaotically between dynamical regimes corresponding to "zonal" and "blocked" flow. The latter regime is also of practical interest because blocking patterns interfere with the normal flow of weather patterns and often give rise to extreme conditions downstream. The synchronization leads to a testable prediction about correlations between blocking activity in the two sectors, which compares favorably with observations and with general circulation model (GCM) output. Second, we show that the same computational results, with different assigned interpretations for the two channels, have implications for the

[1] email: duane@ima.umn.edu

CP676, *Experimental Chaos: 7th Experimental Chaos Conference,*
edited by V. In, L. Kocarev, T. L. Carroll, B. J. Gluckman, S. Boccaletti, and J. Kurths

selection of new observed data required to update weather prediction models.

EDDY-INDUCED SYNCHRONIZATION OF COUPLED PARALLEL CHANNELS

The Single-Sector Channel Model

The gross features of the midlatitude circulation, including the irregular vacillation between zonal and blocked flow regimes, can be captured in a 2-layer, quasigeostrophic, reentrant channel model with a jet upstream of the region where blocking occurs [2, 3]. The model is given by a prognostic equation for potential vorticity q:

$$\frac{Dq_i}{Dt} \equiv \frac{\partial q_i}{\partial t} + J(\psi_i, q_i) = F_i(q_i) + D_i(\psi) \tag{1}$$

where the layer $i = 1, 2$, ψ is streamfunction, and the Jacobian $J(\psi, \cdot) = \frac{\partial \psi}{\partial x} \frac{\partial \cdot}{\partial y} - \frac{\partial \psi}{\partial y} \frac{\partial \cdot}{\partial x}$ gives the advective contribution to the Lagrangian derivative D/Dt. Eq. (1) states that potential vorticity is conserved on a moving parcel, except for forcing F_i and dissipation D_i. The discretized potential vorticity is

$$q_i = f_0 + \beta y + \nabla^2 \psi_i + R_i^{-2}(\psi_1 - \psi_2)(-1)^i \tag{2}$$

where $f(x,y)$ is the vorticity due to the Earth's rotation at each point (x,y), f_0 is the average f in the channel, β is the constant df/dy and R_i is the Rossby radius of deformation in each layer. Periodic boundary conditions are imposed in the longitudinal x-dimension. Following Vautard et al. [2], we use an expression for dissipation of the form

$$D_i = (-1)^i v^{\text{int}} \nabla^2 (\psi_1 - \psi_2) - \delta_{i,2} v^{\text{Ek}} \nabla^2 \psi_2 + \alpha \frac{\partial^8 q_i}{\partial x^8} + \alpha' \frac{\partial^8 q_i}{\partial y^8} + D_i^p \tag{3}$$

which includes terms for internal friction between the two layers, Ekman damping at the surface, superviscosity, and extra damping terms (collectively denoted by D_p) for modes with zonal wave numbers 0,1,2, and 3. This form of the dissipation term, model parameter values, and other details are based on those of Vautard et al. [2], except that the width of the channel is half the length and a second channel (not shown in the figures), with flow in the opposite direction, is used to join the upper and lower latitudinal boundaries, to enforce the free slip boundary conditions used by Vautard et al.. If the forcing is chosen to be a relaxation term

$$F_i = \mu_0(q_i^* - q_i) \tag{4}$$

the flow will tend to a jet-like form near the beginning of the channel, for q_i^* corresponding to the choice of ψ^* shown in Figure 1a. The model vacillates chaotically between two relatively stable flow regimes that naturally divide state space, illustrated for instance by

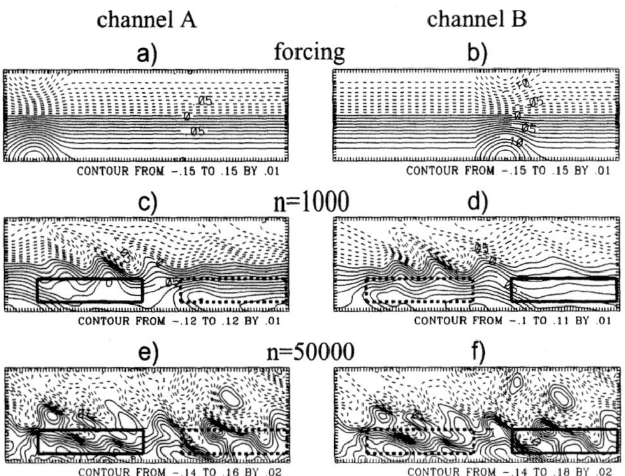

FIGURE 1. Streamfunction (in units of $1.48 \times 10^9 m^2 s^{-1}$)describing the forcing and evolution of a coupled parallel channel model with longitudinally skewed forcing jets $\psi^{A*} \neq \psi^{B*}$ (a,b) and advective coupling (Eqs. (5) and (6)), with $c = 1/2$. Near-identical synchronization occurs by the last time step shown (e,f). The solid-line boxes designate the regions in the two channels used to label a given flow as "blocked" or "zonal". Blocking activity in the dashed-line box in channel A, which is nearly the same as in the solid-line box in channel B after synchronization, anticorrelates with blocking in the solid-line box in channel A. (Similarly for the dashed-line box in channel B.)

the flows in Fig. 1d and Fig. 1c, which are identified with the "zonal" and "blocked" phases of the midlatitude index cycle, respectively. The periodic boundary conditions are taken here to represent the actual topology of the midlatitude region. However, since there is only one jet and one blocking center, the model represents either an active Pacific sector adjoined to a passive Atlantic sector, or vice versa.

The Two-Sector Model

A two-sector model is easily constructed by using a potential vorticity field q^* in (4) derived from a streamfunction ψ^* (via(2)) that would describe a flow with two jets upstream of corresponding blocking centers in each sector, as shown in Fig. 2, in place of the single-jet flow shown in Fig. 1a. However, the relationship of such a two-sector model to a pair of single-sector Vautard-Legras models is not immediately clear, since there is no natural way to impose boundary conditions at a midline in the two-sector model so as to partition it into two coupled sectors. We therefore describe an alternative construction of the two-sector channel model, achieved by coupling two single-sector channels of the type described above, with jets in opposite sectors.

We couple two models of the form (1), imagining a physically unrealizable configuration in which each point in a given layer of one model is coupled to the corresponding point in the corresponding layer of the other model. The coupling is introduced in the

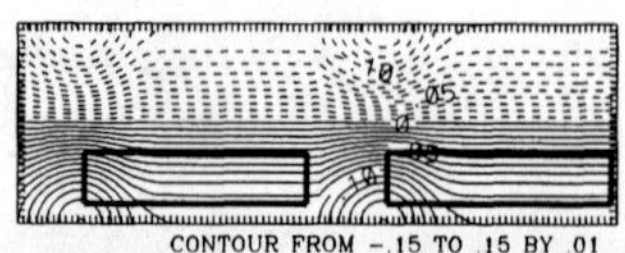

FIGURE 2. The forcing streamfunction $\hat{\psi}^*$ for the two-sector model (8), with one jet in each sector. At any given time, the flow in each sector is labeled as "blocked" or "zonal", depending on whether the minimum, taken over all longitudes x in the box in that sector, of the difference in streamfunction ψ latitudinally across the box, is less than or greater than .01 (in the units of Fig. 1), respectively.

advection terms of the two models:

$$\frac{Dq^A}{Dt} + cJ(\psi^A, q^B - q^A) = F^A(q^A) + D(\psi^A)$$

$$\frac{Dq^B}{Dt} + cJ(\psi^B, q^A - q^B) = F^B(q^B) + D(\psi^B) \tag{5}$$

with wave-number-dependant forcing terms defined by:

$$F_{\mathbf{k}}^{A,B} = (\mu_0 - \mu_{\mathbf{k}}^c)[q_{\mathbf{k}}^{A,B} - q_{\mathbf{k}}^{(A,B)*}] \tag{6}$$

where the flow has been decomposed spectrally and the subscript $\mathbf{k}$ on each quantity indicates the wave number $\mathbf{k}$ spectral component. The wave number dependence in the magnitude of the forcing is introduced so as to avoid constraining the small-scale eddies as the large-scale components of the flow are caused to maintain forms resembling Fig. 1a,b. Specifically, we set

$$\begin{aligned} \mu_{\mathbf{k}}^c &= 0 && \text{if } |k_x| \le k_{x0} \text{ and } |k_y| \le k_{y0} \\ \mu_{\mathbf{k}}^c &= \mu_0[1 - (k_0/|\mathbf{k}|)^4] && \text{otherwise} \end{aligned} \tag{7}$$

defining a slightly smoothed step function, with wave number cutoffs $k_{x0} = 3$ and $k_{y0} = 2$ (in units of waves per channel length or width, resp.) to distinguish between "eddies" and the large-scale flow, following Vautard and Legras [3]. The superscripts A and B in (5) and (6) designate the two separate channel models, each of which has two layers. The layer index i has been suppressed in (5) and (6), and will be suppressed henceforth.

The advective coupling is used because the average $\hat{q} = (q^A + q^B)/2$ of the solutions of (5a) and (5b), for strong coupling $c = 1/2$, is the solution of a model with the average forcing:

$$\left(\frac{D\hat{q}}{Dt}\right)_{\mathbf{k}} = 1/2(F_{\mathbf{k}}^A + F_{\mathbf{k}}^B) + \hat{D}_{\mathbf{k}} = (\mu_0 - \mu_{\mathbf{k}}^c)[\hat{q}_{\mathbf{k}} - 1/2(q_{\mathbf{k}}^{A*} + q_{\mathbf{k}}^{B*})] + \hat{D}_{\mathbf{k}} \tag{8}$$

since the advective coupling terms in (5) combine to give the proper nonlinear advective term in (8). Thus, solutions of the two-sector model defined by a forcing streamfunction such as that shown in Fig. 2 can be obtained from the coupled channel model (5) for judiciously chosen F^A and F^B. As the coupling is increased from $c = 0$ to $c = 1/2$, the two separate Vautard-Legras models are merged continuously. This construction has introduced extra, non-physical degrees of freedom, corresponding to the "passive" sector in each channel. It is shown in the following section that these nonphysical degrees of freedom are dynamically constrained, and essentially disappear in the fully coupled model.

ANTICORRELATION IN BLOCKING ACTIVITY FROM SYNCHRONIZATION OF THE FLOW

The flow fields in the coupled channels governed by (5) are found to synchronize, regardless of differences in initial conditions, as seen in Fig. 1. The two channels cannot exhibit *identical* synchronization because the forcing terms on the right hand sides of the two equations (5) are different. Close inspection of Fig. 1e,f indeed reveals persistent small differences between the flow fields in the two channels. The situation is that of *generalized* synchronization [4]. In the present case, the deviation from identical synchronization is small, so at $c = 1/2$ we have $\psi^A \approx \psi^B \approx \hat{\psi}$. That is, the flow in either channel approximates the flow in a channel with two jets.

As the coupling between the channels is increased from $c = 0$ to $c = 1/2$, the dynamics of each channel changes so as to incorporate an approximate, "virtual" counterpart of the dynamics of the sector that is forced in the other channel.

The relationship between the two sectors can be conveniently characterized in terms of blocked and zonal flow regimes. Blocking can be defined in each sector of the two-jet channel analogously to the definition for a single-jet channel (see Fig. 2). It is found that blocking in one sector anticorrelates with blocking in the other sector, as shown in Fig. 3a for several variants of the model. Anticorrelation is seen to be robust against changes in the shape and position of the jets, unless the jets are skewed latitudinally, a case that will be discussed in the next section.

The anticorrelation in blocking activity follows from generalized synchronization of the two constituent single-jet channels. Generalized synchronization can be obtained from identical synchronization by a change of variables in one of a pair of identically synchronized systems, as in the original examples given by Rulkov et al. [4]. In the present case, the correspondence between the two channels is near to the identity in the representation suggested by Fig. 1, but corresponding states in the two channels are described very differently. While the state of channel A is described as blocked or zonal depending on the flow in the first half of the channel, in the area denoted by the solid box in Fig. 1, just downstream of the channel's driving jet, the state of channel B is similarly described depending on the flow in the second half of the channel. It turns out that blocking activity is weakly anticorrelated in the two halves of either channel, whether or not the channels are synchronously coupled, although blocking occurs less frequently in the passive sector in the uncoupled case. For $c = 0$, anticorrelation in a single channel with one jet is suggested by an early detailed study of the one-sector model, which showed that the eddies induced by a blocked flow pattern tend to maintain that pattern, but that the eddies weakly inhibit blocking at other locations. In the single-jet channel, the smallness of the anti-correlation arises from the time-lag in the coupling of the two sectors, which smears the impact of the active sector

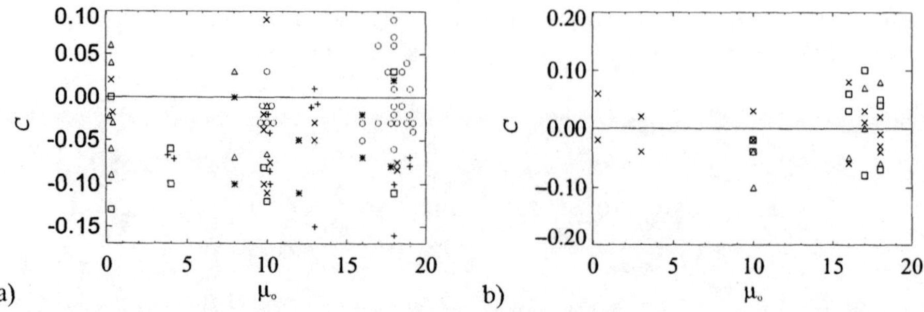

FIGURE 3. a) Correlation between binarized (blocked vs. zonal) states of the two sectors of the model illustrated in Fig. 2, with the binarization as defined in the caption thereto, in numerical integrations of (8) over 4×10^5 time steps with arbitrarily chosen initial conditions. The horizontal axis is the forcing strength μ_0 (in units of $7.3 \times 10^{-5} s^{-1}$). x's denote correlations for the two-sector model as previously defined, squares are for a variant with one sector two-thirds as long as the other, triangles are for one sector half as long as the other, *'s are for shortened jets in both sectors, +'s are for lengthened jets in both sectors, and circles are for a variant in which the latitudinal positions of the two jets are skewed arbitrarily by shifts ranging from a channel half-width to a channel width. Significant anticorrelation is found in all model variants, for μ_0 not too small, except in the case of latitudinally skewed jets. b) Correlation between binarized states of the sectors of the two-sector model, as in a), but with the large scales coupled by using (13) in place of (6), in order to allow the strength of the jets to vary. The sectors appear uncorrelated.

on the passive sector in a long integration over history. Synchronously coupled channels therefore exhibit a small inter-channel anticorrelation as well. (Since blocking is defined nonlinearly in terms of the flow field, anticorrelation in the average field $\hat{q}$ does not follow generally from anticorrelation within the two parallel channels separately, but does follow in the synchronous case where $\hat{q} \approx q^A \approx q^B$.) A change of variables in one channel corresponding to the 180° shift in the zonal direction that is appropriate to describe the rough Atlantic/Pacific symmetry turns near-identical synchronization into a correspondence far from the identity that is manifest as anticorrelation in blocking activity.

The anticorrelation appears to be due simply to the exchange of the small-scale eddy components of the flow, since it is the small-scale components that are left unconstrained by the forcing (6) with the wave number dependence (7). If a **k**-independent forcing is used instead of (6), as in the original Vautard-Legras model, so that the eddies are not free to mediate a sufficiently strong coupling, then both the synchronization between the channels and the anticorrelation between the sectors are found to disappear. In that situation, there is still anticorrelation between the sectors of each separate channel, but without synchronization there is not sufficient coherence between the sectors of the two-jet channel to give anticorrelation in the average field $\hat{\psi}$.

Eddy-induced synchronization is seen more directly for two channels coupled in a simpler manner, by setting $c = 0$ in (5) and introducing the coupling in the forcing terms instead:

$$F_{\mathbf{k}}^A = \mu_{\mathbf{k}}^c [q_{\mathbf{k}}^B - q_{\mathbf{k}}^A] + (\mu_0 - \mu_{\mathbf{k}}^c)[q_{\mathbf{k}}^{A*} - q_{\mathbf{k}}^A]$$
$$F_{\mathbf{k}}^B = \mu_{\mathbf{k}}^c [q_{\mathbf{k}}^A - q_{\mathbf{k}}^B] + (\mu_0 - \mu_{\mathbf{k}}^c)[q_{\mathbf{k}}^{B*} - q_{\mathbf{k}}^B] \tag{9}$$

<h2>n = 1000</h2>

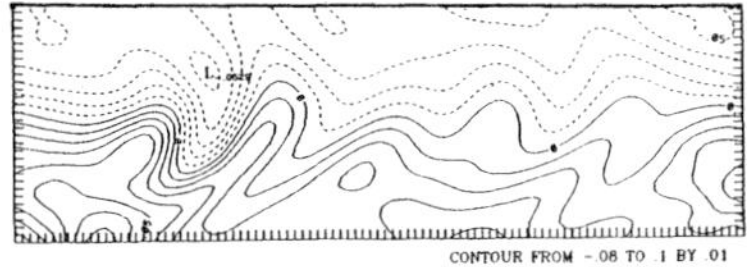 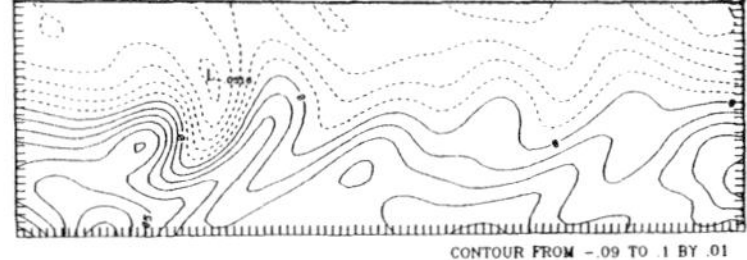

FIGURE 4. Streamfunction describing the flows in two coupled parallel channels, as in Fig. 1, at time step $n = 1000$, but for the dissipative coupling scheme (9) with $k_{noise} = 15$ in (10), so that the smallest scales are decoupled. Forcing is as in Fig. 1a,b. The flows are generally synchronized by the time shown.

We can arrange to couple a select range of frequency components (the "eddies"), by setting:

$$\mu_{\mathbf{k}}^c = \quad 0 \qquad\qquad \text{if } |k_x| \le k_{x0} \text{ and } |k_y| \le k_{y0}$$
$$\mu_{\mathbf{k}}^c = \quad \mu_0[(k_{noise}/|\mathbf{k}|)^4] \quad \text{if } |\mathbf{k}| > k_{noise}$$
$$\mu_{\mathbf{k}}^c = \quad \mu_0[1 - (k_0/|\mathbf{k}|)^4] \quad \text{otherwise} \tag{10}$$

defining a slightly smoothed characteristic function. In previous work [5], synchronization was shown for the case $k_{noise} = \infty$, but it is possible to identify the eddy components that are essential to synchronization/anticorrelation still more specifically. Since chaos synchronization is commonly robust against noise, it might be thought that the highest spatial frequency components, which include numerically-induced noise, would not be an essential part of the coupling required to synchronize the systems. If these smallest-scale components are in fact decoupled, then the channels still synchronize, as shown in Fig. 4 for the skewed forcing case where $q^{A*} \ne q^{B*}$. Similarly, synchronization is obtained with advective coupling if only medium spatial scales are retained in the difference $q^A - q^B$ within the two Jacobian coupling terms with coefficients c in (5).

In fact it is found that differences in the smallest-scale components are exactly the terms that compensate the differences in forcing in the two channels (via derivatives of the differences in the Jacobian coupling terms in (5)), so as to produce near-identical synchronization. These components need not be coupled. It is thus found that there are three ranges of wave numbers of interest: low wave numbers of modes that are ineffective in synchronizing the channels, medium wave numbers defining the synchronizing "eddies", and high wave numbers defining modes that are effectively slaved to the lower modes and need not be directly coupled. That synchronization occurs without coupling the highest spatial frequencies used in the model also establishes that the model is fully resolved for the purposes of the present study and that the inclusion of more modes would not change the results.

That the smallest scales need not be coupled indeed follows from the theory of inertial manifolds [6]. In the case of an infinite-dimensional system possessing a low-dimensional attractor, it might be supposed that the system could be described by a relatively small number of independent dynamical variables, with other dynamical variables effectively slaved to these. That is, the flow field can be decomposed as follows:

$$q = q_{master} + q_{slave} \tag{11}$$

with $q_{master} \in \mathcal{H}_1$, a finite-dimensional subspace, $q_{slave} \in \mathcal{H}_2$, an infinite-dimensional subspace, and $\mathcal{H}_1 \perp \mathcal{H}_2$. A functional relationship

$$q_{slave} = \mathcal{F}(q_{master}) \tag{12}$$

defines an *inertial manifold* within the state space of flows. If we perform the decomposition (12) in each of two coupled channels, then it seems reasonable that we should only have to couple the dynamical variables corresponding to the components q^A_{master} and q^B_{master} to effect synchronization. Provided the coupling of all independent dynamical variables is sufficiently strong, coupling of derived variables, in this case corresponding to q^A_{slave} and q^B_{slave}, should not be required. Likewise, when the forcing terms F^A and F^B are different in form, the the master-slave functions $\mathcal{F}^A$ and $\mathcal{F}^B$ are also different, and so $q^A_{slave} \neq q^B_{slave}$, in agreement with our finding of large differences in the high-frequency components.

COMPARISON WITH OBSERVATIONS

The prediction that exchange of eddies between the Atlantic and Pacific sectors causes anticorrelation of blocking activity cannot be directly compared with observations because the treatment of the large-scale flow in the model is unrealistic. In reality, the strength and positions of the upper tropospheric jets vary considerably. The relationships between the large-scale jet flow in the two sectors must therefore be considered, before predictions can be made about correlations in blocking activity.

Variations in jet strength can be incorporated in the channel model by modifying the forcing term so that only the form, but not the magnitude of the large-scale flow is affected. In place of (6), we use

$$F^{A,B}_{\mathbf{k}} = (\mu_0 - \mu^c_{\mathbf{k}})\Big[\frac{q^{A,B}_{\mathbf{k}}}{\|q^{A,B}\|} - \frac{q^{(A,B)*}_{\mathbf{k}}}{\|q^{(A,B)*}\|}\Big] \tag{13}$$

so as to induce a jet-like flow, of indeterminate strength, in each of the two sectors. The strength of the jet in each sector is thus influenced by the strength of the jet in the opposite sector, as might be implied by the conservation of mass. Resulting correlations are displayed in Fig. 3b. The anti-correlation observed in the previous cases is absent. The interaction between the magnitudes of the two jets has a correlating effect that competes with the anticorrelating effect of the eddy exchange. Apparently, the correlation between the strengths of the jets in the two sectors leads to a correlation in blocking activity. This correlating effect does not involve any detailed synchronization between states.

Variations in jet position can also be readily incorporated in the channel model. Flows with the two jets in latitudinally skewed positions can be used to define the forcing streamfunction and corresponding potential vorticity q^* in (6). Blocking correlations for a random assortment of skew values are plotted as circles in Fig. 3a. In contrast to other changes in model configuration with results that were shown in Fig. 3a, it is seen that introducing a latitudinal skew in jet position tends to destroy the anticorrelation effect.

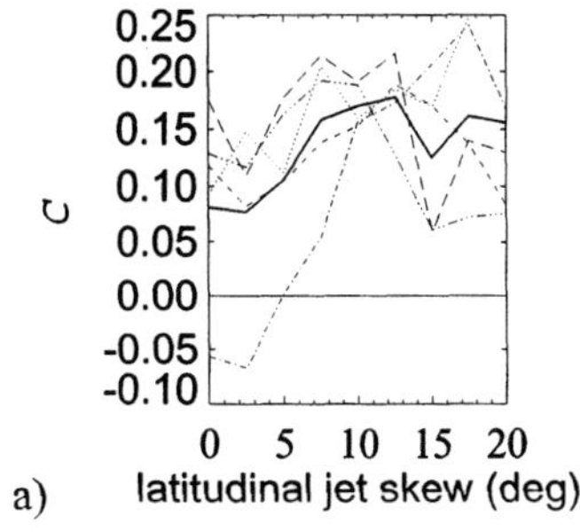

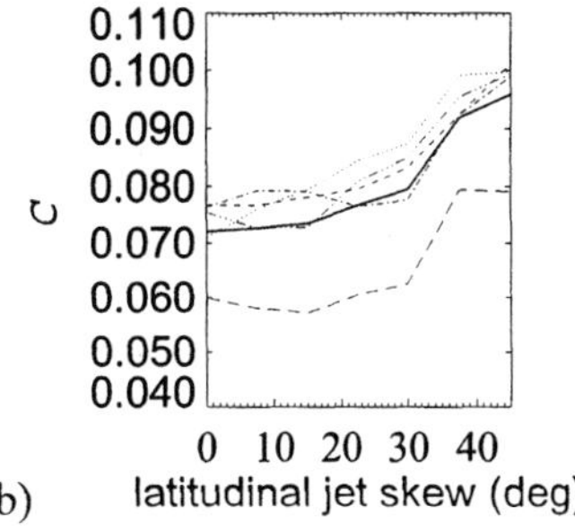

FIGURE 5. a) Correlation between blocking activity in the Atlantic ($100°W - 100°E$) and Pacific ($100°E - 100°W$) sectors, in observed meteorological data, on days of given latitudinal skew between Atlantic and Pacific jet positions. The data and blocking definitions are as in Ref. [9], except Northern Hemisphere winters for the 40 yr. period 1958-98 are considered. Daily jet position is defined as the position of maximum eastward component of wind, in the same observational data set, at an altitude corresponding to a pressure level of 300mb, at non-blocked longitudes. Correlations are shown for the full 40-yr. period (solid line) and for five 8-yr. segments (5 broken lines). The partitioning demonstrates that the increase in correlation as the skew increases from 0 to 10 degrees is statistically significant.

b) Correlation between blocking activity in the Atlantic and Pacific sectors as in a), but for 10^6 days (2700 years) of data produced by the Community Climate Model GCM, run under perpetual winter conditions. Correlations are shown for the full 2700-yr. period (solid line) and for five 540-yr. segments (5 broken lines). The partitioning demonstrates that the increase in correlation over the range of skew values depicted is significant.

The size of the correlating effect due to coupling of the large-scale flow in the Atlantic and Pacific sectors of the real midlatitude circulation is not known. Assuming that this effect, related to conservation of mass, is independent of skew, we hypothesize that anticorrelation is weakened during periods of jet skew. This hypothesis can be compared directly to observed data. Using a standard diagnostic for blocking described in (Duane et al. 1999), and obtaining the position of each of the two jets from the maximum of 300-mb zonal wind speed at non-blocked longitudes (i.e. where the jet is not divided by a blocking high-pressure center), we computed the intersectorial correlation on days of given skew between jet positions, for boreal winters over a 40-year period. The results plotted in Fig. 5a show significant increase in correlation as the skew increases from 0° to 10°, as expected.

To achieve greater confidence in the observed relationship between blocking correlation and jet skew, the analysis was repeated with data obtained from a general circulation model (GCM), simulating 1000 years of perpetual winter. Results are shown in Fig. 5b. While the increase in correlation with jet skew occurs over a different range of skew values than in observations, this difference might be attributed to the fact that the GCM reproduces observed blocking behavior in the Pacific sector poorly. The general trend still confirms the hypothesis and the synchronization theory on which it was based.

POTENTIAL APPLICATION OF CHANNEL SYNCHRONIZATION TO DATA ASSIMILATION

The synchronization of abstract parallel channels suggests a more immediate application in a different area of meteorology. In numerical weather prediction, it is well known that a climate model must be regarded as a semi-autonomous system that generates a useful version of reality, different from, but influenced by incoming observations. That is, the model variables are never completely re-set to the values of new observed data, since both sets of quantities are typically erroneous. Rather, the usual procedure is to form a weighted average of observations and a previous forecast to re-initialize the model, which therefore never completely loses track of its own history (e.g. [7]). The goal is to induce the model to synchronize with the objective world.

If one of the two abstract channels is taken to represent "truth" and the other to represent "model," then our results on synchronization have implications for numerical weather prediction. The two channels would here be coupled unidirectionally, but the results presented above for bidirectional coupling are expected to carry over. In particular, the finding that the smallest scales need not be coupled to effect synchronization of the channels implies that observations are only necessary within a limited spatial bandwidth. If the channel model configuration is indeed relevant, then only a limited number of observation stations are needed. The addition of further stations would not improve the synchrony or the predictive power of the model. Of course, this suggestion agrees with the usual practice in weather forecasting.

Synchronization thus provides a novel framework in which strategies for data assimilation might be evaluated and compared. With the goal of making Lyapunov exponents, or finite-time Lyapunov exponents, as negative as possible for directions transverse to the synchronization manifold, one could vary strategies for choosing the positions and timing of new data to be coupled unidirectionally to the predictive model. Both the choice of data and the coupling scheme would be subject to variation. In particular, truncations of different bases that have been proposed to expand the flow state (e.g. bred vectors and singular vectors) could be compared for efficacy in synchronizing the two channels when coupled. It would be hoped that general insights regarding synchronization would assist the optimization process.

PHILOSOPHICAL UNDERPINNINGS

The use of synchronization to describe the relationship between objective reality and computational model transcends the application to weather prediction. One wonders whether the synchronization paradigm may indeed be applicable to the mind-matter relationship that has been the subject of much philosophical discussion. At least in one commonly held worldview, the human mind is extremely effective in modeling and predicting the future behavior of the objective world based on rather limited input. Could our proficiency be viewed as the result of a synchronizing process?

A role for chaos synchronization in the mind-matter relationship is in accord with ideas about *synchronicity* due to the psychologist Carl Jung and to his collaboration

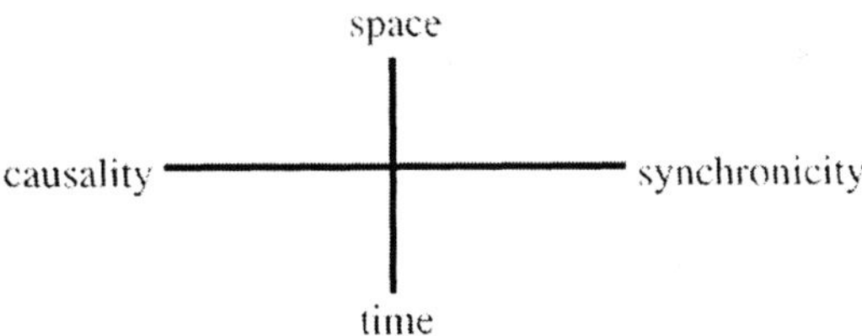

FIGURE 6. Diagram constructed by Carl Jung, later modified by Wolfgang Pauli, to suggest relationships based on synchronicity as an "acausal connecting principle", existing alongside causal relationships.

with Wolfgang Pauli (resulting in a jointly authored book in 1955 [8]). Jung and Pauli jointly constructed diagrams such as the one shown in Fig. 6 to suggest that an "acausal connecting principle" exists alongside causality. Jung invoked such a principle to explain surprising coincidences between mental and physical events, such as dreams and related actual occurrences. He was supported by Pauli, who early recognized the need for connections of a new type to describe quantum phenomena.

Pauli believed it important that the new connections were inconstant. The focus was on events so surprising as to appear to defy rational scientific explanation. He spoke of a need to return to irrationality in our understanding of the world, and tended to separate these philosophical speculations from his scientific work.

Chaos synchronization may offer a way to reconcile Jungian synchronicity with a rational scientific ontology. If synchronization is a sound approach to cryptography, as often claimed, then two effectively unpredictable systems may exhibit significant correlations because they are linked by a signal that would appear meaningless to an interceptor. The limited connections between matter and mind may then effectively carry more information than is apparent. Systems that are weakly coupled may exhibit very brief periods of synchrony amidst apparently de-synchronized behavior. The world may be rife with surprising coincidences to which meaning can be attached. All of this can be said because there is not much that is special about the abstract coupled channel models described above - many other forced-dissipative systems are expected to exhibit similar synchronization phenomena.

SUMMARY AND CONCLUSIONS

The one observational prediction of this study, that exchange of eddies results in an anticorrelating tendency in blocking activity between the Atlantic and Pacific sectors that weakens with latitudinal jet skew, has been borne out in observed data and in GCM data.

While this confirmation of the synchronizing effect in observations is preliminary, the ubiquity of synchronization in quasi-two-dimensional fluid dynamical models has here been firmly established. In particular, it is now clear that synchronization due to exchange of medium-scale eddies occurs independently of the resolution of the model. Synchronization in truncated models such as the one discussed by the author previously

[9, 10] is therefore relevant to the behavior of fully resolved models. Synchronization is observed in the models for different forms of coupling. Coupling two quasigeostrophic channel models via their advective terms results in a synchronizing tendency that implies a coherence between different sectors of the same channel.

To find scenarios in which the synchronization tendency would lead to larger correlations, one might seek a pair of systems that are functionally distinct but not geographically distinct. A candidate is the pair consisting of the Indian monsoon system and the El Niño-Southern Oscillation (ENSO) system. Correlations between these two systems have indeed been reported [11]. The possibility that detailed synchronization (perhaps intermittent) underlies these correlations merits investigation.

There is also room for theoretical development regarding the eddy-induced synchronization phenomenon. In particular, it is not known what characterizes the range of medium scales required for synchronization. It may be possible to make some statements about the upper and lower limits of this range without relying purely on empirical studies.

The results on synchronization of parallel channels point decidedly to an application in the area of numerical weather prediction and data assimilation. The idea would be to use synchronization tendencies to optimize data assimilation strategy, by targeted observation and optimized coupling of observations to the numerical model. However, questions remain about how to use synchronization *efficiently* to pick an optimal data assimilation strategy. Exploring the space of all strategies for the fastest synchronization would be computationally expensive. New theoretical tools might conceivably produce some *a priori* knowledge about expected synchronization rates. The promise of a useful general framework for data assimilation in meteorology and elsewhere would then be fulfilled.

ACKNOWLEDGMENTS

The author thanks his collaborators on previous work leading up to the present exposition: Joe Tribbia at the National Center for Atmospheric Research, and Jeff Weiss and Peter Webster at the University of Colorado.

REFERENCES

1. L.M. Pecora, T.L. Carroll, G.A. Johnson, and D.J. Mar, *Chaos* **7**, 520-543 (1997).
2. R. Vautard, B. Legras, and M. Déqué, *J. Atmos. Sci.* **45**, 2811-2843 (1988).
3. R. Vautard and B. Legras, *J. Atmos. Sci.* **45**, 2845-2867 (1988).
4. N.F. Rulkov, M.M. Sushchik, and L.S. Tsimring, *Phys. Rev. E* **51**, 980-994 (1995).
5. G.S. Duane and J.J. Tribbia, *Phys. Rev. Lett.* **86**, 4298-4301 (2001).
6. R. Temam, *Infinite-Dimensional Dynamical Systems in Mechanics and Physics*, Springer-Verlag, New York, 1988, pp.408-445.
7. R. Daley, *Atmospheric Data Analysis*, Cambridge Univ. Press, Cambridge, 1991.
8. C.G. Jung and W. Pauli, *The Interpretation of Nature and the Psyche*, Pantheon, New York, 1955.
9. G.S. Duane, *Phys. Rev. E.* **56**, 6475-6493 (1997).
10. G.S. Duane, P.J. Webster, and J.B. Weiss, *J. Atmos. Sci.* **56**, 4183-4205 (1999).
11. P.J. Webster and S. Yang, *Quart. J. Roy. Met. Soc.* **118**, 877-926 (1992).

An Equivariant 3D model for the long-term behavior of the solar activity

C. Letellier[♥], J. Maquet[♥], L. A. Aguirre[♦]& R. Gilmore [♣]

[♥] *CORIA / CNRS UMR 6614 - Université et INSA de Rouen, Av. de l'Université, BP 12,
F-76801 Saint-Etienne du Rouvray cedex, France*
[♦]*Universidade Federal de Minas Gerais, Av. Antônio Carlos 6627, 31.270-901
Belo Horizonte, MG., Brazil*
[♣]*Physics Department, Drexel University, Philadelphia, PA 19104, USA*

Abstract. Modeling dynamics underlying the sunspot numbers is an important problem because such data indicate the relative activity of the Sun. A key point in modeling sunspot data, which follows an 11-year cycle, is the need to take into account the reversal of the Sun's magnetic field, which follows a 22-year cycle. This can be done using an appropriate coordinate transformation applied to the phase portrait reconstructed from the sunspot numbers. Such a transformation introduces symmetry in the phase portrait and has the advantage of unfolding the structure of the dynamics. Global models have been obtained from such data. It is shown that the models capture the basic dynamical structure underlying the data which appears to be the structure of a Rössler attractor with an additional half twist.

INTRODUCTION

Solar activity is produced by the emergence of magnetic flux through the photo-sphere forming active regions which include sun-spots. The most characteristic feature of solar activity is its basic 11-year cycle discovered by Schwabe in 1843[1]. Schwabe also described an irregular behavior with fluctuations in the cycle duration as well as in the individual shape and maximum intensity. In 1919, Hale and co-workers discovered that every 11-years the polarity of the Sun's magnetic field reverses[2].

Although the global aspects of the solar cycle are well explained by dynamo theory, it remains doubtful whether the observations favor an explanation in terms of nonlinear (chaotic) dynamics or stochastic processes. Indeed, none of the studies involving nonlinear dynamical systems theory[3-5] provides convincing evidence of a chaotic Sun since no universal method was able to discriminate between colored noise with power law spectra and underlying dynamical processes in data[6]. On the other hand, since the end of the eighties, several different global modeling techniques have been developed for constructing sets of ordinary differential equations or discrete maps [7-11]. These are particularly powerful techniques for providing a global model from a very limited amount of data. Integrating or iterating these global models may generate synthetic data with the same underlying dynamics. When a satisfactory global model is obtained, clear evidence for a nonlinear deterministic component of the dynamics is thus provided. To the best of our knowledge, a single global model

CP676, *Experimental Chaos: 7th Experimental Chaos Conference,*
edited by V. In, L. Kocarev, T. L. Carroll, B. J. Gluckman, S. Boccaletti, and J. Kurths
© 2003 American Institute of Physics 0-7354-0145-4/03/$20.00

has been reported[12] for the sunspot data. On the one hand, such a model captures a few characteristics of the sunspot number dynamics, but on the other hand presents some differences from the solar dynamics.

SUNSPOT NUMBERS AND PHASE SPACE RECONSTRUCTION

Two reasons for the dynamical mismatch may be conjectured. First, it has been noted that the data are not of uniform quality[13]. The sunspot number series was built as monthly means from 1749. Thus, not more than 23 cycles are available (Fig. 1). This is definitely not enough for using standard algorithms for searching signatures of a low-dimensional chaotic attractors in time series (like Lyapunov exponents, correlation dimension and so on). Moreover, the sunspot numbers before 1850 were reconstructed by Wolf and are somewhat unreliable since some characteristics of the underlying dynamics are significantly different for the data recorded before and after 1850[14]. Even after 1850, the dynamics appears to be non-stationary, that is, there is still some change in the dynamics which cannot be explained in terms of a low-dimensional deterministic system[14]. Since known attempts to construct a global model used the full time series available, this could explain the limited success of previous attempts to obtain global models from the sunspot data, as reported in the literature.

Second, the reversals of the Sun's magnetic field have been introduced manually by changing the algebraic sign of the even cycles. Such a procedure presents the disadvantage of forcing the trajectory to pass near the origin of the reconstructed space when switching from one cycle to the next. In that domain, the noise contamination is sufficient to hinder any successful global modeling. Moreover, it will be shown below that this procedure is not adequate.

The first step in obtaining a global model from a time series like the sunspot numbers (Fig. 1) is to reconstruct the phase space using delay coordinates[15]. The phase space is thus spanned by $u_1 = R(t), u_2 = R(t + \tau), u_3 = R(t + 2\tau),\ldots$ (Fig. 2) where τ is a delay equal to 15 months. This is close to the value used by Mundt *et al* [3]. Since we are mainly concerned with the long-term behavior of the solar activity, the monthly averaged sunspot numbers used were slightly smoothed to remove short-term fluctuations. This does not significantly affect the global organization of the trajectory in the reconstructed phase space but does simplify modeling.

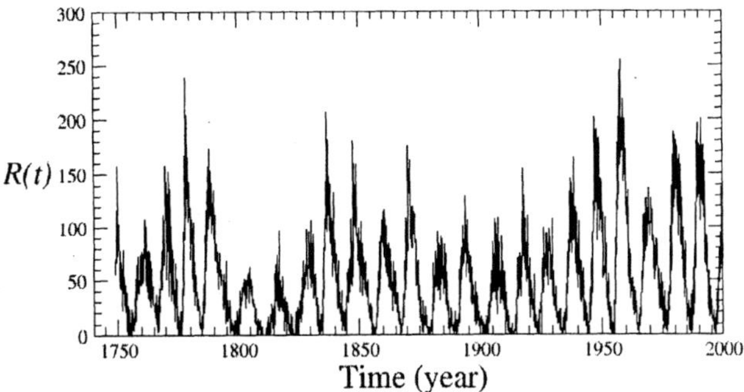

FIGURE 1. Monthly averaged sunspot numbers using Wolf's index $R=k(10g+f)$, where g counts the number of sunspot groups and f counts the individual sun-spots. The factor k was introduced to allow a "normalization" among the different observers who contributed. For these reasons, only the data after 1850 were used in this work for obtaining global models. The time series used here is available on the web-site of the National Geophysical Data Center (NGDC) in Boulder, Colorado, USA[16].

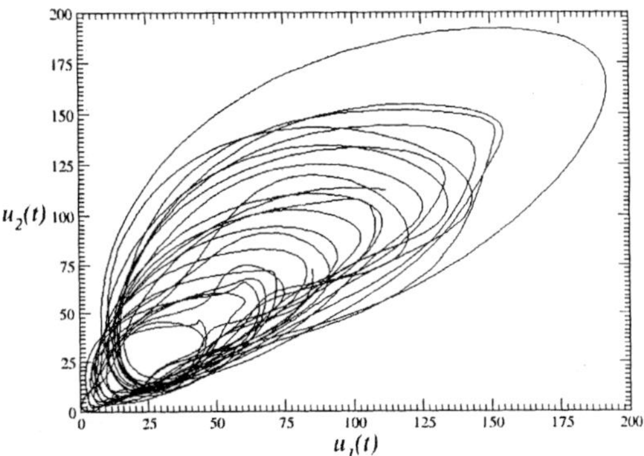

FIGURE 2. Plane projection of the phase portrait reconstructed using the delay coordinates $u_1 = R(t)$ and $u_2 = R(t+\tau)$ where $\tau = 15$ months. Note that the 23 cycles recorded since 1749 are shown here. This explains why some drift may be observed in this phase portrait.

In order to estimate the number of dynamical variables required for spanning the reconstructed phase space, a false nearest neighbors technique was used[17,18]. Note that the small number of cycles and the non stationarity of the data precludes finding a clear threshold over which an unambiguous description of the dynamics can be provided. Our computations suggest that the embedding dimension could be 3. Although a part of the data is not very reliable, such embedding dimension is a first clue that the dynamics underlying the long-term solar activity might be low-dimensional. One may reasonably assume that the three dimensional phase space $R^3(u_1, u_2, u_3)$ would be sufficient to obtain a trajectory without any self-intersections.

The 11-year sunspot cycle is driven by the 22-year magnetic field cycle. That is, $n(t) \approx B(t)^2$, where $n(t)$ is some measure of sunspot activity and $B(t)$ is some measure of magnetic field activity. Our goal is to construct a model relating sunspot and magnetic field activity. To be explicit, we construct a double cover[19] of the sunspot attractor (Fig. 4) in which the magnetic field reverses polarity every 11 years. A useful double cover is obtained using the transformation

$$\psi = \begin{vmatrix} u_1 = \Re(x+iy)^2 = x^2 - y^2 \\ u_2 = \Im(x+iy)^2 = 2xy \\ u_3 = z \end{vmatrix} \tag{1}$$

where $i = \sqrt{-1}$, $\Re$ and $\Im$ indicate the real and imaginary parts, respectively. Every double cover obtained from this transformation is symmetric under rotation by π radians about the z-axis in the phase space $R^3(x,y,z)$. In order to guarantee Hale's polarity law, the rotation axis must pass through the hole in the image portrait (Fig. 3). If it does not - if the symmetry axis is taken at or near the origin in Fig. 2 - the double cover attractor will be similar to the Lorenz attractor and the sign of the magnetic field will change "randomly" from one 11 year cycle to the next.

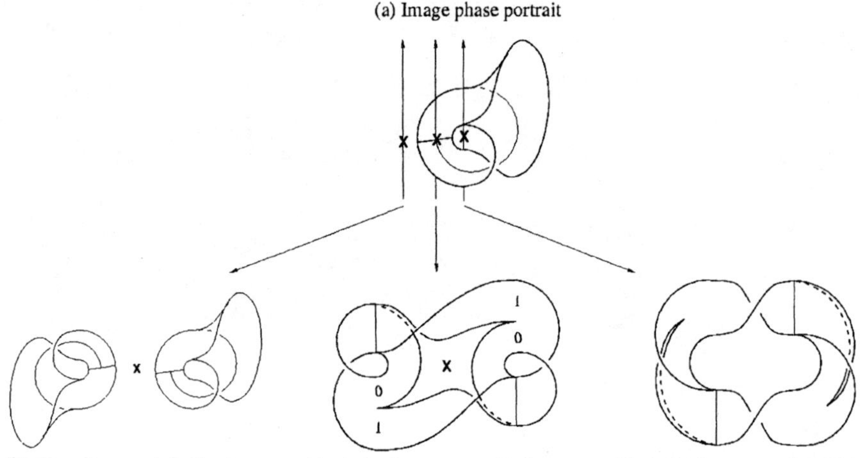

FIGURE. 3. Depending on the location of the singularity (the rotation axis or the center) designated by a cross in the image, three types of non-equivalent covers may be obtained. The sign of the dynamical variables in the cover space are reversed at each cycle in the image only for case (d), that is, when the singularity is located in the center of the image phase portrait. Otherwise, the sign is reversed irregularly (c) or never (b). Note that the branched manifold for the image phase portrait (d) corresponds to the one identified on the phase portrait generated by the global model.

Fig. 4(a) shows an x-y plane projection of the double cover obtained from the image sunspot attractor in Fig. 2 when the symmetry axis is parallel to the u_3 axis and passes through $(u_1, u_2) = (30,30)$, that is, the singularity is located in the center of the image sunspot attractor. In this representation of the dynamics the x- and y variables exhibit 22-year cycles while the z variable exhibits an 11-year cycle and is never negative $(z \geq 0)$.

In order to model the dynamics represented by this double cover, it is necessary to model three functions, one each for $\dfrac{dx}{dt}$, $\dfrac{dy}{dt}$ and $\dfrac{dz}{dt}$. To simplify the modeling effort, we instead introduce a differential embedding based on the variable x according to $X=x$, $Y=\dot{x}$, $Z=\ddot{x}$. This embedding requires the estimation of only one function[11]. The behavior of the variable $X(t)$ obtained from this model will be compared with the behavior of $X(t)$ [cf. Eq. (1)] obtained directly as the double cover of the sunspot data.

Using global modeling techniques, it has been possible to obtain models. In what follows the results obtained with a three-dimensional continuous-time model are presented. Integrating this model generates a chaotic attractor (Fig. 4(b)) topologically equivalent to the branched manifold[20] shown in Fig. 3(d). Consequently, the dynamics associated with the phase portrait reconstructed from the sunspot number (Fig. 2) is characterized by a reverse horseshoe branched manifold, that is, there is an additional half twist compared to the Rössler attractor. This structure is displayed by global models (continuous and also discrete). This nontrivial agreement is taken as strong evidence that the topology of the model attractor corresponds to that of the dynamics underlying the sunspot data. Indeed, this attractor compares very favorably with the experimental data shown in Fig. 4(a), in the same differential embedding. The model was generated from the 2.5 cycle length of data shown in Fig. 4(b).

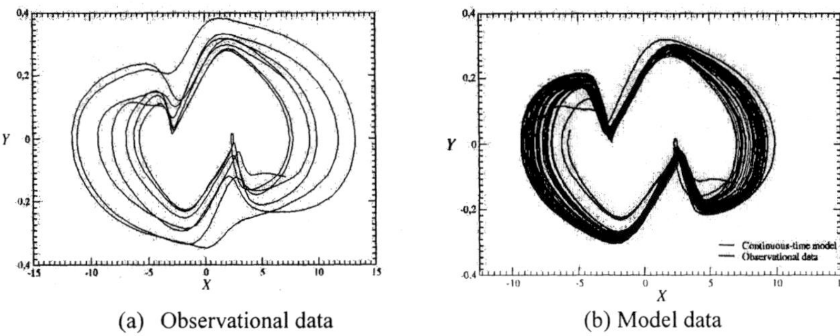

<table>
<tr><td style="text-align:center">(a) Observational data</td><td style="text-align:center">(b) Model data</td></tr>
</table>

FIGURE. 4. $X\text{-}Y$ plane projection of the differential embedding for the recorded sunspot numbers (a) and for the three-dimensional continuous-time model. The global model is estimated using two and half cycles shown in (b). The model is able to reproduce the fluctuations over a range not so large as for the observational data. This results directly from the non-stationarity of the data that the model cannot capture.

The three-dimensional continuous-time model captures the main characteristic of the dynamics underlying the sunspot number. The chaotic attractor generated by the model is a little bit less developed than the observational data suggest. This is a common result when a global modeling technique is used as extensively discussed in Letellier *et al* [11]. Usually, such a feature results from the noise contamination which cannot be captured by the model. In addition to that, the lack of stationarity of the solar activity or, at least, of the dynamics underlying the recorded sunspot numbers, contributes to such a departure. This explains also why it is not possible to obtain a global model from the whole data set available. Nevertheless, numerical integration from any point of the observational data provides a trajectory which settles down onto the chaotic attractor shown in Fig. 4(b).

CONCLUSION

Here we show that, by using an appropriate coordinate transformation for introducing the symmetry properties in the phase portrait reconstructed from the sunspot numbers, it has been possible to obtain a three-dimensional global model generating a   chaotic attractor. The model dynamics is very close to the solar dynamics. The unavoidable non-stationarity contained in the data precludes obtaining a model that produces fluctuations in the same amplitude range. This may come from the fact that sunspot numbers are only semiquantitative indicators of solar activity[4]. The global model obtained here therefore provide a first evidence of a nonlinear deterministic component in the long-term solar activity. Moreover, the topological mechanism has been identified as being a reverse horseshoe, that is, a Rössler-like attractor with an additional half twist. Similar results were obtained with a NARMA polynomial model to be described elsewhere.

ACKNOWLEDGMENTS

Collaboration between C. L. and L. A. is supported by CNRS and CNPq. R. G. is supported in part by NSF Grant PHY 9987468. A part of this work was done during a stay by C. L. at Drexel University.

REFERENCES

1. Schwabe H., *Astron. Nach.*, **495**, t. XXI, 235 (1843).
2. Hale G. E., Ellerman F., Nicholson S. B. & Joy A. H., *Astr. J.*, **49**, 153-178 (1919).
3. Mundt M. D., Maguirre II W. B. & Chase R. R. P. , *J. Geophys. Res. A*, **96** (2), 1705-1716 (1991).
4. Kremliovsly M., Solar Physics., **151**, 351-370 (1994).
5. Jinno K., Xu S., Berndtssen R., Kawamura A. & Matsumoto M, *J. Geophys. Res. A*, **100** (8), 14773-14781 (1995).
6. Theiler J., Eubank S., Longtin A., Galdrakin B. & Farmer J. D, *Physica D*, **58**, 77-94 (1992).
7. Crutchfield J. P. & McNamara B. S., *Complex systems*, **1**, 417-452 (1987).
8. Giona M., Lentini F. & Cimagalli V., *Phys. Rev. A*, **44** (6), 3496-3502 (1991).
9. Gouesbet G., *Phys. Rev. A*, **46** (4), 1784-1796 (1992).
10. Brown R., Rul'kov N. F. & Tracy E. R. *Phys. Lett. A*, **194**, 71-76 (1994).
11. Letellier C., Le Sceller L., Dutertre P., Gouesbet G., Fei Z. & Hudson J. L., *J. Phys. Chem.*, **99**, 7.016-7027 (1995).
12. Lainscsek C., Schürrer F. & Kadtke J. B, *Int. J. Bif. and Chaos*, **8** (5), 899-914 (1998).
13. Eddy J. A., *Science*, **192**, 1189-1202 (1976).
14. Carbonell M., Oliver R. & Ballester J. L., *A. & A.*, **290**, 983-994 (1994).
15. Packard N. H., Crutchfield J. P., Farmer J. D. & Shaw R. S., *Phys. Rev. Lett.*, **45** (9), 712-716 (1980).
16. www.ngdc.noaa.gov.
17. Abarbanel H.D.I. & Kennel M.B., *Phys. Rev. E*, **47**, 3057-3068 (1993).
18. Cao L., *Physica D*, **110**, 43-52 (1997).
19. Letellier C. & Gilmore R., *Phys. Rev. E*, **63**, 016206 (2001).

APPROXIMATE SELF-AFFINE MODELS OF ATMOSPHERIC TIME SERIES

Oleg I. Yordanov

American University in Bulgaria, Blagoevgrad 2700, Bulgaria

Abstract. We construct phenomenological models for the two-point statistics of various atmospheric quantities. The latter are measured under the Atmospheric Radiation Measurement (ARM) Program and made it accessible to the scientific community. The models are defined in the spectral domain as two-segment, power-law functions, explicitly account for the finite-size effects, and the crossover phenomena between different (approximate) self-affine regimes. Pseudo-periodic components observed in the data are also incorporated in the models.

INTRODUCTION

The atmosphere is the most dynamic component among those forming the global climate. Because it is permanently into a state of fully developed turbulence, the characterization of the atmosphere even when restricted to its second order statistics is a formidable task. The theory of statistical turbulence [1] can be used for this purpose with a limited success only. There are mainly two hurdles in this. First, the theory is built around the assumptions of local homogeneity and isotropy, which in field experiments is violated by the impact of the environment conditions and anisotropic formations that exist in the real atmosphere; e.g. clouds, fronts, etc. Second, the theory mostly pertains to the so-called Kolmogorov's inertial (similarity) range [2], namely for turbulence scales between the characteristic (for the flow) large-scale down to the small scale where the energy dissipates into a heat; for refinement and recent status of similarity theory see for example [3]. Outside this range a practitioner has to relay on phenomenological models. In terms of the autocovariance function(ACF), an empirical approach is still indispensable both for very short and very long (representing the tail) lags.

The main purpose of this report is to present a phenomenological model for the turbulent velocity field represented by a time series measured at a fixed point, 5 meters above the ground. More specifically, we focus on modelling the power spectra and ACF of the velocity pulsations. In parallel, spectra and ACFs of temperature, pressure and relative humidity are constructed and compared with those of the velocity.

All time series we analyse in this study are measured under the Atmospheric Radiation Measurement (ARM) Program of US Department of Energy, using The Temperature, Humidity, Wind, and Pressure System (THWAPS), see `http://www.arm.gov/docs/instruments/static/thwaps.html`, and made freely and easily accessible to the world-wide scientific community. The sampling rate of all time series is 10 Hz, the recorded data represent averages over period of 5 minutes. The particular data set that we retrieved pertains to a 5 days period between May 02 and May 06 of

CP676, *Experimental Chaos: 7ᵗʰ Experimental Chaos Conference*,
edited by V. In, L. Kocarev, T. L. Carroll, B. J. Gluckman, S. Boccaletti, and J. Kurths
© 2003 American Institute of Physics 0-7354-0145-4/03/$20.00

2000, and is recorded at the Southern Great Plain' (SGP) Facility. Each of the time series comprises $L = 1440$ equidistant (with time step of $\Delta t = 300$ sec) points. All quantities except the wind are measured at about 1 meter above the ground.

A still ongoing effort to develop a general methodology for building spectral and ACF models of data that exhibit self-affine (self-similar) symmetry is briefly described at its current level of maturity in the next section. In the third and the last section, we present and discuss the main result of our work – the statistical models of all four atmospheric quantities of interest – and draw conclusions.

GENERIC FORM OF THE MODELS

The model we suggest for all time series described in the Introduction is constructed in the frequency domain as a generic multi-segmented power-law function. Let $\omega_0 > 0$, ω_1, ..., $\omega_n = \omega_{(N)}$ be a partition of the frequency domain, $\omega_{(N)} = \pi/\Delta t$ being the Nyquist angular frequency. We define (n-segmented) spectral model $S(\omega)$ as a piece-wise continuous function, which in any of the subintervals $\omega_k \leq \omega \leq \omega_{k+1}$, $k = 0, 1, \ldots n - 1$ is represented by a power-law $S^{(k)}(\omega) = A_k \omega^{-\alpha_k}$, $A_k > 0$. The continuity conditions imposed at the splitting frequencies ω_k, namely $S^{(k-1)}(\omega_k) = S^{(k)}(\omega_k)$, can be used to express all spectral constants A_k in terms of A_0. This reduces the number of the independent parameters of the model to $2n + 1$: n spectral exponents α_0, α_1, ..., α_{n-1}; n splitting frequencies ω_0, ω_1, ..., ω_{n-1}; and A_0. It should be stressed, that we do not put restriction on the values of α_k. Thus, by having $\alpha_k < 0$, we allow the model to approximate accenting branches of the spectrum. The corresponding ACF, $\mathcal{A}(t)$, can be calculated in a closed form, a consequence of proposed spectrum simplicity. Since the contribution to the ACF of the segments is additive, for this essentially we need to evaluate the ACF corresponding to a single segment. The latter were obtained in [4, 5] and expressed in terms of type $_1F_2$ hypergeometric function. We note here only that these expressions are analytic functions with no singularities in the entire complex plane; they are easily and fast computed from their series, or for large arguments, from their full asymptotic expansions [5]. The latter is especially important since the ACF needs to be computed for arguments extending over several orders of magnitude.

The random processes that stem from the multi-segmented spectra $S(\omega)$ have interesting self-affine properties which are inferred from the asymptotic form of $\mathcal{A}(t)$. They will be presented in detail elsewhere, here we note only that if $\omega_{k+1}^{-1} \ll t < \omega_k^{-1}$, to a very good approximation

$$\mathcal{A}(t) \sim \sigma^2 - \chi^{3-\alpha_k} t^{\alpha_k - 1} + o(\omega_k^2 t^2) + \text{oscillatory terms}, \tag{1}$$

where σ^2 is the variance of the process and χ is called *chronotesy*. Explicit expressions in terms of the spectral parameters for these as well as for $o(\omega_k^2 t^2)$ and the oscillatory terms can be found in [5]. In the typical case of $1 < \alpha_k < 3$, the dominant term in (1) is the power-law term. If we assume, as it is often encountered in the experiments, that one of the segments, say the k-th, extends over a wide range of frequencies, the ACF exhibits an approximate self-affinity with a scaling exponent of $\alpha_k - 1$. In the case of $\alpha_k > 3$, the process shows saturated self-affinity, and if $\alpha_k < 1$, the self-affinity is blurred by the oscillatory terms in (1). These different regimes of self-affinity are shown to be caused by finite size effects (in this case the finite extent of the segment) which for different values of α_k have a qualitatively different impact [5]. We note also that if the spectrum have another prominent segment with different spectral exponent, the behavior of the ACF crosses over to a scaling behavior determined by the value of this exponent.

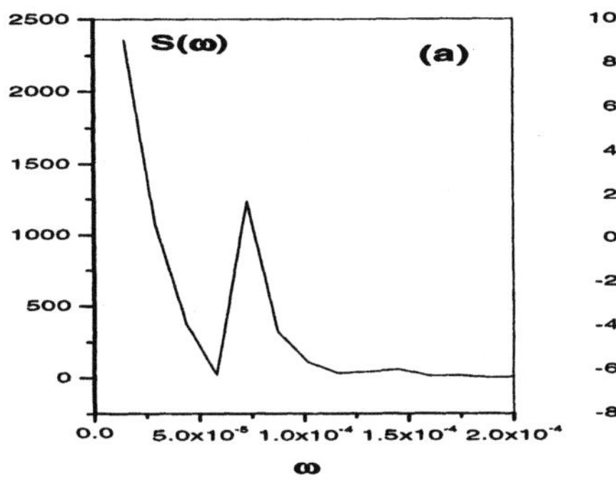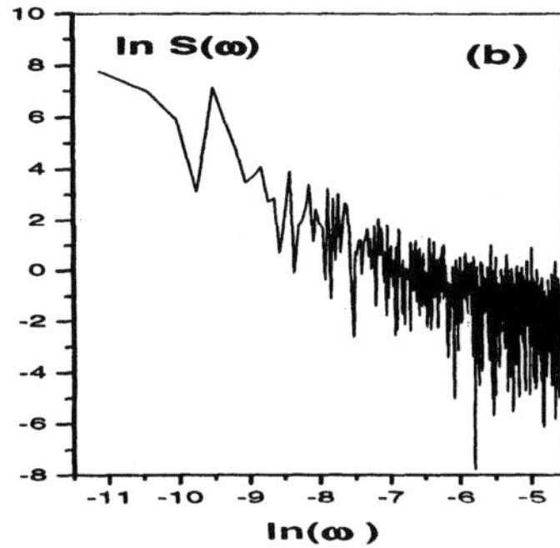

FIGURE 1. Sample spectrum of the wind velocity time series: (a) low-frequency part shown in detail to illustrate the one day periodicity peak at $\omega = 7.27 \times 10^{-5}$ sec^{-1} ; (b) the entire spectrum in log-log scale, suggesting two power-law segments at most.

Frequently, the chaotic time series include (pseudo-)periodic components, which are manifested as peaks in the spectrum. This is the case with all four time series we study. An example is presented in Fig. 1, where in (a) the lower frequency part of the wind time series spectrum is depicted, and in (b) the entire spectrum is plotted in log-log scale. The peak clearly seen in (a) at $\omega = 5 \times (2\pi/T) = 7.27 \times 10^{-5}$ sec^{-1}, corresponds to a period of one day; $T = L\Delta t$, the length of the time series. To model a pseudo-periodic behavior in general we augment the spectrum by a sum of finite width delta-like functions centered at K frequencies $\omega_k^{(p)}$:

$$S^{(p)}(\omega) = \sum_{k=1}^{K} s_k \left[\frac{\sin T(\omega + \omega_k^{(p)})}{T(\omega + \omega_k^{(p)})} + \frac{\sin T(\omega - \omega_k^{(p)})}{T(\omega - \omega_k^{(p)})} \right]. \tag{2}$$

The spectrum (2) leads to an ACF of the form [6]:

$$\mathcal{A}^{(p)}(t) = \sum_{k=1}^{K} a_k \cos \omega_k^{(p)} t, \tag{3}$$

where the amplitudes a_k are related to heights of the spectral peaks s_k by $a_k = 2\pi s_k/T$. Either $\omega_k^{(p)}$ or s_k or more often both are considered independent parameters adding $2K$ additional parameters in the fitting procedure.

The procedure of model validation which we employ involves fitting the sample ACF rather than the sample spectrum. This choice is dictated by the lower sample variances of the ACF estimate. It also avoids dealing with spectral leaks, aliasing effect and other drawbacks of the periodogram estimate of the spectrum, e. g. see [6] and [7]. The procedure of sample ACF fitting, however, requires to account for the bias of the latter. The details of this technical but important aspect of the method will be presented in upcoming publication [8]. However, the significance of the bias corrections for all models is readily demonstrated below.

RESULTS AND DISCUSSION

We begin the expose of the main results by constructing the approximately self-affine model of the time series directly related to the atmospheric turbulence – the wind velocity data. The first decision to make, how many segments the model spectrum should contain, is usually the most ambiguous one, especially when as in our case the periodogram is blurred by strong fluctuations. We test two segments, since at most two segments can be discerned in Fig. 1(b). No other periodicity than the one discussed in relation with Fig. 1(a) is observed and hence we take $K = 1$ in (3). Thus, our test model includes 7 free parameters.

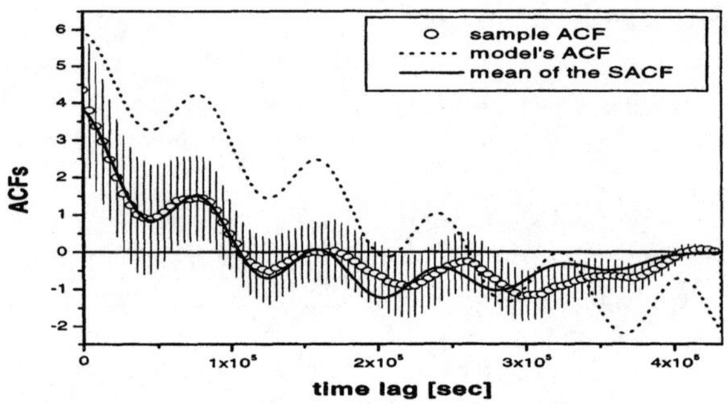

FIGURE 2.   Sample ACF (circles) of the wind velocity, theoretical mean of the sample ACF (solid line) used to fit the sample ACF and the model's ACF (dashed line). The bars illustrate the estimated variances of the SACF.

Next, to ensure a meaningful fit, one has to find "good" initial values for the parameters. These can be obtained in the following manner: Either from the slope of the periodogram or better from the slope of the sample structure function (SF) in log-log scale estimate α_0 and if possible α_1. (We recall, that the SF, $\Delta(t)$, is related to the $\mathcal{A}(t)$ by $\Delta(t) = (\sigma^2 - \mathcal{A}(t))/2$ and it is always positive.) In the case of wind data, we estimate $\alpha_0 = 2.1$. The initial value of α_1 is set to 1.89, rather arbitrary. The same applies to ω_1, which is set at the mid point of the frequency range. The value of the sample spectrum at $\omega = 2\pi/T$ should be close to $A_0\omega_1^{-\alpha_0}$, which gives $A_0 \approx 0.2 \times 10^{-5}$. We take $\omega_1^{(p)} = 7.27 \times 10^{-5}$ sec^{-1} and from $s_1 \approx 1233$, see Fig. 1(a), we evaluate $a_1 \approx 0.18 \times 10^{-1}$. Starting with these values, a fairly standard nonlinear least squares procedure yields the best fit values of the spectral parameters enlisted in the first row of Table 1. The quality of the fit is illustrated in Fig. 2 where the SACF is presented by the open circles; for the sake of clarity one of every fifteen points is depicted. The technique of bias correction involves fitting the data by the *ensemble* mean of the SACF, $\left\langle A^{(s)}(t) \right\rangle$, rather than by the model's ACF, $\mathcal{A}(t)$, itself [8]. The former is evaluated using $\mathcal{A}(t)$ as an approximation of the "true" ACF. It is seen that the fit is reasonably good, especially

TABLE 1. Values of the best fit parameters for all four data sets

data set	ω_0	α_0	A_0	ω_1	α_1	a_1	$\omega_1^{(p)}$	MPPAD
wind	3.64×10^{-6}	2.20	9.24×10^{-7}	7.24×10^{-3}	1.81	0.89	7.27×10^{-5}	0.22
pressure	3.64×10^{-6}	1.70	4.71×10^{-4}	1.05×10^{-3}	4.81	0.22	7.27×10^{-3}	0.27
temp.	3.64×10^{-6}	2.06	6.56×10^{-6}	1.05×10^{-3}	4.80	14.4	7.27×10^{-3}	0.84
humidity	0.5×10^{-4}	2.46	1.52×10^{-5}	1.26×10^{-3}	5.3	199	7.27×10^{-3}	16.3

for relatively short time lags. The mean per point absolute discrepancy (MPPAD) over the entire range of time lags is 0.22, see the last column in Table 1. The bars, however, indicate estimated ACF variances [8] and not 'true' experimental errors. The model's ACF is shown by the dashed line, the difference between the two lines represents in fact the values of the bias corrections. Clearly, they are significant which underlines the importance of the bias accounting technique.

The results need few comments: First, the spectrum essentially has a single segment, ω_1 is close to $\omega_{(N)}$ (with large confidence limits, which we do not present). This indicates a range of turbulence scales smaller than thus determined by Δt. Second, $\omega_0 < \omega_{(F)}$ which is an evidence that the process is not stationary, something that is expected on physical grounds. Third, the value of the spectral exponent α_0 is greater than that expected on the basis of the Kolmogorov's theory (of 1.67). This discrepancy in principal can be attributed to the impact of the earth surface [1].

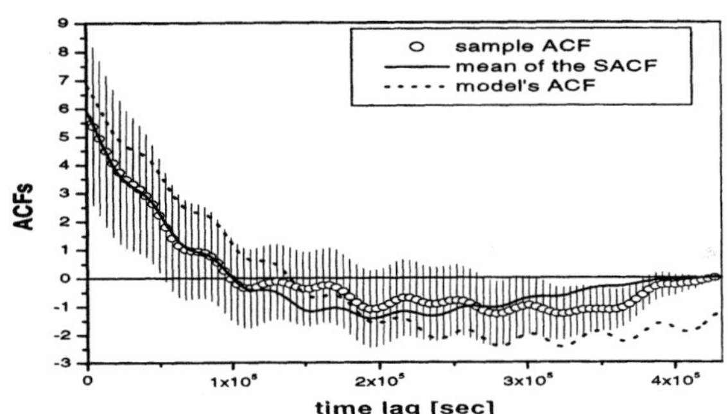

FIGURE 3. The same as Fig. 2 for the pressure ACFs.

Similar good fits are obtained with the other data, see Figs. 3-5. The values of the retrieved parameters, see Table 1, show that the pressure, the temperature and the relative humidity are not effected by the small scale turbulence: observe that $\omega_1 \ll \omega_{(N)}$ for them and $S(\omega) \approx 0$ for $\omega_1 < \omega < \omega_{(N)}$. Also, their principal spectral exponents α_0 are different from the wind's exponent. Finally, in all cases the one day periodicity is readily and well approximated by the spectrum (2) with $K = 1$.

ACKNOWLEDGMENTS

I am very grateful to the ARM administration for their open data policy. This work has been supported by the FIRE grand of the American University in Bulgaria.

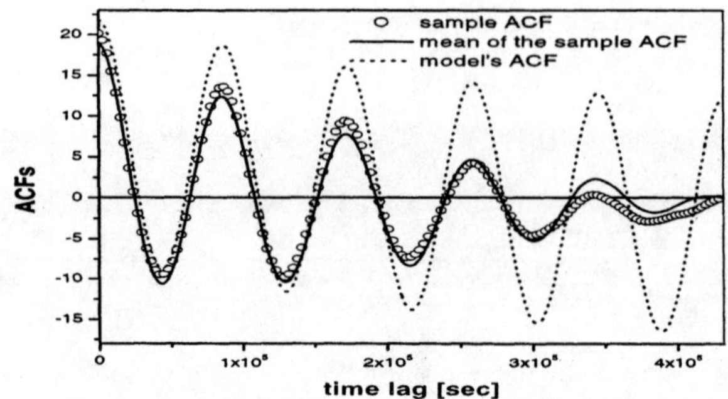

FIGURE 4.   The same as Fig. 2 for the temperature ACFs.

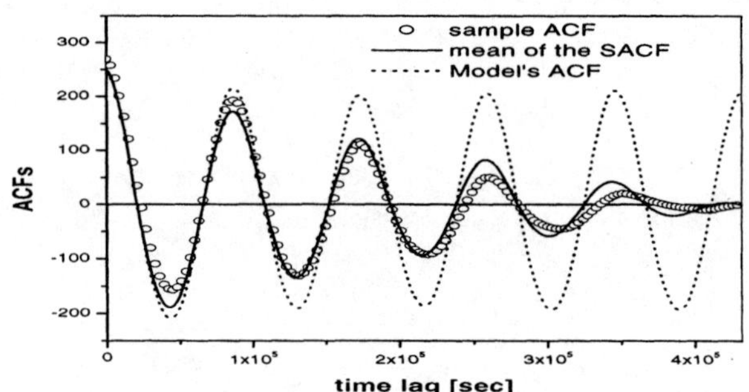

FIGURE 5.   The same as Fig. 2 for the relative humidity ACFs.

REFERENCES

1.   Monin, A. S., and A. M. Yaglom, A. M., *Statistical Fluid Mechanics*, MIT Press, Boston, 1971; S. Panchev, *Random Functions and Turbulence* Pergamon Prees, Oxford, 1971.
2.   Kolmogorov, A. N., *C.R. Acad. Sci. USSR*, **30**, 299-302 (1941).
3.   Shivamoggi, B. K., *Nonlinear Dynamics and Chaotic Phenomena*, Kluwer Academic Press, Dordrecht, 1997, chapter 7.
4.   Yordanov, O. I., and Nickolaev, N. I., *Phys. Rev. E*, **49**, R2517–R2520 (1994).
5.   Yordanov, O. I., and Nickolaev, N. I., *Physica D*, **101**, 116–130 (1997).
6.   Priestley, M. A., *Spectral Anal. and Time Series*, Academic Press, London, 1981.
7.   Press, W. H., Flannery, B. P. S. A. Teukolsky, & Vatterling W. T., *Numerical Recepies*, Cambridge Univ. Press, Cambridge UK, 1986.
8.   Atanasov I. S., and Yordanov O. I., (to be published).

Local Low Dimensionality and Relation to Effects Of Targeted Weather Observations

D. J. Patil*, Istvan Szunyogh*, Aleksey V. Zimin*, Brian R. Hunt*, Edward Ott*, Eugenia Kalnay* and James A. Yorke*

*University of Maryland

Abstract. A statistic, the BV (bred vector) dimension, is introduced to measure the effective local finite-time dimensionality of a spatiotemporally chaotic system. It is shown that the Earth's atmosphere often has low BV-dimension. The implications for improving weather forecasting through data assimilation and targeted observations are discussed.

INTRODUCTION

A key question in the understanding of local instabilities that lead to the breakdown in forecasts, is the complexity of their dynamics (behavior). In this paper, we demonstrate that, in spite of the atmosphere's high dimensionality, in a suitable sense the *local finite-time* atmospheric dynamics is often low dimensional, and we conjecture that these low dimensional regions have a strong relationship to dynamical instabilities.

This paper outlines results that are detailed in [1, 2, 3, 4, 5] as well as several preliminary results which we plan to include in forthcoming papers. Section 2 introduces the method of breeding [6, 7] to produce an ensemble of forecasts from different initial conditions (see [8] and [9] and references within for a review of different methods to generate the initial conditions). Section 3 describes the bred vector ensembles that we utilize. Results regarding the local dimensionality from a similar ensemble (National Weather Service's National Centers for Environmental Prediction's Ensemble Forecasting System) have been previously discussed in [1, 2]. The major difference between the two ensembles is that the results presented here are based on an ensemble that uses many (six times) more bred vectors. A main aim of this work is to investigate the robustness of the results of [1, 2] to changes in the ensemble size. In Section 4 we discuss the statistic, which we call the *Bred Vector dimension* (BV-dimension), which effectively determines the dimension of the subspace spanned by members of the ensemble over a geographically localized region. Using our statistic we investigate the Earth's atmospheric dynamics for the Northern Hemispheric winter. A large membership ensemble run is used to assess the robustness and saturation of the BV-dimension. In Section 6 we provide preliminary results from a study on the relationship between the BV-dimension and the propagation of the impact of targeted observations from the 2000 Winter Storm Reconnaissance Program. We conclude with a discussion about the potential implications of our finding for weather forecasting and data assimilation.

CP676, *Experimental Chaos: 7th Experimental Chaos Conference*,
edited by V. In, L. Kocarev, T. L. Carroll, B. J. Gluckman, S. Boccaletti, and J. Kurths
© 2003 American Institute of Physics 0-7354-0145-4/03/$20.00

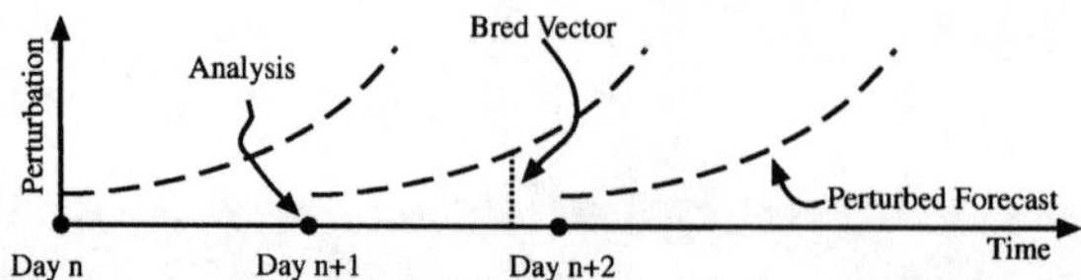

FIGURE 1.   Schematic of the breeding cycle for ensemble forecasts

BREEDING

The procedure for breeding can be outlined in the following steps: (a) add a perturbation to the base state (usually the atmospheric analysis) at a given time t_0; (b) integrate both the base state and the perturbed state forward in time for a period of $t_1 - t_0$ (for this paper $t_1 - t_0 = 24$ hours); (c) at time, t_1, subtract the base state from the perturbed state and rescale the difference, so that it has the same norm (for the NWS ensemble system, rotational kinetic energy at approximately 500hPa is used) as the initial perturbation; (d) this rescaled difference is then used as a new perturbation to the base state and the process (a)-(c) are sequentially repeated. The process is illustrated in Figure 1. The iterated difference between the perturbed and the main solution is called the *bred vector*(see appendix for a theoretical discussion of the breeding process).

DATA

The study we report in this paper is based on ensemble forecasts generated by a replica of the operational Global Ensemble Forecasting System at the National Centers for Environmental Prediction (NCEP). We produce six times the number of ensembles that are currently available at 0000 UTC from NCEP by generating 30 pairs of perturbed forecasts by the breeding process for the period of January 1, 2000 to February 28, 2000. Since the perturbed forecasts are integrations of positive and negative initial perturbations, we average these pairs of perturbed forecasts to yield 30 bred vectors. The bred vectors are generated at a T62 spectral resolution and output on a latitude-longitude spatial grid with 2.5°x2.5° resolution at 21 levels. The perturbed forecasts are rescaled to a regionally varying mask every 24 hours (see [10],[7], and [11] for operational implementations). The rescaling is meant to initialize the magnitudes of the bred vectors to roughly reflect the uncertainties in the analysis of the atmospheric state.

BRED VECTOR DIMENSION

In this section, we outline how the bred vectors can be utilized to obtain a useful measure of the dimensionality of the space in which instabilities in the forecasts are likely to grow the fastest. We begin with considering regions at a fixed pressure level of 5 by 5 grid points by choosing a grid point as the center of the region as well as an array of 24 neighbors so that the array best covers a 1100 km x 1100 km square (at high latitudes some points are skipped in longitude to keep an approximate uniform distance). Given N fields at each point (e.g., temperature, wind speeds, etc.), the values at these 25 points, for each bred vector, are ordered to form a $25N$ dimensional column vector which we call a *local bred vector*.

The issue we want to address is the linear independence of the k local bred vectors. That is, we want to determine the effective dimensionality of the subspace spanned by the local bred vectors. To do this we use empirical orthogonal functions (EOF) [also known as principal component analysis] [12]. The underlying concept is to find the lowest dimensional subspace that, in a least squares sense, optimally represents the majority of the data.

The k local bred vectors form the columns of a $25N \times k$ matrix, B. The $k \times k$ covariance matrix of B is $C = B^T B$, where B^T is the transpose of B. Since the covariance matrix is nonnegative definite and symmetric, its k eigenvalues λ_i are nonnegative ($\lambda_i \geq 0$), and its eigenvectors, after multiplying by B and normalizing, form an orthonormal set of vectors v_i which span the column space of B. We order the eigenvalues by $\lambda_1 \geq \lambda_2 \geq \ldots \geq \lambda_k$. The singular values of B are $\sigma_i = \sqrt{\lambda_i}$. The eigenvalues λ_i are a measure of the extent to which the k column vectors making up B point in the direction v_i, and each $\sigma_i^2 = \lambda_i$ is said to represent the amount of variance in the set of the k unit vectors that is accounted for by v_i. By identifying how many of the vectors of v_i represent most of the variance of B, we can identify an effective dimension spanned by the k local bred vectors. For example, if two out of five singular values are zero, then the subspace spanned by the k local bred vectors is three dimensional. However, if some of the singular values are nonzero but small, the issue becomes more difficult. One option is to use thresholding (as is often done when trying to isolate the dominant modes in EOF), but there is difficulty in determining the "best" value of the threshold. Instead we opt to define the following statistic on the singular values which we call the *Bred Vector dimension* (BV-dimension):

$$\psi(\sigma_1, \sigma_2, \ldots, \sigma_k) = \frac{\left(\sum_{i=1}^{k} \sigma_i\right)^2}{\sum_{i=1}^{k} \sigma_i^2}. \tag{1}$$

As examples of the statistic, we describe several cases with $k = 5$ vectors of unit length. If the k local bred vectors comprising B were all the same, then the singular values would be $\sqrt{5}, 0, 0, 0, 0$. This would yield a statistic of $\psi(\sqrt{5}, 0, 0, 0, 0) = 1$. If the k local bred vectors were equally distributed between two orthogonal unit vectors v_1 and v_2, in the sense that each one accounts for half the variance, then the singular values would be $\sqrt{5/2}, \sqrt{5/2}, 0, 0, 0$, and our statistic would yield $\psi(\sqrt{5/2}, \sqrt{5/2}, 0, 0, 0) = 2$. If the local bred vectors again lie in the two dimensional subspace spanned by unit vectors v_1 and v_2, but the two are not equally represented, then this could give $2 \geq \psi \geq 1$. For example, if 4 of 5 unit local bred vectors were pointing in the same direction while the other pointed in an orthogonal direction, then the singular values would be $2, 1, 0, 0, 0$ and $\psi(2, 1, 0, 0, 0) = 1.8$. While the dimension of the space spanned by the local bred vectors is 2, our statistic gives an intermediate value reflecting the degree of dominance of one direction over the other. In general our statistic returns a real value between 1 and k. Note that while small perturbations due to noise or numerical error will typically cause the dimensionality of the space spanned by the k local bred vectors to be k, the effective dimension ψ may be substantially lower and is insensitive to small changes in σ_i due to noise or numerical error.

REGIONS OF LOW BV-DIMENSION

In this paper we use two types of local bred vectors. The first is constructed using the east-west and north-south wind components at 250 hPa, 500 hPa, and 850 hPa. Hence, $N = 3 \cdot 2$ and the local bred vectors are $25 \cdot 6 = 150$ dimensional. The second type of bred vector we use is constructed by an *energy rescaling* in order to transform different physical variables (wind, temperature, and surface pressure) to comparable quantities of the same physical dimension [13]. The dimension of these bred vectors is 1525.

An example of the BV-dimension applied to a set of 30 bred vectors from February 6, 2002 is shown in Figure 2. The shading represent the effective dimension of the subspace spanned by the local bred vectors (BV-dimension). Several large regions of low dimensionality (BV-dimension less than 8) are evident. This indicates that in these regions, the local bred vectors effectively span a space of substantially lower dimension than that of the full space (and the number of bred vectors). The regions of low BV-dimension can be shown to be statistically significant and not due to random fluctuations, by applying the BV-dimension to *surrogates* of the data [14] that consist of bred vectors chosen from different days which are substantially far apart (removing temporal correlations). When this is done we observe no regions with low BV-dimension (visual examples for operational data are provided in [1, 2]). Furthermore, we find the regions of low BV-dimension are independent of the choices of variables used to construct the local bred vectors. For example, the local bred vectors constructed from vorticity at 500 hPa identify similar regions of low BV-dimension to the methods described earlier in this section. The primary difference is the magnitude of the BV-dimension in the respective regions of low dimensionality.

It is empirically found that regions of low BV-dimension are robust to increasing the size of the ensemble. More specifically, as the ensemble size is increased, the minimum BV-dimension begins to saturate, while the maximum dimension does not. Figure 3 illustrates this property by plotting the average, over days, of the minimum and maximum BV-dimension for the northern midlatitudes. For comparison, results are also shown for random vectors consisting of Gaussian white noise. While the maximum BV-dimension rapidly increases as the ensemble size is increased, it remains at a large distance from the hypothesis of noise. This result indicates the robustness of the regions of low BV-dimension to the size of the ensemble.

Consistent with [2], large regions of low BV-dimension develop and persist on the order of less than a week and move eastward at about $10°$. An example of the time evolution of several regions of low BV-dimension over the northern Pacific ocean are presented in Figure 4. The plots on the left show results from a forecast initiated on February 9, 2000 and figures on the right display results for the corresponding verifying analyses.

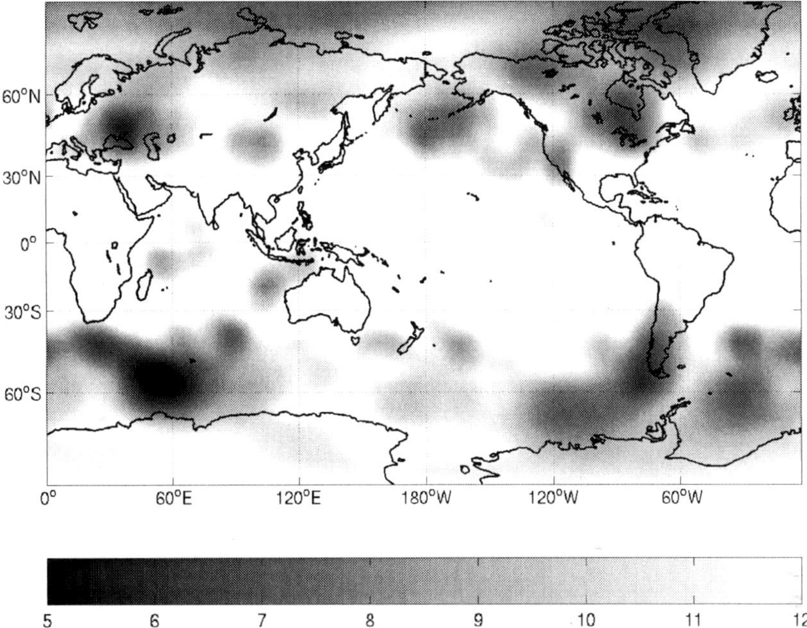

FIGURE 2. An example of the spatial variation of BV-dimension for February 6, 2002. The shades represent the effective subspace spanned by the local bred vectors (BV-dimension).

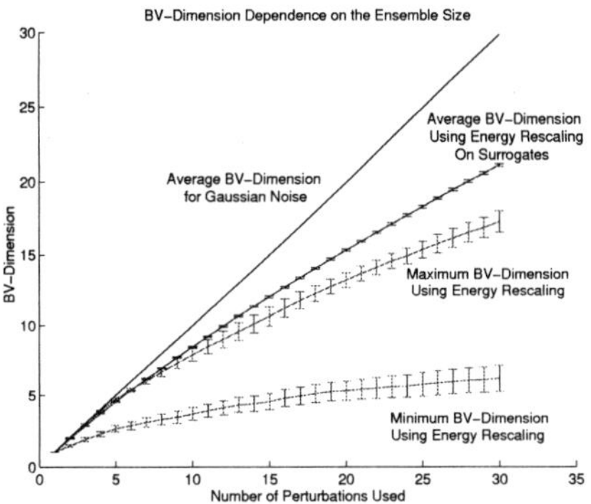

FIGURE 3. The time average of the minimum and maximum BV-dimensions for 30N to 65N as a function of the number of members in the the ensemble. The effective saturation of the minimum BV-dimension suggests robustness of regions of low BV-dimension. The vertical bars represent one standard deviation. For comparison, saturation plots for Gaussian noise and surrogates are shown.

RELATIONSHIP BETWEEN THE EFFECTS OF TARGETED WEATHER OBSERVATIONS AND BV-DIMENSION

The National Centers for Environmental Prediction/National Weather Service (NCEP/NWS) implemented an operational targeted weather observation program

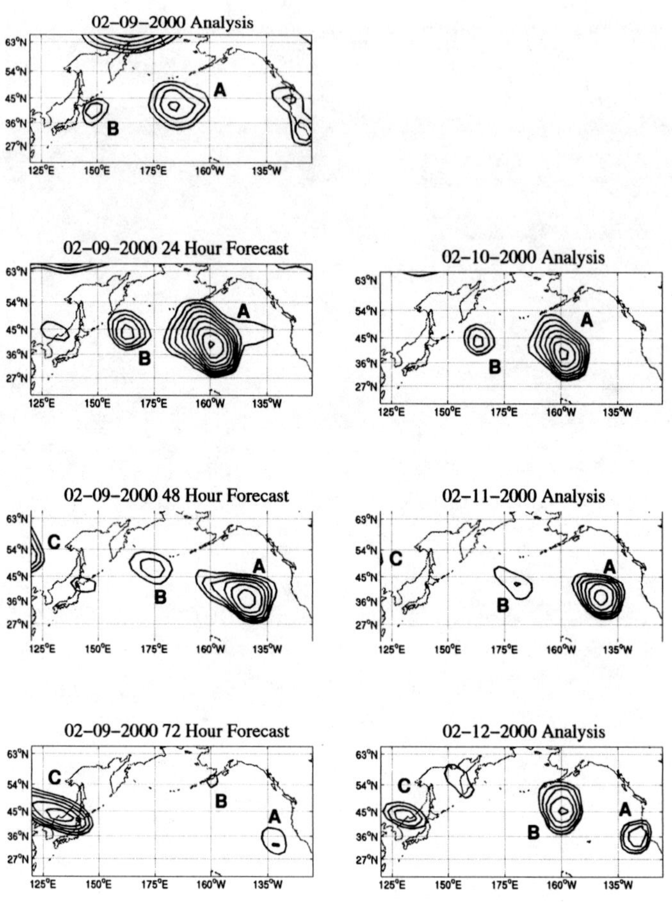

FIGURE 4. An example of the development and propagation of several regions of low BV-dimension from a forecast initiated on February 9, 2002 and the corresponding verification times. Contours of BV-dimension 9 to 5 are used to identify the regions labeled A, B, and C.

with the aim of reducing the risk of forecast failures in the prediction of severe winter storms with potentially large impact on society [15]. At NCEP the Ensemble Transform Kalman-Filter (ETKF, [16, 17]) is used to determine the future observational time, t_o, and observational region from where extra observations taken at t_o are most likely to considerably reduce the error in the prediction of the selected storm at a future verification time, t_v ($t_v > t_o$).

While the forecast verification results [18, 19, 11, 15] have provided convincing evidence that the ETKF technique, and its predecessor the Ensemble Transform (ET) technique [20], are reliable practical tools to target observations, the underlying assumptions of these algorithms have not been rigorously tested. These assumptions are that (i) the analysis error covariance matrix can be well approximated by linearly combining a small ensemble of short-term forecasts valid at the future observational time, t_o, and (ii) the forecast error covariance matrix for the grid points within the verification region can be reasonably approximated by applying the same weights to the ensemble of forecasts valid at t_v. Given these assumptions, the ETKF technique provides an estimate of the forecast error variance reduction (formally, the reduction in the trace of the forecast error covariance matrix) due to the hypothetical extra observations. The main goal of this paper is to explore whether a small ensemble can represent the uncertainty in the dynamical evolution of the atmospheric flow both at observational and verification times. In other words, we ask: Does the atmosphere show finite-time local low-dimensional chaotic behavior in the regions, where the influence of the targeted data propagates?

To answer the question above, the BV-dimension is computed based on the two components of the wind vector at the 250, 500 and 850 hPa pressure levels. This choice was made because the same variables are used in the operationally implemented version of the ETKF technique. In what follows, the relationship between the effects of targeted observations and local low dimensionalities in the atmosphere is shown for a typical example from the 2000 Winter Storm Reconnaissance program [21].

The BV-dimension shown in Figure 5 was computed based on the initial perturbations of the experimental ensemble. In this figure the small crosses mark the dropsonde locations (panel a) and the circles mark the verification regions (panels c and e) for a targeting case from the 2000 Winter Storm Reconnaissance program. The figure shows that the targeted observations were taken in an isolated region of local low dimensionality. The dimension of this region, about 5-6, is extremely low considering that the largest possible value of the BV-dimension for the given ensemble is 30. To track the propagation of the targeted low dimensional spot, an additional forecast ensemble was created. This ensemble was identical to the 30-pair research ensemble, except for that the initial perturbations were set to zero outside of the targeted low dimensional region. The region of local low-dimensionality propagates eastward with the corresponding wave. The propagation of the selected low-dimensional spot is shown in Figure 6. The local dimensions shown in this figure were computed by first adding a small magnitude random noise component to the ensemble perturbations. This was needed to obtain large dimension values in the regions where the magnitude of the ensemble members was smaller than the noise level due to the use of a spectral model. Although this approach slightly increased the local dimension, it has made the tracking of the selected low dimensionality straightforward. After 24 hours the targeted low dimensional feature propagated to the west coast verification region. This is a key finding of our study. This implies that *the targeted data had a great potential to reduce the forecast error in the verification region: both the analysis and the forecast uncertainties were confined in low dimensional subspaces, and these low dimensional subspaces were dynamically connected.* The contour lines in Figure 1 confirm that along the regions, where the targeted low-dimensionality propagates, the forecast was mainly improved. We have recently developed an algorithm to track the propagation of the error reduction in forecasts due to increased observations in regions of low BV-dimension and we plan to provide details in a future paper.

Upper Tropospheric Wave Packets

The propagation of the local low dimensionality cannot explain how the impact of the dropsonde data propagated to the east coast verification region at 48hr forecast lead time. Szunyogh et. al [19, 21], argued that the impact of the targeted data was propagated from the Pacific to the Atlantic region by packets of short upper tropospheric Rossby waves. In Figure 7 the wave packet envelope function, $A(x)$, is shown, where $A(x)$ is computed using our algorithm described in [4] which assumes that the meridional wind component, $v(x)$ can be expressed as

$$v(x) = A(x)cos(\phi(x)), \tag{2}$$

where $\phi(x)$ is the spatially varying phase.

Figure 7 shows that an atmospheric wave packet has developed about 12 hours after observational time. The east coast verification region was located at the leading edge of the eastward propagating wave packet at verification time, t_v. In order to show that the dropsonde signal propagated as an atmospheric wave packet, the signal, $s(x)$, was also demodulated by substituting $s(x)$ for $v(x)$ in Equation 2. The signal, $s(x)$, is formally defined as the difference between a forecast that was initiated by assimilating all targeted and standard observations and a forecast that was initiated by assimilating only the standard observations. The results demonstrate that after an initial transient, which is not longer than 12 hours, the leading edge of the targeted data impact propagated with the atmospheric wave packet.

A comprehensive analysis of all targeting cases from the WSR00 field program is in progress.

CONCLUSIONS

In this paper our main result is a means of identifying local low dimensional behavior (the BV-dimension) of the atmospheric system. We have provided evidence that these low dimensional, dynamical instabilities (regions of low BV-dimension) cover a significant portion of the globe, that they are intrinsic to the dynamics of the atmospheric system, and that they typically last for several days.

At any given time t_0, there is inevitably a discrepancy $\vec{\Delta}(t_0)$ between the true atmospheric state and its representation in the computer model (analysis). Now consider a later time $t_1 > t_0$, and suppose that in a region of interest there is a low BV-dimension at time t_1. This implies that any local discrepancy $\vec{\Delta}(t_1)$ between the true state and its representation in the computer model (forecast error) lies predominantly in the "unstable subspace", the space spanned by the few vectors that contribute most strongly to the the low BV-dimension. We conjecture that in many cases this information can yield a substantial improvement in forecasting. In particular, the implication is that the data assimilation algorithm should correct the computer model state by moving it closer to the observations along the direction of the unstable subspace since that is where the true state most likely lies [22]. Current data assimilation techniques (e.g., that used by the NWS) do not take this into account. Based on these results we have developed a data

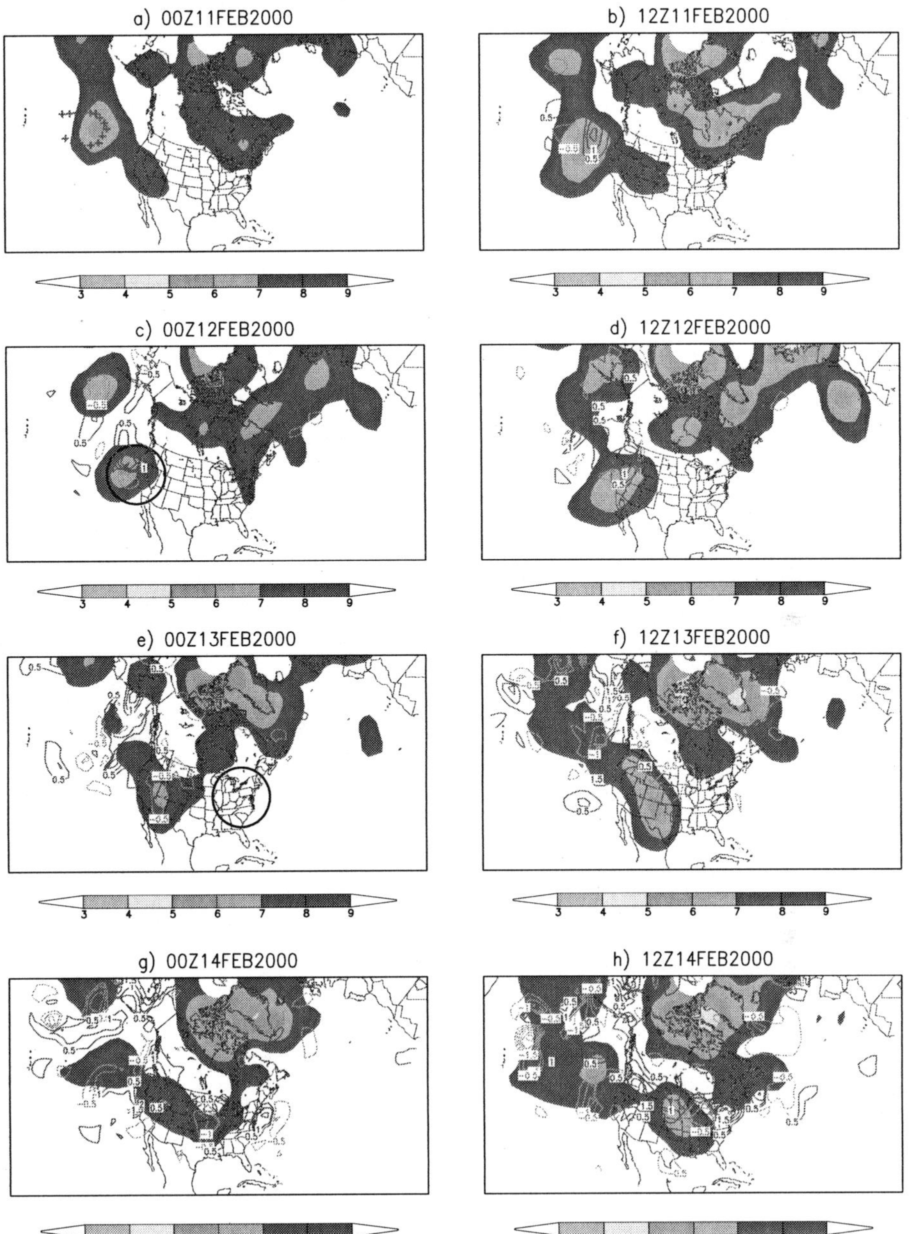

FIGURE 5. Shown by shades is the local BV-dimension. Contour lines indicate improvement(degradation) in the surface pressure forecast initiated at February 11 0000UTC. The contour interval is 0.5 hPa. The circle represents the time and geographical regions in which the targeted observations were intended to improve the forecasts.

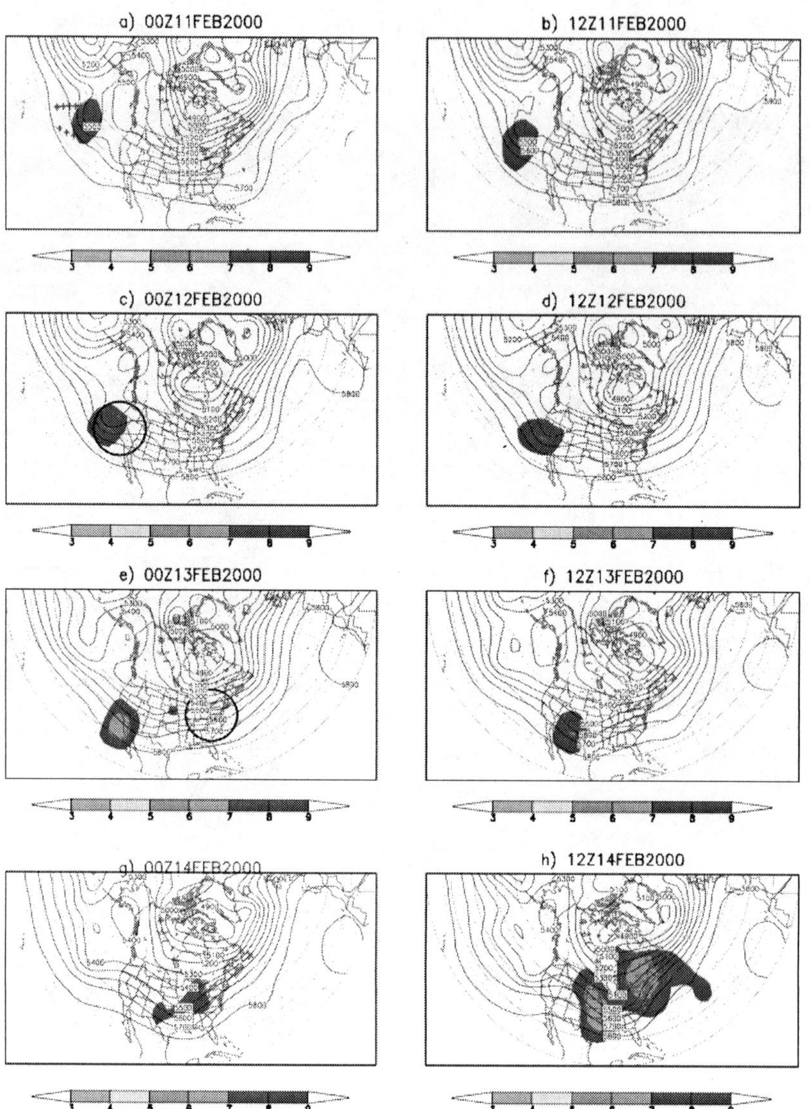

FIGURE 6. Propagation of the targeted low dimensional region. The geopotential height at 500 hPa pressure level is shown by contour lines. The contour interval is 100 gpm. The circle represents the time and geographical regions in which the targeted observations were intended to improve the forecasts.

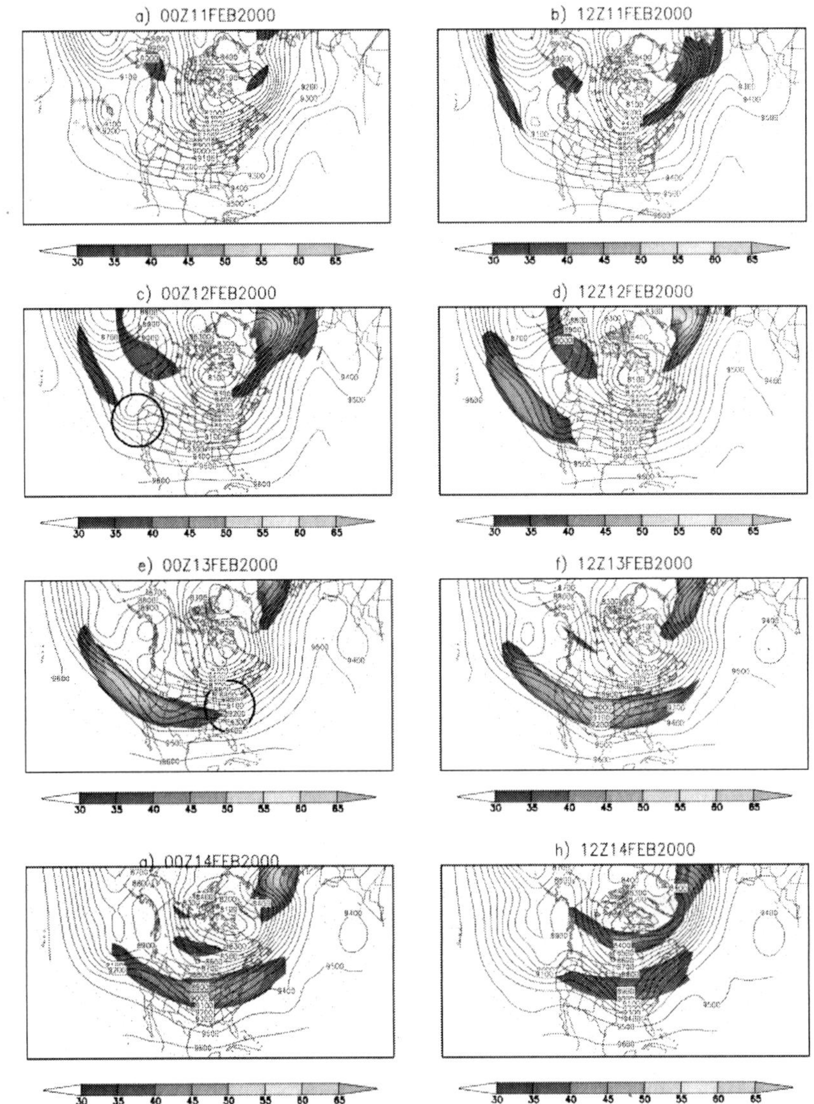

FIGURE 7. Shown by shades is the wave packet envelope function for the atmospheric wave packet at the 300 hPa pressure level. The geopotential height at the 300 hPa pressure level is shown by contour lines. The contour interval is 100hPa. The circle represents the time and geographical regions in which the targeted observations were intended to improve the forecasts.

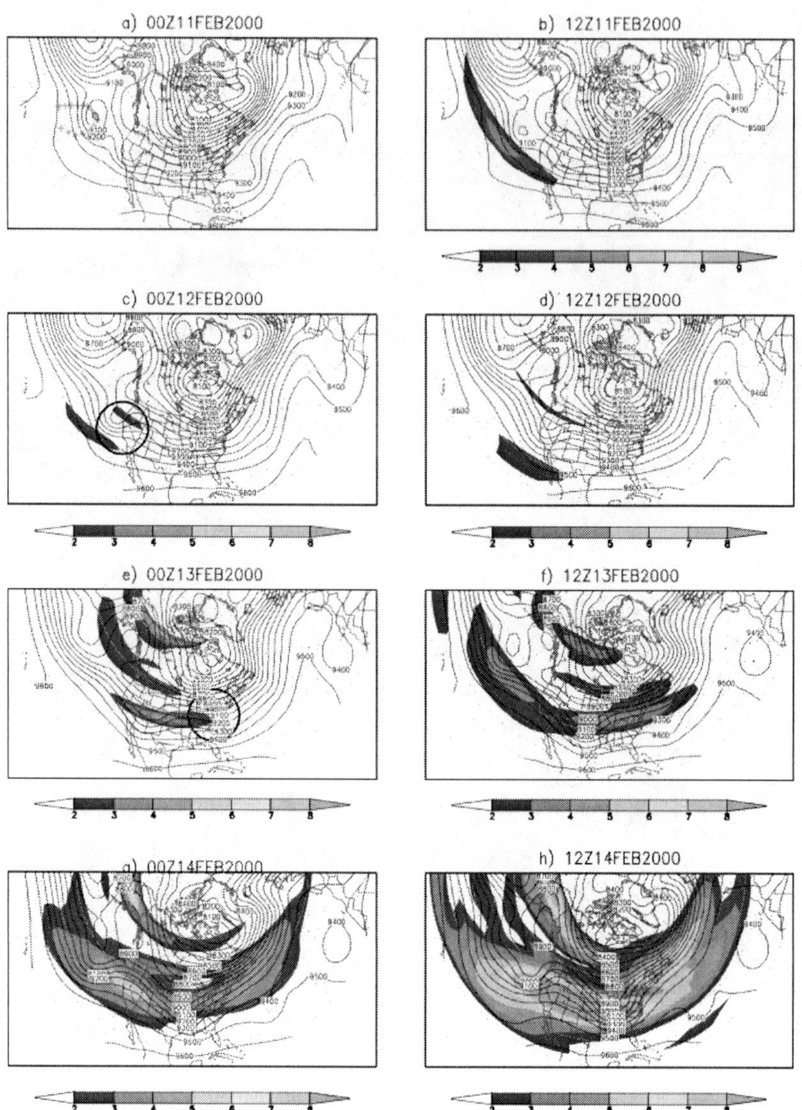

FIGURE 8. Shown by shades is the wave packet envelope function for the signal $s(x)$ at the 300 hPa pressure level. The geopotential height at the 300 hPa pressure level is shown by contour lines. The contour interval is 100hPa. The circle represents the time and geographical regions in which the targeted observations were intended to improve the forecasts.

assimilation algorithm that exploits the local low dimensionality [23, 24]. We plan to report on these results in a later paper.

APPENDIX: THEORETICAL DISCUSSION OF THE BREEDING PROCESS

For a m-dimensional dynamical system

$$\frac{d\mathbf{x}}{dt} = \mathbf{F}(\mathbf{x})$$

important dynamical characteristics can be studied by investigating the infinitesimal displacement between two trajectories, $\mathbf{x}(t)$ and $\mathbf{x}(t) + \delta\mathbf{x}(t)$, as well as the behavior of the associated tangent vector $\delta\mathbf{x}(t)/|\delta\mathbf{x}(0)|$ that evolves according to the equation

$$\frac{d\delta\mathbf{x}}{dt} = \mathbf{DF}(\mathbf{x}(t))\delta\mathbf{x}.$$

Where $\mathbf{DF}$ denotes the Jacobian of $\mathbf{F}$. The exponential separation of the two trajectories, $\mathbf{x}(t)$ and $\mathbf{x}(t) + \delta\mathbf{x}(t)$, is characterized by the Lyapunov exponent:

$$\Lambda = \lim_{t \to \infty} \frac{1}{t} ln \frac{|\delta\mathbf{x}(t)|}{|\delta\mathbf{x}(0)|}.$$

For a given trajectory there are in general m different Lyapunov exponents, Λ_i, corresponding to different choices of the initial perturbation $\delta\mathbf{x}(0)$. The perturbations, $\delta\mathbf{x}(t)$, that correspond to a particular Lyapunov exponent, Λ_i, become concentrated for large t in the direction of a single vector $u_i(t)$, which we call the Lyapunov vector for Λ_i.

From the practical point of view, the utility of Lyapunov exponents and Lyapunov vectors stems from the fact that, in many nonlinear systems, sufficiently small *finite* perturbations, $\Delta\mathbf{x}(t)$, behave like infinitesimal perturbations, $\delta\mathbf{x}(t)$, for finite time, and, over this time interval, these finite perturbations are still large enough to have significant consequences. This situation becomes problematic if in a system like the Earth's atmosphere many spatial scales are present. As an example, consider the spread of a passive scalar in a turbulent fluid with a Kolmogorov energy wavenumber spectrum, $E(k) \sim k^{-5/3}$, where k denotes the magnitude of the wave number. (The presence of many spatial scales in this case is due to the slow algebraic decay of $E(k)$ with increasing k.) For this case the location, $\mathbf{x}(t)$, of a passive particle is given by $dx(t)/dt = \mathbf{v}(\mathbf{x},t)$, where $\mathbf{v}(\mathbf{x},t)$ is the fluid velocity. As shown long ago by [25], on average, $|\Delta\mathbf{x}| \sim t^{3/2}$. This is at odds with the expectation of exponential separation obtained by formally following the Lyapunov analysis and writing $d\delta\mathbf{x}/dt = \delta\mathbf{x} \cdot \nabla\mathbf{v}$. The problem is that if $E(k) \sim k^{-5/3}$ applies for $k \to \infty$ (i.e., for arbitrarily small scales), then $\nabla\mathbf{v}$ does not exist (i.e., the velocity field is not differentiable). This follows from Parseval's theorem which implies $\langle \sum_{\alpha,\beta=1}^{3} (\partial v_\alpha/\partial x_\beta)^2 \rangle = \int_0^\infty k^2 E(k) dk$, where $\langle \cdots \rangle$ denotes spatial average. For $E(k) \sim k^{-5/3}$ the integral $\int_0^\infty k^2 E(k) dk$ diverges at the upper limit of integration. Of course, $E(k) \sim k^{-5/3}$ for a large Reynolds number turbulent flow does not apply for

$k \to \infty$. Rather there is a high k cutoff at $k \sim k_c$ due to viscosity, such that for $k > k_c$, $E(k)$ decreases much more rapidly than $k^{-5/3}$ with increasing k. This indicates that Lyapunov numbers and vectors are only relevant for finite Δx provided that $|\Delta \mathbf{x}| \ll k_c^{-1}$, in which case exponential growth of $|\Delta \mathbf{x}|$ could indeed be expected. However, for $|\Delta \mathbf{x}| \gg k_c^{-1}$, the Lyapunov analysis does not apply, and Richardson's law is applicable. It can be expected that real atmospheric motions contain a large range of scales including, for example, Kolmogorov like turbulence at small scales. This is also indicated by the use of subgrid scale modeling in weather prediction codes. One thus suspects that the use of a Lyapunov type analysis based on infinitesimal perturbations is problematic for scales modeled in these codes (i.e., scales at the grid scale or above). Thus, if one wants to use a weather prediction code to look at how two weather forecasts with differing initial states diverge with time, a linearized Lyapunov type of analysis would not be expected to be valid, and one should look at the nonlinear solutions evolved from the two initial states. Toth and Kalnay 1993, 1997 introduced a procedure based on *finite* perturbations of a computational atmospheric state. They call this the *method of breeding*.

ACKNOWLEDGMENTS

This research was supported by the W. M. Keck Foundation, James S. McDonnell Foundation, Army Research Office, Office of Naval Research, and National Science Foundation (award DMS0104087).

REFERENCES

1. Patil, D. J., Hunt, B., Yorke, J., Kalnay, E., and Ott, E., *Phys. Rev. Let.*, **86**, 5878–5881 (2001).
2. Patil, D. J., *Applications of Chaotic Dynamics to Weather Forecasting*, Ph.D. thesis, University of Maryland (2001).
3. Patil, D. J., Szunyogh, I., Hunt, B., Kalnay, E., Ott, E., and Yorke, J. A., "Using large member ensembles to isolate local low dimensionality of atmospheric dynamics," in *Symposium on Observations, Data Assimilation, and Probabilistic Prediction*, Amer. Meteor. Soc., January 13-17, 2002, pp. 191–196.
4. Zimin, A., Szunyogh, I., Patil, D., Hunt, B., and Ott, E., *Monthly Weather Review, In Press* (2003).
5. Szunyogh, I., Zimin, A. V., Patil, D. J., Kalnay, E., Hunt, B., Ott, E., and Yorke, J. A., "On the dynamical basis of targeting weather observations," in *Symposium on Observations, Data Assimilation, and Probalilistic Prediction*, Amer. Meteor. Soc., January 13-17, 2002, pp. 197–202.
6. Toth, Z., and Kalnay, E., *Bulletin of the American Meteorological Society*, **74**, 2317–2330 (1993).
7. Toth, Z., and Kalnay, E., *Monthly Weather Review*, **125**, 3297–3319 (1997).
8. Ehrendorfer, M., *Meteorol Zeitschrift*, **6**, 147–183 (1997).
9. Kalnay, E., *Atmospheric Modeling, Data Assimilation, and Predictability*, Cambridge University Press, Cambridge, U.K, 2002, chap. 6.
10. Iyengar, G., Toth, Z., Kalnay, E., and Woolen, J., "Are the bred vectors representative of analysis error?," in *11th Conf. on Numerical Weather Prediction*, Amer. Meteor. Soc., Norfolk, VA, 1996, pp. 64–65.
11. Szunyogh, I., and Toth, Z., *Monthly Weather Review*, **130**, 1125–1143 (2002).
12. Scheick, J., *Linear Algebra with Applications*, McGraw-Hill, New York, 1997.
13. Patil, D. J., Szunyogh, I., Hunt, B., Ott, E., Kalnay, E., and Yorke, J. A., *to be submitted* (2003).
14. Theiler, J., Eubank, S., Longtin, A., Galdrikian, B., and Farmer, J., *Physica D*, **58**, 77–94 (1992).

15. *On the use of targeted observations in operational weather prediction*, Fifth Symposium on Integrating Observing Systems, American Meteorological Society, 2001, albuquerque, NM.
16. Bishop, C., Etherton, B., and Majumdar, S., *Monthly Weather Review*, **126**, 420–436 (2001).
17. Majumdar, S., Bishop, C., Etherton, B., and Toth, Z., *In Press* (2001).
18. Szunyogh, I., Toth, Z., Emanuel, K., Bishop, C., Snyder, C., Morss, R., Woolen, J., and Marchok, T., *Quart. J. Roy. Meteor. Soc.*, **125**, 3189–3218 (1999).
19. Szunyogh, I., Toth, Z., Morss, R., Majumdar, S., Etherton, B., and Bishop, C., *Mon. Wea. Rev.*, **128**, 3520–3537 (2000).
20. Bishop, C., and Toth, Z., *J. Atmos. Sci.*, **56**, 1748–1765 (1999).
21. Szunyogh, I., Toth, Z., Zimin, A., Majumdar, S., and Persson, A., *Monthly Weather Review*, **130**, 1144–1165 (2002).
22. Kalnay, E., and Toth, Z., "Removing growing errors in the analysis," in *Preprints; Tenth Conference on Numerical Weather Prediction*, Amer. Meteor. Soc., July 18-22, 1994, pp. 212–215.
23. Ott, E., Hunt, B., Szunyogh, I., Corazza, M., Kalnay, E., Patil, D., and Yorke, J., *Los Alamos Archives* (2002), URL `http://arxiv.org/abs/physics/0203058`.
24. Ott, E., Hunt, B., Szunyogh, I., Zimin, A., Kostelich, E., Corazza, M., Kalnay, E., Patil, D., and Yorke, J., *Under Review* (2003), URL `http://keck2.umd.edu/weather/`.
25. Richardson, L. F., *Proc. R. Soc. Lond. A*, **110**, 709–737 (1926), (also in Collected Papers of L. F. Richardson Vol.1, ed. P. G. Drazin pp. 523-551, Cambridge University Press. 1993).

Can Sea Clutter and Indoor Radio Propagation be Modeled as Strange Attractors?

J.B. Gao, S.K. Hwang, H.F. Chen, Z. Kang, K. Yao, and J.M. Liu

Electrical Engineering Department, UCLA
Los Angeles, CA 90095-1594, USA

Abstract. Sea clutter is the backscattered returns from a patch of the sea surface illuminated by a radar pulse. The amplitude waveforms of sea clutter and indoor radio propagation are very complicated. Can the apparent randomness of these waveforms be attributed to be generated by low-dimensional chaos? Based on the assumption that a chaotic attractor is characterized by a non-integer fractal dimension and a positive Lyapunov exponent, Haykin et al (1992) concluded that sea clutter while Tannous et al (1991) concluded that indoor radio propagation data were chaotic. However, a numerically estimated non-integral fractal dimension and a positive Lyapunov exponent may not be sufficient indication of chaos. Other researchers have also indirectly questioned the chaoticness of the sea clutter. We employ a more stringent criterion for low-dimensional chaos developed by Gao and Zheng (Phys. Rev. E, 1994) to study a two minute duration sea clutter data provided by Haykin, and indoor radio propagation data measured at UCLA, and show that these data are not chaotic. We carry out a multifractal analysis and find that sea-clutter data can be modeled as multiplicative multifractals with a lognormal envelope distribution, while the radio propagation data can be modeled as a weak multifractal in the sense of structure function technique.

INTRODUCTION

Sea clutter refers to the backscattered returns from a patch of the sea surface illuminated by a transmitted radar pulse. The amplitude waveforms of sea clutter and indoor radio propagation are very complicated. The complexity of the former comes from the fact that the data is generated by radar pulses massively reflected from a wavy and possibly turbulent sea surface; the complexity of the latter lies in the (possible) fading of the channel and multipath interference due to massive reflection of electromagnetic waves from objects of vastly different sizes, some visually identifiable, some visually indetectable. The study of sea clutter and radio propagation is not only of theoretical importance, but may also be of practical relevance. For example, study of sea clutter may help us understand the severe limitations on the detectability of radar returns from "point" targets such as low-flying aircraft, small marine vessels, navigation buoys, small pieces of ice, submarine periscope, etc, while study of radio propagation may help us better understand the nature of fading of a channel and multipath interference, thus help us design better signal detection methods, increase the signal-to-noise ratios, and increase the number of users that can be suported by a given channel.

Traditionally sea clutter and radio propagation is often studied in terms of certain simple statistical features, such as the marginal probability density function. For example, in the phenomenological modeling of sea clutter, it has been recognized for quite a while that some sea clutter amplitude signals are approximately log-normally distributed [16].

CP676, *Experimental Chaos: 7ᵗʰ Experimental Chaos Conference*,
edited by V. In, L. Kocarev, T. L. Carroll, B. J. Gluckman, S. Boccaletti, and J. Kurths
© 2003 American Institute of Physics 0-7354-0145-4/03/$20.00

That is, in a logarithmic scale, the amplitude signals are approximately normally distributed. On the other hand, it has been found that some amplitude signals of radio propagation follow a Weibull distribution [10]. Such modeling, however, basically does not shed much light on the true nature of sea clutter or radio propagation. Hence, one does not have any insight as how good the log-normal and Weibull distributions are for the sea clutter and radio propagation signals, respectively, and where is the basis for this log-normality in the sea clutter data, or the Weibullian for the radio propagation.

Since the complicated sea clutter signals are functions of complex (sometimes turbulent) wave motions on the sea surface, while wave motions on the sea surface clearly have their own dynamical features that are not readily described by simple statistical features, it is thus very desirable to understand sea clutter by considering some of their dynamical features. To this end, a natural question to ask is whether sea clutter is low-dimensional deterministic chaos. That is, if the complicated sea clutter wave forms are generated by nonlinear deterministic interactions of a few modes (i.e., number of degrees of freedom). If the answer is yes, then the apparent randomness in the sea clutter signals has a deterministic origin. This would contrast sharply with the modeling of sea clutter based solely on random processes. The same question can be asked about the radio propagation, since it is basically a problem of scattering from objects of very different sizes, some static, some dynamic, such as floating dusts, possibly forming some sort of fractals, thus similar to the scattering of radar pulses from a sea surface.

Chaos is also commonly called a strange attractor. Being an "attractor", the trajectories in the phase space are bounded. Being "strange", the nearby trajectories, on the average, diverge exponentially fast. Mathematically, the latter property can be expressed as follows. Let $d(0)$ be the small separation between two arbitrary trajectories at time 0, and $d(t)$ be the separation between them at time t. Then, for true low-dimensional deterministic chaos, we have $d(t) \sim d(0)e^{\lambda_1 t}$, where λ_1 is positive and called the largest Lyapunov exponent. Due to the boundedness of the attractor and the exponential divergence between nearby trajectories, a strange attractor typically is a fractal, characterized by a simple and elegant scaling law of $N(\varepsilon) \sim \varepsilon^{-D}$, where N represents the (maximal) number of boxes, of linear length not larger than ε, needed to cover the attractor, and D is typically a non-integral number called the fractal dimension of the attractor.

In the past decade, Haykin et al. [11, 12] and Tannous et al. [20] carried out analysis of some sea clutter data and radio propagation data, respectively, using chaos theory, based on the assumption that strange attractors are fully characterized by a non-integral fractal dimension and a positive Lyapunov exponent. By numerical computations, they found a non-integral correlation dimension (which is a tight lower bound for the fractal dimension [9]) and a positive largest Lyapunov exponent from their data. Thus they suggested that sea clutter and radio propagation be modeled by low-dimensional strange attractors.

However, a non-integral dimension together with a positive largest Lyapunov exponent obtained by computational means instead of analysis may not be a sufficient indication of deterministic chaos. A well-known example is the so-called $1/f$ processes. These are random processes with spectral density of $S(f) \sim f^{-(2H+1)}$, where $0 < H < 1$ is sometimes called the Hurst parameter. A trajectory formed by such a process has a fractal dimension of $1/H$. As we shall explain shortly, with most algorithms of estimating the largest Lyapunov exponent, one obtains a positive number for the "Lyapunov exponent", and interpret the random process as originating from deterministic chaos.

Recently, Cowper and Mulgrew [1] , Noga [17], Davies [2], and Haykin et al [13] separately have questioned the chaoticness of the radar sea clutters based on different reasons. Since the approaches of Cowper and Mulgrew [1] , Noga [17] and and Davies [2] are indirect, it would be very desirable to directly re-assess if sea clutter is really chaotic. If the answer is yes, then one can specifically ask how may log-normal distribution and chaos coexist for sea clutter. On the other hand, the radio propagation data used by Tannous et al. [20] is quite non-stationary. Given that wireless communications have been playing an ever growing role in our daily lives, there is an even more urgent need to re-assess whether radio propagation can be modeled as chaos. To answer these questions, in this paper we employ a number of signal processing tools from the more advanced chaos and fractal theories to analyze a sea clutter data set measured on a patch of free sea surface (i.e., without point targets), supplied by Haykin's group, and a number of indoor radio propagation data, measured at UCLA. We shall show that neither sea clutter signals nor radio propagation data (at least for these data sets) are low-dimensional chaos. we shall also show, however, that they are multifractals. The multifractal nature of sea clutter justifies why the sea clutter signals are log-normally distributed.

SEA CLUTTER AND RADIO PROPAGATION DATA

The sea clutter data analyzed here is supplied by Haykin's group. The sampling rate is 1000 Hz. The data set is 2^{17} points long. Fig. 1 shows two short segments of the amplitude signal. We notice that the signal between two successive local maxima is about 10 points long and are reasonably smooth. The waveform between two successive local maxima may be considered to correspond to the gradual turning over of a wave on the sea surface. To have a feeling of how the entire amplitude signal looks like, we have plotted it in Fig. 2(a). Now we notice that the signal is very irregular. Fig. 2(b) has plotted the envelope signal of the data, which is formed by picking up the successive local maxima of the data. Local maxima of the amplitude of the data may be thought of as the square root of the energy of the successive waves of the sea surface. Thus, the irregularity of the data reflects the wide variation of the energy of successive waves turning over on the sea surface. We shall show below that log-normal distributions are more applicable to the envelope signal, and less applicable to the original amplitude data. In other words, the energy of successive waves on the sea surface follows log-normal distribution. This simply follows from that if x follows log-normal distribution, then x^2 also follows log-normal distribution.

None-line-of-sight propagations over approximately 50 feet at 1 GHz were conducted inside a medium size laboratory room with diverse objects. The HP83620A RF Sweep Generator was used as the transmitter, the HP Spectrum Analyzer E4407B was used as the receiver, two high gain transmit/receive antennae were used, and the Tektronix TDS694C 4-channel oscilliscope capable of 25G samples/second with internal interpolation was used for data collection. Four sets of data of 120,000 samples were collected. A low-pass filter at 4 GHz was used on the collected data. The envelopes were obtained from the interpolated values of the peaks of the observed data.

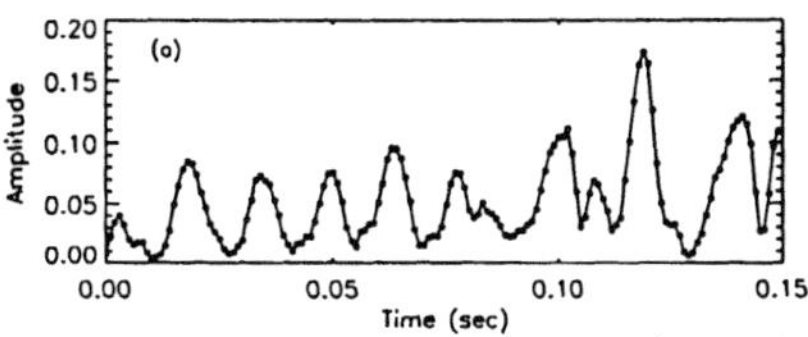

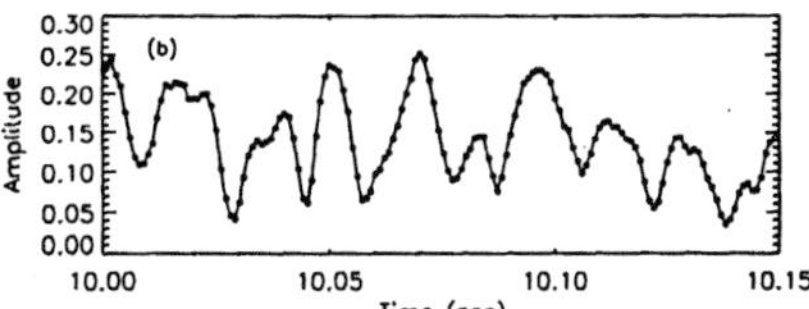

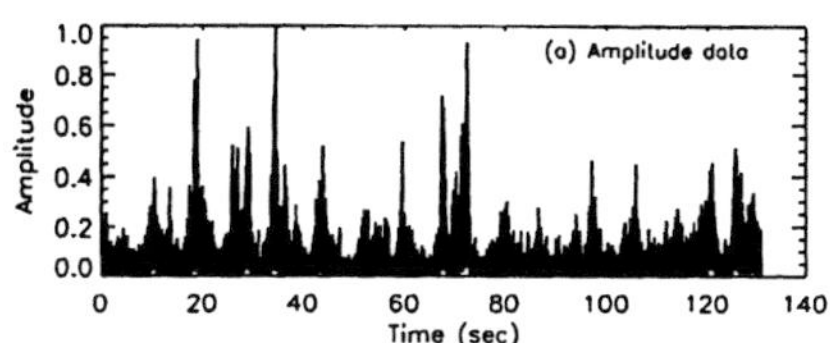

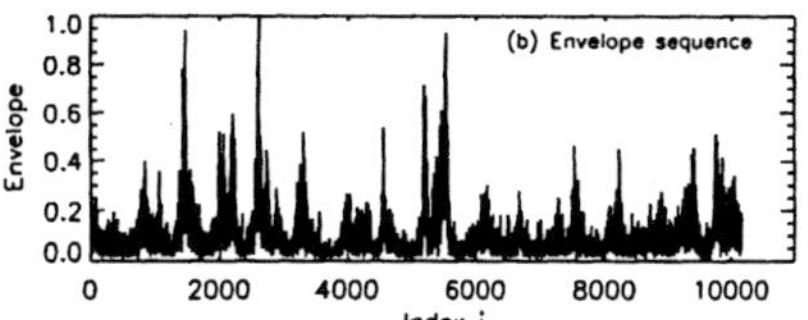

FIGURE 1. Two short segments of the amplitude signal of the sea clutter data.

FIGURE 2. The complete amplitude signal (a) and its envelope (b).

ARE SEA CLUTTER-RADIO PROPAGATION DATA CHAOTIC?

In this section, to study the sea clutter data set, we employ the direct dynamical test for low-dimensional chaos developed by Gao and Zheng [6], which is one of the more stringent test for chaos available in the literature.

Given a scalar time series, $x(1), x(2), ..., x(N)$, one first constructs vectors of the following form using the time delay embedding technique [19]: $V_i = [x(i), x(i+L), ..., x(i+(m-1)L)]$, with m being the embedding dimension and L the delay time. For example, when $m = 3$ and $L = 4$, we have $V_1 = [x(1), x(5), x(9)]$, $V_2 = [x(2), x(6), x(10)]$, and so on. For the analysis of purely chaotic signals, m and L has to be chosen properly. This is the issue of optimal embedding (see [5, 6] and references therein). After the scalar time series is properly embedded, one then computes $\Lambda(k) = \left\langle \ln \left(\frac{\|V_{i+k} - V_{j+k}\|}{\|V_i - V_j\|} \right) \right\rangle$, with $r \leq \|V_i - V_j\| \leq r + \Delta r$, where r and Δr are prescribed small distances. The angle brackets denote ensemble averages of all possible pairs of (V_i, V_j) and k is called the evolution time. A pair of r and Δr is called a shell. The computation is typically carried out for a sequence of shells. For true low-dimensional chaotic systems, the curves $\Lambda(k)$ vs. k for different shells form a common envelope. The slope of the envelope estimates the largest positive Lyapunov exponent. An example is given in Fig. 3(a) for the well-known chaotic Lorenz system. We note the common envelope at the lower left corner of Fig. 3(a). The existence of that common envelope guarantees that a robust positive Lyapunov exponent will be obtained by different researchers no matter which shell they use in the computation, thus ensures determinism. For non-chaotic systems, the common envelope is absent. As an example, Fig. 3(b) shows the $\Lambda(k)$ vs. k curves for a set of uniformly distributed random variables. Here, we note there is no common envelope in the lower left corner of Fig. 3(b). We also note, most other algorithms for estimating the largest Lyapunov exponent is equivalent to compute $\Lambda(k)$ for $r < r_0$, where r_0 is selected more or less arbitrarily, then obtain $\Lambda(k)/k$, for not too large k, as an estimation of the largest Lyapunov exponent. With such algorithms, one can obtain a "positive" Lyapunov exponent for white noises and for $1/f$ processes. However, this positive number critically

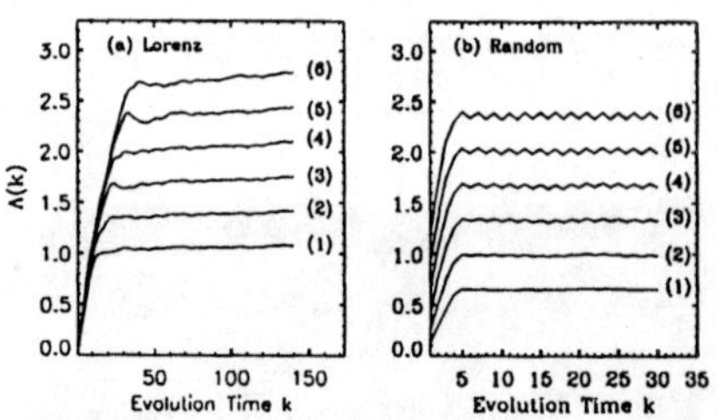
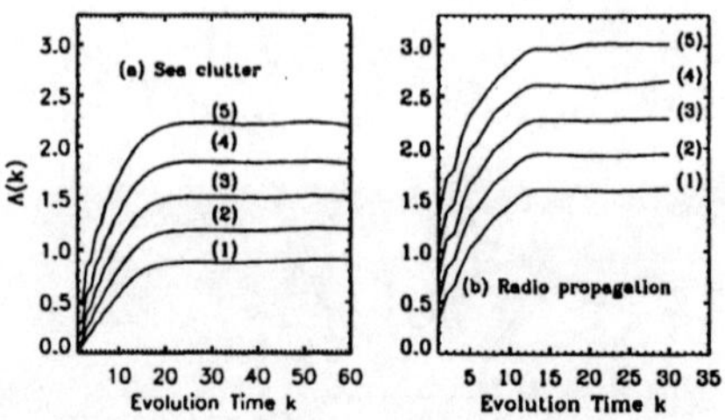

FIGURE 3. $\Lambda(k)$ curves for (a) the chaotic Lorenz attractor and (b) a random process. Numbers $1, \cdots, 6$ represent different shells [6].

FIGURE 4. $\Lambda(k)$ curves for sea clutter and radio propagation data. Numbers $1, \cdots, 5$ represent five different shells.

depends on the parameter r_0 that is selected to use in the computation, hence, typically is different for different researchers. Thus, we are observing the realization of a random process! While there is little danger to interpret white noise as low-dimensional deterministic chaos, the possibility of mistaking $1/f$ processes as "chaotic" may result from using certain computational algorithms that inappropriately yield "positive" Lyapunov exponents.

The $\Lambda(k)$ vs. k curves for the sea clutter and radio propagation signal envelopes are shown in Fig. 4(a) and (b), respectively, with $m = 7$ and $L = 2$ for the sea clutter, and $m = 4$ and $L = 3$ for the radio propagation data. Similar curves were obtained for other choices of m and L. We do not observe a common envelope here. Hence, we have to conclude that these given sea clutter and radio propagation data are not chaotic.

SEA CLUTTER AND RADIO PROP. DATA AS MULTIFRACTALS

To gain deeper insights into sea clutter and radio propagation, we perform multifractal analysis of these data. Mathematically, multifractals are characterized by many or infinitely many power-law relations. Strange attractors, as geometrical objects, are often multifractals, in the sense that, the entire attractor may be partitioned into many (possibly infinitely many) subsets, each subset has its own fractal dimension, and its "weight" in the original attractor is well defined.

We first work with a specific type of multifractal, called random multiplicative process model, to analyze sea clutter. Such models have been used to characterize many other physical problems. In this framework, one computes the generalized moment $M_q(\varepsilon) = \sum_i w_i^q$, where q is a real number, and the positive "weights" w_i, $i = 1, 2, ...$, can be readily computed from a time series, as we shall explain shortly. One then checks if the following scaling law holds, $M_q(\varepsilon) \sim \varepsilon^{\tau(q)}$. If the above relationship holds and $\tau(q)$ is not a linear function of q, then we conclude the waveform is a multifractal. Note that large positive q "emphasizes" large weights, while large negative q "emphasizes" small weights. Also note that $\tau(q)$ being nonlinear in q is equivalent to the weights w_i begin not constants. Thus, when w_i are obviously nonuniform, one only needs to check if the scaling described above holds or not. An example of a multifractal time-series is shown in Fig. 5 for some network traffic data [7]. From this model, one can show that the weights in the generalized moment must have a log-normal distribution with

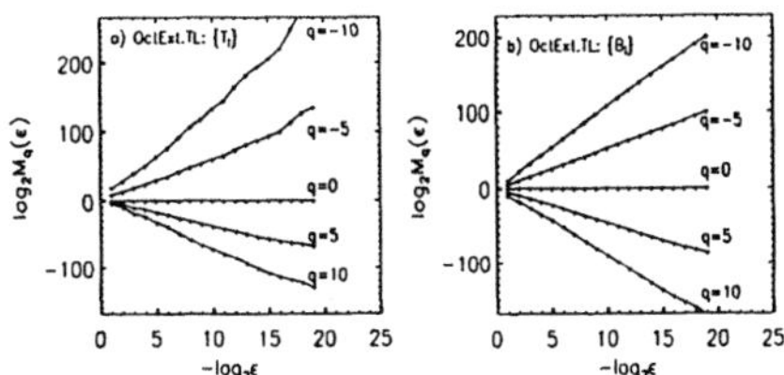

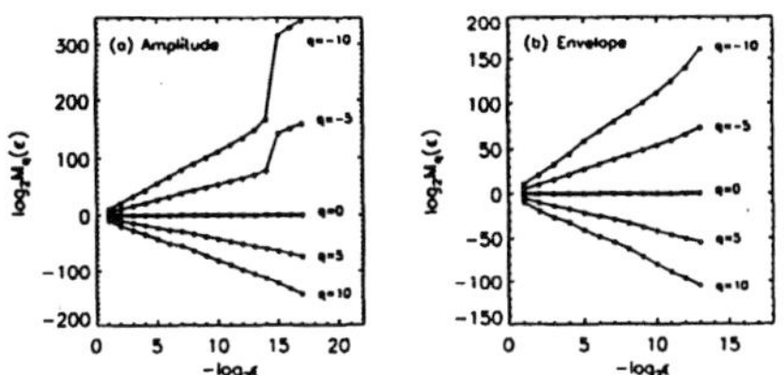

FIGURE 5. $\log_2 M_q(\varepsilon)$ vs. $-\log_2 \varepsilon$ for the inter-arrival time series of WAN traffic. Reproduced from Fig.8(a) of [7].

FIGURE 6. $\log_2 M_q(\varepsilon)$ vs. $-\log_2 \varepsilon$ for the amplitude and envelope signals of sea clutter.

$\tau(q) = -\ln(2\mu_q)/\ln 2$ [7].

Fig. 6(a) shows the result of multifractal analysis on the amplitude signal of sea clutter. We observe that the multifractal scaling is valid for all positive q as well as for negative q on not too short time scales. Since negative q emphasizes the part of the signal with small magnitudes, the above feature thus suggests that the short portion of the data between successive local maxima does not follow this multifractal scaling laws. We are therefore motivated to carry out multifractal analysis on the envelope signal of Fig. 2(b). The result is shown in Fig. 6(b). Indeed, now we see the multifractla scaling holds for all q on all time scales. The results of Fig. 6 suggest that the amplitude signals will only be approximately log-normally distributed, while the envelope signal will be more so. This is indeed the case, as shown in Fig. 7(a) and (b). This simply suggests that the detailed signals for each wave motion on the sea surface, represented by the data between successive local maxima, do not follow log-normal distributions.

Next, we use structure function technique [4] to perform a different type of multifractal analysis of the radio propagation data. The procedure is as follows. Let $x(t), t = 1, \cdots, n$, denote the radio propagation envelope data. Then we remove the mean value from the $x(t)$ time series. Denote the new time series as $x'(t), t = 1, \cdots, n$, where $x'(t) = x(t) - \bar{x}$, where $\bar{x}$ is the mean of $x(t)$. Then we construct a new time series $y(t), t = 1, \cdots, n$, by summing up the $x'(t)$ time series: $y(t) = \sum_{i=1}^{t} x'(i)$. Often, $y(t)$ is called a random walk process of $x'(t)$, while $x'(t)$ an increment process. The essence of the structure function based multifractal analysis is to check whether the following scaling laws $F^{(q)}(m) = \langle |y(i+m) - y(i)|^q \rangle \sim m^{\zeta(q)}$ holds, where $\zeta(q)$ is a function of real value q, and the average is taken over all possible couples of $(y(i+m), y(i))$. A negative q value emphasizes small absolute increments of $y(t)$, while a positive q value emphasizes large absolute increments of $y(t)$. Large absolute increments of $y(t)$ can be thought of as bursts. Hence, large positive q value focuses on the intermittency of the signal (or field). Since our interest is in the variations of $y(t)$, we shall only focus on positive q's. When $\zeta(q)$ is a nonlinear function of q, the data set is said to be a multifractal, in the sense of Parisi and Frisch [18]. Following [3], we also define $H(q) = \zeta(q)/q$. It is clear that $H(q)$ is equivalent to $\zeta(q)$. Note that for a standard Brownian motion (Bm) process, $H(q) = 1/2$ for $q > -1$. The extension of Bm is the fractional Brownian motion (fBm) [14], which has a constant $H(q)$, with $0 \leq H(q) \leq 1, H(q) \neq 1/2$.

$H(2)$ is of special interest to us. It characterizes the correlation structure of the data set $y(t)$, since when the above scaling law for $q = 2$ holds, then the variance of $y(t)$

scales with t as $t^{2H(2)}$. Equivalently, this gives a power-law decaying auto-covariance function, $R(t)$, for $y(t)$, $R(t) \sim t^{2H(2)-2}$. By the Wiener-Khinchin theorem, this gives a decaying power spectral density (PSD) for the data that also follows a power-law, $f^{-(2H(2)+1)}$. Note that when the PSD for $y(t)$ follows $f^{-(2H(2)+1)}$, then the PSD for $x'(t)$ follows $f^{-(2H(2)-1)}$. When $H(2) = 1/2$, the process is called memoryless or short range dependent. The most well-known example is the Bm process. In nature and in man-made systems, often a process is characterized by an $H \neq 1/2$. Prototypical models for such processes are fBm processes. When $0 \leq H(2) < 1/2$, we have negatively correlated increments; a jump up is more likely followed by a jump down and vice-versa. This leads to a process less nonstationary than Bm. When referring to fBm, this is called "anti-persistence" by Mandelbrot [14]. For $1/2 < H(2) \leq 1$, we have positively correlated increments. This means that a jump tends to be followed by another jump in the same direction. This type of noise is more nonstationary than Bm. In fBm, this is called "persistence" [14]. Such processes have long memory properties.

In practice, it is not known whether the time series under study is a multifractal or not, one thus first computes $F^{(q)}(m)$, then checks if $F^{(q)}(m)$ varies with m in a power-law manner. We have carried out analysis on the envelopes of the radio propagation data set measured at UCLA. Fig. 8a shows $log_2 F^{(q)}(m)$ versus $log_2(m)$ and Fig. 8b shows $H(q)$ versus q. These results tend to indicate the data are weakly multifractal, with $H(2)$ having values about 0.5. Due to the complexity and the seeming time and scenario dependency of when and where the data are measured, clearly more data must be collected before any general conclusion on the radio propagation data can be made.

CONCLUDING REMARKS

It is well known that the measurement and characterizing of radar sea-clutter returns are quite complicated. In the first part of the paper, by employing some advanced tools for signal processing from chaos and random multiplicative fractal theory, we have found that (at least for the two minutes of available) sea clutter data set, it is multifractal but not chaotic. The multifractal scaling laws hold well for the envelope signal of the sea clutter amplitude data, but less so for the original amplitude signal on time scales small enough to represent individual waves on the sea surface (which is "emphasized" by negative q). This determines that log-normal distribution describes quite well the envelope signal while only approximately the amplitude signal. In the second part of the paper, we also used the structure function technique of multifractal theory, and showed the measured wideband radio propagation data over a fine time scale is not chaotic and exhibits weak multifractal behaviors. We note these data behave very different from the conventional narrowband fading radio propagation data. Clearly, more data are needed before any general conclusions can be made from the basic understanding and application point of view. For applications, we may exploit possible differences of the multifractal nature of free sea surface clutters from that of sea clutter with point targets. We also may exploit the multifractal properties of the radio propagation data to find adaptive receiver schemes to improve SNR and decrease BER.

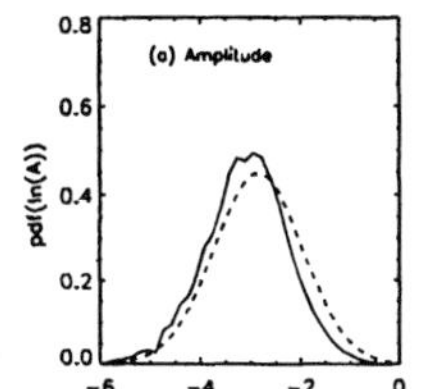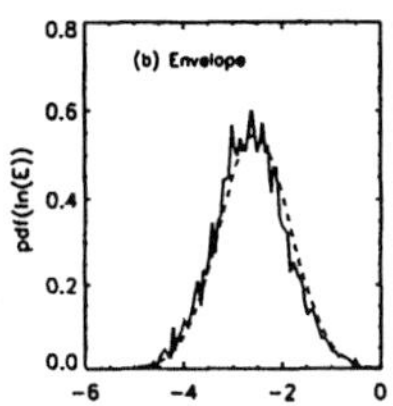

FIGURE 7. Distributions of amplitude and envelope signals (in logarithmic scale) for sea clutter. Dashed lines are normal distributions.

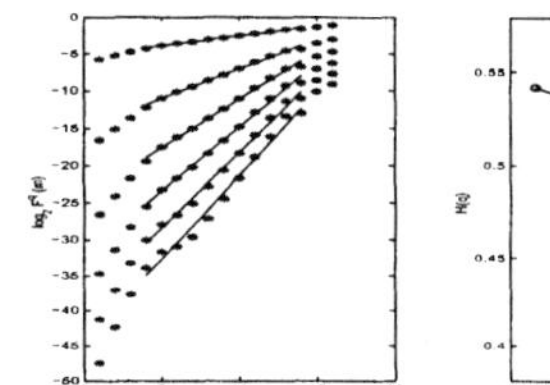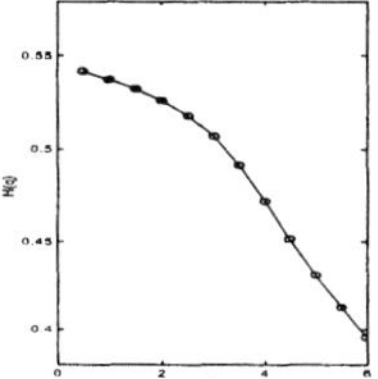

FIGURE 8. Indoor radio propagation data (a) $\log_2 F^{(q)}(m)$ vs. $\log_2(m)$; (b) $H(q)$ vs q.

ACKNOWLEDGMENTS

This work is partially supported by the ARO-MURI grant DAAG55-98-0269 and NASA-Dryden grant NCC2-374.

REFERENCES

1. M.R. Cowper and B. Mulgrew, "Nonlinear Processing of High Resolution Radar Sea Clutter," Proc. IJCNN, vol. 4, July 1999, pp. 2633.
2. M. Davies, "Looking for Non-Linearities in Sea Clutter," IEE Radar and Sonar Signal Processing, Peebles, Scotland, July 1998.
3. A. Davis, A. Marshak, W. Wiscombe, R. Cahalan, 1994: Multifractal char. of nonstationarity and intermittency in geophysical fields-obs., retrived, or sim. *J. Geophys. Res.*, **99**, 8055-8072.
4. U. Frisch, 1995: *Turbulence—The legacy of A.N. Kolmogorov.* Cambridge University Press.
5. J.B. Gao and Z.M. Zheng, "Local exponential divergence plot and optimal embedding of a chaotic time series", *Phys. Lett. A* **181**, 153 (1993).
6. J.B. Gao and Z.M. Zheng, "Direct dynamical test for deterministic chaos and optimal embedding of a chaotic time series", *Phys. Rev. E*, **49**, 3807 (1994).
7. J.B. Gao and I. Rubin, "Multiplicative multifractal modeling of Long-Range-Dependent network traffic", *Int. J. Comm. Systems*, **14**, 783-201 (2001).
8. J.B. Gao and I. Rubin, "Multifractal modeling of counting processes of Long-Range-Dependent network T raffic", *Computer Communications*, **24**, 1400-1410 (2001).
9. P. Grassberger and I. Procaccia, "Characterisation of strange attractors", *Phys. Rev. Lett.*, **50**, 346(1983).
10. H. Hashemi, M. McGuire, T. Vlasschaert and D. Tholl, "Measurements and Modeling of Temporal Var. of the Indoor Radio Prop. Channel". *IEEE Trans. Vehicular Tech.*, **43** 733-737 (1994)
11. S. Haykin and S. Puthusserypady, "Chaotic dynamics of sea clutter". *Chaos*, **7**, 777-802 (1997).
12. S. Haykin, Chaotic dynamics of sea clutter (John Wiley) 1999.
13. S. Haykin, R. Bakker, and B. Currie, "Uncovering Nonlinear Dynamics – The Case Study of Sea Clutter," *Proc. of IEEE*, vol. 90, May 2002, pp. 860-881.
14. B.B. Mandelbrot, *The Fractal Geometry of Nature* (San Francisco: Freeman, 1982).
15. B.B. Mandelbrot, 1997: *Fractals and Scaling in Finance.* New York: Springer.
16. F.E. Nathanson, Radar design principles (McGraw Hill) 1969, pp. 254-256.
17. J. L. Noga, "Bayesian State-Space Modelling of Spatio-Temporal Non-Gaussian Radar Returns," Ph.D thesis, Cambridge University, 1998.
18. G. Parisi, U. Frisch, in *Turbulence and Predictability in Geophysical Fluid Dynamics and Climate Dynamics*, eds. M. Ghil, R. Benzi, and G. Parisi (North-Holland, 1985), pp. 84.
19. F. Takens, in *Dynamical systems and turbulence, Lecture Notes in Mathematics*, Vol. **898**, edited by D.A. Rand and L.S. Young (Springer-Verlag, Berlin) 1981, p.366.
20. C. Tannous, R. Davies, and A. Angus, Strange attractors in multipath propagation. *IEEE Trans. Commun.*, **39** 629-631 (1991).

Lagrangian Structures in Very High-Frequency Radar Data and Optimal Pollution Timing

Francois Lekien*, Chad Coulliette*† and Jerry Marsden*

*Control and Dynamical Systems, California Institute of Technology, Pasadena, CA 91125
†Environmental and Engineering Science, California Institute of Technology, Pasadena, CA 91125

Abstract. Very High-frequency (VHF) radar technology produces detailed surface velocity maps near the surface of coastal waters. The use of measured velocity data in environmental prediction, however, has remained unexplored. In this paper, we uncover a striking flow structure in coastal radar observations along the coast of Florida. This structure governs the spread of organic contaminants or passive drifters released in the area. We compute the Lyapunov exponents of the VHF radar data to determine optimal release windows in which contaminants are advected efficiently away from the coast and we show that a VHF radar-based pollution release scheme using the hidden flow structure reduces the effect of industrial pollution in the coastal environment.

INTRODUCTION

The release of pollution in coastal areas [1, 2, 3] can lead to dramatic consequences for local ecosystems if the pollution recirculates close to the coast rather than being transported out to the open ocean and safely absorbed. We demonstrate that sensitivity to initial conditions in coastal flows can create different patterns of behavior for released contaminants. Depending on their release position and release time, identical parcels of contaminants can have completely different effects on the environment. Using a combination of accurate current measurement [4] and recent developments in nonlinear dynamical systems theory [5, 6], we uncover previously unknown flow structures[1] that govern the mesoscale transport of pollutants. Knowledge of these Lagrangian structures should lead to predictions [7] on a number of phenomena, ranging from the motion of plankton populations to the evolution of oil spills. In this paper we demonstrate how Lagrangian structures can be exploited to reduce the damaging effects of coastal pollution.

We describe fluid particle motion as a dynamical system obeying:

$$\dot{\mathbf{x}} = \mathbf{v}(\mathbf{x}, t) \, . \tag{1}$$

In constrast to earlier approaches to optimal pollution release in simple flow models [8, 9, 10, 11, 12], we rely on real-time data obtained directly from coastal radar antennas. To determine the velocity, the left-hand side of Equation (1), we examine very

---

[1] By flow structure we mean Lagrangian structures, i.e. sets of distinguished fluid particles moving along with the flow.

CP676, *Experimental Chaos: 7ᵗʰ Experimental Chaos Conference,*
edited by V. In, L. Kocarev, T. L. Carroll, B. J. Gluckman, S. Boccaletti, and J. Kurths

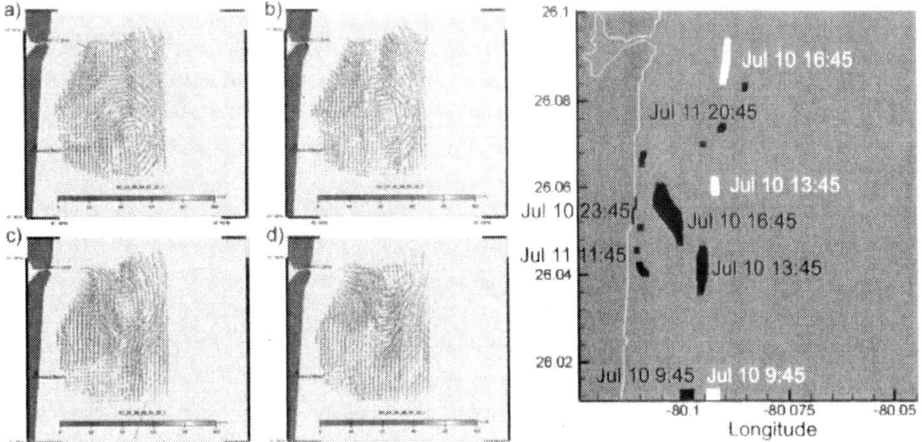

FIGURE 1. On the left, surface current images of the velocity pattern obtained by HF radar along the coast of Florida, near Fort Lauderdale from SFOMC 4-D Current Experiment on June 26, 1999: a) 01:20 GMT; b) 02:20 GMT; c) 04:00 GMT; d) 05:20 GMT. On the right, two parcels of contaminant released at exactly the same time, but at slightly different initial locations on July 10, 1999 at 09:45 GMT. The white parcel leaves the domain quickly and dissolve in the open ocean. The black parcel re-circulates near the coast for more than 36 hours. *Animation: http://www.transport.caltech.edu/florida.html*

high-frequency (VHF) radar measurement of near-surface currents along the coast of Florida [4]. Except for measurement errors, the VHF data shown on Fig. 1 is an exact footprint of the actual velocity as described by Equation (1).

VHF RADAR DATA ALONG THE COAST OF FLORIDA

The OSCR VHF model [13, 14] was deployed for the Southern Florida Ocean Measurement Center (SFOMC) 4-Dimensional Current Experiment from June 25-August 25, 1999. Radio waves are backscattered from the moving ocean surface by surface waves of one-half of the incident radar wavelength. This Bragg scattering effect [15] results in two discrete peaks in the Doppler spectrum. In the absence of surface current, spectral peaks are symmetric and their frequencies are offset from the origin by an amount proportional to the surface wave phase speed and the radar wavelength. If there is an underlying surface current, Bragg peaks in the Doppler spectrum are displaced by the radial component of current along the radar's look direction. Using two radar stations sequentially transmitting radio waves resolves the two-dimensional velocity vector for placing the data into a geophysical context.

The two transmit/receive stations operated at 50 MHz and sent electromagnetic signals scattered from surface gravity waves with 3-m wavelengths. Coastal ocean currents were mapped over a 7 km × 8.5 km domain at 20 minute intervals with a resolution of 250 m. The radars were located in John Lloyd State Park, Dania Beach (Master) and an oceanfront site in Hollywood Beach (Slave) which equated to a distance of 7 km.

NUMERICAL EXPERIMENTS

The temporal complexity of the currents becomes evident from tracking different evolutions of a fluid parcel[2] released at the same time, but at a slightly different location. We show the results of two such numerical experiments on Fig 1. The complete animation and others can be downloaded from `http://www.transport.caltech.edu/florida.html`. We have performed this analysis using two parcels of particles launched at 09:45 GMT on July 10, 1999. Using available VHF velocity data, we advected the fluid particles using a 4th order Runge-Kutta-Fehlberg algorithm (RKF45) combined with a 3rd order tricubic interpolation in both space and time[3].

Note that the black contaminant parcel remains near the coast, whereas the white parcel exits the domain quickly to the North and advects into the open ocean. The latter scenario is highly desirable, because it minimizes the impact of the contaminant on coastal waters, by causing it to be safely dispersed in the open ocean. This observation inspires us to understand and predict different evolution patterns of a fluid parcel, depending on its initial location and time of release. Such patterns are known to be delineated by repelling material lines or finite-time stable manifolds [16, 17, 18, 19, 20, 21]. Here we shall use a recently developed nonlinear technique, Direct Lyapunov Exponent [6] (DLE) analysis, which identifies repelling or attracting material lines in velocity data as local maximizing curves of material stretching.

LAGRANGIAN COHERENT STRUCTURES

The DLE algorithm starts with the computation of the flow map, the map that takes an initial fluid particle position $\mathbf{x}_0$ at time t_0 to its later position $\mathbf{x}(t, \mathbf{x}_0)$ at time t. To perform this analysis, we launched a grid of 200×200 particles at time t_0. Using a 3rd order interpolation of the VHF radar data, we advected the particles of the grid for $t - t_0 = 25$ hours, using a 4th order Runge-Kutta-Fehlberg algorithm. We used these particle trajectories to approximate the flow map. We modeled the coastline as a free-slip boundary, and disregarded particles that crossed the open boundaries of the domain on the northern, eastern and southern edges. All of these numerical algorithms have been compiled into a software package, MANGEN, that is available from the authors upon request.

We then compute the largest singular value $\sigma_t(\mathbf{x}_0, t_0)$ of the spatial gradient of the flow map [21]. More specifically, we compute the largest eigenvalue of the Cauchy-Green strain tensor

$$\Sigma_t(\mathbf{x}_0, t_0) = \left[\frac{\partial \mathbf{x}(t, \mathbf{x}_0)}{\partial \mathbf{x}_0} \right]^T \left[\frac{\partial \mathbf{x}(t, \mathbf{x}_0)}{\partial \mathbf{x}_0} \right], \tag{2}$$

with the superscript T referring to the transpose of a matrix.

---

[2] A fluid parcel is a simplified model for a blob of contaminant.

[3] The resulting local interpollator provides a C^1 velocity field in extended phase space.

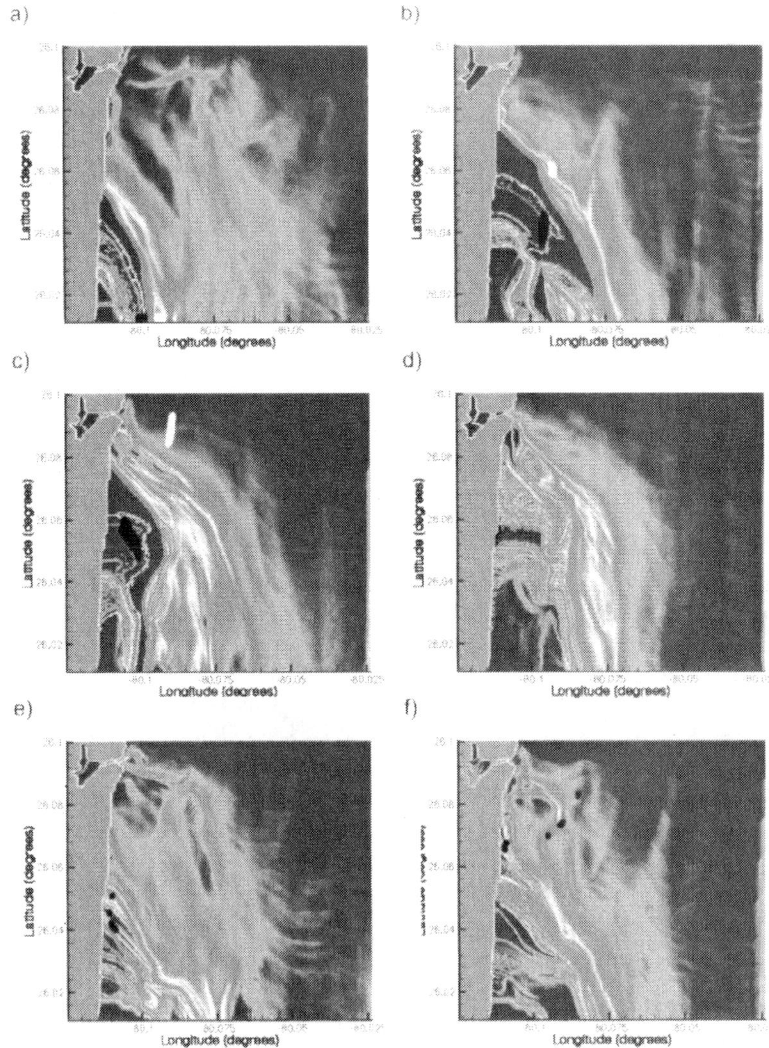

FIGURE 2. Level Sets of Direct Lyapunov Exponents along the coast of Florida on (a) July 10, 1999 09:45 GMT, (b) July 10 13:45 GMT, (c) July 10 16:45 GMT, (d) July 10 23:45 GMT, (e) July 11 11:45 GMT and (d) July 11 20:45. The simulation clearly shows a stable manifold (i.e. repelling material line) attached to the coast near Fort Lauderdale. Superimposed on each figure panel are the respective positions of the two parcels from Fig. 1. Every particle north of the manifold flows through the northern open boundary. It is non-optimal to release contaminants below the branch of the manifold because it will remain between the coast and the manifold for a long time. *Animation: http://www.transport.caltech.edu/florida.html*

Repelling material lines are local maximizing curves or ridges of the scalar field $\sigma_t(\mathbf{x}_0, t_0)$[6, 22]. The same procedure performed backward in time (i.e., for $t < t_0$) would render attracting material lines at t_0 as ridges of $\sigma_t(\mathbf{x}_0, t_0)$.

Composed of fluid particles, these curves remain hidden to naked-eye observations of unsteady current plots, yet they fully govern global mixing patterns in the fluid. Such Lagrangian structures in measured ocean data have previously been inaccessible due to lack of an efficient extraction method.

ANALYSIS OF THE DATA

Selected frames of the contour level sets of the Lyapunov exponents are shown in Fig. 2. During the two months of the experiment, the plot reveals a strong stable Lagrangian structure attached to the coast near Fort Lauderdale, propagating to the southeast. This structure acts as a Lagrangian barrier between the coastal recirculating zone (southwest of the material line) and the Florida Current (northeast of the same material line). The material line is a barrier in the sense that particles cannot cross it. Superimposed on Fig. 2 are the two parcels used in Fig. 1. A quick analysis reveals that any particle northeast of the barrier (white parcel) is flushed out of the domain in only a few hours. In contrast, parcels starting southwest of the barrier (black parcel) typically re-circulate several times near the Florida coast before they finally rejoin the current. The animation corresponding to the panels of Fig. 2 can be downloaded at http://transport.caltech.edu/forida.html. It is important to realize that without the use of DLE or a similar method, the Lagrangian structure would still be there, but could not be seen or made use of in this way. We prefer to think of the currents as not influencing particle paths directly, but rather that the currents influence the Lagrangian structure, such as causing transport barriers and alleyways, and the Lagrangian structure directly influences the particle paths.

MINIMIZATION OF THE EFFECT OF POLLUTION

We remark that the location of the base of the structure (on the coast) can be used as a criteria to minimize the effect of coastal pollution. We will refer to the intersection of the coastline and the Lagrangian structure as the barrier point. Factories and sewing centers along the coast should not release anything if the barrier point is located North of them. To illustrate how an efficient pollution release algorithm can be set up, we imagined a source of pollution with a fixed position along the coast. Using the DLE plots of Fig. 2, we identify zones of (light gray) favorable release[4] and (dark gray) dangerous release[5].

To minimize the effect of coastal pollution, we propose using a holding tank that stores contaminants during dangerous release zones. The tank stores pollution during

---

[4] when the manifold is below the position of the factory

[5] when the manifold is above the factory

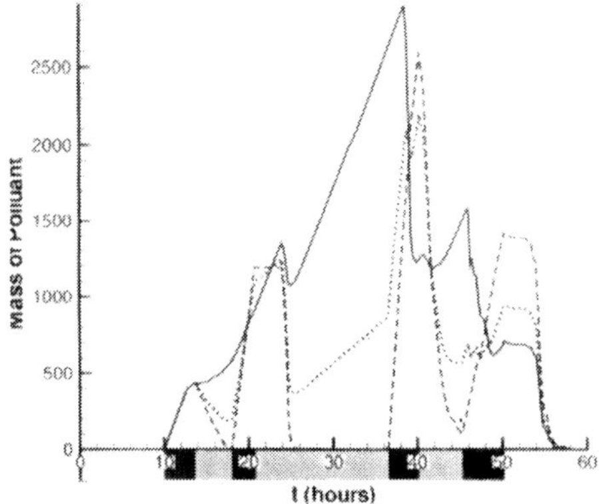

FIGURE 3. Total mass of contaminant in the coastal area for different source of pollution. The solid line is the result of pollution at a constant rate. The light and dark gray blocks on the x-axis are the zones identified by the DLE algorithm as respectively good release zones and bad release zones. The dashed line and dotted line are the result of optimized pollution release with respectively $\alpha = 0\%$ and $\alpha = 30\%$.

the half-period of the barrier point oscillation, during which contaminants should not be released. The contents of the tank are released once the barrier point passes south of the source of pollution. On Fig. 3 is the total mass of contaminant for the two release mode. Clearly, by getting information from the DLE plots, the white factory (dashed curve) is able to reduce by a factor of three the local contamination caused by the same amount of pollution.

Also shown on Fig. 3 is a 3rd experiment. In many case the damage to the environment is a function of the maximum concentration of contaminant. From this point of view our algorithm (releasing nothing during "dark gray" zones and as much as possible during "light gray" zone) does not seem to be efficient. The peak of maximum concentration for the white factory has only decreased by a small amount. To set up a more elaborate algorithm, we define a new degree of freedom α, the percentage of incoming created contaminant that will be released during "dark gray " zones. From this point of view, the black curve of Fig. 3 corresponds to $\alpha = 30\%$, the dashed curve to $\alpha = 0\%$ and the purple curve to $\alpha = 33\%$. Fig. 3 shows that a significant reduction of the peak of maximum concentration can be obtained using an appropriate partial release during zones that are marked dangerous by the DLE algorithm.

ACKNOWLEDGMENTS

The VHF radar data used in this paper was collected and analyzed at the Southern Florida Ocean Measurement Center (SFOMC) by Arthur Mariano and Edward Ryan. The authors are grateful to the Office of Naval Research and particularly to Manuel Fiadero for his support and advices. This research was funded by ONR grant N00014-97-1-0071.

CONCLUSION

We have shown the existence of a set of repelling material lines in radar data obtained from the Fort Lauderdale area on the Florida coast. We have also shown how these material lines can be used to minimize the effect of coastal pollution by determining optimal release times. This approach can be used for making predictions about the trajectories of buoyant contaminants or the trajectories of nearly Lagrangian tracers. The data source can be VHF radar data or any other current data source, such as data-assimilated ocean models that approximate the near-surface velocity field to some reasonable level of accuracy. The other advantage of using ocean models is that the velocity provided is 3D+1, thus we can explore the Lagrangian structures that develop at various depths. Since these types of models are becoming more readily available and configured for various areas, we are currently in the process of extending MANGEN to compute Lyapunov exponents for 3D+1 velocity data. A real-time experimental realization of our pollution release will be most important, and is planned for future work.

REFERENCES

1. Prahl, F. G., Crecellus, E., and Carpenter, R., *Environ Sci Technol*, **18**, 687–693 (1984).
2. Rice, D. W., Seltenrich, C. P., Spies, R. B., and Keller, M. L., *Environmental Pollution*, **82**, 79–91 (1993).
3. Verschueren, K., *Handbook of Environmental Data on Organic Chemicals*, Van Nostrand Reinhold Co., New York, 1983.
4. Shay, L., Cook, T., Haus, B., Martinez, J., Peters, H., Mariano, A., VanLeer, J., Edgar An, P., Smith, S., Soloviev, A., Weisberg, R., and Luther, M., *EOS, Transactions, American Geophysical Union*, **81** **(19)**, 209–213 (2000).
5. Stirling, J. R., *Physica D*, **144**, 169–193 (2000).
6. Haller, G., *Physica D*, **149**, 248–277 (2001).
7. Coulliette, C., Lekien, F., Marsden, J., Haller, G., and Paduan, J., *Physical Review Letters* (2002).
8. Webb, T., and Tomlinson, R. B., *J. Env. Eng.*, **118**, 338–362 (1992).
9. Smith, R., *IMA J Appl Math*, **51**, 187–199 (1993).
10. Bikangaga, J. H., and Nussehi, V., *Water Res*, **29**, 2367–2375 (1995).
11. Giles, R. T., *Water Res.*, **29**, 563–569 (1995).
12. Smith, R., *J. Hydraulic Eng.*, **124**, 117–122 (1998).
13. Prandle, D., *J. Phys. Oceanogr.*, **17**, 231–245 (1987).
14. Shay, H. C., L. K.and Graber, Ross, D. B., and Chapman, R. D., *J. Atmos. Ocean Tech.*, **12**, 881–900 (1995).
15. Stewart, R. H., and Joy, J. W., *Deep-Sea Res*, **21**, 1039–1049 (1974).
16. Miller, P. D., Jones, C. K. R. T., Rogerson, A. M., and Pratt, L. J., *Physica D*, **110**, 1–18 (1997).
17. Poje, A. C., and Haller, G., *J. Phys. Oceanogr*, **29**, 1649–1665 (1999).
18. Coulliette, C., and Wiggins, S., *Nonlinear Proc. Geophys*, **8**, 69–94 (2001).
19. Ridderinkhof, H., and Zimmerman, J. T. F., *Science*, **258**, 1107–1111 (1992).
20. Lapeyre, G., Hua, B. L., and Legras, B., *Chaos*, **11**, 427– (2001).
21. Haller, G., *Phys. Fluids A*, **13**, 3368–3385 (2001).
22. Haller, G., *Phys. Fluids A*, **14**, 1851–1861 (2002).

HYDRODYNAMICS, TURBULENCE, AND PLASMAS

Experiment and Model for the Dynamics of Vortex Ripples in Sand

M. Abel[*], K.H. Andersen[†], C. Ellegaard[**], J. Krug[‡], L.R. Søndergaard[**§]
and J. Udesen[**§]

[*]Institut für Physik, Universität Potsdam, D-14415 Potsdam, Germany
[†]3shape, Copenhagen, Denmark
[**]Niels Bohr Institute, University of Copenhagen, DK-2100 Copenhagen, Denmark.
[‡]Fachbereich Physik, Universität Essen, D-45117 Essen, Germany.
[§]Department of Mathematics and Physics, University of Roskilde, Box 260, DK-4000 Roskilde,
Denmark.

Abstract. Vortex ripples in sand are studied in a one-dimensional setup experimentally. The results are compared with a simple coarse-grained particle-like model. In this model, ripples are considered as interacting particles with nearest neighbor interaction. The key ingredient is a nonlinear interaction function which incorporates the main dynamical aspects of the pattern dynamics. With a multivariate nonparametric regression the model is evaluated and the interaction function is found explicitly. The resulting model is integrated numerically and its behavior is compared with the experimental runs. The stable band of ripples is measured independently.

INTRODUCTION

Patterns in sand are not only fascinating to watch but as well highly complex in the underlying physics. The best-known examples for sand patterns are probably dunes and ripples in the deserts. Underwater patterns can be observed readily on the beach as ripples under the oscillating flow of the water. Other patterns are formed in river beds and far from coastal regions. The dimensions of the objects can reach from hundreds of kilometers for sand waves in the sea or draas in the desert over several magnitudes of meters for typical desert dunes in air to centimeters for rolling grain ripples in water [1, 2].

Laboratory experiments for many of these objects are not straight-forward to perform. The physics is very complex, involving granular matter theory and fluid dynamics at the same time - two of the most involved research areas of modern physics. Among the existing sand patterns, vortex ripples are interesting objects due to their reasonable size and the possibility of controlled laboratory experiments [3, 4, 5]. In recent works the physics of vortex ripples has been investigated by detailed numerical fluid dynamical simulations and simple models have been proposed [6, 7]. In this article, some findings are summarized and updated results concerning the data analysis are shown.

For a two dimensional setup, new bifurcations have been found [5] and stability of the patterns has been investigated [8]. Since the interaction of the ripples has not been fully understood, an experiment has been devised to investigate a one dimensional pattern. The model presented in [6] can be checked by analyzing experimental data [9]. To this

CP676, *Experimental Chaos: 7th Experimental Chaos Conference,*
edited by V. In, L. Kocarev, T. L. Carroll, B. J. Gluckman, S. Boccaletti, and J. Kurths
© 2003 American Institute of Physics 0-7354-0145-4/03/$20.00

end we use the approach proposed in [10] for spatio-temporal data analysis.

We first present details about the experimental setup, subsequently the model under consideration is explained and formulated, the results of the data analysis are presented in the following section; the article concludes with a short discussion.

THE EXPERIMENT

The main component of the experiment is a channel between two concentric plexiglass cylinders. The channel is 11 mm wide and 15 cm high. This channel is partly filled with sand, then filled completely to the lid with water. The sand consists of glass beads with a diameter of 250 ± 50 μm. This channel can be oscillated by driving it with an excentric on a motor through a long connecting rod. The length of this rod is sufficient (1.5 m) to ensure very little deviation from a sinusoidal motion, see Fig. 1.

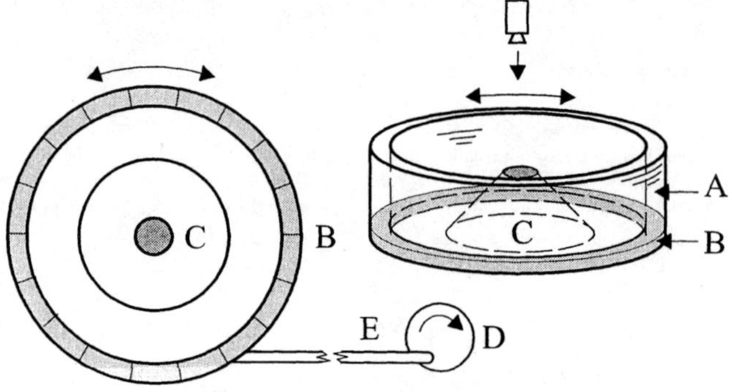

FIGURE 1. Sketch of the experimental setup

The channel with the sand is viewed from above with a CCD camera by a conical mirror in the centre. The camera has a resolution of 640×480 pixels and is triggered precisely at the extreme of the sinusoidal motion. The view seen by the camera is shown in Fig. 2, an unfolded view in Fig. 3.

The profiles are extracted from such pictures and recorded as a function of time. An example of a time series is shown in Fig. 4. We create initial conditions with small ripples by oscillating at a small amplitude. The experiment starts when amplitude and frequency are changed to their desired values. Throughout this paper, we only use the condition $a=6$ cm, $v=0.6$ Hz. There is nothing special about this condition, the same qualitative results are obtained for other parameters; clearly this point deserves more detailed and quantitative analysis. Growing the ripples from initial conditions with small homogeneous ripples results in a final number of 19–21 ripples with a length (averaged over 16 realizations) of $\tilde{\lambda}/a=1.25$. The uncertainty in the length due to periodicity is around 5 %.

For the later data analysis, we must have information about the ripple lengths and their temporal change. The first are found after an unfolding of the images by a fit to triangles with fixed slope (Fig. 3). The second is obtained by consecutive pictures

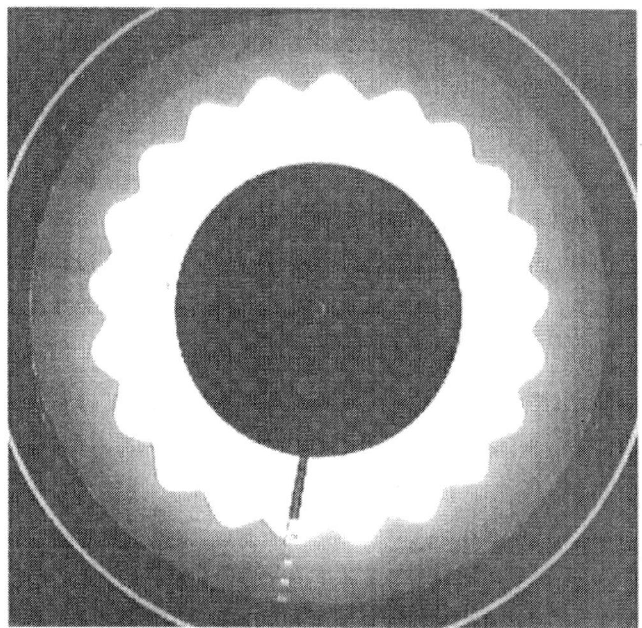

FIGURE 2. The view seen by the camera in the conical mirror

FIGURE 3. The unfolded profile from Fig. 2. On top, one sees triangles, fitted to the slopes, which are used for the data analysis.

of the camera. Fig.4 shows the evolution starting from a small initial wavelength with subsequent annihilation of ripples. Around each annihilation the ripples have not been analysed, as it becomes inaccurate to fit the triangles during these events.

The experiment is run with different values of a until a homogeneous pattern occurs. This way a certain number of ripples, N, are created as initial conditions. The actual experiment is started from this pattern with the above parameters and run for 10000 driving periods. We have created initial conditions with $N \in [17;24]$ and observed that for $N > 22$ ripples are annihilated, while for $N < 18$ one or more new ripples are created. We therefore conclude that there exists a stable band $N \in [17.5;22.5]$ (for $a = 6$ cm, $v = 0.6$ Hz), which corresponds to $\lambda_{min}/a = 1.13 \pm 0.03$ and $\lambda_{max}/a = 1.45 \pm 0.05$.

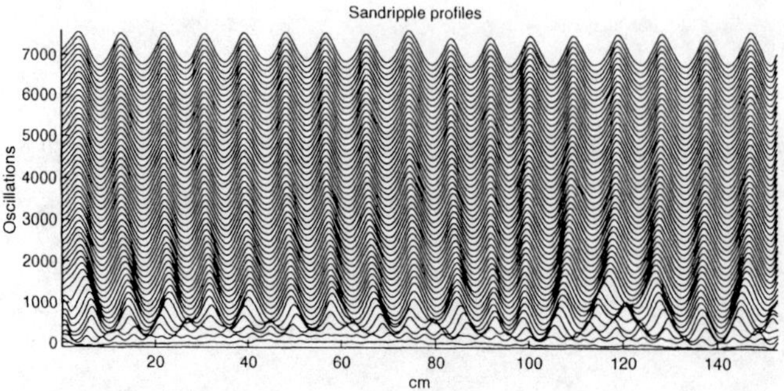

FIGURE 4.   Timeseries of sand profiles

THE MODEL

We consider the pattern dynamics of fully developed vortex ripples. In this case, the flow is dominated by vortices on the lee sides of the ripples (hence the name), the so called separation bubbles [11, 12].

Parameters and dimensional analysis

The typical parameters for the flow are driving amplitude a, frequency f, viscosity v, and the density ρ_w. For the sand, one has density ρ_s, settling velocity w_s and grain diameter d. From these, one can calculate the dimensionless numbers

$$Re = \frac{a^2 f}{v} \, , \quad s = \frac{\rho_s}{\rho_w} \, , \quad \tilde{w}_s = \frac{w_s}{af} \, , \quad \Theta = \frac{\tau_{bed}}{\rho_w(s-1)gd} \, , \tag{1}$$

with g the gravity and τ_{bed} is the maximum shear stress on a flat the bed. Typical values in the experiment are $a = 5$ cm, $f = 3$ s^{-1} yielding a Reynolds number Re of the order of 10^3. The Reynolds number is a measure of the degree of turbulence of a fluid; with the value given above the flow is turbulent and large scale structures, like the separation bubble are independent of the Reynolds number [13]. For quartz in water $s = 2.65$, the non-dimensional settling velocity, $\tilde{w}_s$, is assumed to be large enough to avoid suspension (in the experiment this is tuned by the grain size). The Shields parameter Θ represents the non-dimensional shear stress on the fluid-granular boundary.

If the shear stress is higher than a certain critical value Θ_c, grains loosen and are transported by the flow. This transport takes place in the thin bed load layer above the bed and is a function of the Shields parameter. For turbulent boundary layers, no analytical expression for the shear stress can be given and one uses an empirical relation between shear stress and fluid parameters. For further analysis we use throughout the article non-dimensional time $t \to ft$ and space $x \to x/a$.

For more details about sand motion, we refer to the literature about sediment transport [2]

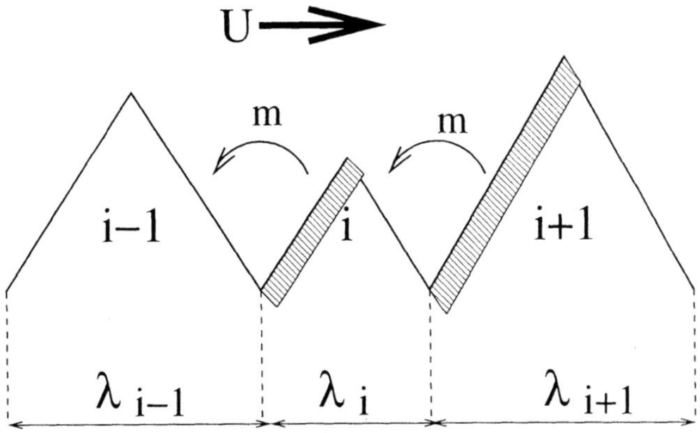

FIGURE 5. Sketch of the mass transport between adjacent ripples during one half period. The direction of mass transport (m) is indicated by arrows, it is opposed to the flow direction (U) and the amount depends on the strength of the separation bubble which in turn depends on the ripple size.

The discrete model

In [6, 7, 14, 15] results from numerical simulations are presented which coincide with ours and others experimental findings [3, 4, 9]. The following qualitative picture emerges:

Let us consider a single ripple. During a half period, a separation bubble is built up, sediment is transported *towards* the ripple which has formed the separation bubble, even upwards, as long as the angle of repose allows for it. The chain of dependencies now goes as follows: the transported amount of sand depends on the shear stress on the bed, which depends on the strength of the separation bubble, which in turn depends on the height of the ripple. The ripple will thus gain mass in each half period on the lee side, while on the "wind" side the neighboring ripple will take mass away from it.

For the second half period the scenario is the same on the opposite side of the ripple. For two neighboring ripples with different size, this means that the net effect over a whole period is a mass transport from the small to the large one (cf. Fig. 5). This holds as long as the ripples are not widely separated. If, e.g., two ripples are larger than the extension of the separation bubble, they effectively do not reach each other and in the trough between them a new, very small ripple can nucleate. From this picture, we build a discrete model for the ripple dynamics. In this model, ripples are assumed to be triangular, symmetric and are characterized by their length, λ. The height h is related to the length by the angle of repose, $\tan\alpha = 2h/\lambda$. The central assumption of the model is that the mass transport during a half period only depends on the size of the ripple that creates the respective separation bubble.

If a ripple gains mass, it gains length. If all ripples are of similar size, one can assume that the ratio of transported mass to gained length $\Delta m/\Delta\lambda$ is equal for all ripples. This is a great simplification which certainly does not hold for ripples which are of very different size. Mass transport can now be recast into the transport of length from one ripple to the other.

The coarse-graining step consists in the accumulation of all the complex effects during one half period in one function, to be determined later. If the flow goes from left to right, ripple i gains material from ripple $i+1$ and looses to ripple $i-1$ (see Fig. 5). For the half period $(t, t+\frac{1}{2})$ one obtains

$$\lambda_i(t+\frac{1}{2}) - \lambda_i(t) = f(\lambda_{i+1}) - f(\lambda_i) , \tag{2}$$

please recall that length and time are scaled. All the physics is contained in the interaction function $f(\lambda)$, whose formal notation is given by the integral of the mass transported through the trough between two adjacent ripples over half a period, see [6] for details. Within the interval $(t+\frac{1}{2}, t+1)$, it follows by the same argument

$$\lambda_i(t+1) - \lambda_i(t+\frac{1}{2}) = f(\lambda_i) - f(\lambda_{i-1}) . \tag{3}$$

Put together, one obtains the model as a mapping from time t to time $t+1$

$$\Delta_t \lambda_i := \lambda_i(t+1) - \lambda_i(t) = -f(\lambda_{i+1}) + 2f(\lambda_i) - f(\lambda_{i-1}) , \tag{4}$$

where Δ_t is the difference between two subsequent discrete times, as defined above. This model conserves the total length of a ripple system, whereas mass conservation is violated. A more refined model can fix this by the cost of being slightly more complicated [6]; basically a ripple is allowed to be asymmetric which requires the measurement of crests and troughs. This however turned out to be experimentally not very accurate and the refined model has not been analyzed further.

Stability analysis and equilibria

We find three equilibria in the model: 1) all ripples are of equal length, i.e., a homogeneous state, 2) period two states, where $f(\lambda_i) = f(\lambda_{i+1})$ and $\lambda_i \neq \lambda_{i+1}$, 3) any juxtaposition of ripples with length λ_a and λ_b and $f(\lambda_a) = f(\lambda_b)$. The latter, spatially inhomogeneous, states are unstable [6] and are not further discussed. Here we present the stability analysis of the homogeneous state which is the physically important one.

The linear stability of the homogeneous state $\lambda_i = \lambda_{eq}$ follows by setting $\lambda_i = \lambda_{eq} + \delta_i$:

$$\Delta_t \delta_i = -f'(\lambda_{eq})(\delta_{i+1} - 2\delta_i + \delta_{i-1}) . \tag{5}$$

This is nothing but a discrete diffusion equation with diffusion coefficient $-f'(\lambda_{eq})$. If $f'(\lambda_{eq})$ is negative (positive), the perturbation will shrink (grow) and the pattern is stable (unstable).

How does the interaction function look like? From simulations one can evaluate the mass transport through the troughs and finds a concave function. The above mentioned observations suggest, too, an unstable branch for small ripples which are still to grow and a stable branch for long ripples which are close to the equilibrium length. For very long ripples one expects again an unstable branch to allow for small ripples to grow in the troughs (cf. [5, 6, 7]). Hence the smallest stable length is at λ_{min} where f has a maximum. Assuming that the model holds for arbitrary small ripples, an infinitesimal ripple will gain mass and grow, if $f(\lambda_{eq}) < f(0)$ and thus the maximum length for which a homogeneous pattern is stable is given by $f(\lambda_{max}) = f(0)$.

If a ripple is continuously loosing mass to its neighbors, it shrinks to zero and does no longer contribute to the dynamics. This is consequently included in the model by removing the ripple from the equations when it becomes smaller than a certain bound. The set of equations is then reduced by one. On the other hand can we include the creation of a ripple by a corresponding rule when two adjacent ripples grow bigger than λ_{max}; this is however not used in the data analysis and the subsequent discussion. We conclude with the observation that there exists a band of ripple lengths for which the homogeneous pattern is stable, outside this band short length instabilities occur [6]. This can be verified experimentally.

RESULTS

In this section, we first present the stable band measurement and show then the results for the interaction function, obtained from a series of 9 independent measurements. The used tool is nonparametric data analysis [10, 16], which works by fitting a generalized additive model to the data [17, 18]. The model as "prepared" for the data analysis reads

$$\Delta_t \lambda_i = -f_l(\lambda_{i+1}) + 2f_c(\lambda_i) - f_r(\lambda_{i-1}) \, . \tag{6}$$

We now allow for different functions f_l, f_c, f_r, in the very end the analysis should confirm the equality of the functions. The functions are obtained numerically by a back-fitting algorithm [17] which determines the functions by minimizing the least-squares error $\chi^2 = \sum_{i,t} \left[\Delta_t \lambda + f_l - 2f_c + f_r \right]^2$. The optimal transformations $f_{l,c,r}$ are found in the least-squares sense varying the functions iteratively in the space of measurable functions. The resulting functions are given nonparametrically, i.e., numerically as points, e.g., $[\lambda_i, f_c(\lambda_i)]$. In a recent publication, only the maximum error on the functions has been given, below we give error bounds as twice the local standard deviations. Please note that the calculation of error bounds with errors in variables is a subtle issue and should be treated with care [16, 19, 20].

As an overall measure of the quality of the regression we use the correlation Ψ of a function with the sum of all the others [16]. For instance for the center function f_c one calculates

$$\Psi_c = |corr(2 \cdot f_c \, , \, \Delta_t \lambda + f_l + f_r)| \, , \tag{7}$$

and accordingly for the other functions Ψ_l and Ψ_r. This measure signals a good result if Ψ is close to unity, if Ψ is close to zero the corresponding function is meaningless for the dynamics in a statistical sense. Most real world applications yield values in between. One disadvantage consists of the disability to distinguish between measurement noise which lowers correlations and "true" deficiencies of the model. With the latter we mean that if parts of the residuals contain deterministic information which is not captured by

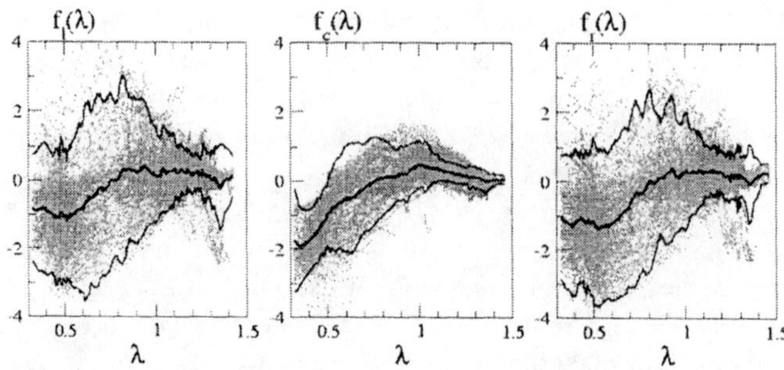

FIGURE 6. Data points from all sets are analyzed in one procedure. The resulting functions f_l, f_c, f_r are plotted with pointwise standard deviations (black lines) on top of the partial residuals (grey dots). The left and right functions show wide scatter. The curves display the expected convex shape. For small ripple length left and right function differ from the center one.

the model, the missing parts then lower the correlations (7). The measurement error can be inferred from the camera resolution to be 3.3%.

The analysis presented here comprises 9 independent measurements which have been analyzed 1) all together to obtain one final result directly 2) separately to be averaged afterwards to the final result. By the second way, one can test instationary behavior during the run and compete with the bias-variance dilemma. If there occur large differences in the final result this is an indication for systematic errors and dependencies in the measurement/analysis procedure. Below, we show results for the first type of analysis and verify consistency by a comparison of the final result.

The result functions are displayed in Fig. 6 with twice the local standard deviation as error indication on top of the corresponding residuals. Clearly there is more scatter in the left and right functions in contrast to a pretty clear picture for the center one. This is reflected by the values $\Psi_l = 0.560$, $\Psi_c = 0.8604$ and $\Psi_r = 0.6376$. Given that the data for each of the functions are of equal precision, one can infer that there may be a deterministic part in the residuals of the left and right function which is not captured by the model. This is clear, if one considers the simplifications assumed in the derivation of the model, further studies will aim at clarifying that point, see the discussion below.

If one plots all three functions together, the coincidence is quite good, cf. Fig. 7. Since in the simple model the functions should be treated as independent, equally weighted estimates for the interaction function, we calculate the final interaction function as the average $f = 1/3(f_l + f_c + f_r)$ which is plotted as a thick line in Fig. 7. We do not present in detail the separate analysis of the different measurement runs, but note that there has been no essential difference in the final result. This is shown in the inset to Fig. 7, where the two results are compared. More details about the analysis are found in [16], together with references to the statistical literature.

The resulting function has been used as input for a numerical integration of the model (4). Qualitatively, the numerics should show coarsening until the final wavelength is reached. Quantitatively, it is interesting, if one can predict the final length from the model, if the stable band is found correctly and if the time and length scales coincide with the experiment, i.e., typical relaxation times and the final length must meet the

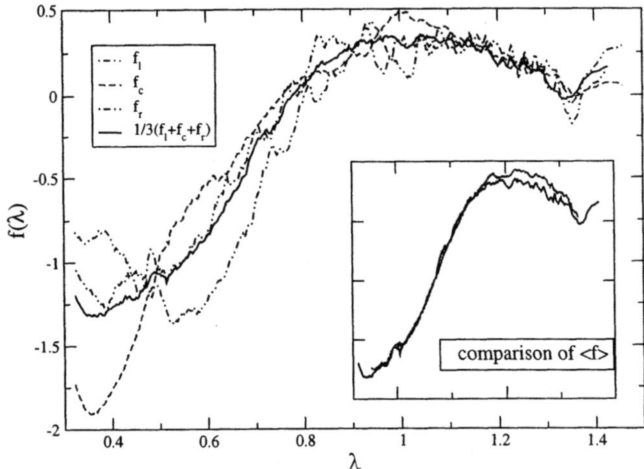

FIGURE 7. The result functions with the final function obtained by averaging.For long ripple lengths -the regime, the model has been developed for- one observes a very good coincidence of sides and center functions. A difference is not expected for the simple model; assuming them to be equal, averaging yields the black line as $f = 1/3(f_l + f_c + f_r)$ as estimate for the true interaction function. If each of the 9 runs is analyzed separately, the functions are found by first averaging over the runs and then as described previously, no essential difference occurs, this is shown in the inset by comparing the latter with the first way of analysis.

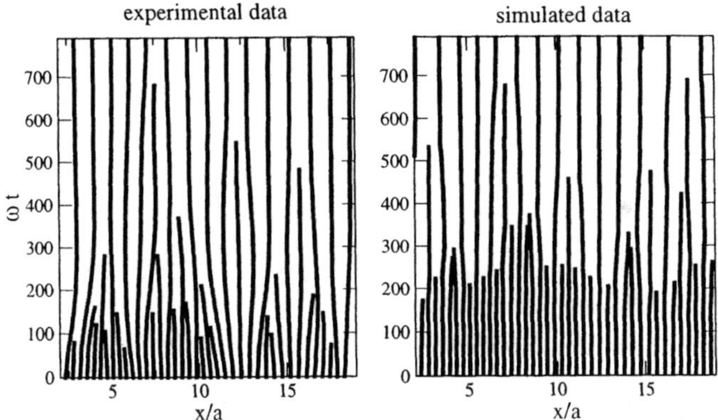

FIGURE 8. Experimental (left) and simulated (right) paths of the ripples. The space-time plot shows the same qualitative behavior, however differences in the final length and the relaxation time towards the final state.

experimental results.

The first observation is, that the model shows the right qualitative features. In Fig. 8 we compare the experimental space-time plot with the simulated one. both show coarsening until a stable pattern is reached. There are, however clear differences in the final length and the time scales to reach the final state: The experiment typically ended with

$N = 20$ ripples, whereas the simulation yields $N = 18$. The relaxation time to reach the final state depends strongly on the slope of the function f. But there is the region $\lambda < 0.3$ for which no data are available and thus the curve has to be extrapolated. Changing the slope of this part yields a different relaxation speed, the final length is, however, conserved. The lower edge of the stable band coincides pretty well with the separate stable band measurement ($\lambda_{min} = 1.13$) with the value $\lambda_{min} = 1.08$. The upper value should be found by $f(\lambda_{max}) = f(0)$. The stable band measurement yields a value of $\lambda_{max} = 1.45$, even with a large uncertainty inherent in an extrapolation it is hard to imagine to find $f(0)$ suitable. For the final length one can find an upper bound by a Maxwell construction [9]. The result depends again on the extrapolated branch of the interaction function, experimenting with varying slopes gave at best a value of $\tilde{\lambda} = 2.2$ which is far outside the definition range of $f(\lambda)$.

Why this discrepancy in the quantitative comparison? The model (4) has been formulated for ripple patterns which do not deviate too much from the homogeneous pattern and thus lengths are close to the final one. The creation of a ripple is, however, far from this situation: typically a very small ripple nucleates between two larger ones, a model to compete with this should then depend not only on the ripple creating the separation bubble but as well on its neighbors. The data analysis becomes then harder, but numerical work hints to some possible candidates for a generalized interaction function of the type $f(\lambda_i, \lambda_{i+1})$.

DISCUSSION AND CONCLUSION

We have performed an accurate experiment to test the discrete model, which describes ripple formation close to the homogeneous state. Experimental data have been processed by an advanced data analysis method to yield the basic ingredient of the model, a nonlinear interaction function, in a nonparametric form.

The result of the data analysis gives a model which shows the required qualitative features. Quantitative coincidence is not reached to a fully satisfactory degree. This can be explained by the assumptions made when deriving the model from theoretical considerations, some are violated such that the model is not expected to hold for all realized situations. In order to obtain a more accurate description, one should generalize and modify the model for small ripple lengths.

The formation and evolution of ripples and underwater patterns of sand in general is an important topic for technical applications. In this field, this work is nothing but a first, small step towards understanding or description of the interesting patterns formed by the fluid-granular interaction. An extension of the presented experiment to two dimensions should be possible, it is however hard to formulate a system of equations which describes the rich phenomenology the system displays [5]. As a connecting link, data analysis can be used to verify or reject models, whose direct justification is not possible.

In this sense, we show that it is possible to join numerical techniques with modelling principles from nonlinear dynamics, statistical data analysis and experiments. Doing so, a consistent way of analyzing and improving complex models has emerged. We think that our approach is promising and look forward to future tasks.

ACKNOWLEDGMENTS

We thank T. Bohr, M. van Hecke and H. Voss for fruitful discussion. M.A. and J.K are grateful to NBI and DTU for hospitality. M.A. is supported by DFG Project number 52059903).

REFERENCES

1. Bagnold, R. A., *The Physics of Blown Sand and Desert Dunes*, Chapman and Hall, Methuen, London, 1941.
2. Fredsœ, J., and Deigaard, R., *Mechanics of Coastal Sediment Transport*, World Scientific, Signapore, 1992.
3. Stegner, A., and Wesfreid, J. E., *Phys. Rev. E*, **60**, R3487–R3490 (1999).
4. Scherer, M. A., Melo, F., and Marder, M., *Phys. Fluids*, **11**, 58–67 (1999).
5. Hansen, J. L., van Hecke, M., Haaning, A., Ellegaard, C., Andersen, K. H., Bohr, T., and Sams, T., *Nature*, **410**, 324 (2001).
6. Andersen, K. H., Chabanol, M. L., and v. Hecke, M., *Phys Rev. E*, **63**, 66308 (1999).
7. Andersen, K. H. (1999), ph. D. thesis, Copenhagen University, http://www.nbi.dk/ kenand/Thesis.html.
8. Hansen, J. L., van Hecke, M., Ellegaard, C., Andersen, K. H., Bohr, T., Haaning, A., and Sams, T., *Phys. Rev. Lett.*, **87**, 204301 (2001).
9. Andersen, K. H., Abel, M., Krug, J., Ellegaard, C., Soendergaard, L. R., and Udesen, J., *Phys. Rev. Lett.*, **88**, 4302 (2002).
10. Voss, H. U., Buenner, M., and Abel, M., *Phys. Rev. E*, **57**, 2820–2823 (1998).
11. Bagnold, R. A., *Proc. R. Soc. London, Ser. A*, **187**, 1 (1946).
12. Andersen, K. H., *Phys. Fluids*, **13**, 58–64 (2001).
13. Holmes, P., Lumley, J., and Berkooz, G., *Turbulence, Coherent Structures, Dynamical Systems and Symmetry*, Cambridge University Press, Cambridge, 1996.
14. Krug, J., *Adv. Compl. Systems*, **4**, 353–362 (2001).
15. Hellen, E. K. O., and Krug, J., *Phys. Rev. E*, **66**, 011304 (2002).
16. Abel, M., *Int. J. Bif. Chaos* (2002), accepted for publication; http://arxiv.org/abs/nlin.PS/0202058.
17. Hastie, T., and Tibshirani, R., *Generalized Additive Models*, Chapman and Hall, London, 1990.
18. W., H., *Applied Nonparametric Regression*, Cambridge University Press, Cambrige, 1990.
19. Fan, J., and Truong, Y. K., *Ann. Stat.*, **21**, 1900–1925 (1993).
20. Carroll, R. J., Maca, J. D., and Ruppert, D., *Biometrika*, **86**, 541–554 (1999).

Experimental characterization of space-time chaos in nonlinear optics

L. Pastur, U. Bortolozzo, P.L. Ramazza

Istituto Nazionale di Ottica Applicata, Largo E. Fermi 6, 50125 Firenze, Italy

Abstract. We report on the characterization of two different states of space-time chaos based on correlation functions and on Karhunen-Loève decomposition. The model system is an optical feedback loop closed through a nonlinear Kerr-like medium, in which different symmetries can be excited at threshold.

INTRODUCTION

We present an experimental model system for the study and comparison of various kinds of STC. The system consists in an optical feedback loop closed through a nonlinear medium. Different symmetries can be excited at the threshold of pattern formation — and therefore different variety of STC can be generated at higher pump level — by simply modifying the spatial features of the feedback pattern. To illustrate the possibilities of the system, we studied two different STC regimes by applying to each of them different characterizing tools, namely space- and time-correlation functions and the decomposition of pattern time-series on the base of empirical eigen-patterns, better known as the Karhunen-Loève decomposition.

THE MODEL SYSTEM

The experimental setup is an optical feedback loop closed through a nonlinear medium provided by a liquid crystal light valve (LCLV) (Fig.1). The input plane wave is generated by an Ar^+ laser operating at 514 nm. A couple of lenses in the feedback loop provides a one-to-one image of the LCLV front face onto a plane located at a distance l from a coherent fiber bundle that, in turn, relays its input intensity distribution to the back face of the LCLV. Over a range of feedback intensities of the order of 5 mW/cm^2, the LCLV acts as a Kerr medium: the phase retardation at the front side of the LCLV is proportional to the intensity impinging on its back side. In these conditions, if the input signal at the rear of the valve is a single spectral component signal $\gamma \sin(k_0 x)$, the device will typically produce in the transmitted wave higher spectral components in the form:

$$e^{i\gamma sin(k_0 x)} = \sum_n J_n(\gamma) e^{ink_0 x}, \tag{1}$$

CP676, *Experimental Chaos: 7th Experimental Chaos Conference,*
edited by V. In, L. Kocarev, T. L. Carroll, B. J. Gluckman, S. Boccaletti, and J. Kurths
© 2003 American Institute of Physics 0-7354-0145-4/03/$20.00

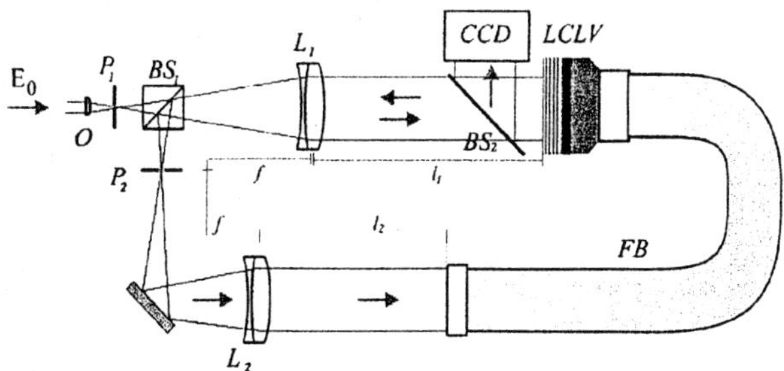

FIGURE 1. Experimental setup. An extended laser beam is closed through a non-linear Kerr-like medium (liquid crystal optical valve). Instabilities develop in the transverse plane of the beam. O: microscope objective; P_1, P_2: pinholes; BS_1, BS_2: beam splitters; LCLV: liquid crystal light valve; L_1, L_2: lenses of focal lens f; FB: fiber bundle; CCD: video camera. In our experiment, $2f - (l_1 + l_2) = -240$ mm is the effective free propagation length

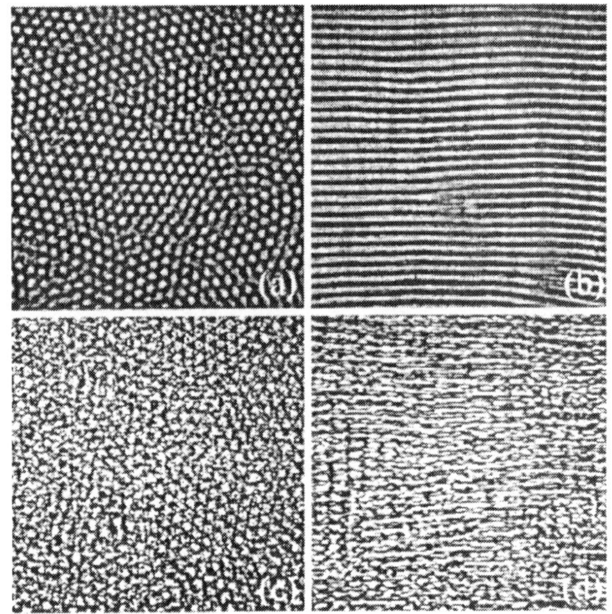

FIGURE 2. Hexagons and stripes at onset ((a) and (b) resp.), in the spatio-temporal regime ((c) and (d)).

where $J_n(\gamma)$ is the nth-order Bessel function. After propagation, the phase modulations transform into intensity distribution due to diffraction, and a negative feedback loop establishes for some spatial frequencies, resulting in pattern formation. The nonlinear phase to amplitude encoding expressed by (1) transfers energy from lower to higher scales, possibly resulting in the excitation of turbulent-like patterns.

Previous studies in this system have shown the occurrence of a rich variety of spatio-temporal patterns [1]. In some cases, characterization of the transition from hexagonal structures to space-time chaos has been given [2] . In this paper, we study the dependence on the pump level of the patterns observed in the cases of local or global (translated) feedback. We keep fixed the free propagation length at $l = -24$ cm, and the spatial frequency cutoff for the feedback field (given by the pinhole P_2) at a value allowing the existence of only the first unstable wavenumber among the possible set $q_n = 2\pi\sqrt{(n+1/2)/(\lambda_{Ar}l)}$ in the feedback signal (n odd and λ_{Ar} the light wavelength).

The control parameter that leads the transverse beam from the uniform state, to pattern formation at I_c (with a characteristic scale $\lambda \sim 0.28$ mm), to space-time chaos, is the incident light intensity I. When the feedback loop is closed and local, the natural pattern beyond I_c has an hexagonal symmetry (Fig.2.a). When the feedback is translated over a few λ along the x-axis, the hexagons disappear to the benefit of stripes that develop parallelly to x (Fig.2b).

Snapshots of the optical transverse plane at the front side of the LCLV are obtained by digitizing the output of a 768×576-pixels CCD video camera. At any value of the pump intensity, time series of $T = 30$ patterns $u(\mathbf{x},t)$ are realized at a sampling rate of 300 ms. The estimation of the dynamical indicators is performed on the fluctuations with respect to the time averaged pattern, $v(\mathbf{x},t) = u(\mathbf{x},t) - \langle u(\mathbf{x})\rangle_t$.

CORRELATION LENGTHS

For a proper description of space-time chaos, correlation functions must be determined both on space and time. The spatial correlation functions, defined as

$$C(\Delta\mathbf{x},t) = \int v(\mathbf{x},t)v(\mathbf{x}+\Delta\mathbf{x},t)d\mathbf{x}, \tag{2}$$

are actually computed by inverse Fourier transforming the spectral density of v, and then averaged over the time series. For hexagons, a further average over the azimuthal angle θ gives the correlation function depending only on the two-point distance Δr. For any experimental situation, we evaluate the *correlation length* ξ as the spatial shift Δx after which the envelope of the correlation function is 10 % lower than its maximum.

In Fig.3a is reported the evolution of ξ *vs.* the control parameter I, for the case of a local feedback. As expected the characteristic length decreases when the intensity increases, meaning that more spatial coherence is lost as the level of chaos is increased. Note that the maximum correlation length observed close to threshold ($\sim 6\lambda$) is roughly in agreement with the estimation one can have from a visual inspection of Fig.2a about the average domain size ($\sim 8\lambda$). As "domain" we mean here a region of space within which the hexagons have a uniform orientation. This result was not obvious *a priori* since the correlation function of a two-dimensional patterned signal is sensitive to local features, as discussed in [4]. In fact, a demodulation of the pattern before evaluating the correlation function would probably bring a better information about the domain size.

In the stripes shown in Fig.3b, the correlation length ξ_y perpendicularly to the stripes is always slightly greater than the correlation length ξ_x in the direction of the stripes, but

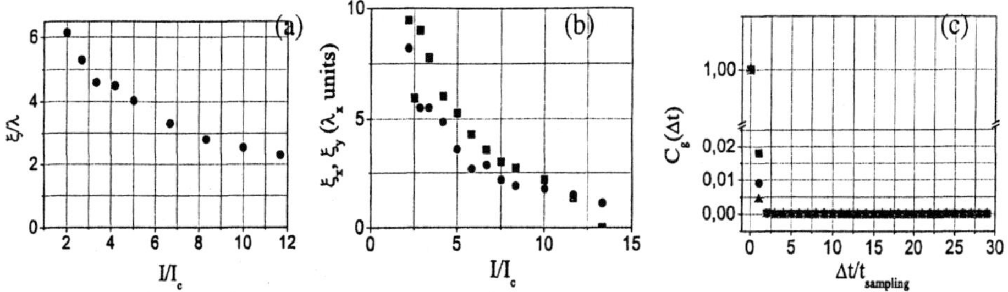

FIGURE 3. Spatial correlation length ξ vs. the light intensity I for Hexagons (a) and horizontal stripes (b); circles: perpendicularly to the rolls (ξ_y), squares: parallelly to the rolls (ξ_x). (c) Glogal time-correlation function at $I/I_c \sim 3$ (squares), $I/I_c \sim 6$ (circles), $I/I_c \sim 9$ (triangles), for stripes.

both decrease as I increases, confirming a loss of order in both direction as the level of chaos is increased.

The usual global time-correlation function is defined as:

$$C_g(\Delta t) = < \int v(\mathbf{x},t)v(\mathbf{x},t+\Delta t)\, d\mathbf{x} >_t, \qquad (3)$$

and gives an indication on how fast the pattern as a whole loses its correlation in time. The time-correlation τ was found to be lower than the sampling rate as soon as the STC threshold (Fig.3c)[1]. Note that the hardware parformances do not allow much lower sampling rates. This result however indicates that the pattern is actually non-stationary and its dynamics fastly decorrelated in time, which is characteristic of a spatio-temporal disorder. It is worthwhile here to note that a "local" time-correlation function can be defined as:

$$C_l(\Delta t) = < \int v(\mathbf{x},t)v(\mathbf{x},t+\Delta t)\, dt >_\mathbf{x}, \qquad (4)$$

that gives an information on how quickly a point in the system loses coherence in time. In the general case $C_g(\Delta t) \neq C_l(\Delta t)$. This point will be discussed elsewhere.

THE KARHUNEN-LOEVE DECOMPOSITION

The Karhunen-Loève (KL) decomposition is a long used technique in signal analysis and processing [3]. While usual techniques of decomposition deal with fixed basis functions (*e.g.*, cos(·) and sin() for the Fourier transform), this technique is aimed to extract from a pattern series the empirical eigen-pictures that best fit the statistical ensemble given by the data. By this technique, it is often possible to encode the information contained

---

[1] Note however that $C_g(\Delta t)$ seems to be sensitive to a contrast renormalization on v: slightly different results were obtained by stretching the contrast of each pattern between the two common extreme values $v_{min} = -128$ and $v_{max} = 127$.

in the pattern series within only a few modes of the KL basis. This "adaptive basis" procedure has found applications in the analysis of turbulent signals in fluids [5], and was later extended to several other physical situations, including the study of space-time chaos [6, 7, 8, 9].

Let us briefly recall how the Karhunen-Loève technique works. Given a time series of images $v(\mathbf{x},t)$, the aim is to find two sets of orthonormal functions $\phi_n(t)$, $\psi_n(\mathbf{x})$ such as

$$v(\mathbf{x},t) = \sum_{n\geq 1} \mu_n \phi_n(t)\psi_n(\mathbf{x}), \qquad (5)$$

where the functions $\{\psi_n(\mathbf{x})\}$ describe the data information content in the spatial domain, and are therefore sometimes called the "topos". Analogously, the $\{\phi_n(t)\}$ are named the "chronos" [7]. The set of function $\phi_n(t)$ are the solutions of the eigenvalue problem having as a kernel the time correlation matrix [8]:

$$K(t',t'') = \int v(\mathbf{x},t')v(\mathbf{x},t'')d\mathbf{x}. \qquad (6)$$

The $\{\psi_n\}$ are further determined by projecting each data picture on the $\{\phi_n\}$ and averaging on time:

$$\psi_n(\mathbf{x}) = \frac{1}{\mu_n} < v(\mathbf{x},t) \cdot \phi_n(t) >_t . \qquad (7)$$

The expansion coefficients μ_n in (5) result to be given by

$$|\mu_n|^2 = \lambda_n, \qquad (8)$$

the $\{\lambda_n\}$ being the eigen-values associated to $\{\phi_n, \psi_n\}$. Since the $\{\psi_n\}$ and $\{\phi_n\}$ are normalized, this amounts to say that the eigenvalue associated to each function represents its "energy contribution" to the data series.

Having obtained the KL spectrum, an approximated reconstruction of the pattern is obtained by truncating the expansion (5) up to the n_0^{th}-mode (Galerkin projection):

$$v(\mathbf{x},t) \simeq v_{n_0}(\mathbf{x},t) = \sum_{n=1}^{n_0} \mu_n \phi_n(t)\psi_n(\mathbf{x}), \qquad (9)$$

where $n_0 \leq T$.

A typical eigenvalue spectrum (usually called *singular spectrum* [9]) is illustrated on Fig.4a. For increasing intensity I, the spectrum tail level grows, while the energy contained in the first few modes decreases; see e.g. Fig. 4c, showing the fraction of the total energy contained in the first mode only. The spectrum becomes flatter and flatter, indicating an energy equipartition tendency over the whole spectrum as the level of chaos is increased. At that point, the Karhunen-Loeve decomposition technique may become of little utility, the statistical ensemble being not reducible to a few modes anymore. However, one can see that the number of modes involved for recovering 3/4 of the energy remains lower than 10 up to $I/I_c \sim 4$ (Fig.4b).

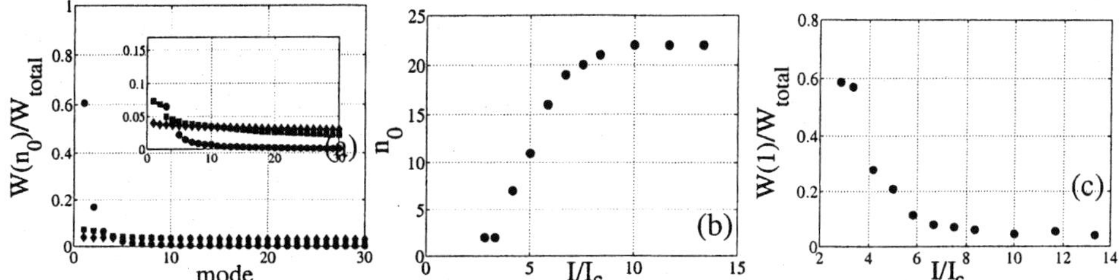

FIGURE 4. (a) Typical singular spectrum, from the Karhunen-Loeve decomposition of stripes at $I/I_c \sim$ 2 (circles), $I/I_c \sim 6$ (squares), $I/I_c \sim 9$ (diamonds). The tail level increases with I while the energy contained in the first mode decreases. (b) number of modes required for recovering at least 3/4 of the total energy. Averaged over 30 different times. (c) Fraction of the total energy contained in the first mode.

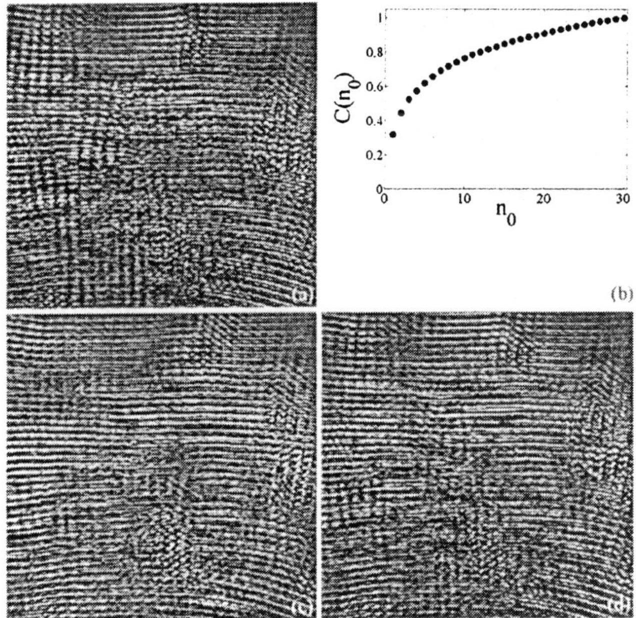

FIGURE 5. Fluctuation-pattern at the output of the LCLV at $I/I_c \sim 6$ (a). KL reconstruction using $n_0 = 1$ mode (c) and $n_0 = 15$ modes (d). (b) Correlation between the initial pattern and the reconstructed pattern *vs.* the number of modes involved for the reconstruction (averaged over 30 different times).

Another indicator we introduced to quantify the efficiency of the method is the average cross-correlation $C(n_0)$ between the input pattern v and the pattern v_{n_0} reconstructed using n_0 eigen-patterns:

$$C(n_0) = < \int v_{n_0}(\mathbf{x},t)v(\mathbf{x},t)d\mathbf{x} >_t \qquad (10)$$

From Fig. 5d it can be seen that the cross-correlation between the real pattern and its reconstruction using the only first eigen-pattern well inside the turbulent regime ($I/I_c \sim 6$) is still very good ($\sim 30\%$). On Fig.5 are reported the reconstructed patterns using $n_0 = 1$ mode (5b) and $n_0 = 15$ modes (5c), to be compared to the real pattern (5a), recovered with $n_0 = 30$.

CONCLUSION

We have studied two different states of STC, easily excited in an optical feedback loop closed through a nonlinear Kerr-like medium. The correlation functions are fastly decreasing both on space and time, revealing the spatio-temporal disorder in the chaotic regime. The Karhunen-Loeve decomposition revealed to be efficient for extracting the relevant information from a pattern time-series over a satisfying range of the control parameter. This last result is encouraging for the development of control algorithms based on the time-evolution of the few first KL modes amplitude.

REFERENCES

1. See e.g. P.L. Ramazza, S. Boccaletti, U. Bortolozzo, S, Ducci, E. Benkler and F.T. Arecchi, Phys. Rev. **E65**, 066204 (2002) and references therein.
2. R. Neubecker, B. Thüring, M. Kreuzer, T. Tschudi, *Chaos, Solitons & Fractals* **10** (1999) 681.
3. see for example *Fundamentals of Digital Image Processing*, A.K. Jain, Prentice-Hall International, Inc (19) p.163.
4. D.A. Egolf, I.V. Melnikov, E. Bodenschatz, *Phys. Rev. Let.* **80** (1998) 3228.
5. N. Aubry, P. Holmes, J.L. Lumley, E. Stone, *Physica D* **37** (1989) 1.
6. L. Sirovich, J.D. Rodriguez, *Phys. Lett. A* **120** (1987) 211; A.M. Fraser, *Physica D* **34** (1989) 391; L. Sirovich, *Physica D* **37** (1989) 126; S. Ciliberto, B. Nicolaenko, *Europhys. Lett.* **14** (1991) 303; S.M. Zoldi, J. Liu, K.M.S. Bajaj, H.S. Greenside, G. Ahlers, *Phys. Rev. E* **58** (1998) R6903; S.M. Zoldi, H.S. Greenside, *Phys. Rev. Lett.* **78** (1997) 1687.
7. M.P. Chauve, P. Le Gal, *Physica D* **58** (1992) 407.
8. L. Sirovich, *Quarterly of App. Math.* **45** (1987), 561.
9. R. Vautard, M. Ghil, *Physica D* **35** (1989) 395.

Fluctuations and Pinch-Offs Observed in Viscous Fingering

Mitchell G. Moore[*], Anne Juel[*†], John M. Burgess[*], W. D. McCormick[*]
and Harry L. Swinney[*]

[*]*Center for Nonlinear Dynamics and Department of Physics,
The University of Texas at Austin, Austin, Texas, 78712*
[†]*Present address: Dept. of Mathematics,
University of Manchester, Manchester M13 9PL, United Kingdom*

Abstract.
Our experiments on viscous (Saffman-Taylor) fingering in Hele-Shaw channels reveal several phenomena that were not observed in previous experiments. At low flow rates, growing fingers undergo width fluctuations that intermittently narrow the finger as they evolve. The magnitude of these fluctuations is proportional to $Ca^{-0.64}$, where Ca is the capillary number, which is proportional to the finger velocity. This relation holds for all aspect ratios studied up to the onset of tip instabilities. At higher flow rates, finger pinch-off and reconnection events are observed. These events appear to be caused by an interaction between the actively growing finger and suppressed fingers at the back of the channel. Both the fluctuation and pinch-off phenomena are robust but not explained by current theory.

Viscous fingering occurs when a less viscous fluid displaces a more viscous fluid in a Hele-Shaw channel (a quasi-2D geometry in which the width w is much greater than the channel thickness b); the interface between the fluids is unstable and forms a growing pattern of "fingers". A single finger forms at low flow rates; more complex branched patterns evolve at high flow rates. This phenomenon is the simplest example of the class of interfacial pattern forming systems which includes dendritic growth and flame propagation. Viscous fingering thus continues to receive attention for the insight it provides into these important problems [1, 2, 3, 4].

Saffman and Taylor first studied the problem in 1958 [5] by injecting air into oil in a Hele-Shaw cell. They observed the formation of a single, steadily moving finger whose width decreased monotonically to $1/2$ of the channel width as the finger speed was increased. Subsequent experimental [6], numerical [7], and theoretical [8, 9] work found that the ratio of finger width to channel width, λ depended on a modified capillary number, $1/B = 12\,(w/b)^2\,Ca$, which combines the aspect ratio, w/b, and the capillary number, $Ca = \mu V/\sigma$, where μ is the dynamic viscosity of the liquid, V is the velocity of the tip of the finger, and σ is the surface tension. A transition to complex patterns of tip-splitting occurs at large $1/B$ values [6, 10, 11, 12]. Our experiments have revealed two phenomena that were not reported in prior experiments [5, 6, 10, 11] or predicted theoretically: fluctuations in the width of the evolving viscous fingers [13], and finger pinch-off events.

CP676, *Experimental Chaos: 7th Experimental Chaos Conference,*
edited by V. In, L. Kocarev, T. L. Carroll, B. J. Gluckman, S. Boccaletti, and J. Kurths
© 2003 American Institute of Physics 0-7354-0145-4/03/$20.00

EXPERIMENTAL METHODS AND DATA ANALYSIS

We conducted experiments in a 254 cm long channel formed from 1.9 cm thick glass plates. The spacing between the glass plates was set by stainless steel strips with thicknesses $b = 0.051$ cm, 0.064 cm, 0.102 cm, or 0.127 cm; the channel width w between the spacers was varied between 19.9 cm and 25.1 cm. Both glass plates were supported and clamped at the sides, and the channel was illuminated from below. For aspect ratios under 150, we used a smaller channel of length 102 cm and width 7.4 cm. Interferometric measurements revealed that the root-mean-square variations in gap thickness were typically 0.6% or less in the large channel and 0.8% in the small channel. Mechanical measurements of the bending of the glass due to the imposed pressure gradient revealed that such deflections were typically 0.2% or less; the maximum deflection (2.2%) was measured in the widest channel close to the oil reservoir at the highest flow rates. Experiments were conducted with air penetrating a Dow Corning silicone oil whose surface tension and dynamic viscosity was either $\sigma = 19.6$ dyne/cm, $\mu = 9.21$ cP or $\sigma = 20.6$ dyne/cm, $\mu = 50.8$ cP at laboratory temperature (22°C). The oils wet the glass completely. We withdrew oil at a uniform rate using a syringe pump attached to a reservoir at one end of the channel; an air reservoir at atmospheric pressure was attached to the other end. The entire experiment was placed on a floating optical table to minimize vibrations and allow precise levelling of the channel.

We obtained images of up to $1200 \times 10,000$ pixels at a resolution of 0.25 mm/pixel using a camera and a rotating mirror. The camera captured up to 11 overlapping frames which were then concatenated, background subtracted, and corrected for perspective effects. The interfaces were then digitally traced, yielding finger width values accurate to 0.1% in the larger channel and 0.3% in the smaller channel. For each flow rate, up to four time sequences of 20-30 digital interfaces were recorded. Finger widths determined in consecutive sequences agreed within the measurement accuracy. Mean width values agreed within 0.5% for data sets repeated after channel disassembly, cleaning, and reassembly. For each experimental run, we determined the time average $\langle \lambda \rangle$ and the root-mean-square (rms) fluctuation from the mean δ_λ (as described in [13]). Each data set was analyzed for flow rates up to the point of tip splitting, beyond which the finger width λ was no longer well defined [1].

FLUCTUATIONS

Typical interface image sequences are shown for $w/b = 158$ and 490 in Fig. 1. The finger width λ (relative to the channel width w) fluctuates visibly at low flow velocities for both aspect ratios (Fig. 1 (a),(c)). In the smaller aspect ratio system the width appears to become steady as the finger velocity is increased (Fig. 1(b)), appearing exactly like the classic "half-width finger" of Saffman and Taylor. However, with sufficient resolution,

---

[1] The first tip instabilities observed with increasing $1/B$ are asymmetric "shouldering" modes, not actual tip splitting [1, 2, 3, 4, 6]; because the first instabilities are not obviously distinguishable from fluctuations, our data at high $1/B$ include $\langle \lambda \rangle$ values averaged over such instabilities.

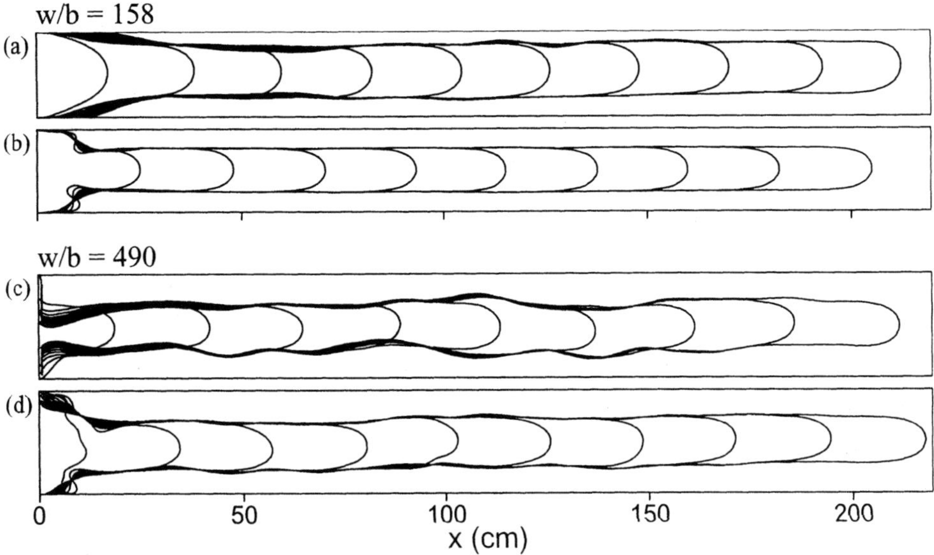

FIGURE 1. Finger images recorded at regular time intervals for different values of aspect ratio w/b and modified capillary number $1/B$. Experimental values for each run: (a) $1/B = 116$, the rms fluctuation of the finger width $\delta_\lambda = 1.30 \times 10^{-2}$, time between successive curves $\Delta t = 270$; (b) $1/B = 1757$, $\delta_\lambda = 3.96 \times 10^{-3}$, $\Delta t = 18$; (c) $1/B = 408$, $\delta_\lambda = 3.32 \times 10^{-2}$, $\Delta t = 810$; (d) $1/B = 2869$, $\delta_\lambda = 1.26 \times 10^{-2}$, $\Delta t = 108$.

fluctuations can still be measured for all velocities up to the onset of tip instabilities. In the higher aspect ratio system the width fluctuates visibly for all flow rates (Fig. 1(d)). We find that the rms fluctuation of the finger widths (relative to the channel width w) is described by $\delta_\lambda = A(\mathrm{Ca})^\beta$ with $A = (1.1 \pm 0.3) \times 10^{-4}$ and $\beta = -0.64 \pm 0.04$, for all the aspect ratios studied (Fig. 2).

The fluctuations in finger width are accompanied by a substantial deviation from the expected relation between finger width and velocity [13]. For low aspect ratios, our finger width measurements are in accord with previous results, but for aspect ratios $w/b \gtrsim 250$, which were not examined in previous work, we find that the mean finger width exhibits a surprising maximum as the tip velocity is decreased. However, the maximum value of the fluctuating finger width observed during finger evolution, $\lambda_{\max}$, still exhibits the classical scaling with $1/B$ for all aspect ratios. This suggests that the fluctuations represent an intermittent narrowing of the fingers from their "ideal" width [13].

The finger width fluctuations and the peak in $\langle \lambda \rangle$ versus $1/B$ have proven robust under many variations of experimental conditions, allowing us to rule out many possible explanations for the phenomena. We obtained identical results with differences in the oil viscosity, the pinning properties at the back of the channel, the pumping method, and the magnitude of the channel's intrinsic gap variations [13]. An instability of the film wetting layer is also unlikely, because the film is very thin at low capillary numbers, where

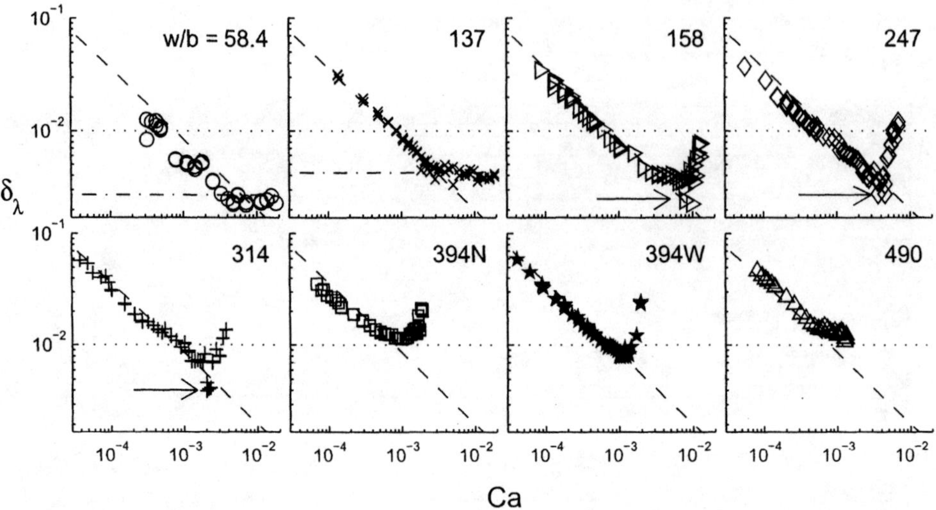

FIGURE 2. The rms fluctuation of the relative finger width δ_λ as a function of capillary number Ca, where the dashed lines describe the best fit to all of the data sets within the observed scaling region, $\delta_\lambda = (1.1 \times 10^{-4})\text{Ca}^{-0.64}$. (The data for 394N and 394W have the same aspect ratio but different widths; the channel for 394N was 20.0 cm wide, 394W was 25.0 cm.) The first two graphs are for data taken in the smaller channel; the horizontal dash-dotted lines correspond to the limits of measurement accuracy for that channel. Fluctuations with $\delta_\lambda \lesssim 10^{-2}$ are not obvious visually, which may explain why previous experiments at lower aspect ratios did not observe them. An upturn in δ_λ occurs at high Ca, signalling the onset of the secondary instabilities in the tips of the fingers. The arrows in the graphs for $w/b = 158$, 247, and 314 point to data from runs where pinch-off events (see fig. 3) occurred, demonstrating the delay of the onset of secondary instabilities in these cases.

the fluctuations are largest. Though the fluctuation power law we observed remained unchanged for *all* experimental variations, we did discover that measurements on two channels with $w/b = 394$ but different values of w and b yielded slightly different results [13]. This difference suggests that either the finger width for a given w/b and Ca is not unique, or perhaps a third parameter, as yet unknown, is necessary to fully describe the problem.

PINCH-OFF EVENTS

In addition to finger width fluctuations, we have observed another unexpected phenomenon: finger pinch-off events followed by reconnection to a different finger, an example of which is shown in fig. 3. For sufficiently high flow rates, several fingers compete during the early stages until one gets ahead, suppressing the growth of the other initial fingers. However, if the initial competition stage was long, the suppressed fingers will continue to interact with the active finger. As the active finger continues to grow, it narrows at the back, and the adjacent short finger grows towards it. This process is

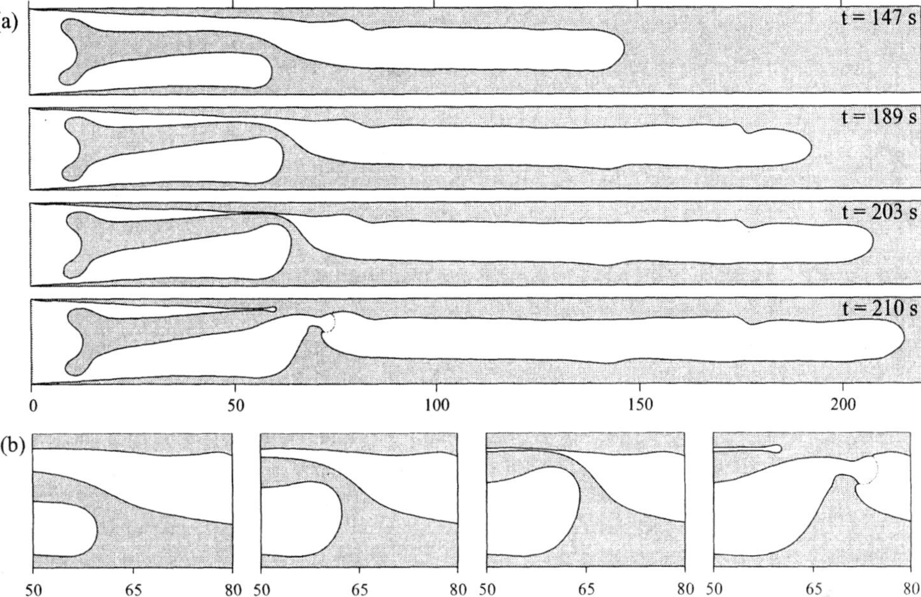

FIGURE 3. (a) A time sequence from a run where a pinch-off event occurred; the times since the beginning of the pumping are shown. The back of the growing finger narrows while the adjacent suppressed finger grows. The process accelerates as it continues; the pinch-off and subsequent reconnection happens in less than two seconds (which is faster than the time resolution we had for this data). The gray line in the last interface shows where the interface between the middle finger and the disconnected bubble finally broke; this point could be clearly seen on the image as a thick film left behind at that point. (b) A close up view of the pinch-off at the same times. (The distance scales shown are in cm.)

very slow at first and gradually accelerates; the final pinch-off and the connection of the previously suppressed finger to the resulting bubble occurs very quickly.

The dynamics of the finger tip appear unaffected during these events, though the onset of secondary instabilities is apparently slightly suppressed. This can be seen in fig. 2; all of the data points indicated by arrows had pinch-off events, and they continue to follow the fluctuation power-law when other runs at similar flow rates without pinch-off events depart that curve, beginning the upturn that marks the onset of the secondary instabilities. At higher flow rates, pinch-offs at the back of fingers and tip instabilities both occur, apparently independently of one another.

Pinch-offs may or may not occur on a particular run at a given flow rate, since this depends on the configuration of the initial finger competition, which can be different each time. In particular, the levelling of the cells is crucial: a small tilt around the long axis will give fingers on the higher side of the cell an early advantage; the shorter dormant fingers resulting from this do not interact as quickly.

Pinch-off events are not mentioned in any of the previous literature; we believe that we observe it because our cell has an unusually high ratio of length to width, allowing sufficient time for evolution towards the pinch-off to occur before the finger reaches the end of the channel. Also, seemingly unimportant asymmetries across the width of the cell, like the levelling mentioned above, can disrupt the long period of finger competition and thus suppress the pinch-off events. We have observed the pinch-off phenomenon in different cells at both high and low aspect ratios. These events can occur at flow rates where the plate bending due to the imposed pressure gradient is unmeasurably small. The cause of this phenomenon is still unknown.

CONCLUSIONS

In conclusion, we have discovered fluctuations that intermittently narrow evolving single fingers; the magnitude of these fluctuations follows a power law with the capillary number for all aspect ratios studied. This is accompanied by a departure from the classic scaling of finger width versus $1/B$ for large aspect ratios ($w/b \gtrsim 250$). In addition, we observe finger pinch-off and reconnection events at the back of the channel; these are apparently due to an interaction between adjacent fingers that has not yet been explored. Neither the fluctuation nor the pinch-off phenomena are predicted by existing viscous fingering theories, nor have they been observed in simulations.

ACKNOWLEDGMENTS

We thank J. B. Swift for frequent guidance and advice. This work was funded by the Office of Naval Research.

REFERENCES

1. Bensimon, D., Kadanoff, L. P., Liang, S., Shraiman, B. I., and Tang, C., *Reviews of Modern Physics*, **58**, 977–999 (1986).
2. Homsy, G. M., *Annual Review of Fluid Mechanics*, **19**, 271–311 (1987).
3. Pelcé, P., *Dynamics of Curved Fronts*, Academic Press, San Diego, CA, 1988.
4. Couder, Y., "Viscous Fingering as an Archetype for Growth Patterns," in *Perspectives in Fluid Dynamics*, edited by G. K. Batchelor, H. K. Moffatt, and M. G. Worster, Cambridge University Press, 2000.
5. Saffman, P. G., and Taylor, S. G., *Proceedings of the Royal Society*, **A245**, 312–329 (1958).
6. Tabeling, P., Zocchi, G., and Libchaber, A., *Journal of Fluid Mechanics*, **177**, 67–82 (1987).
7. DeGregoria, A. J., and Schwartz, L. W., *Journal of Fluid Mechanics*, **164**, 383 (1986).
8. McLean, J. W., and Saffman, P. G., *Journal of Fluid Mechanics*, **102**, 455–469 (1981).
9. McLean, J. W., *The Fingering Problem in Flow Through Porous Media*, Ph.D. thesis, California Institute of Technology (1980).
10. Park, C. W., and Homsy, G. M., *Physics of Fluids*, **28**, 1583–1585 (1985).
11. Kopf-Sill, A. R., and Homsy, G. M., *Physics of Fluids*, **31**, 242–249 (1988).
12. Maxworthy, T., *Journal of Fluid Mechanics*, **177**, 207–232 (1987).
13. Moore, M. G., Juel, A., Burgess, J. M., McCormick, W. D., and Swinney, H. L., *Physical Review E*, **65**, 030601 (2002).

Lagrangian Chaos: Transport, Coupling and Phase Separation

Thomas H. Solomon, Nathan S. Miller, Courtney J. L. Spohn, and Jeffrey P. Moeur

Bucknell University
Lewisburg, PA 17837 USA

Abstract. We present results of experimental and numerical studies of Lagrangian chaos (chaotic advection) in an oscillating vortex chain. We first review previous experiments on enhanced diffusive and superdiffusive transport in this flow. Superdiffusion occurs only as a transient and only at certain well-defined resonant frequencies. At these frequencies, long "flights" are found numerically in the trajectories with lengths with scaling similar to that of Lévy flights. We then discuss on-going experiments of the effects of Lagrangian chaos on more complicated dynamical processes occurring in this flow. Specifically, we study the dynamics of coupled Belousov-Zhabotinsky chemical oscillators in the oscillating vortex chain. We analyze the system as a network with each vortex and its contents forming the nodes of the network. Coupling between the nodes then is via Lagrangian chaos. With Lagrangian chaos, there is either nearest neighbor coupling (enhanced diffusion) or long-range coupling (superdiffusion and Lévy flights) which could make this system an experimentally-realizeable "small world" network. Three different modes of synchronization are found: unsynchronized oscillations, traveling waves, global "blinking" states, and a state in which lobes responsible for enhanced transport mediate the synchronization.

INTRODUCTION

It is well known that simple, laminar fluid flows can have surprisingly complicated mixing properties. Even if the velocity field itself is very well-ordered and deterministic, the trajectories of fluid elements or impurities in the flow may follow chaotic trajectories in the sense that nearby trajectories will separate exponentially in time [1,2]. There have been several studies of the effects of Lagrangian chaos on the mixing of passive impurities (e.g., miscible dyes). An important outstanding question, however, is how the presence of Lagrangian chaos affects *dynamical* processes occurring in a fluid flow.

In this article, we present the results of experimental and numerical studies of Lagrangian chaos in a simple two-dimensional flow composed of a chain of alternating vortices that sways periodically in the lateral direction. We start by discussing the nature of the chaotic mixing in this oscillating vortex chain and how this chaotic behavior affects long-range transport. We then discuss on-going experiments on coupled chaotic oscillators in this fluid flow and how Lagrangian chaos affects the global dynamics and synchronization of these coupled oscillators.

CP676, *Experimental Chaos: 7ᵗʰ Experimental Chaos Conference,*
edited by V. In, L. Kocarev, T. L. Carroll, B. J. Gluckman, S. Boccaletti, and J. Kurths
© 2003 American Institute of Physics 0-7354-0145-4/03/$20.00

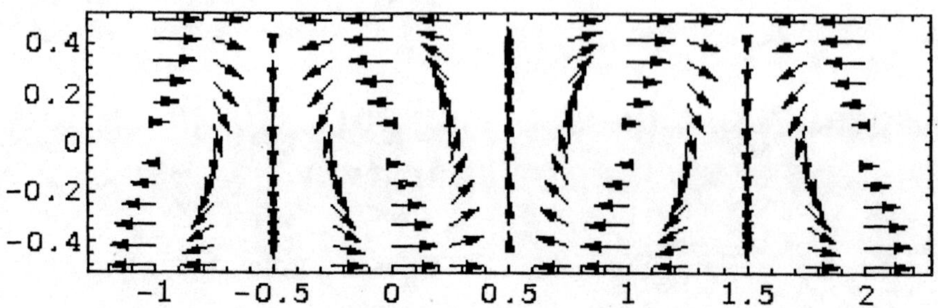

FIGURE 1. Alternating vortex chain for case with free-slip boundary conditions. The no-slip case would look similar, except that the velocities would vanish at the top and bottom boundaries. Time dependence is manifested as lateral oscillation of the entire velocity field with an amplitude B and an angular frequency ω.

MODEL FLOW AND EXPERIMENTAL TECHNIQUES

The flow studied is a chain of alternating vortices, as shown in Fig.1. This system was originally proposed as a two-dimensional cross-section of oscillatory (time-periodic) Rayleigh-Bénard convection [3-5]. In fact, experiments on time-periodic Rayleigh-Bénard convection demonstrated the presence of Lagrangian chaos [4]. Numerically, a simple model was proposed for this flow, based on the following streamfunction [6]:

$$\psi(x,y,t) = \frac{A}{k}\sin\{k[x + B\sin\omega t]\}W(y).\tag{1}$$

The term $B\sin\omega t$ denotes the lateral oscillation of the vortex chain. We will scale the oscillation amplitude by the vortex width w; from hereon, we will denote the amplitude by this non-dimensional $b \equiv B/w$. For a flow with stress-free boundary conditions,

$$W(y) = \sin(\frac{\pi y}{d}),\tag{2}$$

and for a flow with rigid (no-slip) boundary conditions,

$$W(y) = \cos(q_0 y) - A_1\cosh(q_1 y)\cos(q_2 y) + A_2\sinh(q_1 y)\sin(q_2 y),\tag{3}$$

where q_o = 3.973639, q_1 = 5.195214, q_2 = 2.126096, A_1 = 0.06151664 and A_2 = 0.103887 [7]. The velocity field can be obtained by applying Hamilton's equations of motion to this streamfunction:

$$\dot{x} = \partial\psi/\partial y; \quad \dot{y} = -\partial\psi/dx\,.\tag{4}$$

Numerical simulations [4,5] of particle trajectories in this flow demonstrate Lagrangian chaos denoted by trajectories that separate exponentially in time.

We have recently assembled an experimental configuration that generates this flow using a magnetohydrodynamic technique [8,9], as shown in Fig. 2. An electrical current passing through a thin (2 mm) electrolytic solution (salt water or dilute sulfuric acid) interacts with an alternating magnetic field produced by a chain of Nd-Fe-Bo magnets underneath the fluid layer. Lorentz magnetic forces resulting from this interaction produces horizontal forcing in the fluid layer which, in combination with the plexiglass side walls, produces the alternating chain of vortices. Time dependence is imposed externally with a plunger that oscillates slowly up and down, displacing the fluid laterally across the fixed pattern of vortices.[1] This system has an advantage experimentally over time-periodic Rayleigh-Bénard convection in that it allows independent control of both the amplitude and frequency of the lateral oscillations.

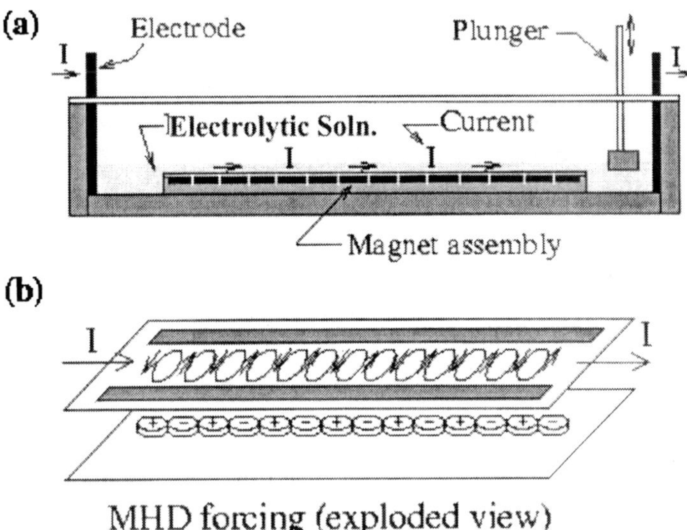

FIGURE 2. Diagram of experimental apparatus. (a) Side view; (b) exploded view showing magnetohydrodynamic (MHD) forcing technique.

For transport measurements, fluorescent dye is injected into the center vortex, and the system is illuminated with black light. The entire apparatus is imaged from above with a CCD video camera, and the images are digitized and analyzed on a PC-compatible computer.

---

[1] The period of oscillation is chosen to be several times the viscous diffusion time (~4 s) for the fluid layer. This is important to avoid "frozen" oscillations of the vortices which would be different kinematically from the model in Equations (1) – (3).

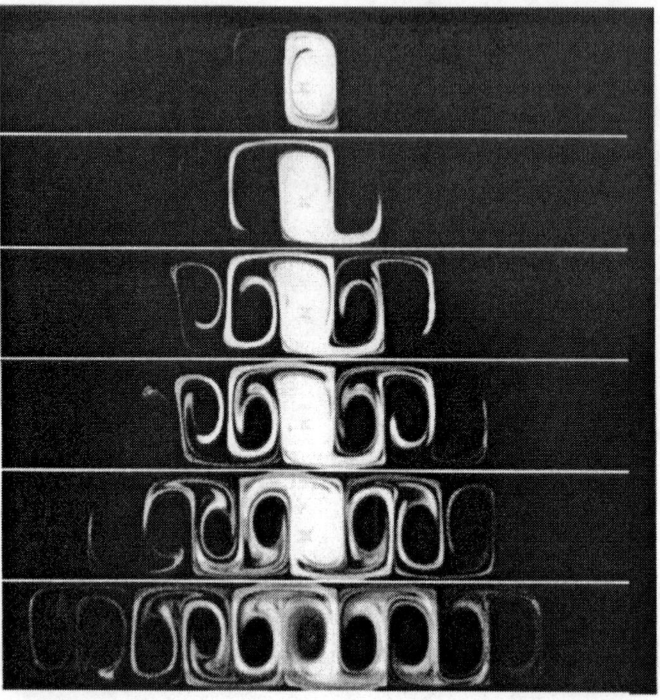

FIGURE 3. Sequence showing the mixing of dye due to Lagrangian chaos in the oscillating vortex chain; amplitude of oscillation $b = 0.12$; oscillation period = 19 s; time (from the top): 0, 1, 2, 3, 4 and 10 periods of oscillation.

ENHANCED TRANSPORT OF PASSIVE TRACERS

A sequence of images showing transport of a miscible dye is shown in Fig. 3. Several important features are visible: (a) A significant amount of stretching and folding is seen, typical of chaotic processes. (b) "Lobes" of dye (and clear fluid) are exchanged between adjacent vortices in the chain. This is the primary mechanism for mixing between the vortices [9,10]. (c) Ordered regions are visible in the vortex centers in the later images. This is typical of Lagrangian chaos (and chaotic systems in general): the phase space – real space in this case – is divided into ordered and chaotic regions, and any tracers initially in one of these regions will not cross into the other. The boundaries between the ordered and chaotic regions are therefore transport barriers.[2]

---

[2] A hole develops in the center vortex as well, even though dye that is initially within an ordered region should never leave. The cause of this hole is a weak, three-dimensional secondary flow that carries fluid up through the vortex centers. In this experiment, the dye that was injected into the center vortex was initially primarily near the top of the fluid layer, so the secondary flow circulated clear fluid up from below.

Quantitatively, a complete description of this process requires solution to the advection-diffusion equation:

$$\partial c / \partial t = D\nabla^2 c - \vec{u} \cdot \vec{\nabla} c, \tag{5}$$

where $c(x,y,t)$ is the concentration field, $\vec{u}$ is the velocity field describing the oscillating vortex chain, and D is the molecular diffusion coefficient. Equation (5) cannot be solved in general, so other techniques must be used to analyze transport in the presence of fluid flows in general, and in situations with Lagrangian chaos in particular.

In advection-diffusion processes, it is common to consider the long-time growth of the variance of the concentration field and compare this growth to a power law of the form:

$$<r^2>\sim t^\gamma. \tag{6}$$

If $\gamma = 1$, then the process is termed "normal" (although enhanced) diffusion, in which case Equation (5) is approximated by the effective diffusion equation:

$$\partial c / \partial t = D^*\nabla^2 c, \tag{7}$$

where D^* is the effective diffusion coefficient. Previous experiments [9,11] have shown that at many typical frequencies in this system, the transport can be approximated quite well by normal diffusion with an effective diffusion coefficient that grows linearly with the oscillation amplitude B for small B.

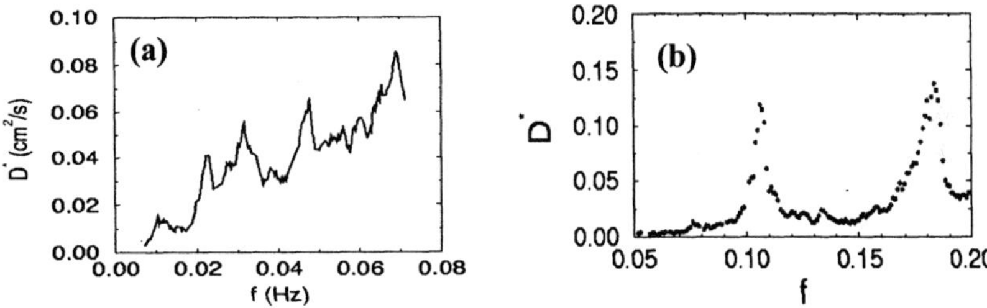

FIGURE 4. Simulations of the frequency dependence of the long-time limit of the enhanced diffusion coefficient D^*. (a) No-slip boundary conditions; (b) free-slip boundary conditions.

The dependence of D^* on the oscillation frequency $f = \omega/2\pi$ is significantly more complicated. Numerically [12], the dependence is dominated by well-defined resonance peaks, as seen in Fig. 4. (Experimentally, it is very difficult to map out the frequency dependence because of the extreme sensitivity to small changes in f.) Furthermore, for the frequencies near the resonance peaks, there is a transient period of superdiffusion where the variance grows faster than linearly – $\gamma > 1$ – although the long-time behavior is still diffusive ($\gamma = 1$). The duration of the transient gets longer and longer the closer the frequency is to one of the resonant frequencies; in fact, it has been shown theoretically [13] that the *long-time* behavior is superdiffusive precisely at these resonant frequencies.

The explanation for these resonance peaks is found by considering the motion of individual tracers in the flow. For most frequencies, Lagrangian chaos results in particle trajectories that have the appearance of random, Brownian-like (diffusive) motion, as seen in Fig. 5(a) for the case with $f = 0.095$ (lower curve). For frequencies near the resonance peaks, however, the trajectories include long "flights" (jumps) – some as long as tens or even hundreds of vortex widths – as seen in Fig. 5(a) for the $f = 0.106$ (upper curve). Quantitatively, these jumps can be described by a Lévy distribution [14,15]

$$p(L) \sim L^{-\mu}, \qquad (8)$$

with $2 < \mu < 3$, as seen in Fig. 5(b). This power law behavior is valid up to a cut-off length that grows (and presumably diverges) as the frequency nears the resonant value.

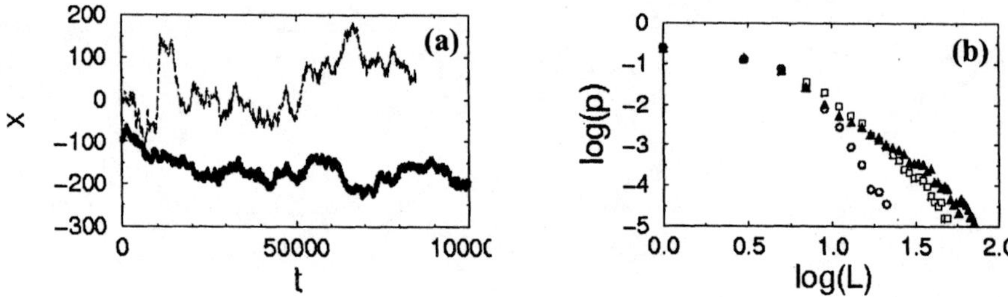

FIGURE 5. Evidence of Lévy flights in the free-slip simulations. (a) Sample trajecotories. The lower curve corresponds to a frequency $f = 0.095$, away from the resonance peaks. The upper curve corresponds to $f = 0.106$, near one of the resonance peaks. (b) Probability distribution functions for the flight lengths L. The frequencies are $f = 0.106$ (filled triangles), 0.103 (open squares), and 0.095 (open circles).

It has been well-established [16-20] that if tracers follow trajectories with flights that scale as in Equation (8) that the transport will be superdiffusive. In this situation, the transport is superdiffusive only for a transient period, consistent with the fact that the flights follow a Lévy distribution only up to a maximum length. The cause of the (truncated) Lévy flights is a resonance between the period of oscillation and typical times for fluid elements to circulate within a vortex, as shown in the cartoon in Fig. 6. At these resonant frequencies, a tracer that has crossed from one vortex to another loops around and finds itself at the opposite corner just at the right time to cross over again. The result is a snake-like trajectory which crosses several vortices in a short period of time. Another way of explaining these flights is with the use of "tangle islands" [21] that appear within the chaotic regions (inside the lobes) for frequencies near the resonant ones [12].

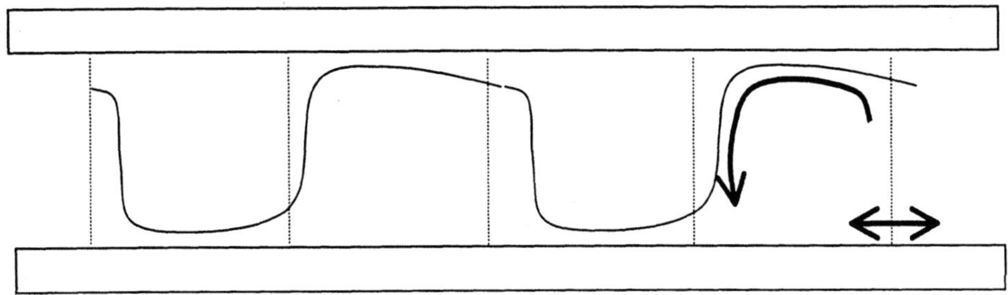

FIGURE 6. Cartoon showing resonance between circulation times and oscillation periods, responsible for the flights.

We have, then, a system where the transport can be either diffusive or (transiently) superdiffusive, depending on the frequency of oscillation.

LAGRANGIAN CHAOS AND COUPLED CHEMICAL OSCILLATORS

Up to this point, we have been discussing kinematical issues; namely, given a flow and some impurities, we have discussed how the impurities move in the flow. But Lagrangian chaos can have a significant effect on *dynamical* processes as well. Different scenarios are possible: (a) The velocity field itself can be partially advected with the flow (although it is more common to discuss advection of vorticity, which is derived from the velocity field), in which case the transport can feed back onto the flow itself. (b) There could be interactions between the various impurity molecules causing deviations from the advection-diffusion equation (Eq. 5). For instance, if the impurity and the surrounding fluid are immiscible, surface tension can inhibit the amount by which the impurity can be stretched by the flow. (c) The "concentration" field might be an active, reacting field, rather than a passive field that is merely advected. This last situation classifies as an advection-diffusion-reaction problem, which can be described by the equation

$$\partial c / \partial t = D\nabla^2 c - \vec{u} \cdot \vec{\nabla} c + f(c),$$ (9)

where *f(c)* is some function denoting reaction of the impurity either with itself of with some additional chemicals in the system.

We have initiated experimental studies of advection-diffusion-reaction processes in the oscillating vortex flow. Specifically, we are looking at dynamics of the Belousov-Zhabotinsky (BZ) chemical reaction [22-25] in the flow and how the global behavior of the system is affected by Lagrangian chaos.

The procedure is quite straightforward: the apparatus is filled initially with chemicals typically used to demonstrate the BZ reaction in its oscillatory state[3]. Using ferroin as an indicator, the reaction oscillates between a red and blue color. The typical oscillation period for this mixture in a well-stirred batch reactor is about 40-50 seconds. The CCD video camera used for the imaging is fitted with a blue filter; consequently, in all the images, the blue phase appears white while the red phases appear black.

FIGURE 7. Snapshots of network behavior. In all cases, white (dark) regions correspond to the blue (red) phase of the BZ reaction. (a) Time-independent flow; no Lagrangian chaos. The chemical reactions in the vortices oscillate independently of each other. (b) Time-periodic forcing, small amplitude (b = 0.1), resonant frequency for lateral oscillations (f = 0.050 Hz). Lobe structures are clearly seen marked by the red (dark) phase of the BZ reaction.

The behavior of the network in the absence of any time dependence is shown in Figs. 7a and 8a. Since there is no Lagrangian chaos for the time-independent flow, the only communication between vortices is via molecular diffusion, a very weak mixing mechanism. Initially (before the starting time in Fig. 8a), the chemical oscillations in the different vortices are in phase since the system is initially well mixed (from pouring the fluid into the system). But after a moderate amount of time, the oscillations drift with respect to each other, as seen both in snapshots (Fig. 7a) and in a space-time plot (Fig. 8a). This drifting can be used to determine a characteristic "coherence time" τ_c, which is on the order of 400 s.

---

[3] The solution used in the experiments is composed of 400 ml of 1M sulfuric acid, 11.44 g of malonic acid, 4.18 g of potassium bromate, 0.44 g of cerium ammonium nitrate and 4 ml of 0.025 M ferroin.

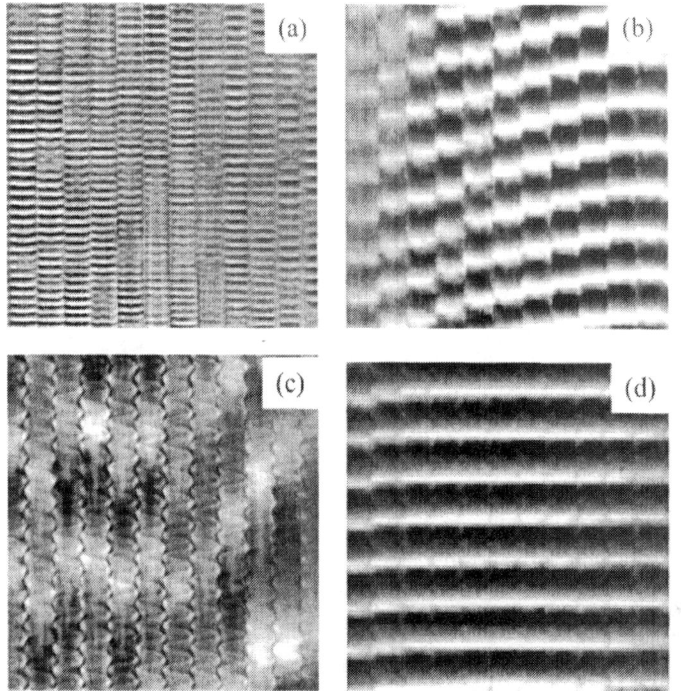

FIGURE 8. Space-time plots showing the dynamics of the networks; the horizontal direction is the distance across the vortex chain, and the vertical direction is time. (a) Time-independent flow (same conditions as Fig. 2a). (b) Time-periodic flow; small amplitude (b = 0.1), f = 0.037 Hz. (c) Time-periodic flow; small amplitude (b = 0.1), f = 0.050 Hz. (d) Time-periodic flow; larger amplitude (b = 0.2), f = 0.050 Hz. Total time spanned is 2500 s for (a) and 300 s for the other three.

For studies of the dynamics with time-periodic (lateral) oscillations, two different initial conditions are used: (1) Initially-synchronized: the entire apparatus is swept, mixing the contents of the entire vortex chain and starting off all the vortices with the same phase. (2) Initially unsynchronized: the oscillations are allowed to proceed for the time-independent flow until the oscillations are out of phase, after which the time-dependent forcing is turned on.

At most frequencies studied, if the oscillations start out synchronized, they drift in phase and become unsynchronized within 20 periods of oscillation. If the pattern starts out unsynchronized, it tends to remain that way at these frequencies. In many cases, patterns will appear within vortices that are clear ghosts of the lobe structure seen in the passive transport. (Compare Figure 7b with Figure 3.) In studies of transport of dye in this flow, the lobes are only visible for a short period of time; basically, as long as there are strong concentration gradients. For the network of coupled chemical oscillations presented here, though, it is quite remarkable that the lobes remain visible indefinitely as *all* the lobes are tagged (and remain tagged) with the same chemical phase. Figure 8c shows a space-time plot for the same conditions as Fig. 7b; the lobes can be seen in this plot as dark oscillatory structures within the vortices but synchronized throughout the chain.

At a couple of well-defined frequencies, the system synchronizes forming global traveling waves, as seen in Figure 8(b). These traveling waves appear regardless of whether the system starts as initially synchronized or unsynchronized. At these frequencies, traveling waves are observed propagating left-to-right, right-to-left, outward from a source in the middle of the vortex chain and inward toward the middle. And in a couple of the experimental runs, the system remained globally-synchronized (each vortex oscillating almost perfectly in phase) for 60 oscillation periods, as seen in Fig. 8(d).

DISCUSSION

In analyzing the results of the advection-reaction-diffusion experiments described in the previous section, it is necessary to consider transport effects, since coupling between the vortices is determined by transport of the contents between vortices. As discussed earlier in this article, at most frequencies transport in the vortex chain is diffusive (although enhanced). In the vicinity of certain well-defined frequencies, though, the transport is characterized by transient superdiffusion and flights. If we think of our system as a network, then comparisons can be made between regular and "small world" networks discussed in recent studies [26]. The main ideas are illustrated in Figure 9 below:

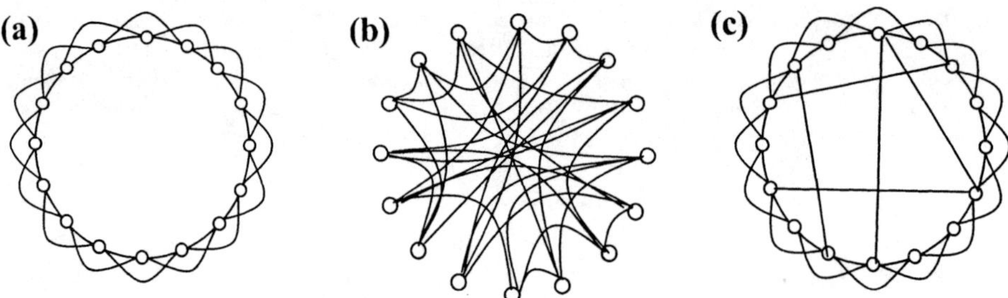

FIGURE 9. Coupling scenarios for networks. (a) Regular, nearest-neighbor coupling. (b) Random coupling. (c) "Small-world" coupling of Watts and Strogatz.

Figure 9(a) shows regular, nearest-neighbor coupling where each node is connected to one or two of its nearest neighbors but does not communicate with any other node directly. This type of coupling leads to "clustering" where local groups of nodes tend to have similar behavior. By contrast, in a randomly-coupled network (Fig. 9b), clustering is not observed; however, this type of network has a property referred to as the "Small World Effect" whereby the number of jumps needed – on the average – to connect any two arbitrary nodes grows logarithmically with the total number of nodes (rather than growing linearly). A third type of network was proposed recently by Watts and Strogatz [26] which captures both clustering and the small-world effect with a mostly-regular network with a few random "short-cut" connections.

Much has been said about these types of network connections, and these classifications have been applied to a wide variety of network systems. But little is known about the effects of these different coupling schemes on the *dynamics* of processes occurring on these networks. The experiments presented here on coupled chemical oscillations in an oscillating vortex chain might provide an experimental system whereby the effects of coupling schemes on the dynamics can be investigated. The coupling here is via transport, which at most frequencies is diffusive, which is inherently a nearest-neighbor coupling mechanism (mixing occurs primarily between adjacent vortices). On the other hand, at or near the resonant frequencies for transient superdiffusion, the flights that appear allow mixing between distant vortices in a short period of time. These flights are, in some sense, analogous to the short-cuts in the Watts/Strogatz small-world model

As to whether the difference in the observed dynamics from the previous section can be explained from this framework, the results are still inconclusive, and more data is needed. In particular, the frequencies for the traveling waves and globally-synchronized states need to be compared carefully with the frequencies for the peaks in the long-term effective diffusion coefficient. These experiments are currently in progress.

In modeling the behavior, the time scales for mixing in this experiment need to be considered. Specifically, how fast must mixing occur to enable synchronization? The longest flight time (from one end of the network to the other) is approximately 100 seconds, whereas typical circulation times are approximately 20 s. The period of the chemical oscillation is around 40 s, which is shorter than the longest flight times. However, our conjecture is that the relevant time scale is the coherence time 400 s, discussed earlier, which is significantly longer than all of the mixing timescales. Furthermore, the model of Ref. 1 needs to be modified, since Lévy flights can connect every vortex with every other vortex, but with a probability distribution that depends both on the separation between the vortices and also on the size of the portion of the chaotic region that gives rise to flights.

ACKNOWLEDGMENTS

This work was supported by the US National Science Foundation (grants DMR-0071771 and REU-0097424).

REFERENCES

1. Aref, H., *J. Fluid Mech.* **143**, 1 (1984).
2. Ottino, J.M., Leong, C. W., Rising, H. , and Swanson, P. D. , *Nature* **333**, 419 (1988).
3. Bolton, E. W., Busse, F. H., and Clever, R. M., *J. Fluid Mech.* **164**, 469 (1986).
4. Solomon, T. H. , and Gollub, J. P., *Phys. Rev. A* **38**, 6280 (1988).
5. Gollub, J. P., and Solomon, T. H., *Phys. Scripta* **40**, 430 (1989).
6. Solomon, T.H., *Transport and Boundary Layers in Rayleigh-Bénard Convection*, PhD thesis, 1990.
7. Chandrasekhar, S., *Hydrodynamic and Hydromagnetic Stability*, Dover (New York), 1961.
8. Willaime, H., Cardoso, O., and Tabeling, P., *Phys. Rev. E* **48**, 288 (1993).
9. Solomon, T.H., Tomas, S., and Warner, J. L., *Phys. Rev. Lett.* **77**, 2682 (1996).
10. Rom-Kedar, V., Wiggins, S., *Physica D* **51**, 248 (1991).

11. Solomon, T.H., Tomas, S., and Warner, J.L., *Phys. Fluids* **10**, 342 (1998).
12. Solomon, T.H., Lee, A.T., and Fogleman, M.A., *Physica D* **157**, 40 (2001).
13. Castiglione, P., Crisanti, A., Mazzino, A., Vergassola, M., and Vulpiani, A., *J. Phys. A* **31**, 7197 (1998).
14. B. D. Hughes, M. F. Shlesinger, E. W. Montroll, *Proc. Natl. Acad. Sci. USA* **78**, 3287 (1981).
15. M. F. Shlesinger, G. M. Zaslavsky, J. Klafter, *Nature* **363**, 31 (1993).
16. Shlesinger, M.F., *J. Stat. Phys.* **10**, 421 (1974).
17. Wang, W.-J., *Phys. Rev. A* **45**, 8407 (1992).
18. Klafter, J., and Zumofen, G., *Phys. Rev. E* **49**, 4873 (1994).
19. Solomon, T.H., Weeks, E.R., and Swinney, H.L., *Phys. Rev. Lett.* **71**, 3975 (1993).
20. Weeks, E.R., and Swinney, H.L., *Phys. Rev. E* **57**, 4915 (1998).
21. Rom-Kedar, V., and Zaslavsky, G., *Chaos* **9**, 697 (1999).
22. I. R. Epstein, *Physica D* **7**, 47 (1983).
23. R. A. Schmitz, K. R. Graziani, J. L. Hudson, *J. Chem. Phys.* **67**, 3040 (1977).
24. A. T. Winfree, *Science* **175**, 634 (1972).
25. K. Showalter, *J. Chem. Phys.* **73**, 3735 (1980).
26. D. J Watts, S. H. Strogatz, *Nature.* **393**, 440 (1998).

Initial Report On The Transition to Turbulence in a Magnetized Plasma Column

M. J. Burin, G. R. Tynan, N.A. Crocker

Center for Energy Research, Fusion Energy Division
Department of Mechanical and Aerospace Engineering
University of California at San Diego, La Jolla Calif. 92093

Abstract. Plasma density fluctuations are being investigated experimentally in CSDX, a magnetized plasma column. We show that an effective Reynolds number of the plasma is directly proportional to axial magnetic field strength **B** and inversely proportional to the ion-neutral collsion frequency, which in our experiment is proportional to the neutral gas pressure. As **B** is increased, plasma density fluctuations grow in amplitude and evolve to either a multi-mode or turbulent state, depending on the gas pressure. In this final state, correlation dimension estimates of the fluctuations are low for the high pressure case (D ~ 2), but are high and unsaturated for the low pressure case. Together these results present a familiar but incomplete picture of a plasma turbulent transition that features the breakup of an attractor.

INTRODUCTION

In this work we characterize the evolution of density fluctuations in a bound plasma from near-quiescence to a turbulent state. After some introductory remarks we show that our control parameters of gas pressure p and magnetic field strength **B** can be viewed in light of a Reynolds number. Then we proceed to discuss the apparatus and data collection. The remainder of this report will discuss the data analysis. Conclusions are then discussed regarding the changes in spectral power distribution and dimensionality with respect to Reynolds number.

In our experiment we believe that the observed plasma density fluctuations are density gradient driven drift waves; they peak near the region of maximum gradient of the plasma density profile and are consistent with expectations from linear theory [1]. Such modes are common in magnetized plasma containment devices, ranging in scale from small linear machines such as ours to large tokamaks. They are analogous to Rossby waves in planetary atmospheres. Attempts to create and study drift wave turbulence in small devices are pertinent because such turbulence is thought responsible for confinement losses that hamper progress towards efficient nuclear fusion. Attempts to estimate a correlation dimension for these turbulent motions have concluded that it is typically high [2], although there have been some exceptions [3]. However, low dimensional dynamics have been identified in the density fluctuations of a number of smaller plasma devices [e.g. 4].

In fluid mechanics one parameter characterizing a transition to turbulence is the Reynolds number, a dimensionless parameter indicating the importance of inertial

CP676, *Experimental Chaos: 7ᵗʰ Experimental Chaos Conference*,
edited by V. In, L. Kocarev, T. L. Carroll, B. J. Gluckman, S. Boccaletti, and J. Kurths
© 2003 American Institute of Physics 0-7354-0145-4/03/$20.00

(nonlinear/convective) forces relative to viscous damping forces. Under the working assumption that the fluctuations are drift waves, we can estimate an effective Reynolds number for our data.

$$\mathrm{Re} \equiv \frac{UL}{\nu} = \frac{V_d L}{r_L^2 \omega_{n-i}} = \frac{L^2 qB}{kT\omega_{n-i}} \sim \frac{B}{p} \qquad (1)$$

Here L is a characteristic scale length, taken to be the density gradient scale length $(\frac{1}{n_p}\nabla n_p)^{-1}$, where n_p is the plasma density. V_d is the diamagnetic drift velocity (proportional to $(\mathbf{B}L)^{-1}$). Also in the equation is the Larmor radius r_L (proportional to $\mathbf{B}^{-1}$) and ion-neutral collision frequency ω_{n-i} (proportional to neutral gas pressure p). We learn from equation (1) that magnetic field and neutral gas pressure may be used as parameters to control the transition to turbulence.

APPARATUS: CSDX

We have constructed the **C**ontrolled **S**hear **De**-correlation Experiment: CSDX, a columnar magnetized plasma machine at UCSD (figure 1). The plasma is powered by a helicon radio frequency source (13.56 MHz) that ionizes a portion of the injected gas (argon in this study). Electromagnetic coils surrounding the chamber generate an axial magnetic field. The magnetic field confines the plasma to about a 10cm diameter.

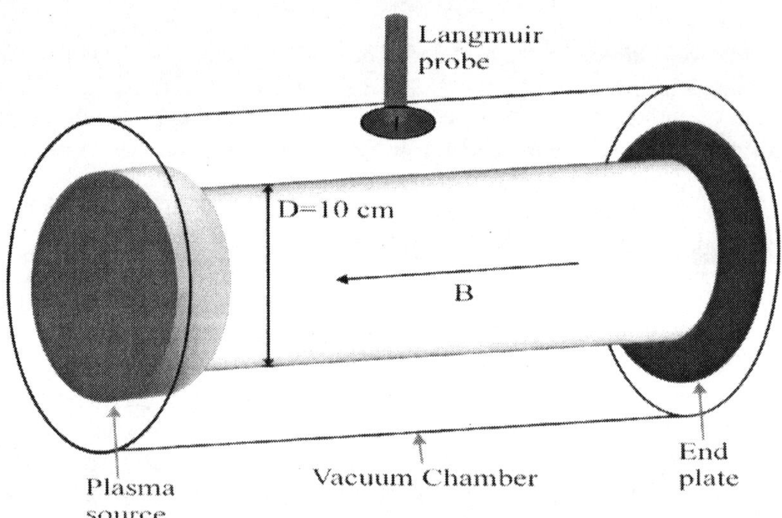

FIGURE 1. The Controlled Shear Decorrelation Experiement, or CSDX. Argon is injected into a cylindrical stainless steel chamber kept near vacuum pressures, with a percentage of this gas becoming ionized by the power ("plasma") source (at left). At the right end are pumps (not shown).

Under normal operating conditions, it can be shown that the CSDX plasma behaves as a magneto-fluid exhibiting drift waves that propagate azimuthally. Both neutral gas pressure p and axial magnetic field strength **B** can be varied, thus allowing for two ways to vary the Reynolds number of the plasma.

INITIAL RESULTS

Time-series data of density fluctuations are obtained with ion saturation current from an electrically biased (-50V) Langmuir probe. Ion current is approximately proportional to plasma density [5]. Sampling was at 200kHz, where the Nyquist frequency is well above the frequency of the drift modes (between 10^3 and 10^4 Hz).

Profiles and Fluctuation Onset

Figure 2 shows density profiles and the onset of fluctuations with increasing magnetic field. Fluctuation measurements were taken near the maximum density gradient 3cm from the center of the plasma column. The increase of fluctuation level with magnetic field is readily apparent. A detailed study of the fluctuations near onset (**B** ~ 600 G) has recently been made [1]. These fluctuations have their greatest amplitude near the maximum plasma density gradient, consistent with their identification as resulting from drift instabilities.

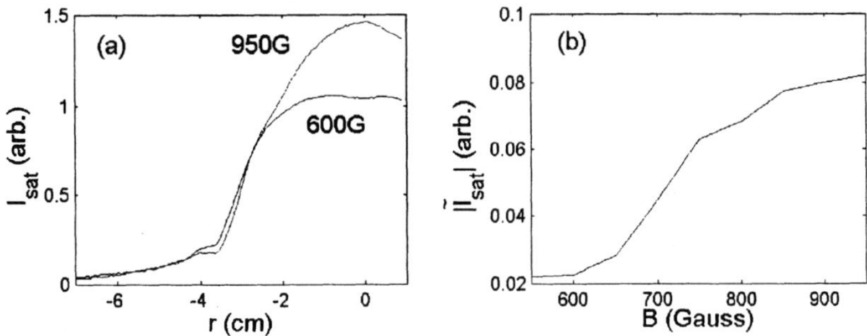

FIGURE 2. Mean density profiles (a) and the onset and increase of drift fluctuation amplitudes with increasing magnetic field (b). Fluctuation amplitudes were obtained near the region of maximum density gradient at r = 3cm.

Power Spectra

The spectral evolution with increasing magnetic field of an argon plasma for two different pressures is given in Figure 3. At the weakest **B** presented (~ 600 G), the plasma fluctuation spectrum consists of few discrete low amplitude modes. As **B** is increased, we observe the expected increase of modal amplitudes as well as the emergence of new ones. The final plasma state (~ 950 G) can be described as either multi-mode (5mTorr case) or turbulent (3mTorr case).

As stated in equation (1), either an increase in **B** or a decrease in p implies an increase in Reynolds number. The power spectra presented in Figure 3 qualitatively agree with this Reynolds number interpretation. A spectral broadening of power is seen for either an increase in magnetic field or a decrease in pressure. The difference between the two pressures at 950G (Figure 3b) is a particularly striking example of a multi-modal plasma which is found to be turbulently broadened at higher *Re*.

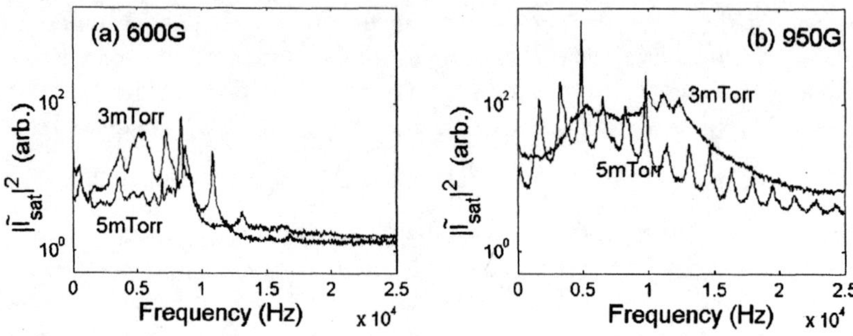

FIGURE 3. Power spectra of plasma density fluctuations for 600G (a) and 950G (b). The probe was positioned near the region of maximum fluctuation activity (see Fig 2a). The power value units are arbitrary but the same for all spectra.

Correlation Dimension Estimates

We now present estimates of the dimensionality of the fluctuations seen in the above spectra. Is the broadband state that develops at high **B** the result of a few nonlinearly interacting modes, and hence of low dimension? Theoretical work has suggested that a trio of nonlinearly interacting drift waves may yield a chaotic broadband spectrum [6], and some drift wave turbulence simulations bear this out [7]. Such a state would be of low dimension, in contrast to the high dimensions in [2] and traditionally thought to describe drift turbulence. To determine the dimension we now turn to phase space reconstructions and, with it, estimates of the correlation dimension.

A system's phase space may be reconstructed from measured time-series of a single variable via the delay-embedding technique [8]. A time-series so construed may exhibit structure (i.e. attractors) in the phase space undetectable by direct inspection of the series itself or it's power spectrum. The correlation dimension, introduced by Grassberger & Procaccia [9], is a measure for quantifying the dimensionality of such structures by looking for power law scaling relationships between points in an embedding space. It is derived from the correlation sum $C(r)$, defined

$$C(r) = \lim_{N \to \infty} \left\{ \frac{1}{N^2} (number\ of\ pairs\ (i, j)\ whose\ separation\ \left| x_i - x_j \right| < r) \right\} \qquad (2)$$

where x_i and x_j are points in the embedding space and r is a threshold distance. N is the number of data points in the time series.

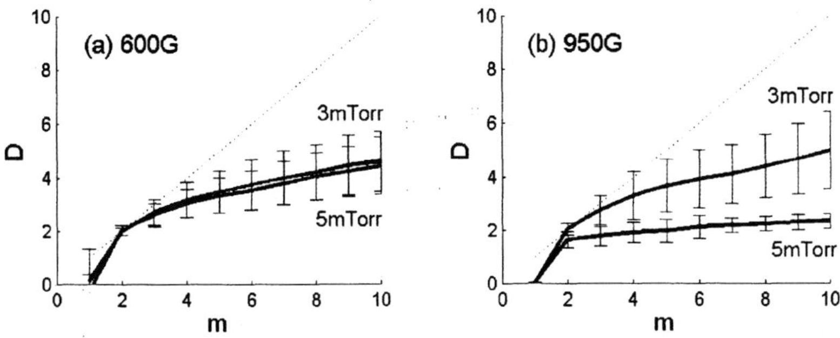

FIGURE 4. Correlation dimension estimates (pointwise) from the density fluctuation spectra shown in figure 3. Time series utilized consisted of 10^5 points. The results above appear to be robust for variations in delay lag time and reference point number.

In a plot of $C(r)$ *vs.* r there may exist a self-similar region where $C(r) \propto r^d$. If d saturates to value D with increasing embedding dimension m then D is the correlation dimension. These dimension estimates are closely related to the degrees of freedom of the underlying dynamical system (if such exists) [11]. A power law scaling region in the proper range of r (typically 1-10% of N [11]) may be indicative of deterministic dynamics underlying the time-series. For such data-sets the amount of data required to resolve a dimension D is formally $(10 \text{ to } 100)^D$. It is predicted that an embedding space of dimension $m = 2D+1$ is sufficient to determine the dimensionality of a D-dimensional attractor [10]. In practice, with data from a deterministic system, the scaling law power should saturate as m is increased above D. Noise will prevent this saturation even in modest levels (relative error <3% [11]).

Figure 4 shows estimates of the (pointwise) correlation dimension for the fluctuation data shown in figure 3. Time series utilized for this analysis consisted of 10^5 points. The results above appear robust for variations in delay lag time and the number of reference points. At 600G, analysis at both pressures reveals that the underlying dynamics are not low dimensional. While both time-series are roughly sinusoidal, leading to the expectation of low dimensional attractors (e.g. limit cycle), the signals are quite noisy with signal to noise ratios (S/N) of only 10-20. It is not surprising then that phase space reconstructions indicate no low dimensional structure. The lack of saturation at both pressures is believed to be the result of insufficient signal power.

At 950G the amplitudes are larger by at least an order of magnitude for both pressures. For the 5mTorr case, we find near saturation yielding D ~ 2. This estimate is consistent with the observation that the power spectrum consists primarily of discrete modes that appear to be harmonics. For the 3mTorr case, which corresponds to higher *Re*, we find no saturation. We believe this result is not due to noise but rather the nature of developed turbulence.

Concluding Remarks

What emerges from this study is a familiar but quite incomplete picture of a transition to turbulence. Density fluctuation onset is accompanied by signal periodicites and (presumably) the formation of simple attractors, but these at present are too noisy to detect. The plasma evolves with increasing magnetic field into a multi-modal or turbulent state, depending on the pressure. These variations with **B** and p are consistent with a transition to turbulence that is characterized by an effective Reynolds number, formulated in equation (1) as proportional to **B**$/p$. If we consider fluctuations at the highest **B** (950G) for both pressures, given the increase of Re from the 5mTorr to the 3mTorr case, we glimpse the breakup of a low dimensional attractor into a much higher dimensional state considered to be turbulent. This scenario has been observed in similar studies in plasma devices that utilized a different type of control parameter [12,13].

Future work on CSDX will include improving signal to noise ratios and a further exploration into parameter space, such as plasma source power and gas type. Additionally, varying the pressure and magnetic field in smaller increments should allow us to discern if the plasma transitions to turbulence or chaos via a traditional route, e.g. by period doubling [14], quasiperiodicity [15-17], or intermittency.

ACKNOWLEDGMENTS

This work was supported by the U.S. Department of Energy under grant number DE-FG03-99ER54553. We thankfully acknowledge Dr. G.Y. Antar for extensive contributions to both the data collection and analysis, as well for providing figure 1. We thank also Mr. Tyler Lynch for making the Langmuir probe utilized on CSDX, and Dr. Gian Mario Maggio for some useful discussions on the correlation dimension.

REFERENCES

1. George, J. MS Thesis, UCSD 2002
2. Sawley, M.L. et al, Phys. Fluids 30, 129-132 (1987).
3. Barkley, H.J. et al, Plasma Phys. Cont. Fusion 30, 217-234 (1988).
4. Qin, J. et al, PRL 63, 163-166 (1989).
5. Hutchinson, I.H. *Principles of Plasma Diagnostics*, C.U.P., 1987, pp 63-65.
6. Biskamp, D. & Kaifen, H. Phys Fluids 28, 2172-2180 (1985).
7. Persson, M. & Nordman, H. Phys Rev Lett 67, 3396-3399 (1991).
8. Packard, N.H. et al, PRL 45, 712-716 (1980).
9. Grassberger, P. & Procaccia, I. PRL 50, 346-349 (1983).
10. Gershenfeld, N., in *Directions in Chaos*, edited by Hao Bai-lin, Singapore: World Scientific, 1988.
11. Kantz, H. & Schreiber, T., *Nonlinear Time Series Analysis*, C.U.P., 1997, pp 69-89.
12. Klinger, T. & Piel, A. Phys. Fluids B 4, 3990-3995 (1992).
13. Gyergyek, T. et al., J. Phys D: Appl Phys. 27, 2080-2094 (1994).
14. Cheung, P.Y. & Wong, A.Y., Phys. Rev. Lett. 59, 551-554 (1987).
15. Chu, J.H., & Lin, I. Phys. Rev. A 39, 233-238 (1989).
16. Weixing, D., et al., Phys. Rev Lett 70, 170-173 (1993).
17. Klinger, T. et al, PRL 79, 3913-3916 (1997).

EXPERIMENTAL CHAOTIC BUBBLING

José Carlos Sartorelli[1], A. Tufaile and V. S. M. Piassi

*Instituto de Física, Universidade de São Paulo, Caixa Postal 66318,
CEP 05315-970 São Paulo, SP, Brazil.*

Abstract. We have studied the air bubble formation in a submerged nozzle in a water/glycerol solution inside a cylindrical tube submitted to a sound wave perturbation. We are reporting the change in the bubbling dynamics when the dissipative pneumatic system was modified. Placing a short tube between the nozzle and the control air valve, we have observed transition from period 1 to period 2 and inverse period doubling bifurcation. Modified the length of the tube, the dynamics has changed. We also obtained the time series of this system subjected to a sound wave.

INTRODUCTION

Bubble columns are widely used in diverse fields as petrochemical, chemical, biochemical, and pharmaceutical industries. Because of this, some authors studied the transition from regular to chaotic behavior in bubble formation in a single nozzle, Tritton and Edgell [1] have observed some attractors by the signal obtained from a hot-film anemometer placed close to the bubble formation using the method of reconstructing a phase-space from a single time series. Mittoni *et al* [2] have given further evidence that bubbling can display chaotic behavior by the characterization of positive Lyapunov exponents from the experiment with non-invasive technique. Rejecting the premise that bubble formation in dense fluid resembles an inverted dripping faucet with period-doubling to chaos, Nguyen *et al* [3] have shown spatio-temporal dynamics in a train of rising bubbles is the dominant source of instability in the bubbling dynamics. On the other hand, studying bubbling from perforated plates with two orifices, Ruzicka *et al* [4] have reported complex behavior in the transition between synchronous and asynchronous regimes.

We have studied the bubble gun experiment [5], which consists of the air bubble formation from a nozzle submerged in a water/glycerol solution inside a cylindrical tube submitted to a sound wave above the fluid. This experiment exhibits transition from quasiperiodicity to chaos, period doubling cascade and Hénon-like dynamics [6] as the same way as in the circle map [7]. We also related these behaviors to the occurrence of subharmonic oscillations and explosion [8]. We are reporting the change in the bubbling dynamics when the dissipative pneumatic system was

---

[1] E-mail : sartorelli@if.usp.br

CP676, *Experimental Chaos: 7ᵗʰ Experimental Chaos Conference*,
edited by V. In, L. Kocarev, T. L. Carroll, B. J. Gluckman, S. Boccaletti, and J. Kurths

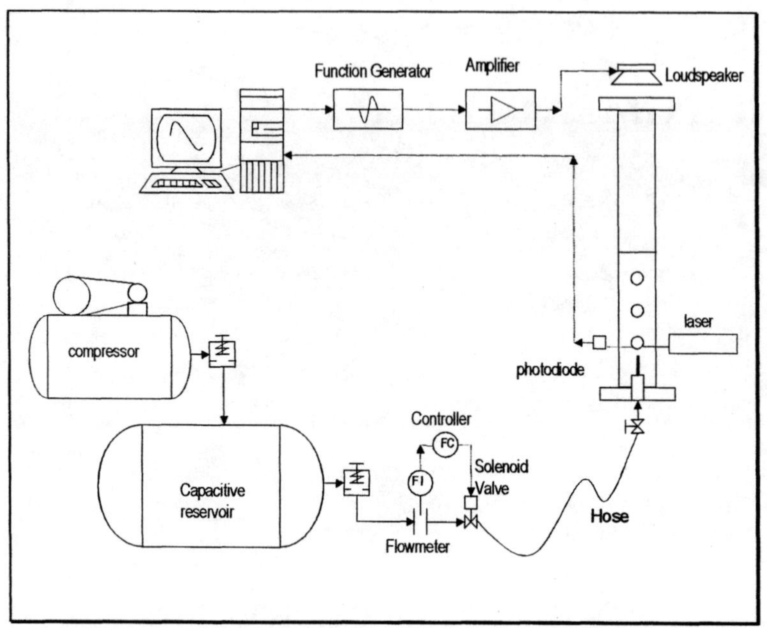

FIGURE 1. Diagram of the experimental apparatus

modified. Placing a short tube between the nozzle and the air control valve, we have observed transition from period 1 to period 2 and inverse period doubling bifurcation, characterizing a *bifurcation bubble*. We also obtained the time series of this system subjected to a sound wave and compared with coupled maps.

EXPERIMENTAL APPARATUS

The experimental apparatus consists of a cylindrical tube with an inner diameter of 11 cm and 70 cm in height. The bubbles are issued by injecting air through a metallic nozzle submerged in a viscous fluid column (20 % water/ 80%glycerol), as is shown in Fig. 1. The inner diameter of the nozzle is 0.78 mm. The nozzle is attached to a capacitive reservoir, and a solenoid valve controlled by a PID controller (BTC-220) sets the air flux. The flow rate is measured by a flowmeter Aalborg (GFM47). A hose was placed between the nozzle and the control valve with an inner diameter of 5 mm. We have used three hose lengths: 50 cm (tube a), 140 cm (tube b) and 300 cm (tube c).

The detection system is the same as the one of the dripping faucet experiment, (see Ref. [9] for details). The delay times between successive bubbles are measured by a time circuitry inserted in a PC slot, with a time resolution equals to 1 µs. The input

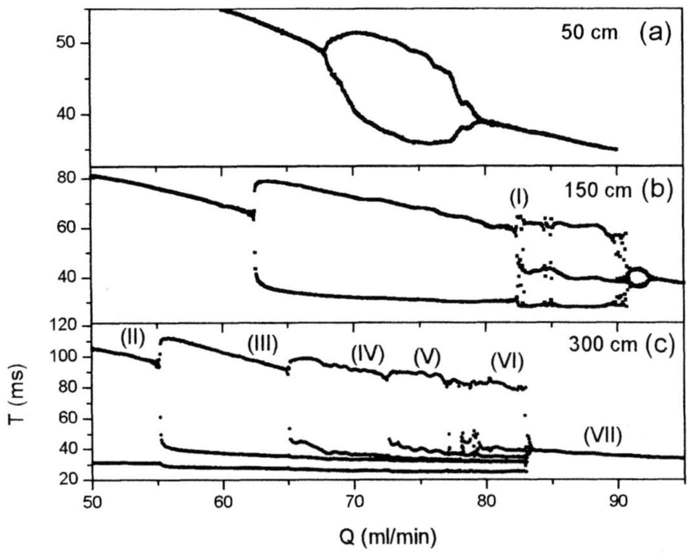

FIGURE 2. Bifurcation diagram using different tube lengths.

signals are voltage pulses, induced in a resistor, defined by the beginning (ending) of scattering of a laser beam. The pulse width is the time interval t_n (n is the bubble number), and the time delay between two pulses defines the crossing time (dt_n) of a bubble through the laser beam, so that the total time interval is $T_n = t_n + dt_n$. Series of time intervals between bubbles {T} were obtained, and then the bubble mean frequency was calculated as $f_b = 1/<T>$. We have estimated the total experimental noise as ~100μs in the period 1 behavior.

Applying a sound wave tuned to the air column above the liquid with a loudspeaker placed at the top of the liquid column we can change the bubble dynamics. Whenever an air bubble crosses an observation point above the needle, it displaces an amount of fluid, and after the bubble passage through the fluid, the initial configuration is restored. Hence, when a bubble train crosses the observation point, it displaces the fluid with the bubbling frequency f_b. This movement is an oscillatory system while the sound wave (frequency f_s) represents the other one that drives the bubbling. In this way, we have constructed a forced oscillator analogous to that occurring in the circle map model [8].

RESULTS AND DISCUSSION

The time interval between bubbles were recorded whilst slowly varying the air injection rate into liquid, as shown in Figs. 2(a-c), in order to observe the importance

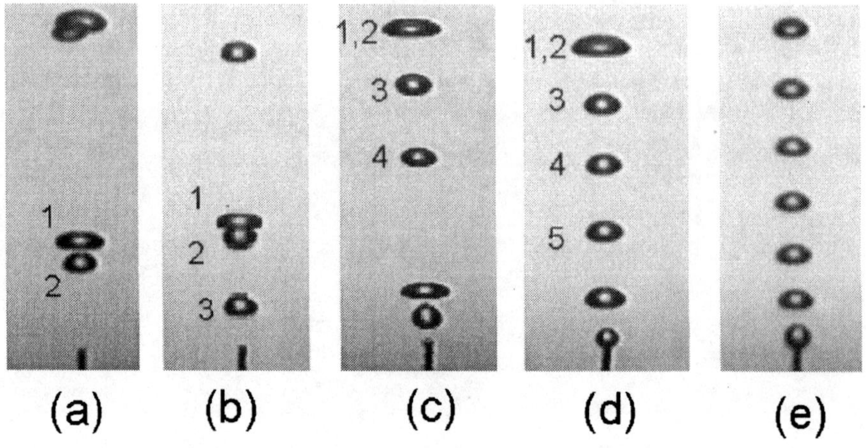

FIGURE 3. Bubble profiles in different bubbling regimes.

of the tube length of the air injection system on the bubbling dynamics. For the shortest tube (tube A), in Fig. 2(a) it is shown a closed loop-like sequence called *bubble* bifurcation. In Fig. 2(b) is shown the bifurcation sequence for the tube B, with a period-1, period doubling, the chaotic region I, followed by a period-3, another period-3 with some chaotic regions inside, and after this we have obtained period-2 and period-1, closing the bubble of bifurcation. The last case in Fig. 2(c) shows the interesting phenomenon of period increase using the longest tube C. For this case the bifurcation system started in a period-2, period-3 (III), period-4 (IV), period-5 (V), followed by a chaotic region, period-6 (region VI), and period-1. In Fig. 3 the bubble profiles represent the behaviors observed in the time series of Fig. 2(c). In Fig. 3(a) the bubble pair shows an example of the periodic dynamics of period-2 observed in the region II of Fig. 2(c), in which the first bubble distorts from an initially spherical shape to a flattened ellipsoidal shape. The wake of this bubble is a low-pressure region that penetrates the following bubble, bringing them into contact. In Fig. 3(b) is shown the bubbling in perio-3 of region III of Fig. 2(c), a bubble pair followed by a single bubble. An image from period-4 is shown in Fig. 3(c), in which the first bubble pair becomes a large single bubble followed by two bubbles (region IV in Fig. 2(c)). In Fig. 3(d) is shown a period-5 (region V), and finally the period-1 in Fig. 3(e) with the bubbles equally spaced representing the dynamics of region VI of Fig. 2(c).

Fixing the air flow rate at 73 ml/min in order to obtain a period-2 with the tube A, and using the applied voltage in the loud-speaker as the control parameter we have obtained the time series shown in Fig. 4. The system behavior can be emulated with a logistic map coupled with the two-dimensional circle map [7]:

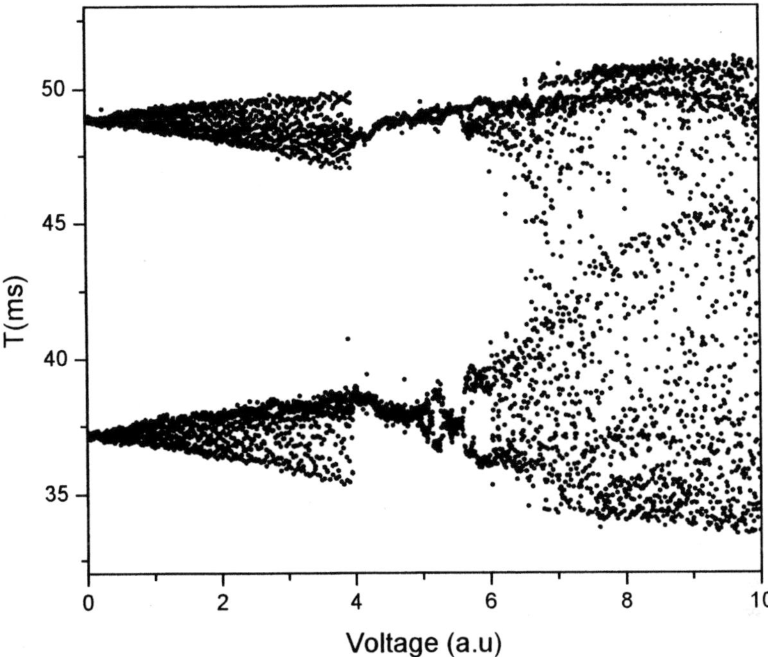

FIGURE 4. Time series using the voltage applied to the loudspeaker as the control from an initial bubbling in a period-2, using the tube A.

$$x_{n+1} = x_n + \Omega - \frac{K}{2\pi} sin(2\pi x_n) + \beta y_n \quad \text{mod} (1),$$

$$y_{n+1} = \beta y_n - \frac{K}{2\pi} sin(2\pi x_n), \tag{1}$$

$$z_{n+1} = 3.32 z_n (1 - z_n) + 0.1 y_n$$

where x_n, y_n and z_n are the dynamical variables, β regulates the damping, Ω is the ratio between two independent frequencies, and K is the intensity of the non-linear external impulse. By using the z variable we obtained a bifurcation diagram as shown in Fig. 5, similar to the experimental one shown in Fig. 4.

CONCLUSIONS

We observed the change of bifurcation branches in the time series of bubble formation when we increased the system dissipative by enlarging the length of the tube. We also have observed features present in the experimental system similar to the

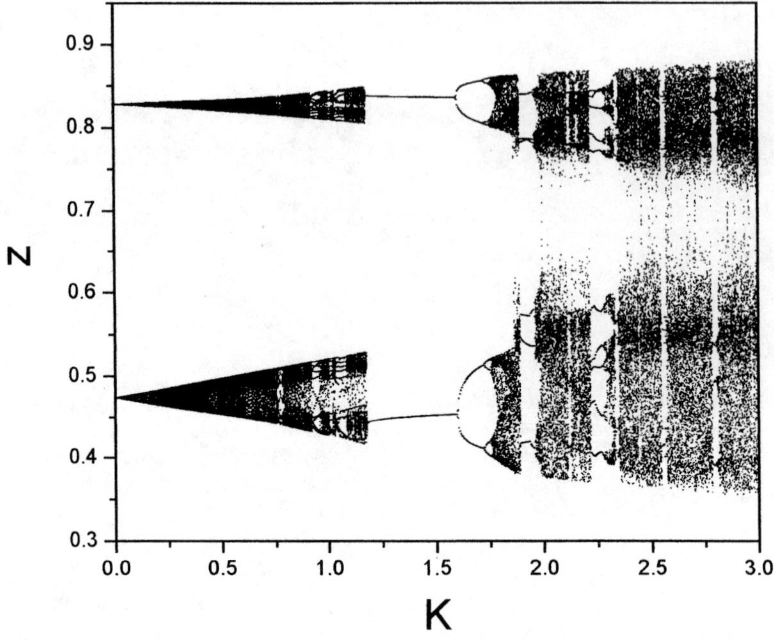

FIGURE 5. Bifurcation diagram from the unidirectional coupling between the two-dimensional circle map and a logistic map placed initially in a period-2.

coupled maps, such as quasiperiodic and intermittent behavior, subharmonic synchronization, period doubling, chaotic attractors and crisis.

ACKNOWLEDGMENTS

This work has been supported by the Brazilian agencies: CNPq and FAPESP.

REFERENCES

1. Tritton, D. J., and Egdell, C., *Phys. Fluids* **5**, 503-505 (1993).
2. Mittoni, L. J., Schwarz, M. P., and La Nauze, R. D., *Phys. Fluids* **7**, 891-893 (1955).
3. Nguyen, K., Daw, C. S., Cheng, M., Bruns, D. D., Finey, C. E. A., and Kennel, M. B., *Chem. Eng. J.* **64**, 191-197 (1996).
4. Ruzicka, M., Drahos, J., Zahradnik, J., and Thomas, N.H., *Chem. Eng. Sci.* **55**, 421-429 (2000).
5. Tufaile, A., and Sartorelli, J. C., *Physica A* **275**, 336-346 (2000).
6. Tufaile, A., and Sartorelli, J. C., *Phys. Lett. A* **275**, 211-217 (2000).
7. Tufaile, A., and Sartorelli, J.C., *Phys. Lett. A* **287**, 74-80 (2001).
8. Tufaile, A., and Sartorelli, J. C., *Physica A* **308**, 15-24 (2002).
9. Tufaile, A., Pinto, R. D., Gonçalves, W. M., and Sartorelli, J. C., *Phys. Lett. A* **255**, 58-64 (1999).

BIOPHYSICS

Dynamics of Single DNA Molecules

Doug Smith

Department of Physics, University of California, San Diego

Abstract. A brief summary is given of experiments in which single DNA manipulation and visualization techniques are used to study problems in polymer dynamics and biophysics. References to original work are provided.

POLYMER DYNAMICS IN FLOW

DNA molecules may be taken as a model system for studying polymer dynamics and rheology questions. Polymer physics concerns the statistical mechanics and hydrodynamics of flexible, chain like molecules and rheology concerns the flow properties of complex fluids and how the macroscopic fluid properties derive from the microscopic dynamics of dissolved macromolecules. Single DNA molecules can be directly manipulated and visualized using optical tweezers and fluorescence microscopy. Of special interest to those in the field of nonlinear dynamics is the conformational dynamics of DNA molecules in elongational and shear flows. Due to the influence of Brownian motion, individual identical molecules in identical high strain rate flow fields experience non-ergodic and heterogeneous dynamics. Polymers in steady shear flow exhibit aperiodic extension fluctuations due to a Brownian-driven conformational instability. For further details on this work please see the references below.

THE WORLD'S STRONGEST MOLECULAR MOTOR

Optical tweezers DNA manipulation has also been used to study the mechanism of packaging of DNA by the bacterial virus phi29. This virus contains a ~10 nm-diameter molecular motor, powered by chemical energy (ATP). The motor grabs onto one end of a single DNA molecule and pulls it into the viral capsid. The motor is thought to work by a rotary mechanism. We have characterized the dynamics and forces generated of this motor by grabbing the free end of the DNA using optical tweezers techniques. Our measurements show that the motor can generate ~60 picoNewtons of force. This force may sound small, but it significantly larger than what other known molecular motors, such as muscle proteins, can generate. The motor is very small (~10 nm sized) and if its force were scaled up with its volume to the size of an automobile motor the force generated would be ~10^{13} Newtons, which is enough to lift more than a thousand aircraft carriers.

CP676, *Experimental Chaos: 7ʰ Experimental Chaos Conference,*
edited by V. In, L. Kocarev, T. L. Carroll, B. J. Gluckman, S. Boccaletti, and J. Kurths
© 2003 American Institute of Physics 0-7354-0145-4/03/$20.00

HOW IS DNA PACKAGED IN HUMAN CELLS?

In physical dimensions, DNA molecules are long, slender threads that are only ~2 nm wide. Amazingly the total length of DNA in a human body is enough to wrap around the earth about 10 million times. An important problem in molecular biology, thus, is to understand how DNA is packaged in such a condensed fashion, yet at the same time is organized so that its stored information remains accessible for dictating essential biochemical processes of life. This topic, of chromatin assembly, is of great current interest to us.

ACKNOWLEDGMENTS

The experiments described were done in Steve Chu's lab at Stanford University and Carlos Bustamante's lab at UC Berkeley.

REFERENCES

1. T. T. Perkins, D. E. Smith, S. Chu, "Optical Manipulation of Single DNA Molecules", *Physics News in 1994* (American Institute of Physics, 1994).

2. T. T. Perkins, D. E. Smith, S. Chu, "Direct Observation of Tube-like Motion of a Single Polymer Chain", *Science* **264**, 819, (1994).

3. D. E. Smith, S. Chu, "Response of Flexible Polymers to a Sudden Elongational Flow", *Science* **281**, 1335 (1998).

4. D. E. Smith, H. P. Babcock, S. Chu, "Single Polymer Dynamics in Steady Shear Flow", *Science* **283**, 1724 (1999).

5. D. E. Smith, S. J. Tans, S. B. Smith, S. Grimes, D. L. Anderson, C. Bustamante, "The Bacteriophage ϕ29 Portal Motor can Package DNA Against a Large Internal Force," *Nature* **413**, 748 (2001).

Nonlinear Synchronization Analysis of Spatiotemporal Heart Data

Jennifer D. Simonotto[§], Michael D. Furman[§], Mark L. Spano[†],
William L. Ditto[§],
Gang Liu[‡] and Katherine M. Kavanagh[‡]

[§]*Department of Biomedical Engineering, University of Florida, Gainesville, Florida, USA*
[†]*US Navy, Carderock Laboratory, W. Bethesda, Maryland, USA*
[‡]*Department of Medicine, University of Alberta, Edmonton, Alberta, Canada*

Abstract. A high-speed video camera and voltage-sensitive dyes were used to acquire high resolution (80x80 pixels) and high-speed (500 μs/frame) optical signals of ventricular fibrillation in a Langendorff-perfused porcine heart. The resulting spatiotemporal dynamics were recorded before and after the application of a defibrillation shock in order to study the mechanism of defibrillation failure. We calculate nonlinear synchronization index measures to qualify the evolution of different types of activity on the heart surface (focal, reentry). We observe changes with time in the spatial distribution of the first Fourier mode, showing that two main types of activity compete on the heart surface during a failed defibrillation.

INTRODUCTION

Ventricular fibrillation (VF) is a serious medical problem; in the western world, VF claims more lives than any other heart disease. But little is understood about how normal, ordered behavior (sinus rhythm) can suddenly change into the complex activity that characterizes VF. It has been established[1,2] that when there are wavebreaks on the surface of the heart (a literal breakup of the smoothly propagating electrical wavefront), spiral waves can result. Further breakdown can lead to VF, but it is unclear exactly how changes in the system can lead to such a breakdown. Currently the only effective therapy for VF is the delivery of a rather large electrical shock (stimulus) across the myocardium with the goal of terminating VF and allowing the sinus node to reestablish sinus rhythm. However, external defibrillation shocks (DS), which for humans can be 200 - 360 Joules in magnitude, are not always successful. It is important to understand why defibrillation failure occurs in order to make DS more effective. Experimentally recorded data from porcine ventricles (taken from a whole heart preparation, but with the main focus on the ventricles) undergoing repeated defibrillation attempts was analyzed to understand the mechanism of defibrillation failure.

CP676, *Experimental Chaos: 7th Experimental Chaos Conference*,
edited by V. In, L. Kocarev, T. L. Carroll, B. J. Gluckman, S. Boccaletti, and J. Kurths
© 2003 American Institute of Physics 0-7354-0145-4/03/$20.00

OPTICAL MAPPING AND DEFIBRILLATION PROTOCOL

High spatial and temporal resolution optical mapping techniques and voltage sensitive dyes were used to visualize the surface excitations of porcine heart. The data were taken at 2000 frames per second by a Red Shirt Imaging CCD camera with an 80x80 pixel resolution covering a 5x5 cm area.

All animal procedures received approval from the Health Sciences Animal Welfare Committee of the University of Alberta. The heart was rapidly removed and submerged in cool perfusion fluid. The right and left coronary arteries were cannulated with DLP arteriotomy canulae. Coronary flow was adjusted between 90 and 120 ml per minute per 100 g to maintain a mean perfusion pressure of 70 – 80 mm Hg. A Millar catheter was used to monitor the perfusion pressure. Perfusate temperature was regulated at 37.5C+/- 0.5C by water jacketed heat exchangers. Temperature was monitored via thermisters located in the water jacket perfusion line and the perfusate inlet port.[3] An intramural temperature probe was inserted near the left ventricular apex.[3] The perfusion connector, on to which the coronary artery canulae were subsequently attached, contained additional temperature control injection ports for infusion of electromechanical dissociation agents and voltage sensitive dye. This proximity of the dye injection site and the coronary artery arterial tree minimizes the dead space for possible dye binding to red blood cell membranes and to blood proteins. The heart was stained with a 100 μM solution of di-4-ANEPPS was infused at 1/10 of the tissue flow rate for an infusion rate of 10 μM / min. The electromechanical dissociation agent Cytochalasin D was infused at 1/10 the arterial flow rate to eliminate movement artifact.

An extracellular mapping system was synchronized with the optical mapping system to continuously monitor atrial and ventricular rhythms during optical mapping. It was also possible to deliver electrical current through these electrodes to facilitate atrial and ventricular pacing. The extracellular electrodes were positioned on the posterior surface of the heart to avoid obstructing optical mapping. Titanium mesh defibrillation patch electrodes were positioned on the right atrium (surface area of 2.3 cm^2) and the apex of the left ventricle (surface area 5.8 cm^2) to facilitate defibrillation.

If the heart was not initially in ventricular fibrillation it was induced with rapid ventricular pacing via the extracellular electrodes. An external defibrillator (Ventritex HVS-02, Ventritex Inc.) was used for defibrillation. A single capacitor biphasic waveform was used in which the leading edge voltage of phase 2 was equal to half the leading edge voltage of phase one. Each phase of the waveform was 6 ms in duration. Defibrillation energies were delivered through external patches, which were placed on the area of the right atrium and LV apex. The initial shock delivered was a low energy shock with first and second phase amplitudes of 200 and 100 volts respectively. If the initial shock was unsuccessful, the strength of the next and subsequent shocks was increased by 50 V until defibrillation was achieved.[4] Defibrillation threshold was defined as the lowest voltage and current that achieved defibrillation.[4]

DATA PREPROCESSING AND GENERAL DESCRIPTION

Due to memory restrictions, the data was limited to 5 seconds of recording at 2000 fps, sufficient to cover the transition from fibrillation $\rightarrow$ DS $\rightarrow$ defibrillation failure. The data was processed with a custom image processing program as described below.

After the camera system acquires the data, it is stored in a proprietary format. Subsequently we read in the raw data in preparation for further processing. At this stage we may average either 2 or 4 frames of the raw data, giving effective frame rates of 1,000 fps or 500 fps. From this data we stretch the dynamic range of each pixel to remove effects due to uneven illumination and non-planar geometry, subtract a global baseline value, and then "debleach" the data by detrending the pixel values and expanding the dynamic range on a frame by frame basis. A typical number of frames used is 9216.

The individual pixel time series presented later in this paper are extracted at this point. For the purpose of making movies of the data, further analysis in the form of a temporal 5-point median filter and, in some cases, a spatial filter was performed.

Movies were made of the processed data, showing electrical excitations over the heart surface. For one movie, four frames of the data were averaged together to give a more well-defined wavefront (wave of excitation in the tissue), but effectively reducing the frame rate to 500 fps. Space-time volume plots were made from the frame-averaged processed data (Fig.1). In these plots one can see several distinct stages in the evolution of the data. Initially the electrical activity was established VF, with multiple waves coexisting on the tissue surface in a complex manner (Fig. 1, upper left). The DS came approximately one second into the data recording, but, before the shock could travel across the entire tissue surface, a spontaneous reentry wave originating in the lower left portion of the tissue collided with the defibrillation wave. We refer to the portion of tissue from which the spontaneous wave originated as the *reentry area*. There was no reentry resulting from this collision, and regular excitation waves propagate across the surface from the bottom edge of the frame to the top edge in normal fashion, but there is a tissue path (from the bottom right to the upper middle) in which waves of excitation travel more quickly compared to the rest of the tissue. After several excitations (one sec), waves are generated from this "fast" area (Fig. 1, upper right); we refer to this area as the *focal area* (as it is a focus of activity generation). The movie then chronicles which type of activity is seen in the rest of the tissue. One can see regular activity induced by the focal area as well as the beginning of reentry activity, seen as "bridges" connecting one whole-surface activation to the next (Fig. 1, lower left). As the movie ends, reentry is the more established type of activity on the heart surface (Fig. 1, lower right). One can see that defibrillation has failed because complex paths of reentry present at the beginning of the movie (Fig. 1, upper left), in which multiple wavelets coexist with each other in a stable manner, begin to reassert themselves at the end of the dataset (Fig. 1, lower right).

Time series were extracted from individual pixels of the processed data for analysis. For this we used the unaveraged dataset (~10000 points at 2000 fps). We sampled the activity with a 13x13 array evenly spaced across the surface of the heart.

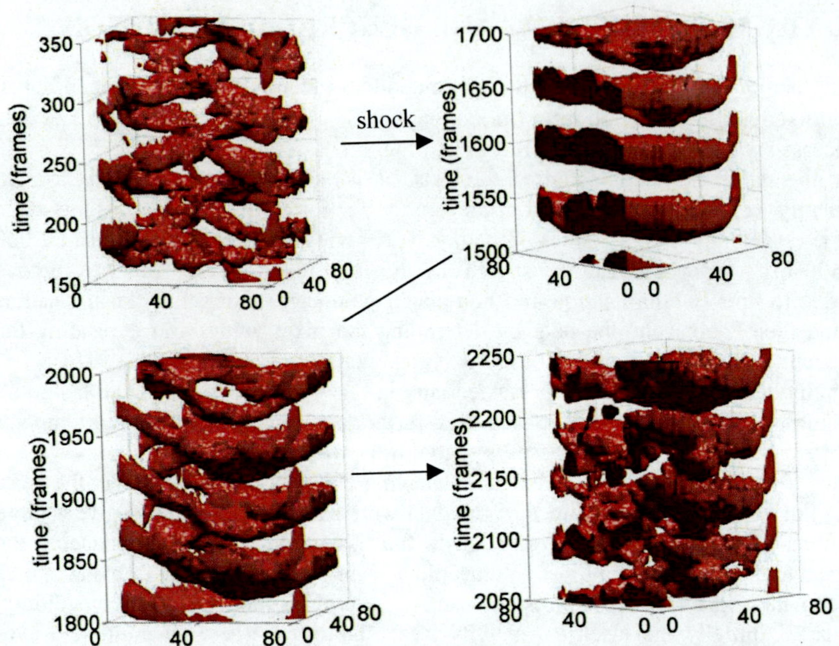

FIGURE 1. Space-time volume plots of processed data as described in the text. The z-axis is time (frames, where 1 frame = 2 msec) and the x- and y-axis are heart surface area (80x80 pixels covering a 5x5 cm area). Note that the viewing angle varies in order to highlight features of interest. *The upper left figure* shows activations from frames 150 to 350 (0.3 to 0.7 sec) showing reentry activity in the form of spatially discontinuous waves. Note the temporal periodicity present in the waves spatially; i.e., some waves occur again and again in the same manner in the same place (meaning that multiple waves coexist in a stable manner with each other). *The upper right figure* shows ordered behavior from frames 1500 to 1700 (3.0 sec to 3.4 sec) which occurs after the defibrillation shock (DS) and is due to excitation wave generation (focal activity) within the frame of visible tissue. *The lower left figure* shows activations from frames 1800 to 2000 (3.6 sec to 4.0 sec) and exhibits reentry in the form of bridges connecting one full-tissue activation to the next. *The lower right figure* shows activations from frames 2050 to 2250 (4.1 sec to 4.5 sec) and emphasizes the breakdown of ordered activity within the full-tissue activations, with reentry connections in some areas of the tissue. Some of the tissue behavior is similar to the spatially discontinuous waves showing temporal periodicity that marked the "chronic" VF at the beginning of the dataset (upper left).

SPECTROGRAMS AND DATA NONSTATIONARITY

A spectrogram[5] was made by calculating the magnitude of the Discrete Fourier Transform of a sliding window on data with a set amount of overlap. This reveals how dominant peaks change in time within a dataset. In calculating the spectrogram for our dataset, the time series is split into overlapping segments each with window length 5000 and 99% overlap. These parameters were a good balance between frequency resolution (larger windows give better results) vs. time resolution of changes (more windows are better).

Fig. 2 shows the spectrogram for time series extracted from different parts of the heart surface. For all points the main feature is the 10 Hz band, which remains significant and stationary for the entire run, as indicated by the warmer colors persistent in this frequency area. The 8 Hz band, however, displays different behavior at different times. At the focal area (Fig.2, right), much of the power in the 8 Hz band at the beginning of the time series shifts down to 7 Hz. For a time series from the reentry area (Fig.2, on left), much of the energy in the 8 Hz band disappears after the shock (which occurred one sec into the dataset). The focus site also has energy in a 1-2 Hz band area which narrows to a band around 2 Hz after the shock and then broadens again by the end of the time series. Spectrograms from other surface areas have some energy in this band at the beginning, but not after the shock.

SYNCHRONIZATION INDEX ANALYSIS

In previous sections we identified areas of tissue which seemed to be the origins of two different types of behavior: reentry and focal. We now investigate how the other tissue areas behave; i.e., whether they display reentry behavior or have more focal characteristics. To more easily identify the type of behavior that a particular area of tissue displays, we use synchronization index measures[6] obtained from the Hilbert transform $x_H(t)$ of individual time series:

$$\zeta(t) = x(t) + ix_H(t) = A(t)e^{i\phi(t)} \tag{1}$$

From the instantaneous phases of individual time series, we can check for n:m synchronization between two different time series by looking at the instantaneous phase difference between the two time series (for simplicity we chose n=m=1).

$$\varphi(t)_{n,m} = n\phi_1(t) - m\phi_2(t) \tag{2}$$

A more robust way of looking for synchronization is to look at the modulus of the phase difference. If the distribution of phase differences modulo 2π is uniform, then the data shows no phase locking and there is no simple relationship between the time series. If there is a significantly large and narrow peak in the phase distribution, then there is phase locking and the time series are connected in some way.[6]

$$\Psi(t)_{n,m} = \phi(t)_{n,m} \bmod 2\pi \tag{3}$$

To quantitatively compare the entire surface of the tissue, we use a synchronization index based on the first Fourier mode of the data[6,7,8] (normalized from 0 to 1):

$$\gamma^2_{n,m} = \left\langle \cos \Psi_{n,m}(t) \right\rangle^2 + \left\langle \sin \Psi_{n,m}(t) \right\rangle^2 \tag{4}$$

Using γ, one can compare the behavior of the activations across the heart surface to the two seminal types of behavior: reentry and focal beating. Two reference points were

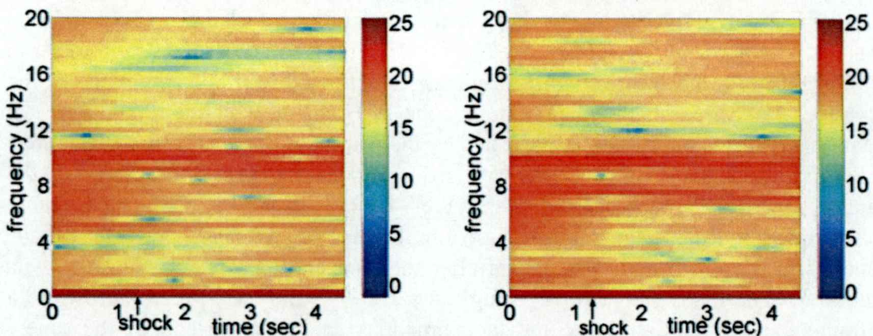

FIGURE 2. Spectrograms, which are the magnitude of the Discrete Fourier Transform of an overlapping sliding window section of data (for these spectrograms, the window length is 5000 points, with 99% overlap) for extracted time series #5020 (figure on left, from reentry area) and #4050 (figure on right, from focal area), showing frequency magnitude changes (colorbar denotes magnitude in dB) as a function of time. The main feature for both reference points, the 10 Hz band, remains steady throughout both time series. At the focal area (right) however, the 8Hz band at the beginning of the time series shifted down to 7Hz by the end of the time series. The reentry area (left) also had the 8 Hz energy band, but it largely disappeared after the defibrillation shock (DS) (which occurred as indicated on the horizontal axes). The focus site (right) also had energy in a 1-2 Hz band which narrowed to a band around 2 Hz after the DS and subsequently reverted to a broader band by the end of the dataset. The reentry site (left) also had a small amount of energy in this band at the beginning of the time series, but it dissipated at the time of the DS.

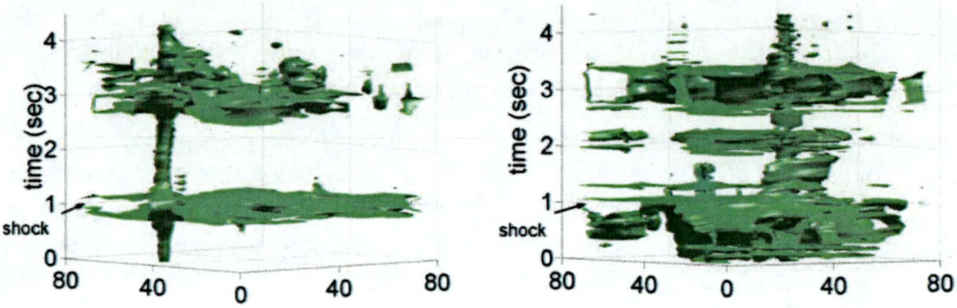

FIGURE 3. Time-space cube of synchronization measures of processed data with different reference points. Overlapping sliding windows of data (2000 points long, with 95% overlap) are used to calculate first Fourier mode changes with time over the tissue. It is referenced to points in the reentry area (pixel #5020, left figure) and in the focal area (pixel #4050, right figure). Values displayed are measures greater than 2/e ($\approx$ 0.74). The z-axis is time (in sec), and the x- and y-axes are position on the tissue surface (extracted from 80x80 pixels covering 5x5 cm area). The shock occurs as indicated on the on the z-axis and is easily visible in the data, since both figures show whole-surface similarity at the same time. The focal area (right) was the more typical behavior (over the entire imaged surface) directly after the defibrillation shock, but reentry behavior slowly reasserted itself over the tissue.

chosen, one within the area previously characterized as focal, and the other within the area previously characterized as reentry, and γ was used to determine which type of beating best characterized each section of the heart. We measured the spatial changes in γ on overlapping boxcar segments (window length 2000, overlap 95%) in order to observe the evolution in time of the tissue behavior. Results are shown in Fig. 3 as a time-space cube, with the x- and y-axes demarking the tissue area, and the z-axis giving the time. Looking at Fig.3, one can see that some tissue areas are indeed more coordinated with the focal area just after the DS (Fig.3, right), while other areas seem to have more in common with the reentry area as time progresses (Fig.3, left). Also, with the exception of the DS (at one sec on z-axis) which only briefly causes all of the tissue to behave similarly, the two reference areas maintain separate areas of influence. One can also see a progression in Fig.3 similar to that of Fig.1; after the DS, regular beating emerges (with the focal area behavior being the most dominant immediately after the shock), but gradually reentry behavior reasserts dominance on the tissue surface toward the end of the dataset.

Thus we have presented the evolution of failed defibrillation and spontaneous focal beating through frequency and synchronization analysis. From the spectrograms, one sees that the focal area displays unique behavior compared to the rest of the tissue (other areas, not shown in this paper, were also examined for confirmation of this fact). This quantifies the tisue pacemaking properties, which changed in frequency and magnitude over time. The synchronization analysis categorizes the tissue behavior into two categories: focal-like behavior or reentry-like behavior. One can clearly see periods in which the entire tissue behaves similarly to the focal area and the gradual breakdown of this type of behavior, with reentry gradually becoming the more dominant type of behavior as time progresses. Future work will focus on the changeover from focal-driven to reentry-like activity, using the spatial distribution of changes in the synchronization index.

ACKNOWLEDGEMENTS

We would like to thank Sheri Prucka for the monetary and scientific support that enabled us to attempt this work. We would also like to thank Dr. Michael Shlesinger and the US Office of Naval Research for continuing support and inspiration.

REFERENCES

[1] Winfree, A.T. *Science* **266**, 1003-1006 (1994).

[2] Witkowski, F.X., Leon, L.J., Penkoske, P.A., Giles, W.R., Spano, M.L., Ditto, W.L., and Winfree, A.T., *Nature* **392**, 78 (1998).

[3] Witkowski, F.X., Plonsey, R., Penkoske, P.A. & Kavanagh, K.M. *Circ Res* **74**, 507-524 (1994).

[4] Feeser, S.A. et al. *Circ.* **82**, 2128-2141 (1990).

[5] Kantz, H. & Schreiber, T., *Nonlinear Time Series*, Cambridge Univ. Press, Cambridge, 1997, Chap. 2.

[6] Rosenblum, M., Pikovsky, A., Schafer, C., Tass, P. & Kurths, J., *Neuroinformatics,* Elsevier, Amsterdam, 2001, pp 279-321.

[7] Dolan, K. & Neiman, A. *Physical Review E* **65**, 026108 (2002).

[8] Quiroga, R. Q., Kraskov, A., Kreuz, T. & Grassberger, P. *Physical Review E* **65**, 041903 (2002).

CONTROL OF OSCILLATION PATTERNS IN A SYMMETRIC COUPLED BIOLOGICAL OSCILLATOR SYSTEM

Atsuko Takamatsu[*†1], Reiko Tanaka[¶§], Takatoki Yamamoto[†], and Teruo Fujii[†]

[*] PRESTO, Japan Science and Technology Corporation (JST), Japan
[†] Underwater Technology Research Center, Institute of Industrial Science, The University of Tokyo, 4-6-1, Komaba, Meguro-ku, Tokyo, 153-8505, Japan
[¶] Dept. of Applied Physics and Physico-Informatics, Keio University, Hiyoshi, Kohoku-ku, Yokohama, 223-8522, Japan
[§] Control and Dynamical Systems, California Institute of Technology, MC 107-81 Pasadena, CA, 91125, USA

Abstract. A chain of three-oscillator system was constructed with living biological oscillators of phasmodial slime mold, *Physarum polycehalum* and the oscillation patterns were analyzed by the symmetric Hopf bifurcation theory using group theory. Multi-stability of oscillation patterns was observed, even when the coupling strength was fixed. This suggests that the coupling strength is not an effective parameter to obtain a desired oscillation pattern among the multiple patterns. Here we propose a method to control oscillation patterns using resonance to external stimulus and demonstrate pattern switching induced by frequency resonance given to only one of oscillators in the system.

INTRODUCTION

Most of biological systems can be modeled with geometrical symmetries. For example, neuronal systems for control of animal locomotion consist of central nervous system and ganglions arranged symmetrically. Such systems could be modeled by a coupled oscillator system that has potential to show a variety of oscillation patterns, namely, a variety of locomotion patterns. One of the most popular parameters to control the patterns would be coupling strength between oscillators. It is, however, ineffective to obtain a desired pattern among multiple patterns if the system has multi-stability. Here we propose a method to control the oscillation patterns by controlling the frequencies of a single oscillator in symmetric coupled oscillators.

We constructed three-oscillator system in a chain with living biological oscillators of plasmodial slime mold, *Physarum polycehalum*. The plasmodial slime mold is an amoeboid multinucleated unicellular organism and shows various oscillatory phenomena in the concentration of intracellular chemicals such as ATP and Ca^{2+} that are considered to lead contraction/relaxation states in the cell, namely, thickness

---

[1] Electronic address: atsuko@iis. u-tokyo.ac.jp

CP676, *Experimental Chaos: 7th Experimental Chaos Conference*,
edited by V. In, L. Kocarev, T. L. Carroll, B. J. Gluckman, S. Boccaletti, and J. Kurths
© 2003 American Institute of Physics 0-7354-0145-4/03/$20.00

oscillation of the cell, induced by complicated mechanochemical reactions among the chemicals, proteins, and intracellular organelles. The thickness oscillation causes protoplasmic streaming inside tube structures of the plasmodial cell, which would relate to interactions between partial bodies of the cell. Thus, this system can be regarded as an assemble of nonlinear oscillators coupled by the tube structure. We have constructed two-oscillator systems and three-, four-, five-oscillator systems in rings by a cell patterning method and have shown that the symmetrically arranged oscillator-systems exhibit multiple oscillation patterns predicted by the symmetric Hopf bifurcation theory using geometrical symmetry proposed by Golubitsky and Stewart [1-4]. In order to obtain more information on the system, we investigated the responses to environmental change. As one of the results, we demonstrate pattern switching between different oscillation patterns induced by frequency resonance in order to control the multi-stable patterns.

METHODS

A three-oscillator system in a chain was constructed with a microfabricated structure of SU-8 (photoresist resin) as shown in Fig. 1(a),(b) by a method previously reported [1,3]. The structure consists of oscillator parts, where thickness oscillation is observed, and channel structures. The coupling strength between the oscillators can be controlled with the channel width in the microfabricated structure [1]. In this experiment, the channel width was fixed at 400µm that corresponds to strong coupling. In this condition, the system could show multiple oscillation patterns [2]. It is known that oscillation frequency in the plasmodium is affected by temperature change [5]. A chip for temperature control was fabricated by patterning a transparent resistant material, ITO (Indium Tin Oxide), on a glass substrate to be used as multi-channel sensors and heaters. Local stimuli in small areas (2×2mm) can be applied to multiple points with small distances (4mm) [6, 7]. The plasmodial oscillator system was placed on this chip. Moreover, transparency of ITO material is useful for optical observation of cells. Stimuli were applied by periodic change of temperature to one of the oscillators located in the center (Oscillator 2; denoted as O2 in Fig. 1 (c)). Temperature of the other oscillators was maintained at 21°C.

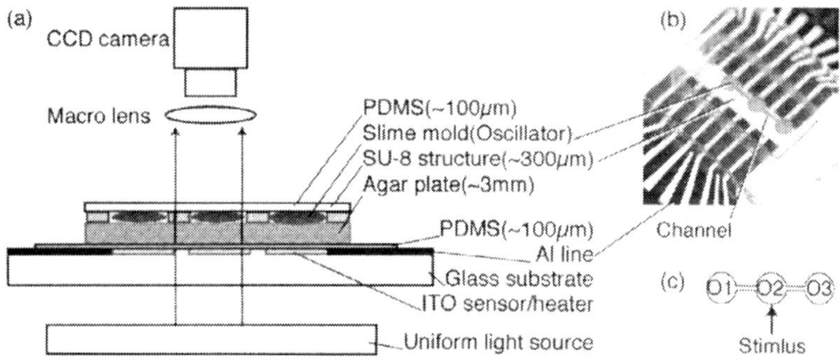

FIGURE 1. (a), (b) Experimental setup. (c) Three-oscillator system in a chain.

OSCILLATION PATTERNS

The system shows multiple oscillation patterns irrespective of existence of stimulus by temperature change. In Fig.2 (a), oscillator 1 (O1) and 3 (O3) show anti-phase oscillation while oscillator 2 (O2) shows double frequency compared with the other oscillators (pattern A). In Fig.2 (b), adjacent oscillators are in anti-phase (pattern B). In Fig.2 (c), the frequencies of all oscillators are doubled (pattern C). As seen in the time series of thickness oscillation in Fig.2 (c), the frequencies are original before the time 200 (sec), and doubled after the time 200 (sec) (denoted with a gray box). In Fig.2 (d), the phases in adjacent oscillators are shifted by 90° then the phase wave propagates from one end to another end (pattern D). The observed patterns were dependent on the sample of the plasmodium even when the coupling strength was fixed, which implies that this system has multi-stability. In addition, spontaneous pattern switching was observed except between the pattern A and B. The most frequently observed pattern was A (41%; in number of periods in 30 minutes observation of 33 samples), secondly C (13%) and D (12%), followed by B (7%), when the coupling strength is fixed at one parameter. The probability of the observed patterns depends on the coupling strength.

By the group theory, this system has Z_2-symmetry that has a reflection denoted with β in Fig. 2 (e). The symmetric Hopf bifurcation predicts the patterns A and in the system with Z_2-symmetry [4]. The pattern A keeps spatio-temporal symmetry under the action of reflection β with a phase shift of 180°. Double frequency of O2 in the pattern A is generated to keep the symmetry when the phase is shifted by 180°, since O2 is located on the reflection axis. The pattern B keeps the symmetry under only the reflection β. The pattern D, however, dose not keep the symmetry under the reflection β with any phase shift. This oscillation pattern would be generated by D_4-symmetry where a hidden oscillator is assumed, whose details will be reported elsewhere. Double frequency of all oscillators in the pattern C cannot be explained by Z_2-symmetry, which could be caused by system specific characteristics of the plasmodium. For simplicity, we focus on only the patterns A and B in this report and, in order to control the patterns A and B, we applied the external stimulus to the center oscillator to keep Z_2-symmetry.

PATTERN SWITCHING BY RESONANCE

We applied the stimulus by change of temperature to the center oscillator O2 and changed the only one control parameter ω_f/ω_0 from 0.5 to 2.0, where ω_f is an angular frequency of stimulus and ω_0 is an original angular frequency of O1 and O3 before stimulus ($\omega_0 \equiv \omega_1 \sim \omega_3$, $\omega_0 \sim \omega_2/2$ in the pattern A, $\omega_0 \sim \omega_1 \sim \omega_2 \sim \omega_3$ in the pattern B).

Fig. 3 shows an example of response to the periodic stimulus. Temperature was changed in a sinusoidal function form ($T_0 + A \sin(\omega_f t + \phi)$; $T_0 = 21.0°C$, $A = 1.5°C$, ϕ is an arbitrary initial phase) from the time around 1900 (sec) as shown in Fig. 3(a). The system showed the oscillation pattern A before starting the stimulus and the pattern B after several periods of stimulation. When the stimulus started, the double frequency of O2 was destroyed, then the dynamical change in phase difference between O1 and O3 from 180° to 0° was observed at the time around 3500 (sec) in Fig. 3 (f). Switching from the pattern B to A was also observed under different condition of stimulus (data not shown).

The mechanism of the switching is considered as follows. Each plasmodial oscillator shows resonance against external periodic stimulus which means the oscillator synchronize with external the stimulus to change its frequency.

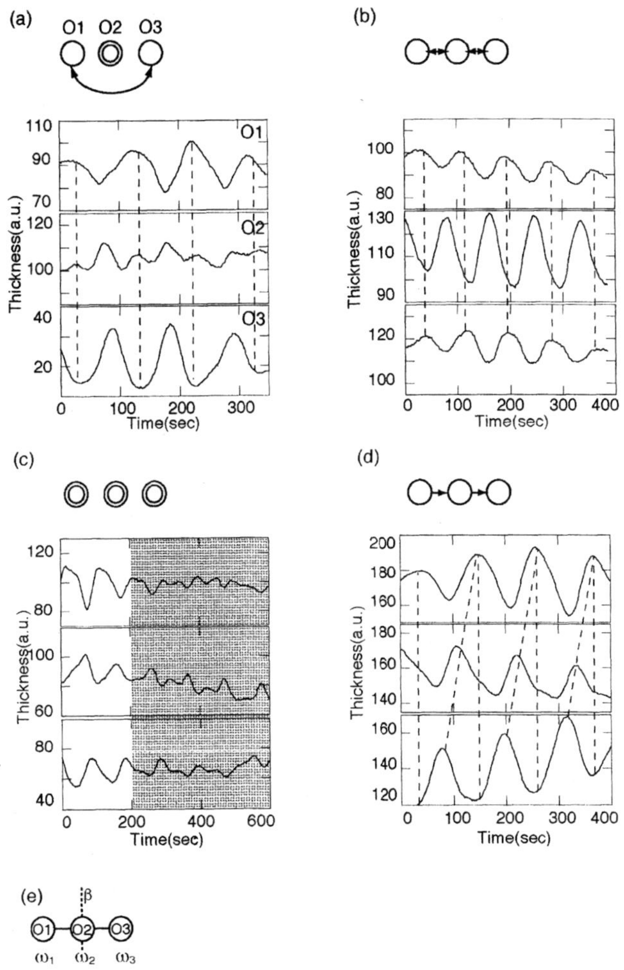

FIGURE 2. (a-d) Schematic diagrams of oscillation patterns and thickness oscillation. Open circles are oscillators. Concentric circles are oscillators have double frequency. Bidirectional and unidirectional arrows mean that phase difference between oscillators is 180° and 90°, respectively. (e) Z_2 symmetry in a chain of three oscillators. β is a reflection.

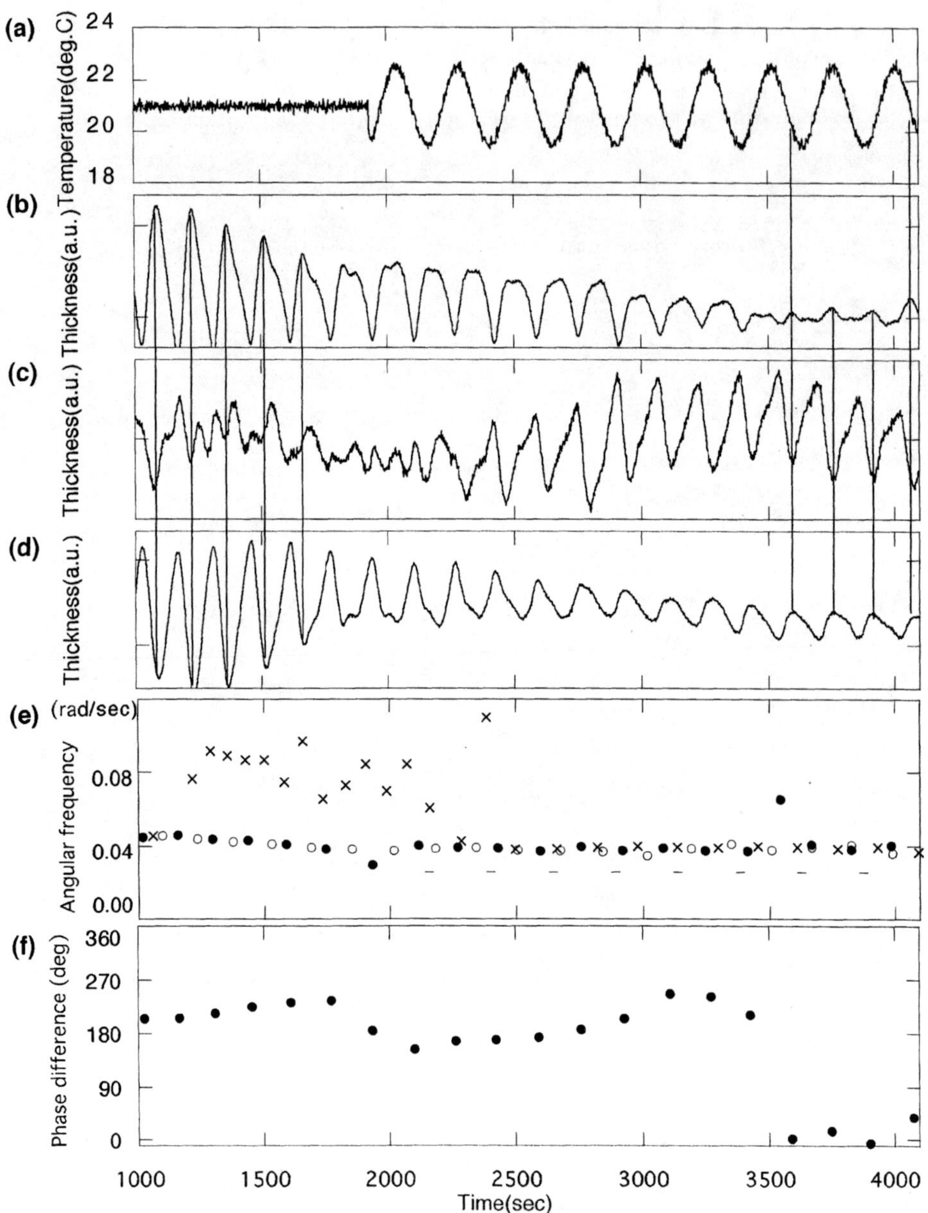

FIGURE 3. Pattern switching by periodic force. (a) Stimulus by temperature change. (b-d) Thickness oscillation in oscillator 1-3. (e) Angular frequency. Bars, filled circles, crosses, open circles are those of stimulus, thickness change in Oscillator 1, 2 and 3, respectively. Original frequencies: ω_1=0.044±0.003, ω_2=0.083±0.010, ω_3=0.043±0.002. Frequency of the stimulus: ω_f = 0.026. Frequencies during stimulus: ω_1=0.040±0.001, ω_2=0.040±0.001, ω_3=0.039±0.002. (f) Phase difference of oscillation between oscillator 1 and 3.

In one oscillator system of the plasmodium, we have observed 1:1, 1:2, 1:3 (ω: ω_f; ω is angular frequency of O1 or O3 during stimulus) resonances and the resonances in other fractional number combination such as 3:2, 4:5, etc., and have found that its phase diagram has Arnol'd tongues [8]. If ω_f /ω_0 is within the tongue, O2 resonates with a different frequency ω from original ω_0. In the case of Fig. 3, $\omega_0 \sim 0.044 \pm 0.003$ and $\omega \sim 0.039 \pm 0.002$ and the system shows 3:2 resonance. It should be noted that the frequency was reduced to 86 ± 9 (%) of the original in all the samples where switching from the patterns A to B was observed. The system showed the resonance in an appropriate ratio in most cases. If the stimulus was applied to increase the frequency, pattern switching occurs from A to B was not observed. On the other hand, the frequency increased when the switching from B to A. Consequently, it is concluded that the control between the pattern A and B can be achieved only by controlling the frequency of the center oscillator that affects on both ends of oscillators symmetrically.

CONCLUSION

We demonstrated the control of the oscillation patterns by using the resonance to external stimulus in the multi-stable system. The multi-stability of the system is commonly observed in other systems of coupled oscillators as well [9]. Thus, the method we demonstrated here could be generally effective to control such system if the oscillation pattern depends on the frequency. Furthermore, we observed the dynamical pattern switching from/to in-phase to/from anti-phase by taking notice of the oscillators in both ends. The anti-phase mode, for example, corresponds to walking mode in bipedal animal and in-phase mode corresponds to hopping mode. This method might also be useful as the model of the control of animal locomotion. Moreover, it should be stated the importance of consideration of the symmetry in the system which generally observed in nature [4, 10].

REFERENCES

1. Takamatsu, A. , Fujii, T., Endo, I. *Phys. Rev. Lett.* **85** 2026 (2000).
2. Takamatsu, A. et al., *Phys. Rev. Lett.* **87** 078102 (2001).
3. Takamatsu, A. and Fujii, T. "Construction of a living coupled oscillator system of plasmodial slime mold by a microfabricated structure" in *Sensors Update* Vol. 10, edited by Wiley-VCH, Weinheim, 2002, pp. 33-46.
4. Golubitsky, M. and Stewart, I., *The Symmetry perspective*, Birkhäuser Verlag, Basel, 2001.
5. Matsumoto, K. et al. *J. Theor. Biol*, **131** 175 (1988).
6. Yamamoto, T. Nojima, T. and Fujii, T. "PDMS-glass hybrid microreactor for cell-free protein analysis" (2002) submitted.
7. Takamatsu, A. and Fujii, T., "Observation of response to stimuli in oscillating cells patterned by microfabricated structure" *Proceedings of The Sixth International Symposium on Micro Total Analysis System (μTASs)*, Nara, Japan, 2002, to appear.
8. Takamatsu, A., Yamamoto, T. and Fujii, T., "Resonant pattern in a plasmodial slime mold by periodic force of temperature." (2002) in preparation.
9. Yoshimoto, M, Yoshikawa, K. and Mori, Y. *Phys. Rev. E* **47** 864 (1993).
10. Tanaka, R., Iwata, S and Shin, S. *Trans. of the Society of Instrument and Control Engineers* 33 441 (1997).

NEUROPHYSIOLOGY

Neurocomputation with Separatrices

Mikhail I. Rabinovich*, Pablo Varona*† and Valentin S. Afraimovich**

*Institute for Nonlinear Science, UCSD. 9500 Gilman Dr. 0402. La Jolla CA 92093.
†GNB. Dpto. de Ingeniería Informática. Universidad Autónoma de Madrid. 28049 Madrid, Spain.
**Instituto de Investigación en Comunicación Óptica, UASLP. A. Obregón 64, 78000 San Luis
Potosí, SLP, México.

Abstract. Information processing in neural networks by computation with attractors (steady states, limit cycles, strange attractors) has been extensively discussed in application to many neural systems: central pattern generators, sensory systems (e.g. visual, olfactory), hippocampus, etc. Computation with attractors in a traditional way faces a fundamental contradiction between robustness and sensitivity. In this paper we discuss a new direction in information neurodynamics based on experiments performed in the locust olfactory system, the orientation sensory system of the marine mollusk *Clione* and the hippocampal place cell networks. This new concept uses the transformation of the incoming spatial or identity information into spatio-temporal output based on the intrinsic switching dynamics of neural networks with nonsymmetric inhibitory connections. This is called the Winner-Less Competition Principle (WLC). The key feature of a network that computes with separatrices is the robustness against noise and the simultaneous sensitivity of the sequence of switching to the incoming information. We present rigorous results about the stability of the sequential switching in the framework of the Lottka-Volterra model. Because of their fast reaction, the discussed neural networks are able to change their intrinsic dynamics to respond to new incoming information and solve many different functional tasks. Computation with separatrices can also be an optimal principle for the design of new paradigms of artificial neural networks.

INTRODUCTION

Plenty of experimental evidence shows that information processing in neural systems, i.e. neurocomputation, can be modeled by dynamical systems. This is possible because two conditions are usually satisfied: (i) time is included in the computation, and (ii) the algorithm of the computation is a set of the deterministic rules. Operations such as associative pattern recognition, sequential learning, generation of spatio-temporal patterns that control behavior, and many others include time in the computation in a natural way. Some computation tasks, like encoding and presentation of sensory information (visual, olfactory, etc.) formally can be done "algebraically", without time, by using "identity" or "spatial" coding (e.g. spatial convergence and divergence and filtering of a spatial message). However, even if it is not necessary in principle, real sensory systems include time in the encoding space to solve non temporal problems such as the encoding and presentation of spatial images. This happens because the spatio-temporal encoding of sensory information has many privileges for the next steps of the information processing (recognition, association with messages from other sensory systems, and decision making for the behavior). Including time in the information processing of spatial images provides such processing with robustness against noise and, at the same time, sensitiveness to the variation of the stimuli. In order to understand

CP676, *Experimental Chaos: 7th Experimental Chaos Conference,*
edited by V. In, L. Kocarev, T. L. Carroll, B. J. Gluckman, S. Boccaletti, and J. Kurths
© 2003 American Institute of Physics 0-7354-0145-4/03/$20.00

and model the process of computation in any neural system, we need to know how sensory signals (or other type of information) have to be fed into and how the result of the computation has to be read out. In the language of dynamical systems that means that we have to know what state of the system corresponds to the final result. It seems very natural to propose that the result of the computation has to be an attractor. Several models of neural systems that compute with attractors are well known. In such models the incoming information usually is a vector of initial conditions and the result of the computation is an attractor. It can be a stable fixed point corresponding to a minimum of the potential function like in the popular Hopfield model [1], a limit cycle [2], or a strange attractor [3].

It is important to emphasize that many models of computation that use the incoming information just as an initial condition for the dynamical system are not flexible and powerful enough. A more general computing approach would take advantage of the fact that the system can change its dynamics depending on the quality and the quantity of the incoming information. Such changing may follow the changing of the stimulus, which is also in many cases strongly time-dependent.

In this paper we discuss a new approach for neurocomputation: computation with separatrices that is based on the WLC principle [4]. According to this approach the computation is a non-stationary stimulus-dependent dynamics of the neural network: a sequential switching from one semi-stable state to another. The mathematical image of the semi-stable state is a saddle fixed point or a saddle limit cycle. For a stable computation with separatrices, each saddle state has to have a one-dimensional unstable separatrix. The sequence of switching is represented in the phase space of the dynamical system by a heteroclinic chain or a heteroclinic contour. Each stimulus has to build just one heteroclinic contour in the phase space of the dynamical system, and each heteroclinic contour corresponds to a specific stimulus. The heteroclinic chain can be open (non-closed).

The questions that we are going to discuss below are: (i) how a dynamical system that computes with separatrices sensitively responses to incoming signals, (ii) what are the conditions for the robustness of such computation, e.g. the topological similarity of the perturbed and original heteroclinic contour, and (iii) how subsystems with closed heteroclinic contours interact with each other. In fact we are going to show in this paper that WLC neural networks that compute with separatrices are able to solve the fundamental contradiction between robustness and sensitivity.

THE MODELS

The activity of many different neural networks [4, 5, 7, 6], can be implemented with the following dynamics:

$$\dot{a}_i \;=\; a_i\left(\sigma(H,S) - \sum_{j=1}^{N} \rho_{ij} a_j + H_i(t)\right) + S_i(t) \tag{1}$$

where $a_i > 0$ represents the instantaneous spiking rate of the principal neurons (PNs) that are making the computation, ρ_{ij}, represents the strength of inhibition in i by j, $H_i(t)$ represents the action from other neural ensembles, and $S_i(t)$ represents the stimuli from the sensors. In many neural networks, the inhibition among PNs is the result of the action of inhibitory local neurons (LNs). Usually LNs also receive an external input and because of this ρ_{ij} can depend the on stimuli.

The dynamical system (1) in the case $\sigma = 1$, $H(t) = S(t) = 0$ is the Lottka-Volterra model. The dynamics of the system is well known when the matrix ρ_{ij} is symmetric ($\rho_{ij} = \rho_{ji}$). In this case the autonomous system has a global Lyapunov function [6, 1] and every trajectory approaches one of the numerous possible equilibrium points. For example, if the inhibitory connections are identical, $\rho_{ij} = \rho$, $\rho_{ii} = 1$, this system has only one global attractor, e.g. $a_i = a_0 = 1/[1+\rho\,(N-1)]$ for $\rho < 1$, and N attractors: $a_i = a_0 = 1$, $a_{j \neq i} = 0$ if $\rho > 1$. No other attractors, e.g. limit cycles, or strange attractors are present in the system. The situation is much more complex and interesting when the inhibition is non-symmetric: $\rho_{ij} \neq \rho_{ji}$. A detailed analysis is only possible in the case $N = 3$ (see references [8, 9, 4]). When $\rho_{ij} > 1, \rho_{ji} < 1$ there exists a heteroclinic contour in the phase space of the system that consists of saddle points and one-dimensional separatrices connecting them. In some regions of the parameter space ρ_{ij}, such heteroclinic contour (or limit cycle in its vicinity) is a global attractor. If ρ_{ij} depend on the stimulus, e.g. as a result of a learning mechanism, the system (1) can generate different heteroclinic contours for different stimuli [4].

Suppose

$$\rho_{ij} = \begin{pmatrix} 1 & \alpha_1 & \beta_1 \\ \beta_2 & 1 & \alpha_2 \\ \alpha_3 & \beta_3 & 1 \end{pmatrix}$$

and $0 < \alpha_i < 1 < \beta_i$ and $\kappa_i = (\beta_i - 1)/(1 - \alpha_i)$. Then the heteroclinic contour is a global attractor if $\kappa_1 \cdot \kappa_2 \cdot \kappa_3 > 1$, and the nontrivial fixed point $A(a_1^0, a_2^0, a_3^0)$ is a saddle point. If $\kappa_1 \cdot \kappa_2 \cdot \kappa_3 = 1$, this fixed point becomes neutrally stable and there exists a family of neutrally stable periodic solutions in the phase space. When $\kappa_1 \cdot \kappa_2 \cdot \kappa_3 < 1$, A becomes a global attractor. The heteroclinic contour exists but looses its stability. It is important to emphasize that in the case $\kappa_1 \cdot \kappa_2 \cdot \kappa_3 > 1$ a small perturbation is able to destroy the heteroclinic contour and then a stable limit cycle appears in its vicinity. This limit cycle is characterized by a finite time period of switching among different states, in contrast with the infinite time of motion along the heteroclinic loop.

When $N > 3$ the dynamics of system (1) can be very complex and even chaotic [5]. We are interested here in the existence and stability of the heteroclinic contours, which are the mathematical image of the winnerless competition behavior. Such contours may exist only in the nonsymmetric case e.g. $\rho_{ij} \neq \rho_{ji}$, when the saddle points (in the heteroclinic contours) satisfy certain conditions.

EXISTENCE AND STABILITY OF THE HETEROCLINIC CONTOUR

In the autonomous case, the system (1) is the canonical Lottka-Volterra model:

$$\dot{a}_i = a_i[1 - (a_i + \sum_{i \neq j}^{N} \rho_{ij} a_j)]. \tag{2}$$

This model, in fact, is a normal form for a neural network with inhibitory connections. Suppose that the dynamics of a network of N inhibitory coupled neurons with M dynamical variables $y_i(t) = (y_i^{(1)}(t), ..., y_i^{(M)}(t)), i = 1, ..., N$ can be described in the form of the following system of ODEs:

$$\dot{y}_i = F(y_i) - \sum_{j=1}^{N} G_{ij}(S)(y_i, y_j) + \tilde{S}_i(t) \tag{3}$$

where F is a nonlinear function that describes the dynamics of an individual neuron with M variables, $G_{ij}(S)(y_i, y_j)$ is a nonlinear operator describing an inhibitory action of the j-th neuron onto the i-th neuron, $S(t) = (S_1(t), ..., S_N(t))$ and $(\tilde{S}_1(t), ..., \tilde{S}_N(t))$ are the vectors representing stimuli to the network. The stimulus here acts in two ways: (i) it adds the perturbation $\tilde{S}(t)$ into (3) as an external force, and (ii) it forms the matrix $G_{ij}(S)$. As a result of averaging, a simplified model that describes the firing rate of the neurons can be written in the form (1), where $\sigma = -1$ when there is no stimulus, and $\sigma = 1$ when the stimulus has a component at neuron i [4]. In the absence of the external force, $\tilde{S}(t)$ the system (2) is just a subsystem of (1) for which all $\sigma = +1$. Thus, we can formulate the stability conditions for the heteroclinic contours in the framework of this model.

"Codimension one" saddle points

A heteroclinic contour consists of finitely many saddle equilibria and finitely many heteroclinic orbits connecting this equilibria. Let's denote by A_1 the equilibrium point $(1, 0, 0, ..., 0)$, by A_2 the point $(0, 1, ..., 0)$, and by A_N the point $(0, 0, ..., 1)$. For the sake of simplicity we assume that there is a heteroclinic orbit r_{ii+1} connecting the points A_i and A_{i+1}, $i = 1, ..., N$ and $A_{N+1} \equiv A_1$. It is simple to see that every point A_i must have only one unstable direction. Otherwise the contour can not serve as an attracting set. By direct verification it can be shown that A_i satisfies this assumption provided that:

$$\rho_{ki} > 1, k \neq i+1, \text{ and} \tag{4}$$

$$\rho_{i+1i} < 1 \tag{5}$$

(Here $i + 1 = i$ if $i = N$).

Moreover, if (4) and (5) are satisfied then the unstable direction at the point A_i is parallel (at that point) to the ort(0...010...0), where 1 corresponds to the ith index. An intersection of hyperplanes $P_{2i} = \cap_{j=1, j \neq i, i+1}^{N} a_j = 0$ is a two dimensional invariant manifold containing points A_i and A_{i+1} such that A_i is a saddle point on P_{2i} and A_{i+1} is a stable node on P_{2i}. The system (2) on P_{2i} has the form:

$$\dot{a}_i = a_i[1 - (a_i + \rho_{ii+1}a_{i+1})], \quad \dot{a}_{i+1} = a_{i+1}[1 - (a_{i+1} + \rho_{i+1i}a_i)] \tag{6}$$

and, from (4) and (5), one has $\rho_{ii+1} > 1, \rho_{i+1i} < 1$.

This implies that there are no equilibrium points in the region $a_i > 0$, $a_{i+1} > 0$ and, since $\dot{a}_{i+1} < 0$ if $a_{i+1} \gg 1$, then it is simple to see that the separatrix, say Γ_i, of the saddle point A_i must go to the attractor A_{i+1}, i.e. there is a heteroclinic connection between A_i and A_{i+1} on the plane P_{2i} (for the case N=3 see the proof in [8]).

Saddle values

The point A_i is a saddle point on P_{2i}. One can write a map from a transversal to the stable separatrix into a transversal to the unstable separatrix along the orbits going through a neighborhood of A_i. In suitable coordinates (ξ, η) it has the form:

$$\xi = c\eta^{\nu_i} \tag{7}$$

where η is a deviation from the stable manifold and ξ is the deviation from the unstable one, c is a constant and

$$\nu_i = -\frac{1 - \rho_{ii+1}}{1 - \rho_{i+1i}} \equiv \frac{\rho_{ii+1} - 1}{1 - \rho_{i+1i}} \tag{8}$$

is the "saddle value" [10]. If $\nu_i > 1$ then the map (7) is a local contraction and P_i is a dissipative saddle. If $\nu_i < 1$ then(7) is a local expansion.

Stability of the heteroclinic contour

The following result tells us that the contour $\Gamma = \cup_{i=1}^{N} \Gamma_i \nu A_i$ can be an attractor.

Theorem 1 *Assume that conditions (4), (5), (6) are satisfied and*

$$\nu = \prod_{i=1}^{N} \frac{\rho_{ii+1} - 1}{1 - \rho_{i+1i}} > 1 \tag{9}$$

(here $i+1 = 1$ if $i = N$). Then there is a neighborhood U of the contour Γ such that for any initial condition $a^0 = (a_1^0, ..., a_N^0)$ with $a_i^0 > 0$, one has $dist(a(t), \Gamma) \to 0$ as $t \to \infty$ where $a(t)$ is the orbit going through a^0.

The proof of the theorem is based on the construction of the Poincaré map along orbits in a neighborhood of the contour Γ (see [11]).

BIRTH OF A STABLE LIMIT CYCLE. ROBUSTNESS

A direct corollary of Theorem 1 is the possibility of the birth of a stable limit cycle in an appropriately perturbed system. Consider the system

$$\dot{a}_i = a_i[1 - (a_i + \sum_{j \neq i}^{N} \rho_{ij} a_j)] + \varepsilon \Psi_i(a) \tag{10}$$

that coincides with (2) for $\varepsilon = 0$, where $a = (a_1, ..., a_N)$ and Ψ_i is a smooth function, $i = 1, ..., N$. For small $\varepsilon > 0$ the system (10) has saddle equilibrium points $A_{i\varepsilon}$ and separatrices $\Gamma_{i\varepsilon}$ (the half of $W_{i\varepsilon}^N$ such that $A_{i\varepsilon} \to A_i$, as $\varepsilon \to 0$ and $lt_{\varepsilon \to 0} \Gamma_{i\varepsilon} \supset \Gamma_{ii+1}$, here lt means the topological limit, i.e. the set of the accumulation points).

Theorem 2 *Assume that the conditions of Theorem 1 are satisfied,*

$$lt_{\varepsilon \to 0}(\bigcup_{i=1}^{N} \Gamma_{i\varepsilon}) = \Gamma \tag{11}$$

and at least one of the separatrices $\Gamma_{i\varepsilon}$ is not a heteroclinic orbit. Then for any sufficiently small $\varepsilon > 0$ the system (10) has a stable limit cycle L_ε (in a neighborhood of Γ) such that $lt_{\varepsilon \to 0} L_\varepsilon = \Gamma$.

The proof of this Theorem can be done in the standard way, i.e., by construction of the Poincaré map and by showing that this map is a contraction in an absorbing region. The condition (11) (or a similar condition) is necessary and sufficient for the existence of an absorbing region [11].

Numerical results show that the system (10) where $\Psi_i(a) \geq 0$ satisfies the condition (11) and has a stable limit cycle (see Figure 1). In this example, the simulations were performed with the following equations:

$$\dot{a}_i = a_i(1 - \sum_{j=1}^{N=6} \rho_{ij} a_j) + \varepsilon a_{i-1} a_{i+2} \tag{12}$$

where $i = 1, 2, ..., 6$ and $i + 3 \equiv i - 3$ if $i > 3$. We used the following values of the connection matrix $\rho_{ij} \neq 0$:

$$\begin{aligned}
\rho_{1,3} &= \rho_{3,5} = \rho_{5,1} = 5; \ \rho_{4,6} = \rho_{2,4} = \rho_{6,2} = 2 \\
\rho_{1,6} &= \rho_{2,1} = \rho_{3,2} = \rho_{4,3} = \rho_{5,4} = \rho_{6,5} = 1.5 \\
\rho_{1,1} &= \rho_{2,2} = \rho_{3,3} = \rho_{4,4} = \rho_{5,5} = \rho_{6,6} = 1
\end{aligned} \tag{13}$$

with $\varepsilon = 0.01$.

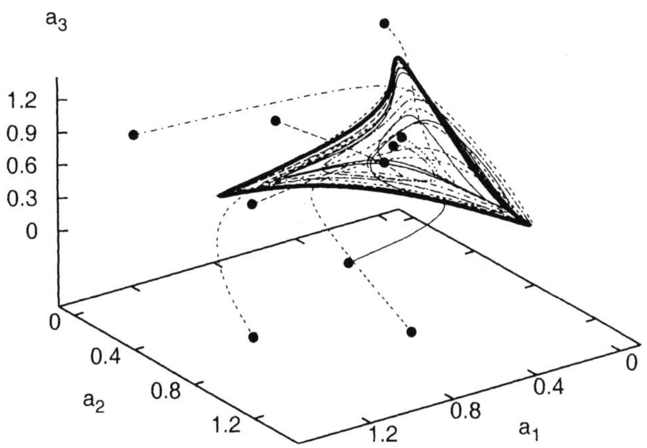

FIGURE 1. 3D projection of the 6-dimensional system (12) showing examples of trajectories falling into the limit cycle from different initial conditions. As the numerical results indicate, this limit cycle in the vicinity of the former heteroclinic contour is a global attractor.

COMPUTATION WITH SEPARATRICES IN THE OLFACTORY SYSTEM

We have used observed features of olfactory processing networks [12] as a guide to the study of computation using competitive networks. In Figure 2 we show the simultaneously recorded activity of three different projection neurons (PNs) in the locust olfactory system, i.e. antennal lobe (AL), evoked by two different odors: despite similar PN activities before the stimulus onset (the result of the action of noise) each odor evokes a specific spatio-temporal activity pattern that results from interactions between these and other neurons in the network [12]. WLC networks produce identity-temporal or spatio-temporal coding in the form of deterministic trajectories moving along the heteroclinic contour. The saddle states in this case correspond to the activity of specific neurons or groups of neurons and the separatrices connecting these states correspond to sequential switching from one state to another.

From the experimental results we infer that a stimulus acts in two principal ways as we hypothesize above: (1) it excites a subset of projector neurons; (2) it modifies the effective inhibitory connections between the projector neurons as a result of the activation of the inhibitory interneurons that connect different PNs. The intrinsic dynamics of these neurons is governed by many variables corresponding to ion channels and intracellular processes. Such detailed description however is not needed to illustrate the principle of "coding with separatrices". We need only to capture the 'firing' or 'not-firing' state of

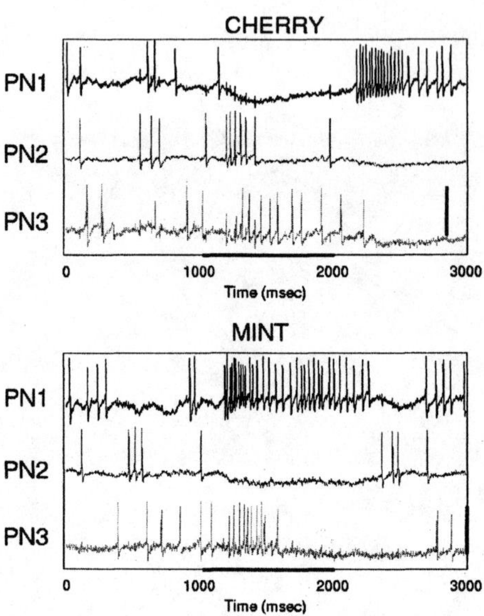

FIGURE 2. The temporal patterns produced by three simultaneously sampled PNs in the locust antennal lobe when two different odors are presented during the time interval from 1000 to 2000 msec. The horizontal bar indicates the time interval when the stimulus was present (see [12] for details, reprinted from [4]).

the component neurons. We thus simplify the model to an equation for the firing rate $a_i(t) > 0$ of neural activity, and thus we arrive to model 1.

When the inhibitory connections are not symmetric, the system with N competitive neurons has different heteroclinic contours depending on the stimulus. The heteroclinic contours are global attractors in the phase space and can be found for a range of values of $\rho_{ij}(\mathbf{S})$. This implies that if the stimulus is changed, another orbit in the vicinity of the heteroclinic contour becomes a global attractor for this stimulus, and guaranties a big capacity of the spatio-temporal representation of the odors [4].

Our numerical simulations [12] show that the network produces different spatio-temporal patterns in response to different stimuli and model the real data from the antennal lobe of the locust.

CHAOTIC DYNAMICS OF COMPETITIVE PATTERNS: HUNTING BEHAVIOR OF CLIONE

Neural networks with WLC dynamics are able to generate new information to answer a simple external signal. Such information can be used for the organization of complex activity and, in particular, chaotic behavior of some animals. Let us consider the hunting activity of a marine mollusk *Clione*. This mollusk is a predator lacking a visual system. It feeds on a small mollusk, *Limacina*. The hunting behavior is a random search for prey: *Clione* "scans" the surrounding space in order to locate and catch the prey. Such behavior is turned on by the smell of the *Limacina*. The main role in the organization of such motion of *Clione* is played by a sensory neural network inside the statocyst (see Fig.3). The statocyst is a special sensory organ responsible for the orientation in the gravitational field [13].

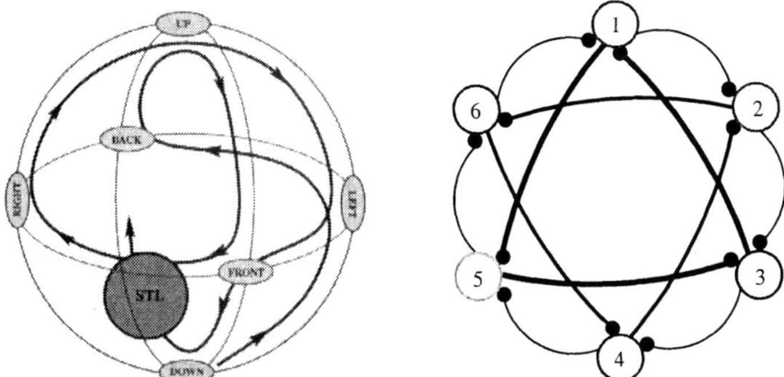

FIGURE 3. Left panel: schematic representation of the statolith motion exciting different receptor neurons inside the statocyst. Right panel: inhibitory connections used in this network (thicker traces mean stronger inhibition).

It is well known from the physiological data that the statocysts have up to 12 receptor neurons (SRNs) that are coupled with inhibitory synapses [13]. These neurons respond to the pressure exerted by the statolith, a stone located inside the statocyst. If no information about a prey (received by the chemical receptors) is present, the receptor neuron D (down, see Figure 3) is excited by the statolith and it inhibits other SRNs. As a result, the information generated by D SRN arrives to the corresponding Central Pattern Generators (CPGs) that control the tail and wing movements. These CPGs establish the habitual "head up" position of *Clione*'s body. If the Hunting Central Neuron (HCN) receives a message from the chemo-sensors about the presence of a prey, HCN excites some SRNs and inhibits others. The behavior of the *Clione* in this case does not depend on the direction of the gravitational field and it moves in a random-like trajectory.

For the phenomenological modeling of the statocyst "hunting" dynamics we can neglect the statolith inertial dynamics and take into account the only key point: the position of the mollusk's body uniquely depends on the message that SRNs are sending to the central neurons that produce the commands to the CPGs. Thus, as a starting point, we consider just a SRN network under the action of the HCN excitation. We suppose that, as a result of the HCN stimulation, all SRNs ("left", "right", "back", "front", "down", and "up") are in the same situation: they receive and send two inhibitory synapses (see Figure 3, right panel).

The dynamics of the SRN's network can be described by model (1) with $N = 6$. In this case, $a_i > 0$ represents the instantaneous spiking rate of the receptor neuron i, $H_i(t)$ represents the stimulus from the hunting neuron to neuron i, and $S_i(t)$ represents the action of the statolith on the receptor that is pressing. When there is no stimulus from the hunting neuron ($H_i = 0, \forall i$) or the statolith ($S_i = 0, \forall i$), then $\sigma(H, S) = -1$ and all neurons are silent; $\sigma(H, S) = 1$ when the hunting neuron is active and/or the statolith is pressing one of the receptors. In our simulations, we have used the values $\rho_{ij} \neq 0$ specified in (13).

247

When there is no activation of the sensory neurons from the hunting neuron, the effect of the statolith ($S_i \neq 0$) in this model is to induce a higher rate of activity on one of the neurons (the neuron i where it rests for a big enough S_i value). We assume that this higher rate of activity affects the behavior of the motoneurons to organize the head up position. The other neurons are either silent or have a lower rate of activity and we can suppose that they do not influence the posture of *Clione*.

When the hunting neuron is active a completely different behavior arises. We assume that the action of the hunting neuron overrides the effect of the statolith and thus $S_i \approx 0, \forall i$. The dynamical system (1) with the ρ_{ij} values specified above (see also Fig. 3) and with a stimuli from the hunting neuron given, for example, by $H_i =$ $(0.730, 0.123, 0.301, 0.203, 0.458, 0.903)$ has a strange attractor in the phase space (see Figure 4). This means that the SRN network generates new information (a chaotic signal with positive Kolmogorov-Sinai entropy) in the presence of the prey, which controls the CPGs and, in fact, organizes the random-like behavior of *Clione*.

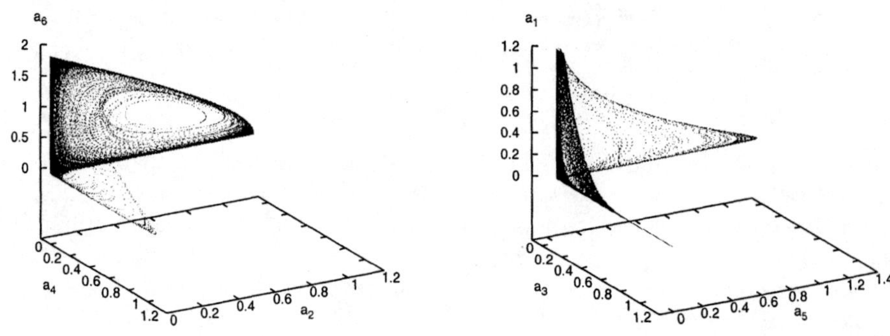

FIGURE 4. Projections of the attractor from the six-dimenional phase space to two different three-dimensional spaces.

The origin of the chaoticity in such dynamical system can be explained in the following manner [5]: due to the diversity in the strengths of the inhibitory connections we may consider the complete network as two weakly coupled WLC triangle networks. Independently each of them has a closed heteroclinic loop (see Figure 1), which becomes a limit cycle under the action of a small perturbation (for example, a stimulus). The periodic oscillations corresponding to these limit cycles have, in general, different frequencies that are extremely sensitive to the distance to the heteroclinic loop in the non perturbed system (such oscillations are strongly non-synchronous). As we showed the weak interaction of these WLC triangles (nonlinear oscillators) generate chaos in wide regions of the control parameter space. New experiments have confirmed the validity of the model [14].

DISCUSSION

We have tried to show that the computation with separatrices based on the WLC principle is a very natural and powerful strategy for information processing in real neural systems. Any kind of sequential activity can be programmed by a network with stimulus dependent nonsymmetric inhibitory connections. It can be the creation of spatio-temporal patterns of motor activity, the transformation of the spatial information into spatio temporal information for successful recognition and many other computations. In addition, we wish to mention that two important computational functions can be successfully implemented by computation with separatrices. These are: (i) sequential memory storage, and (ii) feature binding.

In reference [15] the authors suggest a new biologically-motivated model of sequential spatial memory which is based on the WLC principle. Each stimulus event (visual image, odor, etc...) is represented by a saddle point in the phase space of the system, and a network of one-dimensional separatrices leads the system along the sequence of events in the specific episode. After the learning process, such system is capable of an associative retrieval of the pre-recorded sequence of spatial patterns.

A binding problem occurs when two (or more) different events, e.g. scenes, features, or behaviors are represented by different neural ensembles simultaneously, and for some reason they are all connected with each other. Eventually, these coherent features are integrated by the nervous system of the animal onto a perceptual object, even if the features are dispersed among different sensory systems or subsystems. The binding is ubiquitous and occurs whenever a simultaneous remembrance or representation is important. The most common approach in the modeling of binding is to involve time in operation (von der Malsburg, Singer, and others). The idea is to use the coincidence of certain events in the dynamics of different neural units for binding. This is a dynamic binding. Usually, dynamic binding is represented by synchronous neurons or neurons that are in resonance with and external field. However, dynamical events like phase or frequency variations usually are not very reproducible and robust. It is reasonable to hypothesize that brain circuits that display sequential switching of neural activity [7] use the coincidence of this switching to implement dynamic binding of different WLC networks.

ACKNOWLEDGMENTS

Support for this work came from NIH grant 2R01 NS38022-05A1, Department of Energy grant DE-FG03-96ER14592 and NSF/EIA-0130708. V.A. was supported by Conacyt grant 485100-3-36445-E and by UC Mexus-Conacyt grant.

REFERENCES

1. Hopfield J.J., "Neural networks and systems with emergent selective computational abilities", *Proc. Natl. Acad Sci. USA* **79**, 2554–8 (1982).
2. Li Z., Hopfield J.J., "Modeling of the olfactory bulb and its neural oscillatory processings", *Biological Cybernetics* 61, 379-392 (1989).
3. Skarda, C.A., Freeman W.J., "How brains make chaos in order to make sense of the world", *Behav. Sci.*, **10**, 161–195 (1997).
4. Rabinovich M.I., Volkovskii A., Lecanda P., Huerta R., Abarbanel H.D.I., Laurent G., "Dynamical encoding by networks of competing neuron groups: Winnerless competition", *Physical Review*

Letters **87**(6), 068102–4 (2001).

5. Varona P., Rabinovich M.I., Selverston A.I., Arshavsky Y.I., "Winnerless competition between sensory neurons generates chaos: a possible mechanism for molluscan hunting behavior", *Chaos* **12**(3), 672–677 (2002).

6. Cohen M.A., Grossberg S., "Absolute stability of global pattern formation and parallel memory storage by competitive neural networks", *IEEE Transactions on Systems, Man and Cybernetics*, vol. SMC-13, **5**, 815–26 (1983).

7. Abeles M., Bergman H., Gat I., Meilijson I., Seidemann E., Tishby N., Vaadia E., "Cortical Activity Flips Among Quasi-Stationary States. *Proceedings Of The National Academy Of Sciences Of The United States Of America*", **92** (N19): 8616-8620 (1995).

8. Afraimovich V.S., Hsu S.B., Lin H.E., "Chaotic behavior of three competing species of May-Leonard model under small periodic perturbations", *International Journal of Bifurcation and Chaos* **11**(2), 435–447 (2001).

9. Chi C.W., Hsu S.B., Wu L.I., "On the asymmetric May-Leonard model of three competing species", *SIAM J. Appl. Math.* **58**(1), 211–226 (1998).

10. Kuznetsov, Y.A., *Elements of Applied Bifurcation Theory*, Appli. Math Science, Vol 112. Springer, Berlin (1998).

11. Afraimovich V.S., Rabinovich M.I., Varona P., "Computation with Separatrices: Robustness and Sensitivity", in preparation.

12. Bazhenov M., Stopfer M., Rabinovich M.I., Huerta R., Abarbanel H.D.I., Sejnowski T.J., Laurent, G., "Model of transient oscillatory synchronization in the locust antennal lobe". *Neuron* **30**(N2):553-567 (2001).

13. Arshavsky Y.I, Deliagina T.G., Gamkrelidze G.N., Orlovsky G.N., Panchin Y.V., Popova L.B., "Pharmacologically-induced elements of feeding behavior in the pteropod mollusc *Clione Limacina*. II. Effect of physostigmine", *J. Neurophysiol.* **69**, 522–532 (1993).

14. Levi R., Varona P., Rabinovich M.I., Selverston A.I., Arshavsky Y.I., "Control of hunting behavior in the mollusk Clione: winnerless competition between the statocyst receptor neurons", *SFN Abstracts* **28** 60.1 (2002).

15. Seliger F., Tsimring L., Rabinovich, M.I., "Dynamical model of sequential spatial memory: winnerless competition of patterns". Submitted (2002).

Parameter estimation for neuron models

Isao Tokuda*, Ulrich Parlitz[†], Lucas Illing**, Matthew Kennel** and Henry Abarbanel[‡]

*Department of Computer Science and Systems Engineering, Muroran Institute of Technology, Muroran, Hokkaido 050-8585, Japan
[†]Drittes Physikalisches Institut, Universität Göttingen, Bürgerstraße 42-44, D-37073 Göttingen, Germany
** Institute for Nonlinear Science, University of California, San Diego, La Jolla, CA 92093-0402
[‡] Department of Physics and Marine Physical Laboratory, University of California, San Diego, La Jolla, CA 92093-0402

Abstract. Methods for estimating parameters of the Hindmarsh-Rose (HR) neuron model from a single time series are investigated. Two approaches, (1) synchronization based parameter estimation and (2) adaptive observer, are presented. Both methods are applied to membrane potential data recorded from a single lateral pyloric neuron synaptically isolated from other neurons.

INTRODUCTION

In neuroscience research, nonlinear dynamical systems approach to understanding neuronal activity has recently drawn a great deal of attention [1]. The methods of nonlinear time series analysis [2, 3] applied to isolate neurons from the stomatogastric ganglion of the California spiny lobster *Panuliru interruptus* revealed that the number of degrees of freedom in their membrane potential oscillation ranges typically from three to five [4]. Based on this observation, models of the action potential activity in this biological system have been developed using the framework of Hindmarsh and Rose (HR) [5]. The HR model is represented by three or four dimensional ordinary differential equations (ODEs), in which the complicated current-voltage relationships of the conductance based models [6] are replaced by polynomials in the dynamical variables. The HR model has been further implemented into an analog electronic neuron (EN) [7], whose properties are designed to emulate the membrane voltage characteristics of individual neurons. Using the EN, synchronization and regularization properties between EN and real biological neuron have been reported [8].

Although the numerical and the analog circuit studies have already demonstrated feasibility and consistency of the HR model with the real neurons, the model equations yet have some free parameters. For a systematic construction of the HR model that best matches individual neurons, parameter estimation of the model equations from real data is desired. Here, we present two approaches, (1) synchronization based parameter estimation and (2) adaptive observer, to matching dynamical variable of the HR model to membrane potential data recorded from real neurons.

CP676, *Experimental Chaos: 7th Experimental Chaos Conference,*
edited by V. In, L. Kocarev, T. L. Carroll, B. J. Gluckman, S. Boccaletti, and J. Kurths

SYNCHRONIZATION METHOD

Suppose we have a single time series $s(t)$ measured from an unknown dynamical system. Our goal is to find parameters of a model dynamical system $\dot{\mathbf{y}} = \mathbf{F}(\mathbf{y}, \mathbf{a})$ ($\mathbf{F} : R^d \times R^m \to R^d$) so that the dynamics of $\mathbf{y}$ is commensurate with the unknown dynamics that revealed s. The assumption here is that the general form of $\mathbf{F}$ has been derived as a physiological neuron model but there remain parameter values that are matched to the data s recorded from real neurons. Our approach [9, 10] is to minimize, over free parameters $\mathbf{a}$, $\mathbf{b}$, and $\mathbf{y}_0$,

$$J = \sum_{i=1}^{N} \frac{1}{2}[y_1(t_i) - \Phi_1(s(t_i), \mathbf{b})]^2 \tag{1}$$

subject to the constraints

$$\dot{\mathbf{y}} = \mathbf{F}(\mathbf{y}, \mathbf{a}) + \mathbf{K} \cdot [\Phi(s, \mathbf{b}) - \mathbf{y}], \quad \mathbf{y}(t_1) = \mathbf{y}_0 \tag{2}$$

$$\mathbf{K} = \begin{pmatrix} K & 0 & \cdots \\ 0 & 0 & \cdots \\ \vdots & \vdots & \ddots \end{pmatrix} \tag{3}$$

where $\Phi(s, \mathbf{b})$ is a smooth potentially nonlinear static map with free parameters $\mathbf{b}$.

There is a significant literature in numerical least-squares parameter estimation in ODEs with unperturbed constraint equation, $i.e$ $K = 0$. With long chaotic orbit, this problem becomes progressively ill-conditioned. On account of the exponential separation of trajectories the error surface J grows increasingly bumpy with very large gradients $\partial J / \partial(\mathbf{a}, \mathbf{b}, \mathbf{y}_0)$. Our idea is that, by inducing dynamical synchronization between the signal and the model ($y_1 \approx \Phi_1(s, \mathbf{b})$), the error surface will be smoothed and regularized. With this constrained dynamics, trajectory of the model equations does not run away from the signal even with chaos and numerical process of minimizing the error becomes stable. Among various techniques to minimize the constrained error function J, the quasi-Newton method [11] combined with line search is employed. For computation of the gradients $\partial J / \partial(\mathbf{a}, \mathbf{b}, \mathbf{y}_0)$, variational equations of the constrained dynamics (2) are solved numerically. In order that the model finally reproduces qualitatively similar dynamics as the forcing signal in an autonomous condition, the forcing term of equation (2) must be weakened as $J \to 0$. We realize this by eventually decreasing the coupling strength K nearly to zero in the optimization process.

Let us apply the synchronization technique to 4-dimensional version of the HR equations [7, 8]

$$\frac{1}{\tau} \cdot \frac{dy_1}{dt} = y_2 + b \cdot y_1^2 - y_1^3 - d \cdot y_3 + I + K \cdot s \tag{4}$$

$$\frac{1}{\tau} \cdot \frac{dy_2}{dt} = -y_2 - f \cdot x_1^2 - g \cdot y_4 \tag{5}$$

$$\frac{1}{\tau} \cdot \frac{dy_3}{dt} = \mu \cdot [-y_3 + y_1] \tag{6}$$

$$\frac{1}{\tau} \cdot \frac{dy_4}{dt} = \nu \cdot [-y_4 + y_2]. \tag{7}$$

This HR model has 8 free parameters $\{b,d,I,f,g,\mu,\nu,\tau\}$ (including time scaling parameter τ to match with the real time), where all the 8 parameters were optimized by the synchronization method. For simplicity, the transformation was set to be an identity map, *i.e.* $\Phi_1(s) = s$. As the forcing signal s, membrane potential data $\{s(i{\cdot}\Delta t) : i = 1,\ldots,22000\}$ (sampling frequency: $\Delta t = 0.5$ msec) recorded from a single lateral pyloric neuron synaptically isolated from other neurons were used. As initial condition of the free parameters, parameter values of ref. [8] were used. The coupling strength was initially set as $K = 1$ and with every 250 optimization steps weakened as $K = 0.9$, $K = 0.8$, $\ldots$, $K = 0.1$, $K = 0.1{\times}0.5$, $\ldots$, $K = 0.1{\times}0.5^9$.

Figure 1 shows simultaneous drawing of the real neuron data and the HR dynamics without the forcing term, *i.e.* $K = 0$. Qualitatively, similar behavior has been reproduced by the HR model. The HR dynamics obtained by the synchronization method is chaotic in the sense that it has a positive first Lyapunov exponent $\lambda_1 = 0.0027$ with a Lyapunov dimension $d_{KY} = 2.001$. This is in a good agreement with the real data, which was shown to have a positive Lyapunov exponent by nonlinear time series analysis [4]. Quantitative disagreement between the HR model and the real data such as increasing and then decreasing spike amplitudes present in the data but not in the model might be due to (a) limitation of the simplified model to describe exact physiological process and (b) recording or dynamical noise.

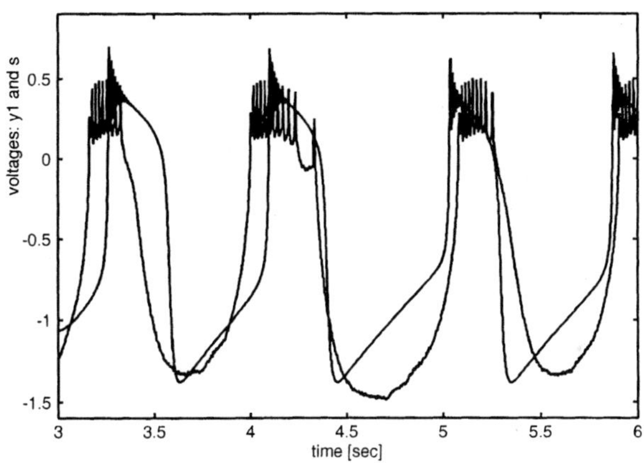

FIGURE 1. Simultaneous drawing of real neuron data (thick line) and HR dynamics (thin line).

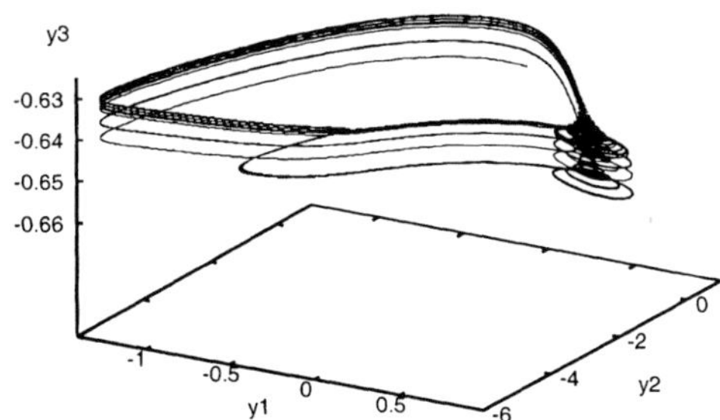

FIGURE 2. Chaotic attractor in (y_1,y_2,y_3)-space obtained by the synchronization method.

ADAPTIVE OBSERVER

In this section we present an adaptive observer, *i.e.* a dynamical system that is driven by the given time series and that possesses additional ODEs governing (slow) variations of model parameters. Once the parameters have converged to the "right" values the model oscillations synchronize with the driving time series.

The observer is based on a 3-dimensional HR model

$$
\begin{aligned}
\dot{v} &= F(v)+y-z \\
\dot{y} &= b_1v+b_2v^2-b_3y \\
\dot{z} &= b_4v-b_5z
\end{aligned}
\tag{8}
$$

where $F(v) = a_0+a_1v+a_2v^2+a_3v^3$. In the above normalized model, redundant parameters present in the original 3-dimensional HR equations [5] have been removed. With this parametrization the model is able to adapt to arbitrary time scale and range of the voltage variable v. To build an observer some extra terms are added to implement driving by an external voltage s

$$
\dot{v} = F(s)+y-z+k(s-v)
\tag{9}
$$

$$
\dot{y} = b_1s+b_2s^2-b_3y+s-v
\tag{10}
$$

$$
\dot{z} = b_4s-b_5z-(s-v)
\tag{11}
$$

and suitable ODEs for all parameters $a_0,...,a_3$ and $b_1,...,b_5$.

$$
\dot{a}_0 = s-v
\tag{12}
$$

$$
\dot{a}_1 = (s-v)s
\tag{13}
$$

$$
\dot{a}_2 = (s-v)s^2
\tag{14}
$$

$$
\dot{a}_3 = (s-v)s^3
\tag{15}
$$

$$
\dot{b}_1 = -0.1(s-v)
\tag{16}
$$

$$
\dot{b}_2 = 0.1(s-v)vz
\tag{17}
$$

$$
\dot{b}_3 = -0.1(s-v)y
\tag{18}
$$

$$
\dot{b}_4 = -0.1(s-v)z
\tag{19}
$$

$$
\dot{b}_5 = -0.1(s-v)yz.
\tag{20}
$$

Note that the right hand sides of all parameter ODEs vanish in the case of perfect synchronization $v = s$. Figure 3 shows the temporal evolution of the measured signal s and the corresponding voltage variable of the HR-observer (9)-(20). The underlying parameter estimation process is shown in Fig. 4. As can be seen in Fig. 3 model variable v converges to the driving neuron voltage s with a small synchronization error $s-v$ (Fig. 3a and section Fig. 3b). The fact that some of the estimated parameters do not converge to fixed values but oscillate with small amplitude may be interpreted as an indication for some limited capability of the HR-model to describe the dynamics of the real neuron quantitatively.

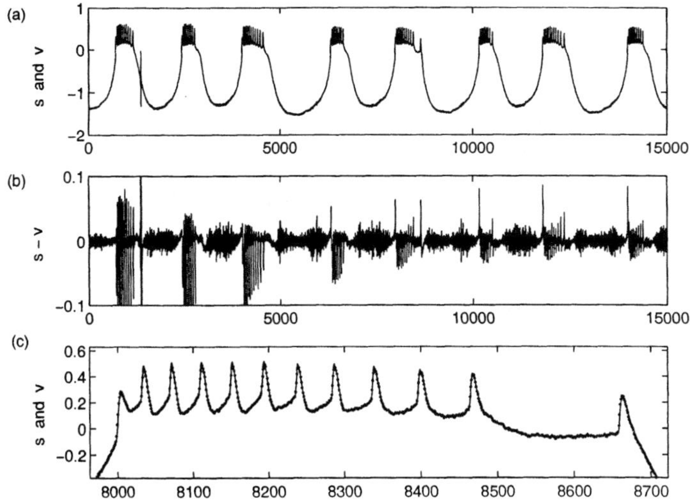

FIGURE 3. (a) Synchronization of the measured voltages s from a real neuron and the corresponding variable v of the Hindmarsh Rose observer (9)-(20). Within the graphical resolution both curves are not distinguishable. (b) Synchronization error $s - v$ vs. sample number. (c) Section of (a) where the driving signal s is plotted using filled circles.

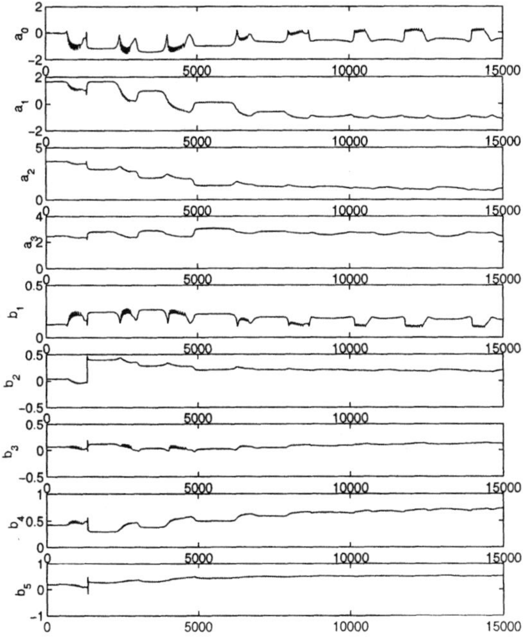

FIGURE 4. Model parameters $a_0, ..., a_3$ and $b_1, ..., b_5$ vs. sample number (time). After some transient the parameters converge to fixed values or oscillate with small amplitude.

CONCLUSIONS

Two approaches have been presented for parameter estimation of the HR models to match with the real neuron data. Both the synchronization technique and the adaptive observer technique utilize synchronized dynamics between the model and the data so as to suppress the modeling error expansion induced by chaotic property inherent in the neuron data. Disadvantages of the synchronization technique are (i) it has a limited basin of attraction that leads to feasible solutions especially when the number of the free parameters is large and (ii) its convergence is rather slow. The adaptive observer, on the other hand, has fast convergence property. There is, however, no systematic way to construct an adaptive observer to optimize all model parameters. Except for some special cases, global convergence of the adaptive observer is not guaranteed.

Alternative approach to the parameter estimation is the multiple-shooting (MS) method [12], which is one of the standard methods for parameter estimation of ODEs. According to our numerical study, as the number of the free parameters is increased, basin of attraction leading to feasible solutions gets much smaller in the MS method and further modifications are necessary in order to apply for the real data. In our future work, combination of the MS method with the idea of synchronization will be examined. Considering the effect of dynamical and observational noise, extended Kalman filtering technique for stochastic nonlinear modeling [13] will be also investigated.

Limitation of the HR modeling of the real neuron data might also be due to an incapability of the simplified HR model to reproduce complicated physiological process. Modified models with (a) nonlinear transformation for adjusting spiking amplitudes to bursting amplitudes and (b) addition of delayed feedback process will be considered in our next study.

REFERENCES

1. H. D. I. Abarbanel and M. I. Rabinovich, Current Opinion in Neurobio. **11**, 423 (2001).
2. H. D. I. Abarbanel, *Analysis of Observed Chaotic Data*, (Springer, New York, 1996).
3. H. Kantz and T. Schreiber, *Nonlinear Time Series Analysis*, (Cambridge University Press, Cambridge, 1997).
4. H. D. I. Abarbanel *et al.*, Neural Comput. **8**, 1567 (1996).
5. J.L. Hindmarsh and R.M. Rose, Proc. R. Soc. Lond. B **221**, 87 (1984).
6. A. L. Hodgkin and A. F. Huxley, J. Physiol. (London) **117**, 500 (1952).
7. A. I. Selverston *et al.*, J. Physiol. (London) **94**, 357 (2000).
8. R. D. Pinto *et al.*, Phys. Rev. E **62**, 2644 (2000).
9. U. Parlitz, L. Junge, and L. Kocarev, Phys. Rev. E **54**, 6253 (1996).
10. H. D. I. Abarbanel *et al.*, Presentation Material from Dynamic Days (2001).
11. D. G. Luenberger, *Linear and Nonlinear Programming* (Addison-Wesley, 1973).
12. E. Baake, M. Baake, H. G. Bock, and K. M. Briggs, Phys. Rev. A **45**, 5524 (1992).
13. D. M. Walker, Phys. Lett. A **249**, 209 (1998).

Heart Motorneuron Dynamics of Leeches

Pietro-Luciano Buono[*] and Antonio Palacios[†]

[*]*Centre de recherches mathématiques, Université de Montréal, Montréal, (Qué) H3C 3J7*
[†]*Nonlinear Dynamics Group, Department of Mathematics, San Diego State University, San Diego, CA 92182-7720*

Abstract. The heartbeat of the medicinal leech consists of two intricate patterns of oscillatory behavior, which are driven by two lateral arrays of motorneurons. On one side of the animal the motorneurons oscillate synchronously, while on the other side they produce a peristaltic wave of oscillations. Then every 20 heartbeats, approximately, the two sides alternate their activity. These two rhythms, including the transitions between them, are known to be initiated by a Central Pattern Generator (CPG) network of neurons. The translation mechanism from CPG dynamics to motorneuron activity, however, is not well understood. In this work, we use symmetric systems of differential equations, accompanied with computational simulations, to investigate such mechanism.

INTRODUCTION

Experiments conducted by R. Calabrese and collaborators [1, 2] reveal that the heartbeat of the medicinal leech is first initiated by a Central Pattern Generator, and then driven by direct contact between two lateral arrays of motorneurons and two lateral arrays of blood vessels. At any given time, all motorneurons on one side of the animal oscillate in phase with each other, while on the other side the motorneurons produce a peristaltic wave of rhythmic activity. Every 20 heartbeats, approximately, the two sides alternate their activity.

Although CPG properties, such as intrinsic dynamics of neurons, network connectivity and type of coupling) have been extensively studied [3, 4, 5, 6, 1, 2], there have been only a few recent attempts at understanding the interplay between CPG dynamics and motorneuron dynamics [7]. In [7], Golubitsky and Stewart address this interplay through a network of coupled cells with $\mathbf{Z}_2$-symmetry, and with a topology that resembles the CPG-motorneurons network. By introducing the concept of "interior symmetries", they are able to reproduce a synchronous-peristaltic state (for left and right sides) as a periodic solution of the entire cell system. A mathematical description of the periodic switching between a synchronous-peristaltic state and a peristaltic-synchronous one, however, has not been addressed yet.

In this paper, we propose an alternative approach to describe not only synchronous and peristaltic rhythms but also the periodic switching problem. More specifically, our approach is based on approximating the network of motorneurons by a coupled cell system with $\mathbf{D}_n$ symmetry. We then show that the CPG output might play a double role. In one role it serves as a switching function to induce periodic transitions between in-phase and peristaltic oscillations, while in the second role, it serves to entrain the frequencies of oscillations of these two motor patterns.

CP676, *Experimental Chaos: 7th Experimental Chaos Conference,*
edited by V. In, L. Kocarev, T. L. Carroll, B. J. Gluckman, S. Boccaletti, and J. Kurths
© 2003 American Institute of Physics 0-7354-0145-4/03/$20.00

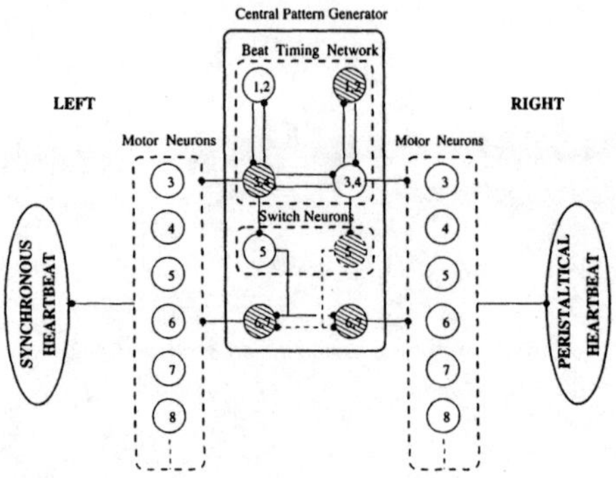

FIGURE 1. The heartbeat of leeches is initiated by a Central Pattern Generator network of neurons whose activity is then translated into synchronous and peristaltic rhythms within two lateral arrays of motorneurons. Shaded neurons fire synchronously with one another and out of phase with those not shaded.

BACKGROUND: EXPERIMENTAL WORK

It has been established also through experimental work that motorneurons behave as identical oscillators [1, 2]. Each oscillator fires tonically when is isolated. Bursting activity in the motorneurons is then induced and controlled by cyclic inhibitory signals produced by a heartbeat CPG. The CPG consists of seven identified pairs of heart interneurons, which also behave as identical oscillators, located in the first seven segmental ganglia, see Figure 1. The experiments also indicate that when the activity of cells HN(3) and HN(4) is synchronized with that of cells HN(6) and HN(7), its ipsilateral side displays the synchronous mode, while when they are out-of-phase the ipsilateral side exhibits the peristaltic mode.

Although synaptic connections between motorneurons do not appear to be explicitly defined in related experimental work [8, 9, 10], it is possible that motorneurons might communicate via antidromic signals. A signal is antidromic if it travels in a direction opposite to its point of origin. In principle, it is then possible that the axons that carry the inhibitory input from heart interneurons in the CPG to motorneurons also carry antidromic signals from one motorneuron to a neighboring one. Evidence of this type of connections can be found in the mastication system of rabbits [11]. Consequently, we adopt the simplifying assumption that motorneurons form linear arrays with synaptic connections among nearest neighbors. Since they also behave as identical oscillators, it is then reasonable that we model each array on each side of the body by systems of coupled identical cells. Epstein and Golubitsky [12] observed that these types of arrays can be studied by embedding them into a circular array with twice the number of cells and with dihedral symmetry. Thus, we consider arrays of n motorneurons with dihedral $\mathbf{D}_n$-symmetry.

NORMAL FORM ANALYSIS

Modeling Assumptions

We then model an interconnected network of motorneurons by a coupled system of $\mathbf{D}_n$ symmetric differential equations of the form

$$\frac{dX_i}{dt} = f(X_i, v, \mu) + \sum_{j \to i} \alpha_{ij}(\mu) h(X_i, X_j), \tag{1}$$

where $X_i = (x_{i1}, \ldots, x_{ik}) \in \mathbf{R}^k$ denotes the state variable of cell i, f governs the internal dynamics of each identical cell (so is independent of i), v and μ are vectors of parameters, h is the coupling function between two cells, and α_{ij} is a matrix of coupling strengths. If n is the total number of cells in the network, then the synchronous mode corresponds to a periodic solution of (1) of the form $X_s(t) = (X_1(t), X_1(t), \ldots, X_1(t))$—all cells fire in phase. Similarly, $X_p(t) = (X_1(t), X_1(t - \phi), X_1(t - 2\phi), \ldots, X_1(t - (n-1)\phi))$, represents a peristaltic wave, where ϕ is the constant phase shift of firing time. Now, since any element of $\mathbf{D}_n$ leaves X_s unchanged, then $\mathbf{D}_n$ is the group of spatial symmetries of X_s. In the peristaltic mode, $\widetilde{\mathbf{Z}}_n$, the group of cyclic rotations of the plane through $0, \phi, \ldots, (n-1)\phi$ coupled with temporal shifts by the same amount, leaves X_p unchanged. Thus $\widetilde{\mathbf{Z}}_n$ is the group of spatio-temporal symmetries of X_p.

Mode Interaction

We assume that both patterns arise, via Hopf bifurcations, when a trivial solution $X = 0$ with $\mathbf{D}_n$-symmetry loses stability at $(v_1, v_2) = (0, 0)$. This implies that $(d_x F)_{000}$ has eigenvalues $\pm\omega_1 i$, $\pm\omega_2 i$, where ω_1 and ω_2 are the frequencies of oscillations of the synchronous and peristaltic states, respectively. In the synchronous mode the eigenvalues are simple because there is no change of symmetry. The peristaltic mode, however, arises via a symmetry-breaking Hopf bifurcation and so the eigenvalues are double. After performing a center manifold reduction on (1), we arrive at a truncated reduced system of ODEs

$$\frac{dz}{dt} = g(z, \lambda_1(v, \mu, \alpha), \lambda_2(v, \mu, \alpha)), \tag{2}$$

where $z = (z_0, z_1, z_2) \in \mathbf{C}^3$, (λ_1, λ_2) are unfolding parameters, and $g(0, \lambda_1, \lambda_2) = 0$. The eigenvalues of $(d_z g)_{0,0,0}$ are the critical eigenvalues of $(d_x F)_{0,0,0}$ on the imaginary axis. We also assume that (2) is in Poincaré-Birkhoff normal form up to any finite order. This introduces an extra $\mathbf{S}^1$ symmetry in each motor pattern, so that g is now $\mathbf{D}_n \times \mathbf{T}^2$-equivariant ($\mathbf{T}^2 = \mathbf{S}^1 \times \mathbf{S}^1$). We can then choose coordinates $z = (z_0, z_1, z_2)$ such that the $\mathbf{D}_n \times \mathbf{S}^1$-action on $\mathbf{C}^3$ takes the following form: $\gamma(z_0, z_1, z_2) = (z_0, e^{\gamma i} z_1, e^{-\gamma i} z_2)$, $\kappa(z_0, z_1, z_2) = (z_0, z_2, z_1)$, $\theta(z_0, z_1, z_2) = (e^{\theta_1 i} z_0, e^{\theta_2 i} z_1, e^{\theta_2 i} z_2)$, where $\gamma = 2\pi/n \in \mathbf{Z}_n$, κ is a fixed element in $\mathbf{D}_n \sim \mathbf{Z}_n$, and $(\theta_1, \theta_2) \in \mathbf{T}^2$ [13]. It follows that $\mathbf{D}_n \times \mathbf{S}^1$ acts trivially on $(z_0, 0, 0)$, so that the synchronous mode now corresponds to a periodic

TABLE 1. Signs of eigenvalues of synchronous and peristaltic motor patterns.

Solution	Sign of Eigenvalues
Synchronous Mode	$0,\ a_\rho^1,\ \sigma_1 = \delta_1\lambda_1$ [four times], $\delta_1 = b_{\lambda_1}^1 - \dfrac{b_\rho^1 a_{\lambda_1}^1}{a_\rho^1}$
Peristaltic Mode	$0,\ b_N^1 + c^1,\ c^1$ [*twice*], $\sigma_2 = \delta_2\lambda_1$ [twice], $\delta_2 = a_{\lambda_1}^1 - \dfrac{a_N^1 b_{\lambda_1}^1}{(b_N^1 + c^1)}$

solution of (2) lying in the invariant subspace $(z_0, 0, 0)$. Similarly, and depending on the direction of the shift, we can identify the peristaltic state as a periodic solution in either subspace $(0, z_1, 0)$ or $(0, 0, z_2)$. Then the general $\mathbf{D}_n \times \mathbf{T}^2$-equivariant mapping $g(z, \lambda_1, \lambda_2) \in \mathbf{C} \times \mathbf{C}^2$ has the form $g(z) = (Az_0, P(z))$, where

$$
P(z) = B\begin{bmatrix} z_1 \\ z_2 \end{bmatrix} + C\begin{bmatrix} z_1^2 \bar{z}_1 \\ z_2^2 \bar{z}_2 \end{bmatrix} + D\begin{bmatrix} \bar{z}_1^{\,m-1} z_2^m \\ z_1^m \bar{z}_2^{\,m-1} \end{bmatrix} + E\begin{bmatrix} z_1^{m+1} \bar{z}_2^{\,m} \\ \bar{z}_1^{\,m} z_2^{m+1} \end{bmatrix}, \tag{3}
$$

and where A, B, C, D, and E, are complex-valued $\mathbf{D}_n \times \mathbf{T}^2$-invariant functions of ρ, N, P, S, T, λ_1 and λ_2, where $\rho = |z_0|^2$, $N = |z_1|^2 + |z_2|^2$, $P = |z_1|^2|z_2|^2$, $S = (z_1\bar{z}_2)^m + (\bar{z}_1 z_2)^m$, $T = i(|z_1|^2 - |z_2|^2)((z_1\bar{z}_2)^m - (\bar{z}_1 z_2)^m)$, $m = n$ if n is odd, otherwise $m = n/2$. Also, the eigenvalue structure of g leads to $A(0) = \omega_1 i$ and $B(0) = \omega_2 i$.

Without loss of generality, we choose the peristaltic state that lies in $(0, 0, z_2)$ and determine its stability properties together with those of the synchronous state through the eigenvalues of $(d_z g)$. The results are listed in Table 1, where $a^1 = \mathrm{Re}A$, $b^1 = \mathrm{Re}B$, $c^1 = \mathrm{Re}C$, $d^1 = \mathrm{Re}D$, and $e^1 = \mathrm{Re}E$. They suggest two possible approaches for modeling transitions between synchronous and peristaltic behavior. In the first approach, we assume that $b_{\lambda_1}^1 > 0$, $a_{\lambda_1}^1 > 0$, $\mathrm{sgn}(p_{\lambda_1}^1) = -\mathrm{sgn}(a_{\lambda_1}^1)$, and $\mathrm{sgn}(\delta_1) = -\mathrm{sgn}(\delta_2)$, so that both branches of solutions are stable, though one bifurcates supercritically and one subcritically. Then a periodic change in the sign of λ_1 would induce the desired motorneuron switching. Details of the second approach will be discussed in future work [14].

Interplay with CPG Dynamics

Let $V_{HN(i)}$ denote the voltage output of CPG interneuron i, i.e. $i = 3, 4, 6, 7$. Assume also $\delta_1 < 0$ and $\delta_2 > 0$. Then under the approach discussed above, we would need λ_1 to be a *switching function* of $V_{HN(i)}$, so that $\lambda_1 = \lambda_1^+ > 0$ if all $V_{HN(i)}$ are in-phase (leading to in-phase heartbeat), otherwise $\lambda_1 = \lambda_1^- < 0$ (leading to a peristaltic heartbeat). A periodic switching between λ_1^+ and λ_1^- can then be realized through a reverse sigmoid function of the form

$$
\lambda_1(V_{CPG}) = s_1 - \frac{s_2}{1 + \exp\left[(V_{CPG} - V_T)/s_T\right]}, \tag{4}
$$

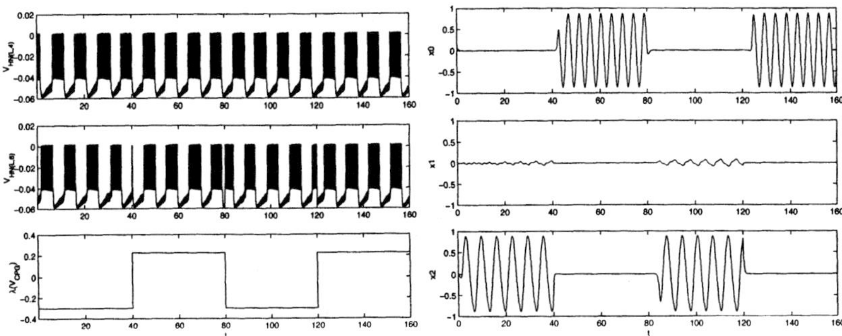

FIGURE 2. (Left) Inhibitory rhythmic output from the CPG. (Right) Normal form response.

where $V_{CPG} = |V_{HN(3)} + V_{HN(4)} - V_{HN(6)} - V_{HN(7)}|$, V_T is a threshold value, s_1, s_2 and s_T are scaling parameters. In particular, $s_1 = \lambda_1^+$ and $s_1 - s_2 = \lambda_1^-$. Choosing s_T very small, we see that $\lambda_1 \to s_1$ if $V_{CPG} > V_T$, while $\lambda_1 \to s_1 - s_2$ if $V_{CPG} < V_T$. Now, since V_{CPG} alternates periodically (approximately every 20 heartbeats) between $V_{CPG} = 0$ and $V_{CPG} = 2|V_{HN(3)}|$ then, according to (4), the motorneuron dynamics will also alternate between in-phase and peristaltic patterns. Finally, we simulate $V_{HN(i)}$ through a Hodgkin and Huxley [15] type of model described in [4, 1].

VISUALIZATION

We now carry out computer simulations of the normal form equations (3) with a network of $n = 4$ motorneurons with $\mathbf{D}_4$ symmetry—the symmetry group of a square. Then Figures 2 show a synchronous mode that appears when the CPG output is in phase, and a peristaltic state that appears when the CPG output is out-of-phase. Observe that the frequencies of oscillations of these two states are not equal. A periodic perturbation of the normal forms, however, can lead to frequency entrainment as is shown in Figure 3. This fact suggests that the CPG output might play a double role. In one role, it would serve as a switching function for λ_1, which which can be associated with a change in coupling strengths. In the other role, the CPG output would also act as a perturbation to induce frequency entrainment.

For visualization purposes, it is possible to map the oscillations in the normal forms to those of a cell system such as (1) without performing a center manifold reduction on (1). Details of this mapping will be provided in [14]. The results are shown in Figure 4.

ACKNOWLEDGMENTS

We thank R. Calabrese, M. Field, and M. Golubitsky for many useful discussions.

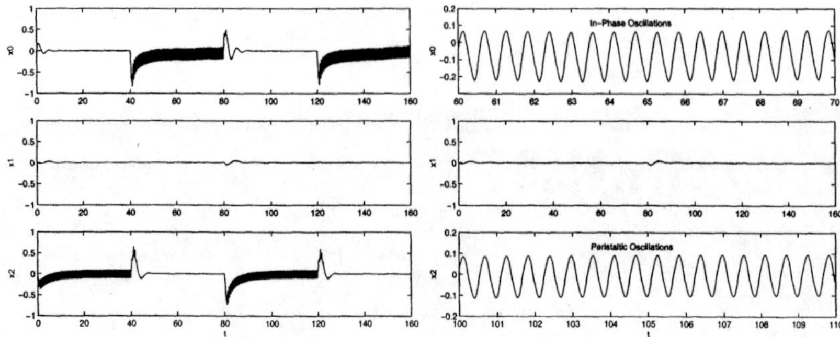

FIGURE 3. (Left) A periodic perturbation of the normal forms can lead to frequency entrainment between peristaltic and in-phase oscillations. (Right) Enlargement of oscillatory states.

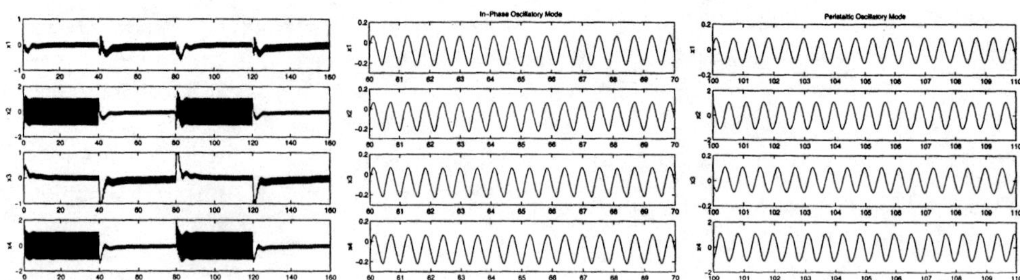

FIGURE 4. (Left) Translation of oscillatory behavior from the normal forms (see Figure 3) to oscillations in (1). (Middle) Enlargement: In-phase CPG perturbations lead to synchronous oscillations, while (Right) Out-of-phase perturbations lead to peristaltic behavior. Frequency entrainment is also achieved.

REFERENCES

1. Nadim, F., Olsen, O., Schutter, E. D., and Calabrese, R., *J. Comp. Neurosci.*, **2**, 216–235 (1995).
2. Schmidt, J., and Calabrese, R., *J. Exp. Biol.*, **171**, 329–347 (1992).
3. Calabrese, R., and Feldman, J., "Intrinsic membrane properties and synaptic mechanisms in motor rhythm generators," in *Neurons, Networks, and Motor Behavior*, edited by P. S. et al., MIT Press, Cambridge, MA., 1997, pp. 119–130.
4. Calabrese, R., Nadim, F., and Olsen, O., *J. Neurobiol.*, **27**, 390–402 (1995).
5. Gramoll, S., Schmidt, J., and Calabrese, R., *J. Exp. Biol.*, **186**, 157–171 (1994).
6. Lu, J., Gramoll, S., Schmidt, J., and Calabrese, R., *J. Comp. Physiol. A*, **184**, 311–324 (1999).
7. Golubitsky, M., and Stewart, I. (2002), preprint.
8. Thompson, W., and Stent, G., *J. Comp. Physiol.*, **111**, 261–279 (1976).
9. Thompson, W., and Stent, G., *J. Comp. Physiol.*, **111**, 281–307 (1976).
10. Thompson, W., and Stent, G., *J. Comp. Physiol.*, **111**, 309–333 (1976).
11. Westberg, K., Kolta, A., Clavelou, P., and Sandstrom, G., *J. Neurosci.*, **12**, 1145–1154 (2000).
12. Epstein, I., and Golubitsky, M., *Chaos*, **3**, 1–5 (1995).
13. Golubitsky, M., Stewart, I., and Schaeffer, D., *Singularities and Groups in Bifurcation Theory*, Appl. Math. Sci., Springer-Verlag, New York, 1988.
14. Buono, P.-L., and Palacios, A., *In Preparation* (2002).
15. Hodgkin, A., and Huxley, A., *J. Physiol. London*, **117**, 500–544 (1952).

CHEMICAL CHAOS

Spatiotemporal Pattern Formation and Chaos in the Belousov-Zhabotinsky Reaction in a Reverse Microemulsion

Irving R. Epstein and Vladimir K. Vanag

*Department of Chemistry and Volen Center for Complex Systems, MS 015,
Brandeis University, Waltham, MA 02454*

Abstract. We study the spatiotemporal behavior of the oscillatory Belousov-Zhabotinsky (BZ) reaction in a reverse microemulsion consisting of water, octane and the surfactant sodium bis(2-ethylhexyl)sulfosuccinate (AOT). By varying the microemulsion composition, we can "tune" its structure, specifically the size and spacing between the nanometer-sized water droplets in which the polar BZ reactants reside. We find a remarkable array of pattern formation as the microemulsion structure and BZ chemistry are varied. Behaviors observed include stationary Turing patterns, traveling and standing waves, spirals, targets, antispirals and antitargets (which travel into rather than out from their center), and spatiotemporal chaos. A simple reaction-diffusion model, which accounts for the BZ chemistry and the differential diffusion of species within water droplets and in the bulk oil phase, is able to reproduce nearly all of the observed behavior.

INTRODUCTION

The class of reaction-diffusion systems constitutes one of the most fertile sources of spatiotemporal pattern formation in general and of experimental chaos in particular [1]. Among these systems, by far the most thoroughly studied, and probably the most versatile, is the Belousov-Zhabotinsky (BZ) reaction [2]. The BZ reaction, which consists of the oxidation of an organic substrate, typically malonic acid, by bromate in aqueous acidic solution catalyzed by a metal ion or complex, gives rise to periodic oscillation, both simple and complex, temporal chaos, a variety of temporally periodic traveling waves, notably spirals and target patterns, and spatiotemporal chaos [1,3]. In recent years, investigators have probed the behavior of the BZ reaction in more structured media such as gels [4], beads of ion exchange resin [5] and membranes [6]. We report here on the remarkable phenomenology of this system in a new medium, a reverse microemulsion consisting of water, oil (octane) and the surfactant sodium bis(2-ethylhexyl)sulfosuccinate, also known as Aerosol-OT or AOT. We summarize the variety of patterns found experimentally in the BZ-AOT system and present a simple model that explains much of the observed behavior. We also suggest some open questions that lend themselves to further research on this fascinating system.

CP676, *Experimental Chaos: 7ᵗʰ Experimental Chaos Conference,*
edited by V. In, L. Kocarev, T. L. Carroll, B. J. Gluckman, S. Boccaletti, and J. Kurths

265

THE BZ-AOT SYSTEM

In unstirred aqueous solution, the BZ reaction gives rise to striking pattern formation phenomena of the sort shown in Fig. 1. When the reaction is run in AOT microemulsion, new features arise that strongly affect pattern formation. All of the key reactants (bromate, malonic acid, catalyst – in this case ferroin (ferrous 1,10-phenanthroline)) are polar and quite soluble in water. In the course of the reaction, a number of key intermediates are produced. Among them are two nonpolar species: bromine, Br_2, which has an inhibitory effect on the reaction, and bromine dioxide, BrO_2, which has an autocatalytic, acceleratory effect. The distributions of these species between the aqueous and oil phases are different and depend on the relative volumes of the phases. Bromine preferentially dissolves in the oil phase, while BrO_2 has approximately equal solubility in either phase. Depending on the composition of the microemulsion, the dominant role in pattern formation may be played either by Br_2 or by BrO_2.

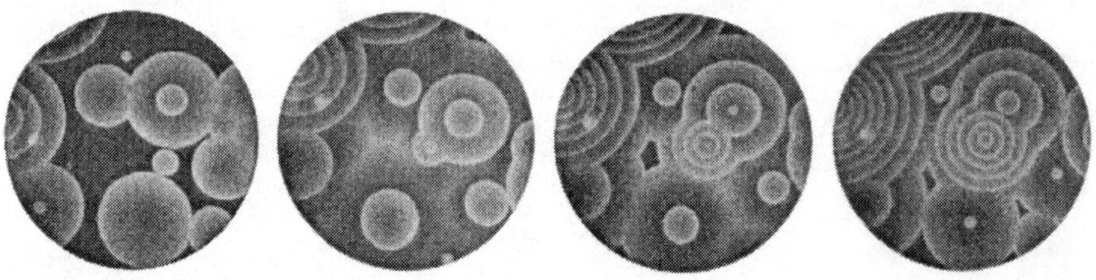

FIGURE 1. Target patterns in the BZ reaction. Shown are four successive snapshots of a petri dish (diameter about 10 cm) containing an initially homogeneous solution of the reactants. Dark background shows high concentration of reduced form (ferroin) of catalyst; light rings show high concentration of oxidized form (ferriin).

A thermodynamically stable AOT microemulsion [7] consists of spherical, nanometer-sized water droplets dispersed in a continuous oil (octane) phase. Each droplet is surrounded by a monolayer of AOT. The BZ reagents are dissolved initially in the aqueous phase, i.e., in the water droplets. When the BZ reaction starts, the nonpolar intermediates Br_2 and BrO_2 are produced in the water droplets and can diffuse into the oil phase. Communication between droplets may occur through the oil phase via diffusion of these small molecules or as a result of mass exchange during droplet collision/fusion/fission. The former process is characterized by typical molecular diffusion constants of the order of 2×10^{-5} cm^2 s^{-1}, while the latter process, which involves the motion of entire droplets, is much slower, with an effective diffusion constant one or two orders of magnitude smaller, depending upon the composition of the system [8].

The structure of the microemulsion is "tunable" [7] in that the radius of the water droplet cores is determined by the molecular ratio $\omega = [H_2O]/[AOT]$ and is roughly equal to 0.17ω (nm) [9], and the spacing between droplets decreases with the volume fraction φ_d of the dispersed phase (water plus surfactant), which is governed by the ratio ([AOT] + [H$_2$O])/[octane]. At a critical value, φ_p, percolation occurs, as channels of water droplets form and the diffusivity of water-soluble and oil-soluble molecules becomes approximately equal [10].

EXPERIMENTAL RESULTS

In Fig. 2, we summarize results obtained over a wide range of reactant concentrations and microemulsion compositions in the BZ-AOT system. In a typical experiment, we first prepare two stock microemulsions, ME I, containing malonic acid and sulfuric acid, and ME II, containing ferroin and bromate, with the same size and concentration of water droplets, by mixing 1.5 M solutions of AOT in octane with aqueous solutions containing the appropriate concentrations of the BZ reagents. The reactive microemulsion is generated by mixing equal volumes of ME I and ME II. The droplet concentration can be varied while keeping the size constant simply by diluting with pure octane.

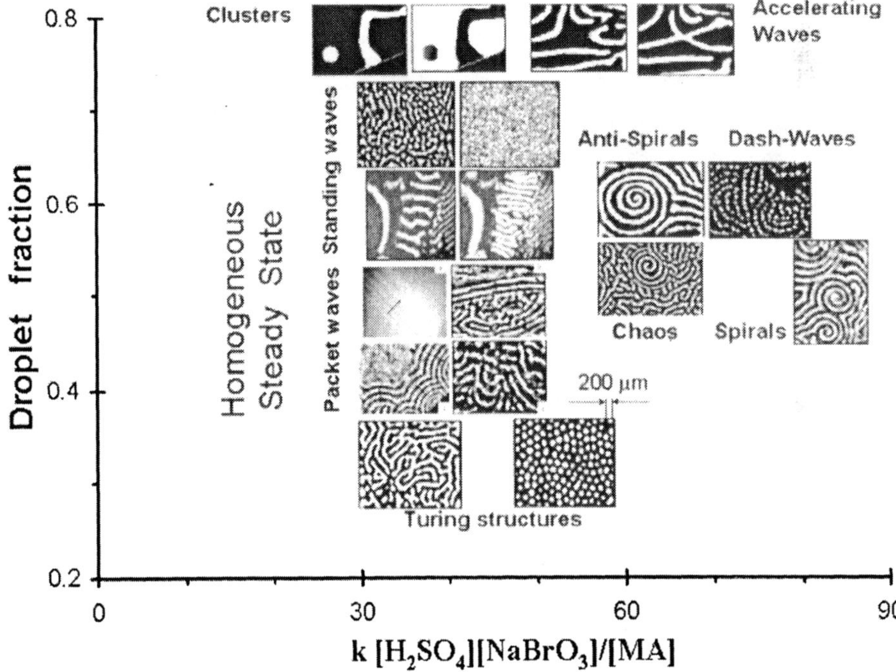

FIGURE 2. "Phase diagram" of patterns observed in the BZ-AOT system. Individual patterns are discussed below.

A small volume of this BZ-AOT mixture is sandwiched between two flat optical windows 50 mm in diameter and separated by a 0.1 mm thick annular telflon gasket with inner and outer diameters of 20 mm and 47 mm, respectively. We observe the patterns, which persist for about 1 h, through a microscope equipped with a digital

CCD camera connected to a personal computer. The reaction layer is illuminated by a 40 W tungsten source passing through a 450 nm interference filter.

Chaos

Two types of spatiotemporal chaos are found in the BZ-AOT system, typically at intermediate values of φ_d. Spiral chaos emerges as spirals initially form and then break up, presumably due to a long-wavelength instability [11]. A snapshot of this form of chaos is shown in Fig. 3a. A second type of chaos occurs when a region of plane waves (Fig. 3b, upper half) serves as a boundary to a set of chaotic wave fragments. This behavior is believed to be associated with the finite wavelength instability and has been simulated with a simple model [21] that we discuss below.

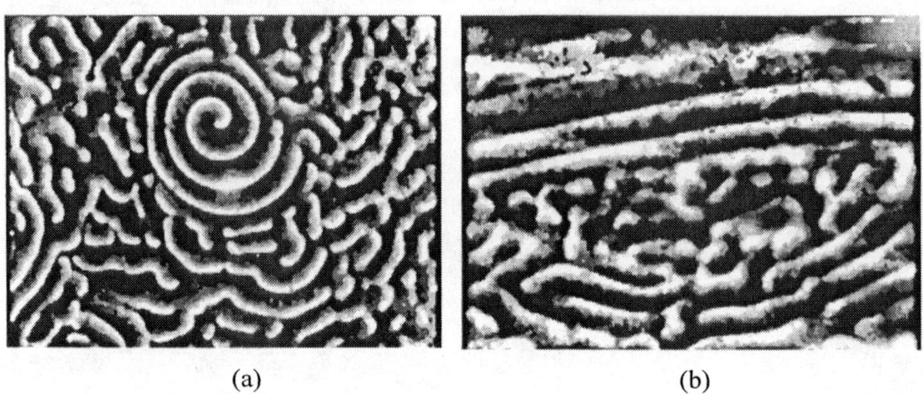

(a)   (b)

FIGURE 3. Spatiotemporal chaos in the BZ-AOT system (a) Spiral chaos; (b) Chaos emanating from plane waves.

Stationary Patterns

One class of phenomena found in the BZ-AOT system consists of patterns in which the spatial structure is fixed. These patterns may be temporally stationary (Turing structures) or they may oscillate periodically (standing waves and clusters). This behavior typically occurs (see Fig. 2) at relatively low ratios of activator to inhibitor.

Turing Structures

A primary motivation for undertaking the study of the BZ-AOT system was the desire to produce Turing structures, patterns which are stationary in time and periodic in space [12], in the BZ reaction. Turing patterns were seen experimentally for the first time [13] in another chemical oscillator, the chlorite-iodide-malonic acid (CIMA) reaction, but efforts to obtain them in a BZ system were unsuccessful, because of the difficulty of generating the required difference in diffusion rates between the activator and inhibitor species. This problem was solved in the CIMA system by carrying out

268

the reaction in a gel impregnated with starch. The starch reversibly forms an unreactive complex with the activator iodine species, effectively slowing their diffusion by about an order of magnitude [14]. The BZ reaction does not lend itself to such an approach.

Instead, by running the BZ reaction in an AOT microemulsion, we are able to utilize the difference in diffusion rates between the polar species in the water droplets and the nonpolar species that diffuse directly through the oil. At relatively low values of φ_d, the inhibitor bromine is found primarily in the oil phase, and hence, as required for producing Turing structures, its effective diffusion rate significantly exceeds that of the activator species, which must diffuse with entire droplets. Under these conditions, Turing patterns are obtained [7], as shown in Fig. 4. Since our system is closed, the Turing patterns are transient, though they persist for an hour or more. In an unstirred flow reactor, it should be possible to sustain them indefinitely.

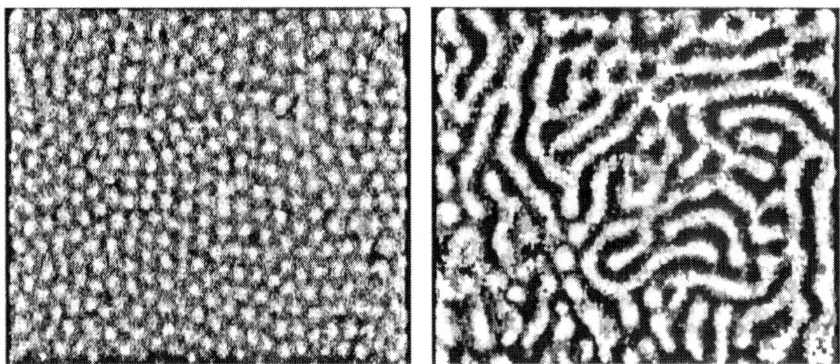

FIGURE 4. Hexagonal (left) and striped/labyrinthine (right) Turing patterns in the BZ-AOT system. Height of each frame is 2 mm.

Standing Waves

At higher values of φ_d, we observe another variety of stationary patterns, standing waves. Like Turing structures, standing waves have an intrinsic wavelength, determined by the reaction and diffusion parameters and independent of the geometry of the system or the initial conditions. Standing waves, however, undergo temporally periodic oscillation everywhere except at a set of fixed nodal surfaces, where the concentrations remain constant. The characteristic wavelengths of the Turing structures and standing waves observed under our conditions are about 0.18 and 0.23 mm, respectively. Standing waves emerge after bulk oscillations cease in this initially oscillatory mixture, and evolve from traveling waves, usually starting near the Teflon gasket, which suggests that they are favored by zero-flux boundary conditions [7]. We believe this phenomenon to be the first experimental example of standing waves in a reaction-diffusion system in the absence of global coupling or external perturbation. An example of a standing wave pattern is shown in Fig. 5.

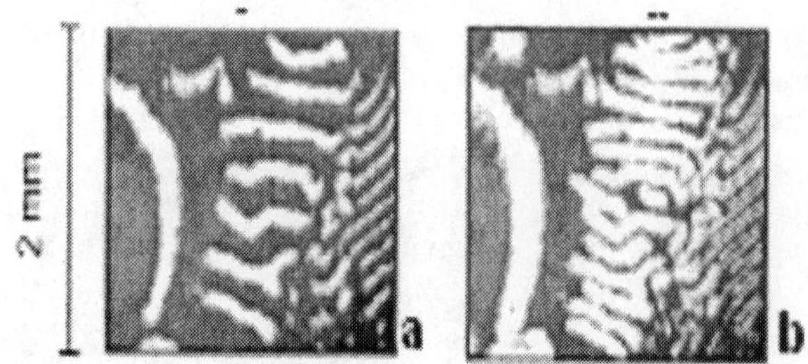

FIGURE 5. Standing waves in the BZ-AOT system. Striped standing waves emerge between traveling waves from the right and fast phase waves from the left. Gray levels quantify [ferroin], with white corresponding to minimum and black to maximum. (b) is summation of snapshot (a) and snapshot of an antiphase pattern that appears 45 s later. Period of oscillation of standing waves is 91 s. Period of oscillation of bulk solution is 60 s. From ref [7].

Clusters

At still higher values of φ_d, above the percolation threshold, we observe another phenomenon, oscillatory clusters. Clusters consist of domains in which nearly all elements in a given domain oscillate with the same amplitude and phase, separated by stationary nodal surfaces. They resemble standing waves, but clusters do not possess an intrinsic wavelength, and their sizes and shapes depend upon the initial or imposed conditions. Clusters have been observed in experiments on aqueous BZ systems subject to a photochemical global feedback [15] and they are found in models of coupled neuron systems [16], but observations of clusters in autonomous reaction-diffusion systems are extremely rare.

Traveling Waves

Perhaps the most frequently studied behavior of the BZ system in unstirred aqueous solution consists of traveling waves, often in the form of target patterns like those shown in Fig. 1 or spirals, which arise when a ring in a target pattern is broken, either by external perturbation or because of internal fluctuations or inhomogeneities. These waves are typically trigger waves, which arise in excitable or oscillatory media when an inhibitor concentration drops below a threshold value, leading to diffusive propagation of a relatively large amplitude wave at a constant velocity determined by the reaction and diffusion rates. In the BZ (and many other systems), the wave is followed by a region of high inhibitor concentration, resulting in a transient refractory zone following passage of a single wave and in annihilation when a pair of waves collide. This latter phenomenon results in the complex structure seen in Fig. 1 when waves emanating from one center meet and annihilate waves from another center. Other geometries of traveling waves, such as nearly planar waves, are also observed. The wave speed depends upon the curvature $1/r$ according to the "eikonal" equation [17]

$$v = v_0 - (1/r)D \qquad (1)$$

where v_0 is the velocity of a planar front. Here we present some examples of traveling waves in the BZ-AOT system that behave quite differently from "normal" BZ traveling waves.

Antispirals and Antitargets

In all examples of spiral and target patterns observed in reaction-diffusion systems in homogeneous media, the waves propagate outward from a center or pacemaker. The BZ-AOT system also supports such patterns. However, at slightly higher φ_d values than those that support "normal" spirals and target patterns, we have observed spirals that rotate inward and target patterns whose rings move toward rather than away from the center [18]. In addition to violating the eikonal equation, "antispirals" differ from their outwardly moving counterparts in several ways. They oscillate at a lower rather than a higher frequency than the bulk; they have a pitch that increases, rather than remaining constant, with distance from the center; and their centers or tips do not "meander" as do normal spiral tips[1] [19]. Apparently similar behavior has been seen in the aggregating slime mold, *D. discoideum* [20], a species that more typically generates normal spirals. Figure 6 shows examples of antispirals and antitarget patterns.

FIGURE 6. Inwardly moving spiral and target patterns in the BZ-AOT system. (**A** to **C**), antispirals, $\phi_d = 0.55$. (**D**), concentric waves (antitarget), $\phi_d = 0.59$. Frame size (in mm): (A) 5.1 by 3.75, (B) 3 by 2.25. (C) 1.8 by 1.5, (D) 2.7 by 2.5. Arrows mark directions of wave movement and anti-spiral rotation. Reprinted with permission from ref. [18]. Copyright 2001 American Association for the Advancement of Science.

---

[1] The tip of the normal spiral in the center of the spiral chaos in Fig. 3a undergoes meandering.

As noted above, ordinary trigger waves move at constant velocity and annihilate on collision. A very different behavior can be observed in the BZ-AOT system above the percolation threshold and with a high ratio of activator to inhibitor concentration [7]. We find waves that speed up as they approach one another and on collision move apart at right angles to the original direction of approach. This surprising behavior is illustrated in Fig. 7. Note how the waves appear to "reach out" toward each other at 110 s and then, after merging at 122 s, move off perpendicularly at 134 s. The initial velocity of the waves is proportional to the square root of the diffusion constant of the water droplets, while the higher velocity just before collision is proportional to the square root of the diffusion constant of small molecules in the oil phase. These observations suggest that, in contrast to the situation when Turing structures are found, it is the activator that diffuses rapidly through the oil phase. Indeed, when we analyze the distribution of nonpolar species between the oil and aqueous phases, we find that at high φ_d it is the activator BrO_2 that predominates in the oil phase.

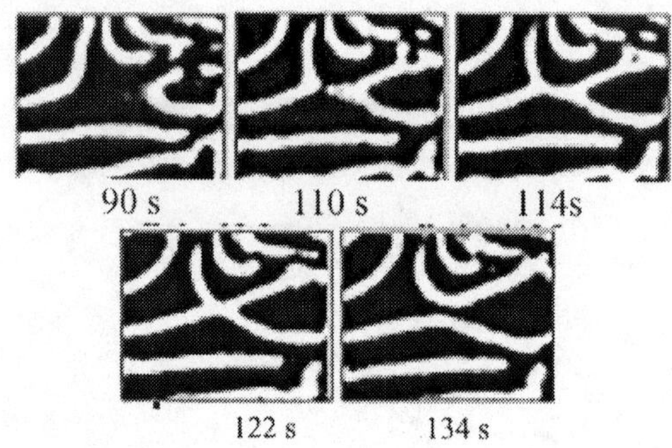

FIGURE 7. Accelerating waves in the BZ-AOT system. Times at which snapshots are taken are indicated. Height of each frame is 4 mm. From ref. [7].

MODELING

In an effort to understand better the origins of the remarkable behavior exhibited by the BZ-AOT system, we have attempted to model it with some relatively simple schemes [7, 21]. Our approach is to start with a description of the BZ chemistry and supplement that model with additional steps corresponding to the exchange of the nonpolar intermediates between the aqueous and oil phases. New variables are added, corresponding to the concentrations of the intermediates in the oil phase, and these species are taken to diffuse much more rapidly than the other variables in the model, which represent concentrations in the droplets. One such model consists of the following equations

$$\partial x/\partial \tau = (x - x^2 - fz(x - q)/(x + q) - \beta x + s)/\varepsilon + D_x\Delta x \tag{2}$$

$$\partial z/\partial \tau = x - z + \gamma u - \alpha z + D_z\Delta z \tag{3}$$

$$\partial s/\partial\tau = (\beta x - s + \chi u)/\varepsilon_1 + D_s\Delta s \tag{4}$$

$$\partial u/\partial\tau = (\alpha z - \gamma u)/\varepsilon_2 + D_u\Delta u \tag{5}$$

where x and z are dimensionless concentrations of $HBrO_2$ and the catalyst (ferriin). If $\beta = \gamma = \alpha = 0$, eqs. (2-3) reduce to the Oregonator model [22]; f and q are parameters, $q \ll 1$, and $1 < f < 3$. The additional variables, s and u, represent an inactive form of the activator (Br_2O_4) and the inhibitor (Br_2), respectively, which are both soluble in the oil phase and diffuse rapidly: $D_s \gg D_x$, $D_u \gg D_z$, $D_u \cong D_s$, $D_x \cong D_z$. They may be thought of as providing long-range positive and negative feedbacks, respectively. The species u produces s (χu term in eq. (4)) *via* a chain of reactions. The corresponding term, $-\chi u$, is omitted from eq. (5), since $\chi \ll \gamma$. Parameters ε, ε_1, and ε_2 specify the ratios between the time scales of x, s, and u, respectively, and z.

This simple 4-variable model reproduces the qualitative features of the bifurcation diagram (Fig. 2) in the BZ-AOT system. A key feature, revealed by linear stability analysis, is that the model possesses a finite wavelength instability, where, as a control parameter is varied, the real part of a complex pair of eigenvalues crosses the wavenumber (k) axis at a finite value k_0, which determines the characteristic wavelength of patterns as $\lambda = 2\pi/k_0$, leading to structures that oscillate in both space and time. Another feature that emerges from the linear stability analysis is the occurrence for some parameters of a negative dispersion, $d\omega/dk$, where ω corresponds to the imaginary part (frequency) of the eigenvalue. The value of $d\omega/dk$ at $k = k_0$ gives the group velocity of wave packets, while the sign of $d\omega/dk$ determines the direction of propagation of phase waves (single waves within a packet). Negative $d\omega/dk$ implies that waves in a group move toward rather than away from an initial perturbation [21], which for spiral or target waves corresponds to antispirals and antitargets.

FUTURE DIRECTIONS

The discovery of such a wide variety of dynamical behavior in the BZ-AOT system opens up a number of attractive directions for future research. While there is still much to be done with this system, it would be worthwhile to insert other reactions that display interesting nonlinear behavior into microemulsions to see whether additional new phenomena emerge. The spatial structures that we have observed to date have length scales on the order of tenths of millimeters, or perhaps 10^5 droplet diameters. It is worth exploring whether structure exists at smaller scales as well. We have employed the ferroin-catalyzed version of the BZ reaction. An appealing alternative is to utilize the ruthenium-catalyzed reaction, which is quite photosensitive [23]. Such a system would afford the experimenter the ability to perturb and control the reaction in a temporally and/or spatially patterned fashion by appropriate illumination. Finally, we note that a typical droplet in the BZ-AOT system has a volume of about 5×10^{-22} L (0.5 zeptoliter). In such a tiny volume, the average number of molecules per droplet, even of species present at relatively high concentrations, is quite small – 300 at 1 M and 0.3 at 1 mM concentrations. Thus, the importance of fluctuations in such systems must be considerable. It would be extremely useful to develop a practical approach to treating systems that contain large numbers of subunits[2] each containing a small number of copies of each species. Such a theory would be applicable not only to reaction-diffusion systems in microemulsions, but also to biological systems in which there may be large numbers of cells that contain only a few copies of key molecules such as enzymes.

ACKNOWLEDGMENTS

This work was supported by grants from the Chemistry Division of the National Science Foundation (CHE-9988463) and the Packard Interdisciplinary Science Program. We thank our colleagues, Anatol Zhabotinsky, Milos Dolnik and Lingfa Yang, for their helpful suggestions on many aspects of these studies.

REFERENCES

1. Epstein, I. R., and Pojman, J. A., *Introduction to Nonlinear Chemical Dynamics. Oscillations, Waves, Patterns and Chaos*, New York: Oxford University Press, 1998.
2. Belousov, B. P., *Sb. Ref. Radiats. Med.*, 145, (1958); Zhabotinsky, A. M., *Biofizika*, **9**, 306-311 (1964).
3. Kapral, R. and Showalter, K. (Editors), *Chemical Waves and Patterns*, Dordrecht: Kluwer, 1995.
4. Tam, W. Y., Horsthemke, W., Noszticzius, Z., and Swinney, H.L. *J. Chem. Phys.*, **88**, 3395-3396 (1988).
5. Maselko, J., Reckley, J. S., and K. Showalter, *J. Phys. Chem.*, **93**, 2774-2780 (1989).
6. Winston, D., Arora, M., Maselko, J., Gáspár, V., and Showalter, K., *Nature*, **351**, 132-135 (1991).
7. Vanag, V. K., and Epstein, I. R., *Phys. Rev. Lett.*, **87**, 228301(2001).
8. Schwartz, L. J., DeCiantis, C. L., Chapman, S., Kelley, B. K., and Hornak, J. P., *Langmuir*, **15**, 5461-5466 (1999).
9. Kotlarchyk, M., Chen, S. H., and Huang, J. S., *J. Phys. Chem.* **86**, 3273-3276 (1982).
10. Baptista, M. S., and Tran, C. D., *J. Phys. Chem. B*, **101**, 4209-4217 (1997).
11. Karma, A., *Nature*, **379**, 118-119 (1996); Zhou, L.-Q., and Ouyang, Q., *Phys. Rev. Lett.*, **85**, 1650-1653 (2000); Ouyang, Q., and Flesselles, J.-M., *Nature*, **379**, 143-146 (1996).
12. Turing, A. M., *Phil. Trans. Roy. Soc. London*, **B237**, 37-72 (1952).
12. Castets, V., Dulos, E., Boissonade, J., and P. De Kepper, P., *Phys. Rev. Lett.*, **64**, 2953-2956 (1990).
14. Lengyel, I., and Epstein, I. R., *Science*, **251**, 650-652 (1991).
15. Vanag, V. K., Yang, L., Dolnik, M., Zhabotinsky, A. M., and Epstein, I. R., *Nature*, **406**, 389-391 (2000).
16. Golomb, D., and Rinzel, J., *Physica D*, **72**, 259-282 (1994).
17. Zykov, V. S., *Biophysics*, **25**, 906-910 (1980).
18. Vanag, V. K., and Epstein, I. R., *Science*, **294**, 835-837 (2001).
19. Winfree, A. T., *Physica D*, **84**, 126-147 (1995).
20. Lee, K. J., Cox, E. C., and Goldstein, R. E., *Phys. Rev. Lett.*, **76**, 1174-1177 (1996).
21. Vanag, V. K., and Epstein, I. R., *Phys. Rev. Lett.*, **88**, 088303 (2002).
22. Keener, J. P., and Tyson, J. J. *Physica D* **21**, 307-324 (1986).
23. Kuhnert, L., Agladze, K. I., and Krinsky, V. I., *Nature*, **337**, 244-245 (1989).

---

[2] In our experiments, there are about 10^{17} water droplets in the sample.

The Onset Of Fluctuations In The Ferroin-Catalyzed Belousov-Zhabotinski Reaction

Harold M. Hastings, Sabrina G. Sobel, Flavio H. Fenton, Stephen Chaterpaul, Claudia Frank, Jordan Pekor and Elizabeth Russell

Departments of Physics and Chemistry, Hofstra University, Hempstead, NY 11549 USA

Abstract. We report on an experimental study of the onset of target waves in the ferroin-catalyzed Belousov-Zhabotinski (BZ) reaction. In the cardiac electrical system, another excitable medium, spontaneous activity can initiate ventricular tachycardia by interacting with normal or other spontaneous electrical activity to generate spiral waves. In normal hearts, these spiral waves generally break down to cause ventricular fibrillation (VF), leading to sudden cardiac death. Our results for the BZ reaction: (1) centers of target waves are spatially correlated, with the correlation likely due to mixing effects and long range modes arising in the "clocking phase", (2) activitations begin after concentrations track an equilibrium through a Hopf bifurcation into an unstable "supercritical" state, (3) allowing very small fluctuations can generate targets.

1. INTRODUCTION

The Belousov-Zhabotinski (BZ) reaction is a widely-studied, oscillatory chemical system with wide variety of spatio-temporal behaviors [1 and references therein, including early work of Belousov and Zhabotinski, 2-5]. The BZ reaction medium is an excitable medium: in the low-excitability regime small perturbations die out and sufficiently large perturbations generate waves of activation. In the high-excitability (auto-oscillatory) regime, spontaneous activations and oscillations generally arise.

Excitable media occur in many areas of biology and chemistry (c.f. [6] and references therein), in particular in the cardiac electrical system [7-10]. At the simplest level, that is, two-variable reaction-diffusion equations, there are both important analogies and important differences between the BZ reaction and the heart. Both involve a fast "activator" variable (the concentration of the autocatalytic species and the membrane potential, respectively) and a slow "inhibitor" variable (the concentration of the oxidized catalyst and a gate variable, respectively). However, BZ activator dynamics involves a quadratic non-linearity, and heart activator dynamics a cubic non-linearity, making spiral waves in the heart less stable (c.f. [6]). It is thus difficult to extend the analogy to spiral wave breakdown in the heart.

Despite these significant differences, the heart and the BZ reaction share standard mechanisms for wavebreak and generation of spiral waves. Our results may thus provide insight into general dynamical mechanisms for spontaneous activity, and the first step in initiation of VF. Finally, chemical models of excitable media are simpler to study than biological models, yet are sufficiently more complex than simple mathematical models to provide a broader range of interesting phenomena, for

CP676, *Experimental Chaos: 7ᵗʰ Experimental Chaos Conference*,
edited by V. In, L. Kocarev, T. L. Carroll, B. J. Gluckman, S. Boccaletti, and J. Kurths
© 2003 American Institute of Physics 0-7354-0145-4/03/$20.00

example, chemical chaos [4, 11, 12], without artificially adding "noise". Chemical chaos is only found over a much narrower parameter range in mathematical models [12]. Also, continuously stirred reactions display surprising complexity and long-range scales of fluctuations [13].

We found that new targets typically arise close (5-10 mm) to existing targets, and propose the following explanation. The onset of activity follows a transition from a stable equilibrium to an auto-oscillatory state as malonic acid is brominated to form bromo-malonic acid (c.f. [5, Ch. 3]); a Hopf bifurcation with the Field-Köros-Noyes "stoichiometric factor" f as bifurcation parameter. As we shall see, dynamics near the bifurcation at f ≈ 0.5 can provide a "chemical amplifier" (c.f. [12]) allowing very small fluctuations to induce targets. (The role of Hopf bifurcations in chemical complexity was recognized ten years ago by Gÿorgi and Field [14]). Moreover, spatial correlation during the transition process then implies spatio-temporal correlation of targets (cf. [15]). Finally, our results are not inconsistent with those of Zhang, Förster and Ross [16], who argue that thermal fluctuations are unlikely to cause spontaneous activity with f fixed and far from the Hopf bifurcation.

2. EXPERIMENTAL WORK

Materials and methods

We generally followed one of the standard recipes for the auto-oscillatory BZ reaction [17], except that the potassium bromate ($KBrO_3$) concentration was lowered to prevent precipitation in the stock solution. This yielded a final reaction volume of 12.1 ml and the following final concentrations: H_2SO_4 0.30 M (making $[H^+]$ = 0.316 M), malonic acid (MA) 0.10 M, $KBrO_3$ 0.25 M, KBr 0.041 M, and ferroin 0.0010 M.

Most reactions were performed in 88 mm diameter glass petri dishes, a few in 58 mm plastic dishes, at room temperature (22 - 25 °C) to minimize temperature gradients and thus convection. The reaction typically "clocked" shortly after the addition of all reagents and started producing wavelike behavior a few minutes later. Significant bubble formation generally occurred 18 - 20 minutes into the reaction, after which reactions were stopped due to wave disturbance. An initial probe of the effects of small perturbations was performed in the excitable (non-auto-oscillatory) regime, at 19 °C with concentrations of three reagents reduced as follows: MA 0.040 M, KBr 0.016 M, $KBrO_3$ 0.10 M, with activity initiated by a silver wire (c.f. [18]).

Reactions were photographed with a Kodak DC-290 digital camera at 90 mm resolution. Images were processed with Adobe Photoshop as follows: the green layer was converted to a gray-scale image, and inverted so that darker areas represent greater concentrations of ferriin (from oxidation of ferroin).

Observations And Analysis Of Clustering

Apparent clustering was found in five replicates, each containing 6 to 16 targets (Figure 1). Clustering was evaluated as follows. First, find the distance d from each

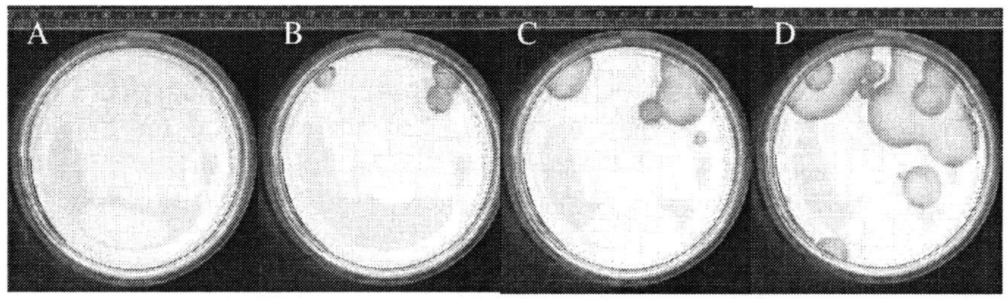

FIGURE 1. <u>Photographs of the BZ reaction</u>, in an 88 mm diameter Petri dish, at the onset of the first target (A), 40 s later (B), 80 s later (C) and 130 s later (D). Note the large-scale, slow "clocking" mode in the bottom of frames (A), (B) and (C), and apparent clustering of targets in frames (B), (C) and (D).

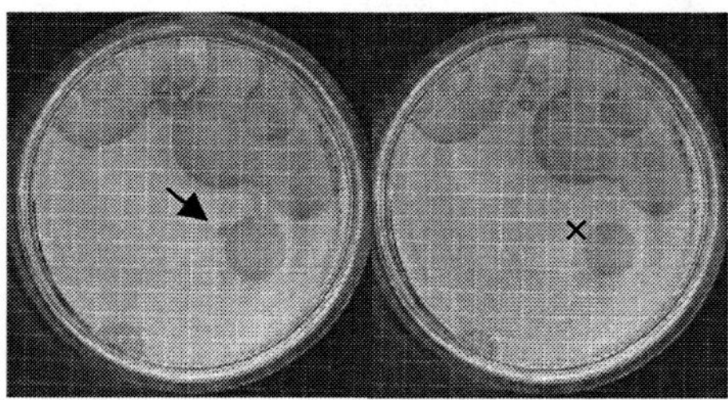

FIGURE 2. <u>Detection of clustering.</u> The target at the head of the arrow in frame (A), 140 s after the onset of activity, first appears in that frame in a burst sequence of photographs. Its location is transferred to the previous frame (B), 130 sec after the onset of activity.

new target (e.g., the labeled target in Figure 2A) to the nearest target in the previous frame (Figure 2B). Then form a strip of width d around the union of all activity in this frame. Let A_{strip} be the area of this strip, and let A be the area in the dish and outside all targets. Then $p = A_{strip} / A$ is the probability that the new target was generated this close to existing targets by chance alone (actually a slight overestimate). Statistical significance is now easily calculated.

The onset of targets displayed significant *spatio*-temporal clustering ($p < 0.01$), with new targets typically arising 5-10 mm from existing wavefronts. On the other hand, at long times targets appeared randomly distributed, with correlation dimension ~ 2 and target numbers in 58 and 88 mm dishes approximately proportional to area. Correlation dimension appeared lower at short times, and target numbers grew more slowly than area (c.f. [15]).

Large-scale correlations were sometimes found in clocking (Figure 1). Restirring the reaction mixture after target formation sometimes yielded broad regions of excitation, sometimes targets with no apparent clustering, occasionally Turing-like patterns.

3. THEORY

The Oregonator

We used the Oregonator model [2,3, c.f. 5] with one modification. The Oregonator represents the rate of bromide production in the reduction of ferriin to ferroin (Fe^{3+} to Fe^{2+}) with a stoichiometric factor f:

$$2\ Fe^{3+} + MA + BrMA \longrightarrow f\ Br\text{-} + 2\ Fe^{2+} + \text{other products.}$$

The stoichiometric factor is considered to depend upon fraction of malonic acid (MA) that has been brominated to form bromo-malonic acid (singly brominated), and even doubly bromonated MA. The initial reagent mixture contains no brominated MA, making f = 0 initially. Bromination of MA during the early stages of the reaction causes f to increase, and auto-oscillatory dynamics requires f > 1/2 (c.f. [5]). We therefore considered f as a dynamically changing variable with simple kinetics (the variable f case was first studied in [14]), yielding the full model (Figure 3):

$$dx/dt = k_3\,[BrO3\text{-}]\,[H^+]^2 y - k_2\,[H^+]xy + k_5\,[BrO3\text{-}]\,[H^+]x - 2k_4 x^2$$
$$dy/dt = - k_3\,[BrO3\text{-}]\,[H^+]^2 y - k_2\,[H^+]xy + f/2\ k_c\,[MA]\,z$$
$$dz/dt = 2\,k_5\,[BrO3\text{-}]\,[H^+]x - k_c\,[MA]\,z$$
$$df/dt = k_f\,(f_\infty - f).$$

Here x = [HBrO$_2$], y = [Br-], z = [ferriin] and the k_i are rate constants (c.f. [5]). We set $f_\infty = 0.70$ and $k_f = 0.012$/s. Qualitative dynamics do not depend upon details of the

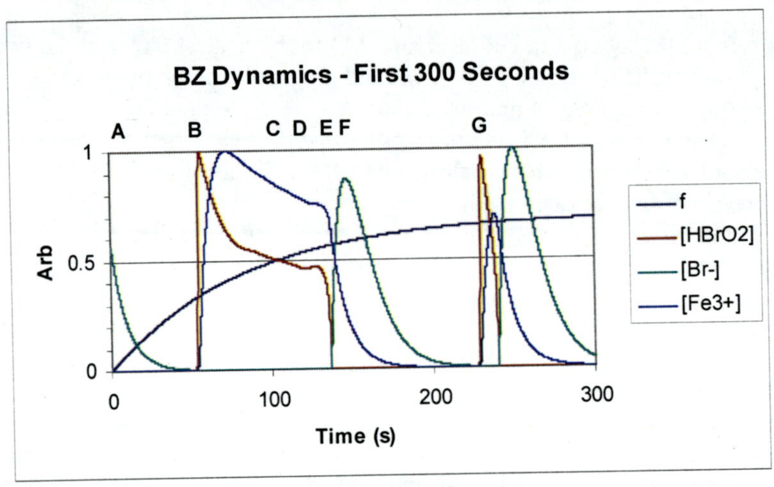

FIGURE 3. <u>The effect of f as a dynamic variable.</u> Simulations with the Oregonator model, initial values [HBrO$_2$] = 0, [Br-] = 0.001 M, [ferriin] = 0, f = 0; f approaches f_∞ = 0.7 with a time scale of 83.3 s. A Hopf bifurcation occurs at t = 104.6 sec, but dynamics closely track the unstable equilibrium until t = 123 s, f = 0.54. At this time, the dynamics are highly unstable, with the real part of the dominant eigenvalue Re(λ) $\approx$ 0.33, making fluctuations double every 2 seconds.

kinetics of f provided that (1) the limiting value of f, f_∞, exceeds 0.5 and (2) f increases slowly compared with other dynamics. This allows concentrations to closely track corresponding equilibrium values as f approaches and passes through the Hopf bifurcation.

Results

Figure 3 above illustrates several important steps in this dynamics: (A) f = 0 at t = 0; (B) clocking (as observed experimentally) after bromide falls to a critical value; (C) bromous acid and ferriin concentrations closely track quasi-steady state values as f is increased to 0.5 (bromide dynamics is always very fast), (D) bromous acid and ferriin concentrations continue to track these values significantly past the Hopf bifurcation at f $\approx$ 0.508; (E,F) a sharp decrease in bromous acid and ferriin concentrations, resp.; (G) the onset of regular, approximately periodic oscillations.

For our choice of concentrations, and f > 0.508, the Hopf bifurcation value, Re(l) $\approx$ 9.52 (f-0.508). This "supercritical" instability allows small fluctuations to nucleate "spontaneous" activations, since the time to exit the relatively small domain of linear oscillations scales with the size of the fluctuation Δx as $-\log(\Delta x / Re(\lambda))$.

In contrast, with f fixed in the auto-oscillatory range (for example, f = 0.75 [16]), the dynamics is relatively stable at most phases of the oscillations. For example, 20 s after the end of an activation, with bromous acid at its low resting value, relatively large (20 %) perturbations cause only small changes in the timing of the next activation pulse. It is thus virtually impossible for thermal fluctuations to cause activations [16].

4. THEORY VERSUS EXPERIMENT

The Oregonator model, with f as a dynamic variable, seemed to provide a good basis for experimental observations. Note however that ferroin is a limiting reagent in experiments, with an initial concentration of 0.001 M, and 5-10 % of the ferroin apparently oxidized to ferriin. In contrast, ferroin is not a limiting reagent in the Oregonator with concentrations reaching 0.01 M. In addition, we used a simple approximation for the bromination of MA. Dynamics near the Hopf bifurcation appear independent of the model, depending only upon the fact that concentrations of bromous acid, bromide and MA have reached (very close to) quasi-steady state values by the time f crosses 0.5. In addition, our model also reproduces clocking and several phenomena of the excitable, non-auto-oscillatory reaction, e.g., a critical value for target production using a silver wire to locally reduce bromide ion concentration (c.f. [18]).

5. DISCUSSION

We found significant spatio-temporal clustering of targets in the onset of activity in the ferroin-catalyzed BZ reaction. Theoretical work suggested a mechanism: the early dynamics (for example, clocking), while bromination proceeds and f increases through the Hopf bifurcation at ~0.5 retains long-term correlations, for example, from early stirring, and leaves parts of the system in unstable "supercritical" states (e.g., Figure 3, at time point D) where small (and possibly random) perturbations can nucleate "spontaneous" activity. This agrees generally with Walgraef et al. [19] and Vidal and Pagola [20]. Further work is needed to better understand and unify (1) statistics of fluctuations and relevant concentrations of intermediates, (2) the role of phase fluctuations [19], (3) the effect of fluctuation size and volume in the unstable range (c.f. [18, 21]) and (4) the role of instabilities as chemical amplifiers in the sense of Ganapathisubramanian and Noyes [12], as well as to quantify the effects of various causes of nucleation (c.f. [16-21]). *Note*: A detailed analysis of microscopic fluctuations will appear in joint work with R.J. Field [22].

ACKNOWLEDGEMENTS

This research was partially supported by National Science Foundation grant DBI-0096692.

REFERENCES

1. Field, R.J., and Burger, M. (eds), *Oscillations and traveling waves in chemical systems*, Wiley, New York, 1985.
2. Field, R.J., Köros, E., and Noyes, R.M., *J. Amer. Chem. Soc.* **94**, 8649-8664 (1972).
3. Field, R.J., and Noyes, R.M., *J. Chem. Phys.* **60**, 1877-1884 (1974).
4. Scott, S.K., *Chemical Chaos*, Clarendon Press, Oxford, UK and Oxford Univ. Press, New York, 1991.
5. Scott, S.K., *Oscillations, Waves, and Chaos in Chemical Kinetics*, Oxford Chemistry Primers 18, Oxford Univ. Pr., Oxford, UK and New York, 1994.
6. Showalter, K., *Nonlinear Science Today* **4**, 1-10 (1995).
7. Glass, L., *Physics Today* **49**, 40-45 (1996).
8. Gray, R.A., and Jalife, J., *Int. J. Bif. and Chaos* **6**, 515-435 (1996).
9. Henriquez, C.S., and Papazoglou, A.A., *Proc. IEEE* **84**, 334-354 (1996).
10. Winfree, A.T., *The Geometry of Biological Time*, 2nd ed., Springer-Verlag, New York, 1999.
11. Hudson, J.L., and Mankin, J.C., *J. Chem. Phys.* **74**, 6171-6177 (1981).
12. Ganapathisubramanian, N. and Noyes, R.M., *J. Phys. Chem.* **76**, 1770-1774 (1981).
13. Menziger, M. and Jankowski, P., *J. Phys. Chem.* **94**, 4123-4126 (1990).
14. Gÿorgi, L., and Field, R.J., *Nature* **355**, 808-810 (1992).
15. Jung, P., *Phys. Rev. E* **61**, 2095-2098 (2000).
16. Zhang, Y.-Z., Förster, P., and Ross, J., *J. Phys. Chem.* **96**, 8898-8904 (1992).
17. Shakashiri, B.Z., *Chemical Demonstrations*, Vol. 2., Univ. of Wisconsin Pr, Madison, WI, 1985, pp. 298-300.
18. Förster, P., Müller, S.C., and Hess, B., *Proc. Nat Acad. Sci. USA* **86**, 6831-6834 (1989).
19. Walgraef, D., Dewel, G., Borckmans, P., *J. Chem. Phys.* **78**, 3043-3051 (1983).
20. Vidal, C., and Pagola, A., *J. Phys. Chem.* **93**, 2711-2716 (1989).
21. Showalter, K., Noyes, R.M., and Turner, H., J. Amer. Chem. Soc. 101, 7463-7469 (1979).
22. Hastings, H.M., Field, R.J., and Sobel, S.G., J. Chem. Phys. (to appear).

INTERDISCIPLINARY

Quantum And Classical Dynamics Of Atoms In A Magneto-optical Lattice

Shohini Ghose, Paul M. Alsing and Ivan H. Deutsch

Dept. of Physics and Astronomy, University of New Mexico, Albuquerque, NM, 87131

Poul S. Jessen and David L. Haycock

Optical Sciences Center, University of Arizona, Tucson, AZ 85721

Tanmoy Bhattacharya, Salman Habib and Kurt Jacobs

T-8, Theoretical Division, MS B285, Los Alamos National Laboratory, Los Alamos, NM 87545

Abstract. The transport of ultra-cold atoms in magneto-optical potentials provides a clean setting in which to investigate the distinct predictions of classical versus quantum dynamics for a system with coupled degrees of freedom. In this system, entanglement at the quantum level and chaos at the classical level arise from the coupling between the atomic spin and its center-of-mass motion. Experiments, performed deep in the quantum regime, correspond to dynamic quantum tunneling. This nonclassical behavior is contrasted with the predictions for an initial phase space distribution produced in the experiment, but undergoing classical Hamiltonian flow. We study conditions under which the trapped atoms can be made to exhibit classical dynamics through the process of continuous measurement, which localizes the probability distribution to phase space trajectories, consistent with the uncertainty principle and quantum "back-action" noise. This method allows us to analytically and numerically identify the quantum-classical boundary.

INTRODUCTION

Coherent control is one of the great challenges in contemporary physics with applications ranging from engineered chemical reactions [1] to electron transport in semiconductors [2-5] . The issues are richest in complex systems with multiple coupled degrees of freedom. At the quantum level, the coupling between the various subsystems can lead to highly entangled states with no classical description. Such entangled states play a major role in various areas of quantum information processing [6] . In the classical limit, systems with coupled degrees of freedom will exhibit nonlinear dynamics and chaotic motion. At the theoretical level, one seeks to better understand the border between the distinct predictions of classical and quantum dynamics, and to perhaps ultimately to control the system's behavior across this boundary.

CP676, *Experimental Chaos: 7ᵗʰ Experimental Chaos Conference,*
edited by V. In, L. Kocarev, T. L. Carroll, B. J. Gluckman, S. Boccaletti, and J. Kurths
© 2003 American Institute of Physics 0-7354-0145-4/03/$20.00

Various studies of quantum systems with a chaotic classical limit have been carried out. While most have focused on static aspects such as wave function scars and energy spectrum statistics [7, 8] , some have investigated the dynamical features of these systems. The experiments of Raizen *et al.*[9, 10] have used ultra cold atoms trapped in standing waves of light (optical lattices) to explore the phenomenon of dynamical localization and the effect of the environment on a kicked rotor system. These experiments demonstrate that the atom/optical realization provides a very clean arena in which to study coherent quantum dynamics versus nonlinear classically chaotic motion.

We have explored a new system in which to study the rich quantum and classical dynamics associated with coupled systems – ultra-cold atoms trapped in far off resonance magneto-optical lattices. Like the kicked rotor, this system has several attractive feature for experiments: the potential can be modeled and designed with great flexibility, and state preparation, controlled unitary evolution and measurement can be performed efficiently through a combination of well established techniques of laser cooling, optical pumping and application of magnetic fields. Coherence times can be very long, and interaction with the environment (dissipation) can be introduced in a controlled manner. Finally, the ability to continuously measure the system enables us to explore the quantum-classical boundary and the emergence of chaos from quantum dynamics. In contrast to the kicked rotor where chaos is due to an externally applied time-dependent perturbation, in our system nonlinearity can arise from two coupled degrees of freedom, namely the atomic magnetic moment and its motion in the lattice. We can thus explore how the quantum-classical boundary is crossed as each of these subsystems varies from microscopic to macroscopic.

In this paper we review our recent experimental and theoretical studies of atomic transport in magneto-optical lattices. In section II we summarize the main features of this trap for alkali atoms. In the classical limit the coupling of spin and motion has the form of the motion of a magnetic moment moving in a spatially inhomogeneous magnetic field. This can lead to chaotic dynamics (section III). We describe the experimental setup and results in section IV. Comparison of the experimental data to quantum and classical theoretical predictions demonstrates the non-classical nature of the observed dynamics. We further show that the motion corresponds to dynamical tunneling through a potential barrier that depends on the internal state of the atom. In section V we explore the key question of measuring classical chaotic trajectories in this system. We study the effect of measurement back-action on our coupled quantum system and the conditions under which classical behavior is recovered. Finally, we end with a summary of our main results in section VI.

ALKALI ATOMS IN A MAGNETO-OPTICAL LATTICE

Optical lattices are formed by the ac-Stark effect arising from the interaction of individual atoms with a standing wave created by a set of interfering laser beams. The physics of this system has been previously described in [11, 12] . Here we briefly summarize the main features. Our one-dimensional optical lattice is formed by two counterpropagating plane waves whose linear polarization vectors are misaligned by a relative angle Θ_L. The resulting field can be decomposed into $\sigma+$ and $\sigma-$ standing waves whose nodes are separated by Θ_L / k, where k is the laser wave number. Consider for simplicity, an alkali atom whose ground state valence electron has spin-1/2. The resulting dipole interaction as a function of the atomic position, z, can be cast as the sum of a scalar part (independent of the atom's spin state) and a vector part which appears as an effective Zeeman interaction [11],

$$U(z,\vec{\mu}) = U_J(z) - \vec{\mu} \cdot \mathbf{B}_{eff}(z) \qquad (1)$$

Here $U_J(z) = 2U_0 \cos\Theta_L \cos(2kz)$, where U_0 is a constant depending on the atomic polarizability and field intensity. The effective magnetic field, $\mathbf{B}_{eff}(z) = B_x \mathbf{e}_x + B_{fict}(z)\mathbf{e}_z$, is the sum of an applied transverse magnetic field, B_x, plus a fictitious field associated with the ellipticity of the optical lattice laser polarization. For our geometry, $\mu_B \mathbf{B}_{fict} = -U_0 \sin\Theta_L \sin(2kz)\mathbf{e}_z$, where μ_B is the Bohr magneton. In actual experiments, we consider alkali atoms interacting with a 1-D lattice far detuned from the $nS_{1/2} \to nP_{1/2}$ resonance (the D2 line). In this far off resonance limit, the form of the potential (Eq. (1)) remains unchanged. The atomic magnetic moment in this case equals $\vec{\mu} = \hbar\gamma\mathbf{F} = -\mu_B\mathbf{F}/F$, where γ is the gyromagnetic ratio and $\mathbf{F}$ is the total angular momentum vector of the atomic hyperfine ground state (including electron and nuclear spin) in units of $\hbar$. We consider here ^{133}Cs, with $F = 4$, the atom used in our current experiments [13]. The eigenvalues of the potential as a function of position result in nine adiabatic potentials (Fig. 1a), the lowest of which exhibits a lattice of double-wells.

The effective magnetic field causes Larmor precession of the atom's magnetic moment direction. Due to the coupling between the atomic position and its magnetic moment through the fictitious magnetic field, the spin precession is accompanied by motion of the wave packet between the double wells. This correlation between the internal state precession and center of mass motion leads to entangled spinor wave packets. The oscillation of the magnetization thus provides a meter through which we can detect the time dependent motion of the packet. Experiments of this type are described in more detail below.

CHAOTIC DYNAMICS

The classical analog to the Hamiltonian associated with the far off resonance magneto-optical potential corresponds to the motion of a massive particle with a magnetic moment moving in a combination of a scalar potential (independent of the moment) plus a spatially inhomogeneous magnetic field [14]. The classical equations of motion then have the form

$$\dot{z} = p/m, \quad \dot{p} = -\frac{d}{dz}\left(U_0(z) - \vec{\mu} \cdot \mathbf{B}_{eff}(z)\right), \quad \dot{\vec{\mu}} = \gamma\left(\vec{\mu} \times \mathbf{B}_{eff}(z)\right), \qquad (2)$$

where $\mathbf{B}_{eff}(z) = B_{fict}(z)\mathbf{e}_z + B_x\mathbf{e}_x$, with fictitious field as given in the previous section. These equations follow from the Heisenberg equations of motion, replacing the quantum operators by their expectation values and neglecting any correlations in the operator products. The effective phase space is four dimensional with two external and two internal variables. The Hamiltonian is in general non-integrable but can be made integrable under two simple physical circumstances: the case in which there is no transverse magnetic field, B_x, [15], and the case of a sufficiently large transverse field so that the motion is adiabatic [16]. We study each regime separately using a set of canonical action-angle variables.

When there is no transverse magnetic field $(B_x = 0)$, $n_z \equiv \mu_z /|\vec{\mu}|$ becomes an additional constant of motion. This results in an integrable Hamiltonian that is identical to that of a simple pendulum

$$H_0 = \frac{p^2}{2m} + C\cos(2kz + D), \tag{3}$$

$$C = U_0\sqrt{4\cos^2\Theta_L + n_z^2\sin^2\Theta_L}, \quad D = \arctan(n_z\tan\Theta_L/2). \tag{3a}$$

whose amplitude and phase depend on the constant z-projection of the atomic moment [17]. The action-angle variables for a pendulum, $(J,\psi)_-$ are well known to be functions of the complete elliptic integrals [18]. For energies close to the bottom of the sinusoidal potential, we can expand the elliptic integrals in a power series, keeping only the first few terms. This enables us to express H_0 as a function of the motional actions J and internal action μ_z / γ. The precession frequencies of the corresponding angle variables ψ and χ can then be computed from Hamilton's equations to be,

$$\omega_1 = \dot{\psi} = \frac{\partial H_0}{\partial J} = \frac{\pi}{2}\frac{\omega_0}{\mathrm{K}(\kappa)}, \quad \omega_2 = \dot{\chi} = \gamma\frac{\partial H_0}{\partial\mu_z} = \frac{\partial C}{\partial n_z}\frac{\gamma H_0}{\mu_B C}, \tag{4}$$

where $\omega_0 = \sqrt{4k^2|C|/m}$ is the oscillation frequency for a harmonic approximation to the sinusoidal potential, $2\kappa^2 = 1 + H_0/|C|$, and $\mathrm{K}(\kappa)$ is the complete elliptic integral of the first kind. The frequency ω_1 represents oscillation of the center of mass in the sinusoidal potential. By moving to a frame that oscillates with the atom, the time dependence in the effective magnetic field is removed, resulting in a *constant* precession frequency ω_2 about the z-axis. The precession angle in this frame is χ. The addition of a transverse magnetic field as a small perturbation to this integrable Hamiltonian couples the oscillations of the two angles, giving rise to nonlinear resonances. The primary resonances occur when the ratio of the unperturbed frequencies is a rational number, and can be calculated for our system using Eqs. (4).

Current experiments operate in the adiabatic regime where the applied transverse field is large and thus cannot be treated as a perturbation as outlined above. We therefore analyze the regime in which the non-adiabatic coupling can be treated as a perturbation. The integrable adiabatic Hamiltonian is obtained by setting the angle α between $\vec{\mu}$ and $\mathbf{B}_{eff}$ to be a constant, so that

$$H_0 = p^2/2m + U_J(z) + \mu_B |\mathbf{B}_{eff}(z)| \cos\alpha. \qquad (5)$$

When $\alpha = 0$ we obtain the lowest adiabatic double-well potential (Fig. 1a). Other fixed values of α correspond to other adiabatic surfaces. The component of $\vec{\mu}$ along the direction of the magnetic field is now a constant of motion and serves as our new action variable. The other action of the system is obtained in the standard way by integrating the momentum over a closed orbit in the double well for a given energy and α. The precession frequencies ω_1 and ω_2, of the conjugate angle variables correspond respectively to the oscillation of the center of mass in the adiabatic double-well potential and precession of the magnetic moment about the local magnetic field direction in a frame oscillating with the atom as described previously. Unlike the previous case however, we cannot obtain analytical expressions for the frequencies and must resort to computing them numerically. Figure 1(b) shows a typical surface of section for motion in this regime. The primary resonance at $n_z = 0.38$ and $\phi = 0$ corresponds to a ratio of the unperturbed adiabatic frequencies of $\omega_2/\omega_1 = 4$. The nonadiabatic perturbative coupling is strong enough to cause the previously stable primary resonance at $n_z = 0.8$ to bifurcate, and secondary resonances to appear around the points $n_z = 0.38$ and $n_z = -0.85$. The secondary resonances result from coupling between the motion around the primary islands to the unperturbed periodic motion. As the energy is increased, the primary resonances eventually disappear and global chaos sets in.

QUANTUM VS. CLASSICAL DYNAMICS

We have recently performed experiments to observe quantum transport of atoms in the magneto-optical lattice in a mesoscopic regime [13] . We prepare a sample of about 10^6 Cs atoms in a well-defined initial quantum state, say $|\psi_L\rangle$, and follow their subsequent quantum coherent evolution. We start out by laser cooling in a standard magneto-optical trap/3D molasses to a temperature $\sim 4\mu K$ and a Gaussian density distribution of ~ 200 _m RMS radius, followed by further cooling in a near-resonance 1D lin-θ-lin lattice. The atoms are then transferred to a far off resonance (detuned 3000 linewidths from resonance) 1D lin-θ-lin lattice and optically pumped to $|m_F = 4\rangle$. In order to select the motional ground state in the corresponding potential, the depth of the lattice is lowered and it is accelerated at $300\,m/s^2$ for 1.5ms so that atoms outside the ground band can escape. Optical pumping and state selection is done in the presence of a large longitudinal external field B_z (-55mG) to lift degeneracies between the optical potentials and prevent precession of the magnetic moment. This state selection procedure prepares the atoms with roughly 90% population in the target state. Once this is achieved, we increase the lattice depth, change the lattice acceleration to free-fall, ramp up the transverse field B_x and finally ramp B_z to zero.

By performing this sequence slowly enough, we adiabatically connect the ground state in the $m_F = 4$ potential to the left-localized state of the lowest optical double well.

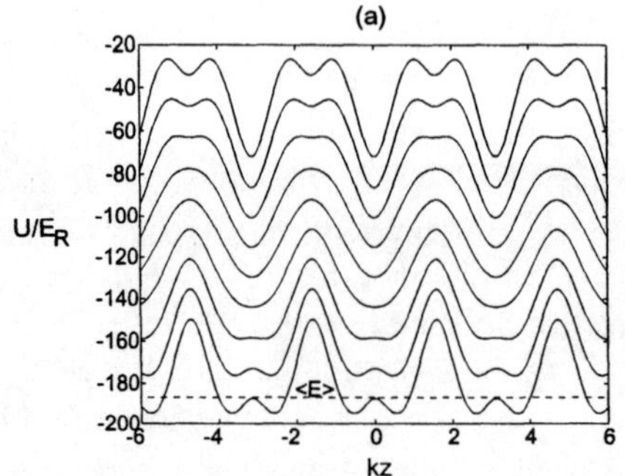
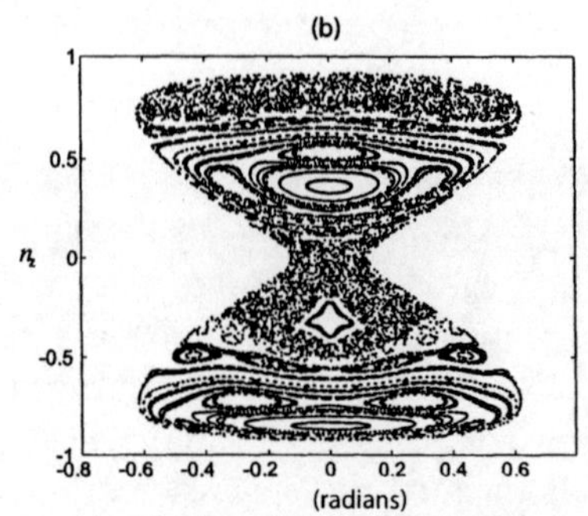

FIGURE 1. (a) Adiabatic potentials for Cesium atoms in an optical lattice with an additional external transverse magnetic field. The mean energy of the state prepared in experiments [13] is just greater than the lowest adiabatic potential barrier energy (horizontal line). The Poincaré surface of section in (b), for $p = 0$ and $dp/dt > 0$ using the parameters of the experiment [13] , with $E = -186.8E_R$ $\left(E_R/h = 2kHz\right)$, shows the effects of the non-adiabatic perturbation term, which makes the classical equations (Eq.(2)) non-integrable.

The motion of the atoms in the double well is measured indirectly by monitoring the time-dependent magnetization of the atomic ensemble. A Stern-Gerlach measurement is performed by releasing the atoms from the lattice, quickly applying a bias field B_z to keep the quantization axis well defined, and letting the atoms fall to a probe beam a few cm below in the presence of a magnetic field gradient of 13G/cm. The magnetic populations can then be extracted from the separate arrival time distributions for different magnetic sublevels.

In order to understand the classical vs. quantum nature of the atomic motion, we compare the experimental data to the predictions of a fully classical calculation as well as a quantum bandstructure calculation [14] . Given the initial state, we compute the joint Husimi or "Q" quasiprobabiluty distribution over both external and internal phase space by employing the familiar motional coherent states as well as the spin coherent states [19] . This initial distribution is evolved both classically and quantum mechanically and compared to the experimental results. The contrast between classical and quantum dynamics is clearly shown in Figure 2(a). Due to the correlation between the atomic magnetic moment and its motion in the well, an oscillation of the mean magnetization between positive and negative values implies motion of the atom from one minimum of the double well to the other. Classical dynamics thus predicts that the mean of the distribution remains localized on one side of the double well. In contrast, the experimental data shows an oscillation between positive and negative values at a frequency well predicted by the quantum model.

A closer look at the reduced classical distribution in the phase space of position and momentum, obtained by tracing over the magnetic moment direction, shows that a part of the distribution does move between wells, but the peak remains localized in one well (Fig. 2b). This implies that transport between the wells is not classically forbidden, but is unlikely for this distribution of initial conditions. This is due to the fact that the classical description of the state corresponding to the Q-function involves a *distribution* of energies. High energy tails of this distribution can classically hop between the left and right wells. However, the experimentally observed oscillations of the mean atomic magnetization are much better described by quantum evolution indicating a *nonclassical* motion of the atom between the double wells. This is not surprising given the fact that for the given experimental parameters, the actions of the system are on the order of $\hbar$ (External center-of-mass action, $I_0 \approx 10\hbar$ and spin $J = 4\hbar$). The only discrepancy with the quantum model is the experimentally observed decay of the oscillations. The probable cause is an estimated ~5% variation of the lattice beam intensities, which is consistent with the observed dephasing times. We estimate the timescale for decoherence due to photon scattering to be of order ~1ms, which is too slow to account for the observed damping. A next generation of the experiment is now underway, in which we hope to increase the dephasing time by an order of magnitude by better control over lattice beam and magnetic field inhomogeneities.

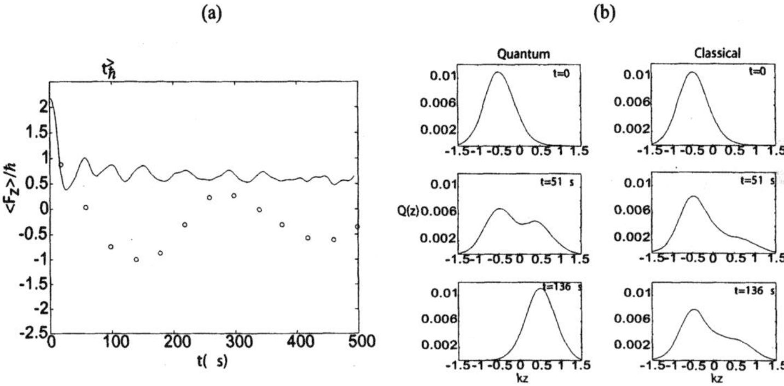

FIGURE 2. (a)Predictions of mean magnetization dynamics. Ideal quantum theory: two-level Rabi-flopping (dashed-dotted); Ideal classical theory: localized at positive $\langle F_z \rangle$ (solid); Experimental: (circles) with a damped sinusoid fit. (b) Reduced Q–distribution in position $Q(z)$, at different times in the quantum versus classical evolution. The quantum distribution oscillates between wells while the classical distribution remains mostly on the left side, with a portion equilibrating between the wells.

We can ask whether the observed non-classical motion can be interpreted as 'quantum tunneling'. For particles with more than one degree of freedom, the identification of tunneling behavior can become ambiguous since the total energy does not uniquely define the particle's classical trajectory. In particular, for the magneto-optical lattice at hand, the potential energy depends not only on the position of the atom, but also on its internal state in a correlated way. The initial state prepared in the experiment mostly populates the lowest adiabatic potential, but at times corresponding to a Schrödinger cat-like superposition in the two wells, there is a small component in the second lowest potential due to the nonadiabatic coupling. Whereas the mean energy of the population in the lowest adiabatic potential is higher than the corresponding double well barrier, the barrier of the next adiabatic potential is much higher than the mean energy of the small population in that potential. The oscillation of the population from right to left well in this adiabatic potential corresponds to tunneling through a classically forbidden regime. The atom therefore sees a population-weighted average of the two lowest adiabatic potential barriers [14] . The non-adiabatic transitions of the internal state cause the tunneling barrier to be dynamical in nature.

QUANTUM-CLASSICAL TRANSITION

Given the disparity between the classical and quantum phase space dynamics, one can ask under what circumstances classical dynamics is recovered? In recent years, it has been widely appreciated that emergent classical behavior can arise when the quantum system is weakly coupled to an environment. Decoherence resulting from tracing over the environment can suppress quantum interference and, in many circumstances, this can lead to an effectively classical evolution of a phase space distribution function [20, 21] . While decoherence can explain classical behavior for *mean values* of observables, it does not succeed in extracting localized "trajectories" from the quantum dynamics. Such trajectories are useful for quantifying the existence of chaos both theoretically and in experiments through the quantitative measure of the Lyapunov exponents. One can recover trajectories from the quantum dynamics when the environment is taken to be a meter that is continuously monitored, leading to an evolution of the system density operator conditioned on the measurement [22] . If one averages over all possible measurement results, the description reverts to that at the level of phase space distributions.

Continuous measurement provides information about the state of the system and thus localizes it in phase space. These localized trajectories have added quantum noise, however, due to quantum measurement backaction. Therefore, in order to recover the desired classical trajectories, the system must be in a regime where the measurement causes strong localization but weak noise [23] .

We have studied both numerically and analytically the conditions for recovering classical behavior in our system given a continuous measurement of the atomic position [24] . We take as our model Hamiltonian,

$$H = \frac{p^2}{2m} + \frac{1}{2} m\omega^2 z^2 + b z \frac{J_z}{J} + c \frac{J_x}{J} \tag{6}$$

which is a harmonic approximation to the magneto-optical lattice potential at a single lattice site. The evolution of the system conditioned on a record of the atoms mean position, $\langle z \rangle + (8k)^{-1/2} dW / dt$, is studied using a stochastic Schrödinger equation [25, 26] ,

$$d|\tilde{\psi}\rangle = \left\{ \frac{1}{i\hbar} H dt - kz^2 dt + \left(4k\langle z \rangle dt + \sqrt{2k} dW \right) z \right\} |\psi\rangle \tag{7}$$

where the tilde denotes an unnormalized quantum state, k is the "measurement strength", and dW describes a Wiener noise process. Here we have assumed perfect measurement efficiency. We can numerically evolve this equation using a Milstein algorithm [27] for the stochastic term. We pick as our initial condition a product of minimum uncertainty states (coherent states) in position and spin, and compare to the classical trajectories centered at the same initial points; we choose the initial spin coherent state in the x-direction though any direction in the x-y plane would be equivalent. We fix $b = -m\omega^2 \Delta z / J$, with $\Delta z \approx 15 z_g$ where z_g is the ground state rms width of the wells. Previous work [23] has established a window of measurement strengths for sufficiently large external (center of mass) actions, which satisfy the dual desire for strong localization and weak measurement back-action. We build upon that result here, choosing $I \approx 1000\hbar$. This puts us in a semi-classical regime where the external degree of freedom is effectively classical but the quantum nature of internal dynamics can be still be important.

We first consider the dynamics of the smallest spin system, $J=1/2$ in the integrable regime ($c = 0$). We see in Fig. 3(a) that the quantum trajectory quickly diverges from the classical trajectory. This can be understood by studying the effect of the position measurement on the spin subsystem. The initial spin state pointing in the x-direction is an equal superposition of spin-up and spin-down states, which move along the wells centered at $z_\uparrow = -\Delta z$ and $z_\downarrow = \Delta z$ respectively (Fig. 4 a,c). The two spin components of the initial spatially localized wave packet spatially thus separate into a left and a right wavepacket, so that the total wave function evolves into an entangled Bell-like state, $|\psi(t)\rangle = |\phi_{left}\rangle|\uparrow\rangle + |\phi_{right}\rangle|\downarrow\rangle$, with $\langle\phi_{left}|\phi_{right}\rangle \neq 1$. This splitting of the wave packet causes an initial rapid increase of the position variance (outer solid curves in Fig. 3a). When the left and right components of the state become spatially resolvable beyond measurement errors, the position measurement, acting as a meter for spin, collapses the wave function into either the left potential (spin-up state) or the right potential (spin-down state). This contrasts the fully classical dynamics, which predicts oscillation about the origin.

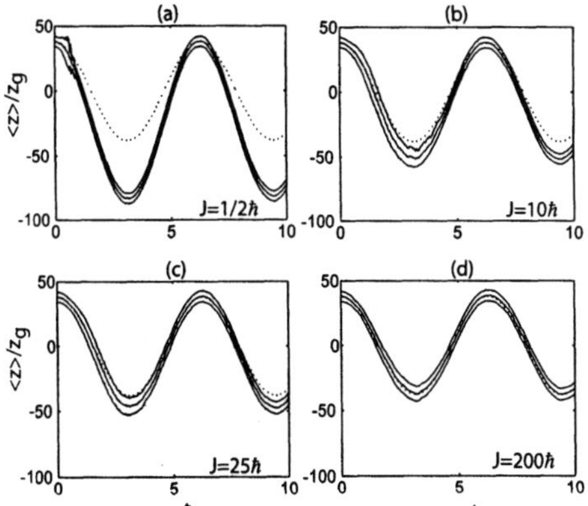

FIGURE 3: Mean position of the measured system (solid) in a single quantum trajectory for different values of spin with $\Delta z \approx 15 z_g, I \approx 1000\hbar, k = \omega/2z_g^2$ Outer solid curves show the variance of the wave function. As J gets larger the mean position evolution approaches the classical (dotted) trajectory.

When the magnitude of the spin, J, becomes much larger than 1/2, an initial spin coherent state in the x-direction is no longer a superposition solely of spin up and down states in the z-basis, but rather, a distribution over all $2J+1$ M_J states, centered around $M_J = 0$. Just as in the spin 1/2 case, an initially localized wave function will spread out in space as the different spinor components move along the different potentials centered at $z_{M_J} = -(M_J / J)\Delta z$, (Fig 4. b,d). However, as J becomes larger, the population distribution becomes more peaked at the $M_J = 0$ state, so that most of the population moves along the potential centered at $z_0 = 0$, which corresponds to the classical potential. The measurement is thus more likely to localize the atom in this classical potential and damp out the tails of the wave function that spread out over the outermost potentials. The key point is that the effective spin measurement in the J_z basis is no longer strongly projective, and therefore the weak noise condition can be met along with the strong localization condition.

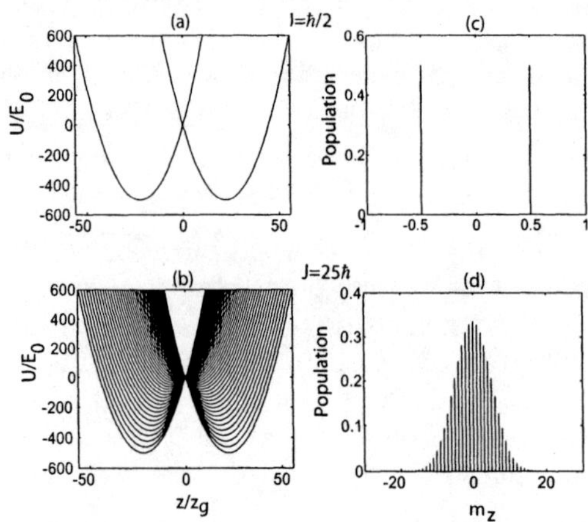

FIGURE 4. The spin up and down components of a spin 1/2 wave function move along 2 different potential wells (a). For J >>1/2, the spin components of the wave function evolve along 2J+1 different potentials (b). Histograms for the population in each mz state (c,d) show that as J gets larger the population becomes peaked around the m_z=0 state (d). The position is thus more likely to localize the wave function in the central (classical) potential well.

The scale of J for which the weak noise and strong localization conditions are satisfied can be analytically determined by following the approach in [23] . The stochastic equations of motion for the mean position and momentum follow from Eq. (7),

$$d\langle z\rangle = \frac{\langle p\rangle}{m}dt + \sqrt{8k}C_{zz}dW$$

$$d\langle p\rangle = -m\omega^2\langle z\rangle dt - b\langle J_z\rangle dt + \sqrt{8k}C_{zp}dW \qquad (8)$$

$$d\langle J_z\rangle = \sqrt{8k}C_{zJ_z}dW$$

where $C_{ab} = (\langle ab\rangle + \langle ba\rangle)/2 - \langle a\rangle\langle b\rangle$ are the symmetrized covariances. The small noise and strong localization conditions applied to these equations can be combined into the condition that the covariance matrix in the ordered basis $\{z,p,J_z\}$ remain small at all times relative to the allowed phase space of the dynamics. We have ignored the x and y components of the angular momentum since we are interested in measuring the position of the particle, which depends only on J_z. Furthermore, we neglect third and higher cumulants in the evolution of the covariance matrix C, since these remain small for large J. Under this approximation, the covariance matrix evolves according to a matrix Riccati equation, $\dot{C}(t) = \alpha + \beta R(t) + R(t)\beta^T + R(t)\gamma R(t)$ [24] . This equation has an analytical solution since the matrices α, β and γ are time independent [28] . However the analytical solutions for the second cumulants are not simple functions of the system parameters. We resort to numerically finding the bounds on the analytical solutions $C(t)$. Our studies indicate that the bounds on $C(t)$ become small relative to the allowed phase space, only when J is much larger than $\hbar$, as expected.

Our numerical and analytical results show that classical dynamics is only recovered in this coupled system when the actions of both subsystems become large relative to $\hbar$. When one subsystem lies in the quantum regime, even a weak measurement of the classical subsystem eventually results in a strongly projective measurement of the quantum subsystem, thus preventing the recovery of classical behavior. Preliminary numerical studies indicate that the same condition holds true in the chaotic regime. When the spin is large enough relative to $\hbar$, the measured trajectory recovers the mixed phase-space of the classical system. However, the analytical solutions of the corresponding Riccati equation for the covariance matrix are much more difficult to obtain since the matrix γ is now time dependent. Work is in progress to find approximate bounds on the covariance matrix and to recover the classical Lyapunov exponent from the measured trajectories.

SUMMARY

Atoms in optical lattices provide a clean testbed in which to study the rich dynamics of coupled systems. Our experiments employ Cesium atoms in a one-dimensional, far-off-resonance, magneto-optical lattice. Preparation of atoms in localized, pure quantum states in the double-potential wells is achieved through a combination of laser-cooling, optical pumping, and state-selection. Because of the correlation between the atomic position and internal state, precession of the atomic spin is accompanied by motion of the center-of-mass wave packet. The entanglement between atomic spin and center of mass motion provides a meter through which dynamical tunneling and other transport phenomena can be directly observed at the microscopic level. We have carried out a detailed study of the classically chaotic dynamics of our atom-lattice system, including a direct comparison between classical predictions and quantum theory/experiment. Our results underscore the profoundly non-classical nature of the observed tunneling Rabi oscillations. We find that under appropriate conditions the atom can "tunnel" through a dynamical energy barrier.

The manifestations of chaos in quantum mechanics and the emergent complexity at the classical level continues to be a problem of fundamental interest. Our numerical and analytical results show that classical chaotic trajectories can be recovered through continuous measurement of this system only when both the external and internal actions are large relative to $\hbar$. In future work we hope to continuously measure the spin of the system via Faraday rotation spectroscopy [29, 30] and develop more tools to explore the quantum-classical boundary.

ACKNOWLEDGEMENTS

SG, PMA and IHD were supported by NSF Grant No. PHY-009569. PSJ and DLH were supported by grants NSF PHY0099582, ARO DAAD19-00-1-0375 and JSOP DAAD19-00-1-0359.

REFERENCES

1. H. Rabitz, *Science* **288**, 824-828 (2000).
2. A. Hache, *et al.*, *Phys. Rev. Lett.* **78**, 306-309 (1997).
3. A. Hache, J. E. Sipe, and H. M. van Driel, *IEEE J. Quantum Electron.* **34**, 1144-1154 (1998).
4. R. Atanasov, *et al.*, *Phys. Rev. Lett.* **76**, 1703-1706 (1996).
5. M. Sheik-Bahae, *Phys. Rev. B* **60**, 11257-11260 (1999).
6. M. A. Nielsen and I. L. Chuang, *Quantum Computation and Quantum Information,* Cambridge University Press, Cambridge, 2000.
7. M. V. Berry, *Proc. R. Soc. London Ser. A* **413**, 183-198 (1987).
8. L. E. Reichl, *The Transition to Chaos in Conservative Classical systems: Quantum Manifestations,* Springer-Verlag, New York,1992.
9. F. L. Moore, *et al.*, *Phys. Rev. Lett.* **73**, 2974-2977 (1994).
10. F. L. Moore, *et al.*, *Phys. Rev. Lett.* **75**, 4598-4601 (1995).
11. I. H. Deutsch and P. S. Jessen, *Phys. Rev. A* **57**, 1972-1986 (1998).
12. I. H. Deutsch, *et al.*, *J. Opt. B.: Quantum and Semiclassical Optics* **2**, 633-644 (2000).
13. D. L. Haycock, *et al.*, *Phys. Rev. Lett.* **85**, 3365-3368 (2000).
14. S. Ghose, P. M. Alsing, and I. H. Deutsch, *Phys. Rev. E* **64**, 056119-056119-8 (2001).
15. D. Feinberg and J. Ranninger, *Physica D* **14**, 29-48 (1984).
16. H. Schanz and B. Esser, *Phys. Rev. A* **55**, 3375-3387 (1997).
17. D. Kusnezov, *Phys. Rev. Lett.* **72**, 1990-1993 (1994).
18. A. J. Lichtenberg and M. A. Lieberman, *Regular and Stochastic Motion,* Springer, New York, 1983.
19. F. T. Arechi, *et al.*, *Phys. Rev. A* **6**, 2211-2237 (1972).
20. W. H. Zurek, *Acta Phys. Pol. B* **29**, 3689-3709(1998).
21. S. Habib, K. Shizume, and W. H. Zurek, *Phys. Rev. Lett.* **80**, 4361-4365 (1998).
22. H. Carmichael, *An Open Systems Approach to Quantum Optics,* Springer-Verlag,Berlin,1993.
23. T. Bhattacharya, S. Habib, and K. Jacobs, *Phys. Rev. Lett.* **85**, 4852-4855 (2000).
24. S. Ghose, *et al.*, *quant-ph/0208064*, submitted to *Phys. Rev. Lett.* (2002).
25. G. J. Milburn, K. Jacobs, and D. F. Walls, *Phys. Rev. A* **50**, 5256-5263 (1994).
26. A. C. Doherty and K. Jacobs, *Phys. Rev. A* **60**, 2700-2711 (1999).
27. G. N. Milstein, *Theory Probab. Appl.* **19**, 557-562 (1974).
28. W. T. Reid, *Riccati Differential Equations,* Academic Press, New York, 1972.
29. A. Kuzmich, L. Mandel, and N. P. Bigelow, *Phys. Rev. Lett.* **85**, 1594-1597 (2000).
30. B. Julsgaard, A. Kozhekin, and E. S. Polzik, *Nature* **413**, 400-403 (2001).

A Chaotic Scattering Experiment

James A. Blackburn

Department of Physics and Computing
Wilfrid Laurier University
Waterloo, Ontario, Canada

Abstract. A table top experiment is described in which a magnetic puck floating on an air cushion was fired at, and scattered from a configuration of three other permanent magnets fixed at the vertices of an equilateral triangle. Both the launch velocity and the lateral displacement of the launch direction with respect to the center line (impact parameter) could be precisely controlled. A ring of ninety Hall effect switches encircling the system was used to detect both when and where the puck entered and exited the scattering region. The resulting experimental observations show extensive evidence of chaotic behavior. A comparison with numerical simulations is also presented.

INTRODUCTION

In a scattering experiment, a particle, or a beam of particles, is fired at a target and the ensuing trajectories are observed. In some systems, the exit angle can be a hypersensitive function of one or more input parameters – a hallmark of *chaotic scattering* [1]. A frequently studied two-dimensional system consists of three scatterers located at the vertices of an equilateral triangle as illustrated in Fig.1.

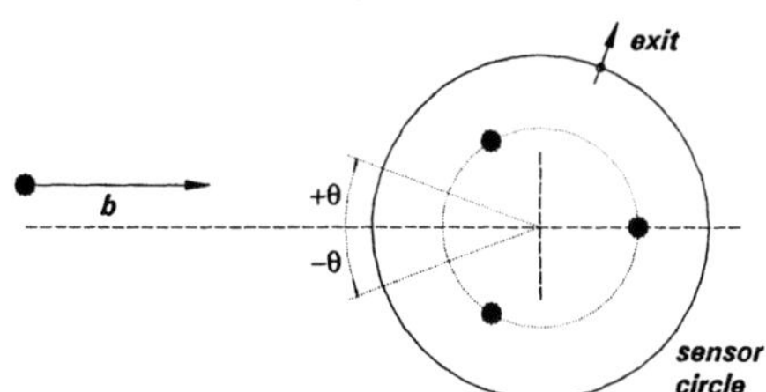

FIGURE 1. System composed of three scattering objects positioned uniformly on a circle of radius R. The incoming particle is defined by its energy β and impact parameter b.

An important special case of this arrangement considers the incoming particle to be a point object, the three scatterers to be disks of specified radius, and the interaction between particle and scatterer to be a perfect reflection at the moment of contact [2]. In contrast, here we assume an inverse square law repulsive force acts between each

CP676, *Experimental Chaos: 7th Experimental Chaos Conference,*
edited by V. In, L. Kocarev, T. L. Carroll, B. J. Gluckman, S. Boccaletti, and J. Kurths
© 2003 American Institute of Physics 0-7354-0145-4/03/$20.00

stationary scatterer and the moving object [3]. The scattering interaction is thus *soft* rather than hard.

THE EXPERIMENT

The experimental apparatus is shown in Fig.2.; its two principal elements are the particle launcher and the scattering array.

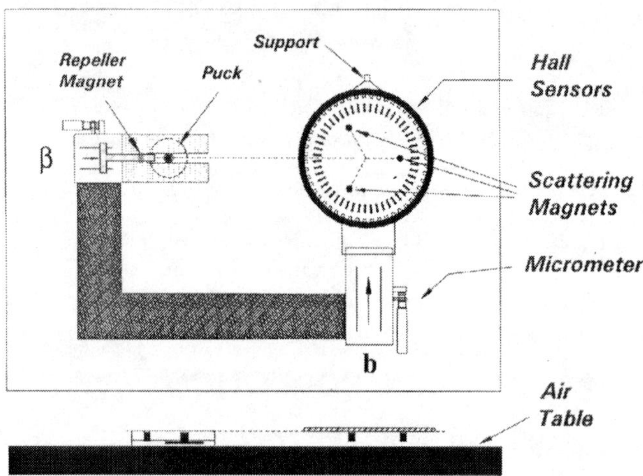

FIGURE 2. The experimental apparatus. The air puck is released at an initial distance from the repeller magnet, and is propelled towards the scattering system that consists of three stationary permanent magnets.

The "particle" was a small pill-shaped high intensity samarium-cobalt magnet (1 cm diameter, 2 cm length) which sat on a thin aluminum disk; this combination formed a puck that floated on a thin air cushion and moved in the plane of the table surface in a frictionless manner. The launcher consisted of a slotted guide block in which was embedded a repeller magnet. The air puck was manually positioned against a mechanical stop (setting the initial distance from the repeller magnet) and released. The β micrometer in the figure determined the initial puck/repeller separation and hence the launch velocity (or, equivalently, the injection energy).

The three fixed scatterers also were samarium-cobalt magnets, oriented so the interactions were all repulsive. They were attached to the underside of a circuit board, hanging down but not quite reaching the table surface. This allowed the skirt of the air puck to slide freely under any scatterer in the event of a close encounter. Because of the exceptionally strong repulsive forces at small separations, no hard physical contact ever occurred between magnets. By means of a second micrometer the entire scattering assembly could be translated along a line orthogonal to the particle's initial launch direction. This allowed precise control of the impact parameter b.

A circular array of ninety Hall effect switches, mounted on the circuit board, was used to detect the passage of the magnetic puck as it entered or left the scattering region. A computer-based data acquisition system recorded these switching pulses and from them determined the final two experimental quantities of interest: *delay time* (how long the particle bounced around within the system before escaping), and *exit angle*.

Figure 3 shows typical experimental results obtained at a fixed value of the injection energy (i.e., for one particular setting of the "β" micrometer). It is apparent that there are intervals of the impact parameter for which the delay time is comparatively small, and that each of these intervals is associated with one of the three possible exit channels that lie between the scattering magnets. Separating these domains are narrower intervals in the impact parameter within which the delay time jumps to much larger values, and the exit angle hops unpredictably between the three channels. This behavior is a reflection of the increased complexity of the particle trajectories within the scattering region.

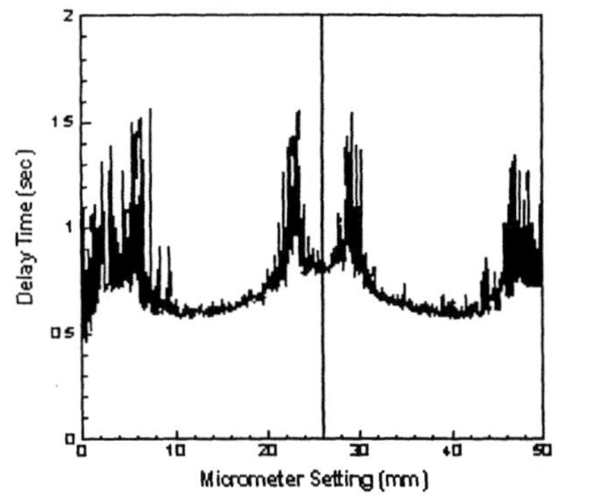
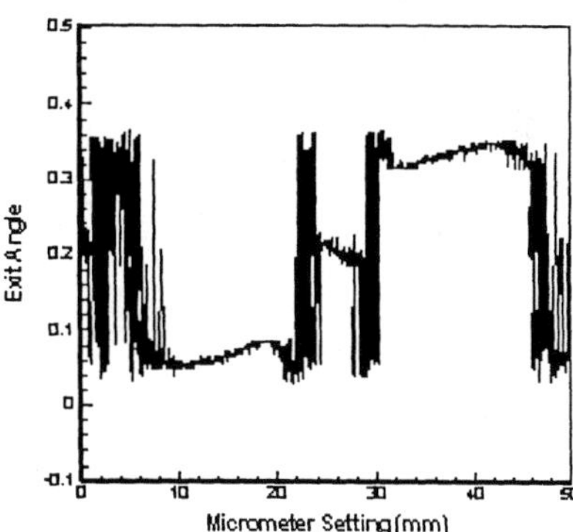

FIGURE 3. Experimental data for Delay Time (left) and Exit Angle (right) as functions of the setting on the "b" micrometer. The vertical line is provided to indicate the micrometer setting that corresponds to the center line of the system (i.e., b=0).

NUMERICAL SIMULATIONS

In dimensionless form, the equations of motion of a particle acted upon by three inverse-law repulsive potentials can be written:

$$\ddot{x} = \sum_{j=1}^{3} \frac{(x - x_j)}{r_j^3} \qquad \ddot{y} = \sum_{j=1}^{3} \frac{(y - y_j)}{r_j^3} \tag{1}$$

where r_j is the distance of the particle at (x,y) from center of the jth potential (x_j, y_j). Spatial coordinates are expressed in units of the radius of the circle on which the three scattering potentials are located, while the normalizing time scale subsumes all the particulars of the physical constants, including the particle's mass, that enter the particle/potential interaction.

Numerical solutions were obtained for this pair of coupled differential equations with various choices of the dimensionless impact parameter b and injection energy β (normalized to the potential energy at the geometric center of the scattering system). The results of a series of simulations covering a range of impact parameters is shown in Fig. 4. The particular value chosen for β was found to give the best match to the corresponding experimental results shown in Fig. 3.

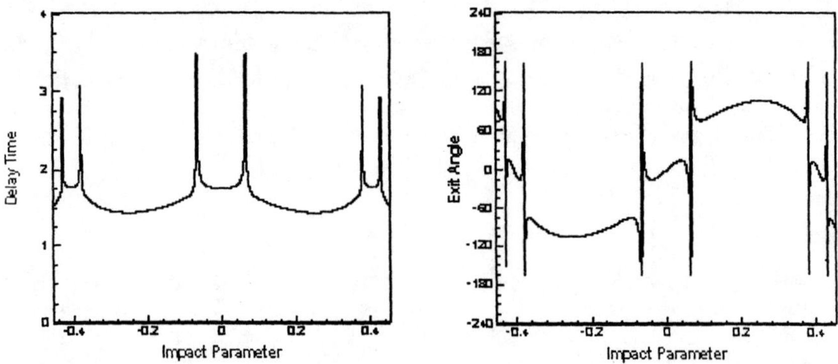

FIGURE 4. Results of a numerical simulation of the scattering equations for β=2.50.

A PROBLEM AND ITS RESOLUTION

In spite of the great care taken in the design and execution of this experiment, the results shown in Fig. 3 seem rather "noisy" when compared with the corresponding numerical simulation presented in Fig. 4. These significant differences remained unexplained until measurements revealed that very slight variations occurred in the launch velocity for repeated trials at exactly the same β micrometer setting. The possible significance of this small effect can be appreciated by examining a complete view of the (β, b, delay time) parameter space (Fig. 5). Scattering functions are actually individual slices from this surface taken at particular values of injection energy. The peaks in any one slice merge to become ridges in the full parameter space, and the ridges are themselves narrow chaotic domains separating the more extensive smooth regions of simple scattering behavior.

A spread in injection energy can be visualized as the bracketed range shown in Fig. 5. The bounding scattering functions are re-plotted in Fig. 6. From these graphical representations, two inferences can be drawn.

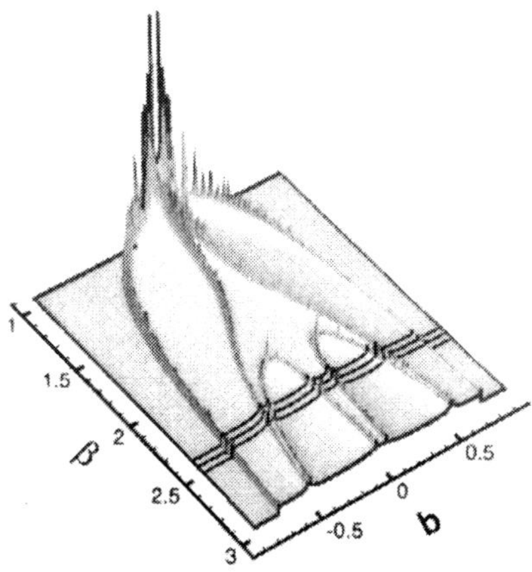

FIGURE 5. Delay time as a function of the impact parameter (b) and injection energy (β), determined from numerical simulations.

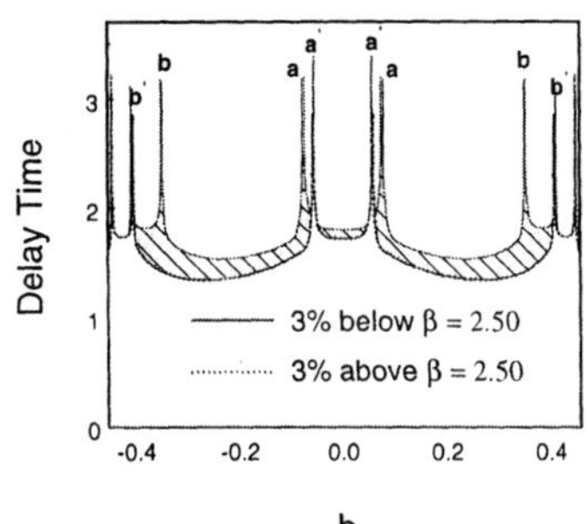

FIGURE 6. Delay time as a function of impact parameter for two values of injection energy. This figure is, in effect, an end view of the surface in Fig. 5 highlighting the slices taken above and below the value β=2.50.

First, because the parameter surface has a downward tilt toward higher energies, there is a corresponding vertical displacement of the non-chaotic portions of the scattering functions, as can be seen in Fig. 6. The effect of an uncertainty in β will be that at each new value of the impact parameter, the delay time will appear as if randomly sampled from between these upper and lower boundaries. This will give the appearance of noise in the smooth portions of the scattering function.

Second, because the chaotic ridges lie along curves in parameter space, the peaks in the scattering function will be displaced horizontally, either outward or inward, as can be seen in Fig. 6. Any variability in β will permit the continuum of peaks lying between these limits to be visited in a random fashion, significantly widening the intervals in b over which enhanced delay times occur.

This evidence prompted a revised set of simulations in which, at each new value of impact parameter, β was randomly selected from within a +/- 3% interval around the mean value of 2.50. The results are shown in Fig. 7.

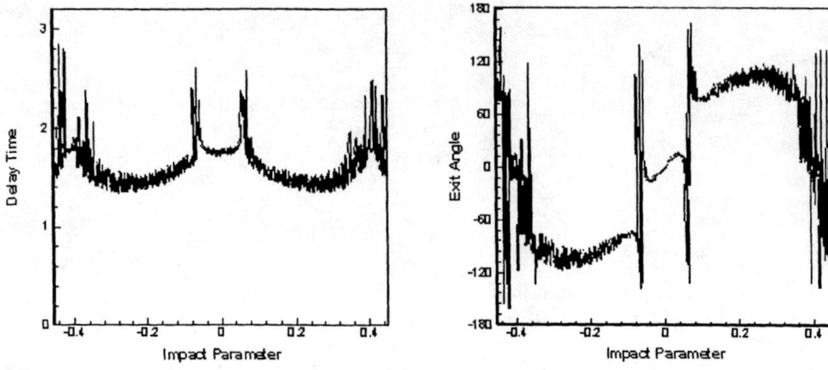

FIGURE 7. Results of a numerical simulation of the scattering equations with a random variation in the injection energy as described in the text.

Now there is a convincing similarity to the experimental observations displayed in Fig. 3, confirming the crucial role of energy blurring on the scattering functions.

These results have potentially profound implications with respect to computationally produced data. In published studies, it is not uncommon to find numerical simulations that focus on extremely narrow intervals of some parameters. As this experiment has demonstrated, it would be unlikely that a real system would exhibit the kind of ultra-ideal scattering functions that are so easily generated with a computer. The very hypersensitivity that is at the heart of chaos means that observed behavior can be grossly modified by unavoidable imperfections and uncertainties.

ACKNOWLEDGMENTS

The work reported here was a collaboration with H.J.T. Smith. Funding was provided by the Natural Sciences and Engineering Research Council of Canada.

REFERENCES

1. E. Ott and T. Tel, *Chaos* **3**, 417-426 (1993).
2. B. Eckhardt, *J. Phys. A.* **20** 5971-5979 (1987); P. Gaspard and S.A. Rice, *J. Chem. Phys.* **90**, 2225-2241 (1989).
3. J.A. Blackburn and H.J.T. Smith, *Int. J. Bifur. Chaos* **12**, 71-85 (2002)

Phase Synchronization in a Plasma Discharge Driven by a Chaotic Signal

Epaminondas Rosa, Jr.[*][†], Catalin M. Ticos[**][†], William B. Pardo[†], Jonathan A. Walkenstein[†], Marco Monti[†] and Jürgen Kurths[‡]

[*]*Department of Physics, Illinois State University, Normal, Illinois 61790*
[†]*Nonlinear Dynamics Lab., Dept. of Physics, University of Miami, Coral Gables, Florida 33146*
[**]*University of Oxford, Department of Engineering Science, Parks Road, OX1 3PJ, UK*
[‡]*Nonlinear Dynamics, Institute of Physics, University of Potsdam, D-14415, Potsdam, Germany*

Abstract. Two different coupled experimental systems (the Chua circuit and a plasma discharge) are demonstrated to be capable of phase synchronizing to each other. Real-time observation of power spectra in association with oscilloscope plotting of the signals of both systems allows the visualization of the transition from unsynchronization to synchronization.

INTRODUCTION

Synchronization is not a new phenomenon. The repeatedly occurrence of simultaneous events has long been observed in nature and also used in many types of devices [1]. It goes back to Huygens [2] who in 1673 reported the synchronous motion of two pendulum clocks hanging on the same wood beam. More recently, in the case of chaotic dynamical systems, synchronization has become an area of great interest. Initially expected not to be able to achieve synchronization [3], chaotic systems can indeed synchronize depending on the type of coupling, parameter values, and other factors. In particular, chaotic systems may be coupled in ways that allow the synchronization of their phases only, keeping amplitudes completely uncorrelated [4]. Phase synchronized chaos has received much attention lately with major theoretical and experimental work [5]. This includes the instance of a chaotic system forced with a periodic oscillator [6] demonstrated experimentally with a plasma discharge tube paced with a sinusoidal function [7], and with a music signal [8]. In all these cases there is a need for a clear understanding of what is meant by phase of the oscillator [9]. In the case of a circular oscillator, phase can defined as an angular variable ϕ increasing monotonically in time on the limit cycle: $d\phi/dt = \omega$, where $2\pi/\omega = T$ is the period of the oscillation. This definition is suitable for chaotic oscillators, such as the Rössler system [10], by introducing a phase angle variable. We take the projection of the attractor on the (x, y) plane and define the phase $\phi(t) = \arctan[y(t)/x(t)]$. For experiments where simultaneous recording of multiple measurements may restrict real time observation of attractors, the Hilbert transform can be useful for studying phase synchronization. It allows the computation of an analytical signal corresponding to each rotation, yielding the phase for that rotation. Using the analytical signal concept introduced by Gabor [11], the instantaneous phase of a signal $s(t)$ is $\phi(t) = \arctan[\tilde{s}(t)/s(t)]$, where $s(t) + i\tilde{s}(t) = A(t)e^{i\phi(t)}$, and

CP676, *Experimental Chaos: 7ᵗʰ Experimental Chaos Conference*,
edited by V. In, L. Kocarev, T. L. Carroll, B. J. Gluckman, S. Boccaletti, and J. Kurths

"

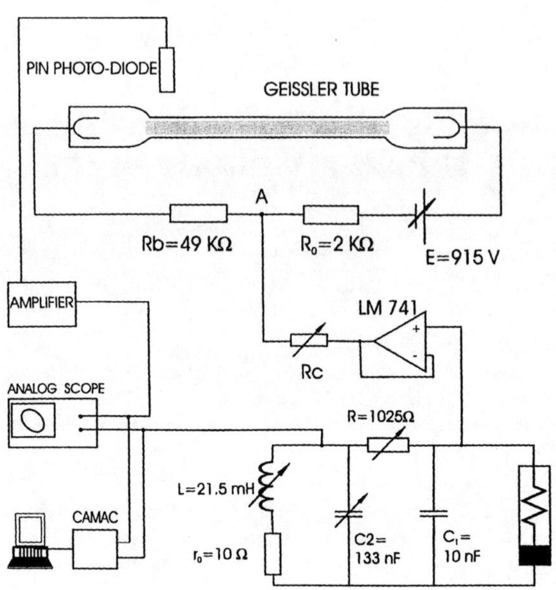

FIGURE 1. Chua-plasma experimental setup.

$\tilde{s}(t) = \pi^{-1}\text{P.V.}\int_{-\infty}^{\infty}\frac{s(\tau)}{t-\tau}d\tau$. Here A(t) is the instantaneous amplitude and P.V. stands for the principal value of the integral.

In this paper we propose to discuss the results of experimental measurements involving a plasma discharge tube coupled to a Chua electronic circuit. The coupling is unidirectional and works with a voltage from the Chua circuit fed into the plasma. We verify that, depending on the strength of the coupling, the two systems can phase synchronize for long time intervals. A special technique allows real time verification of phase synchronization, and proves to be very useful for testing of parameter values associated with different regimes.

EXPERIMENTAL SETUP

The setup of our experimental arrangement is shown in Fig. 1. The unidirectional coupling between plasma and Chua is achieved through an op-amp LM741, and a current limiting variable resistor R_c. This resistor is used to control the strength of the coupling. Increasing the resistance reduces the current with consequent weak coupling, and vice-versa. The Chua circuit [12] has $R = 1025\ \Omega$, $C_1 = 10$ nF, $C_2 = 133$ nF, $L = 21.5$ mH, and $r_0 = 10\ \Omega$. The voltage across the capacitor C_1 is the driving signal injected into the plasma. The plasma is produced in a commercial Geissler tube filled with helium gas. The discharge is powered by a high voltage source set at $E \simeq 900$ V, and sustained by the ballast resistor $R_b = 49$ kΩ. The light intensity oscillations of the plasma column is acquired using a PIN photo-diode.

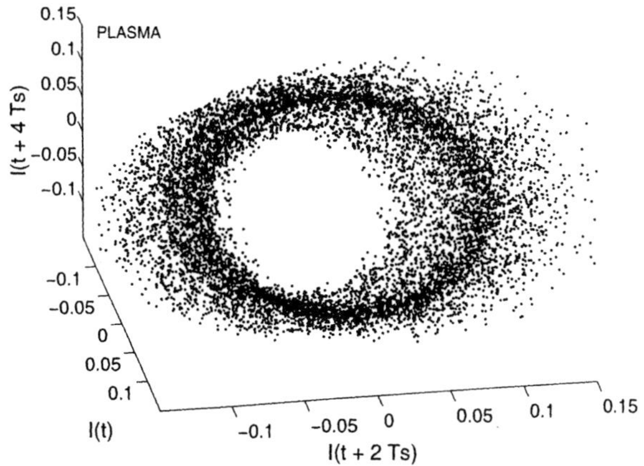

FIGURE 2. (a) Plasma attractor.

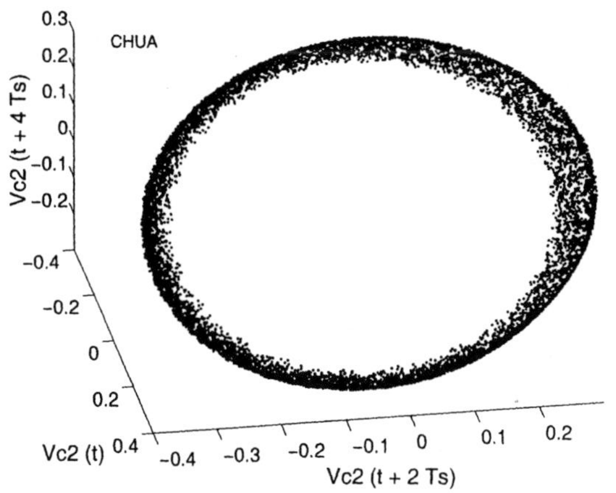

FIGURE 3. (a) Chua attractor.

TUNING THE OSCILLATORS

The signal acquired from the oscillations of the plasma light intensity can be plotted real time on an oscilloscope. If this signal is plotted versus its analog derivative, an attractor can be viewed and used for tuning the system. Alternatively, time delay embedding can also be used to produce the corresponding attractor, as pictured in Fig. 2. The same can be done for the voltage across C_2 of the Chua circuit, with the resulting attractor shown in Fig. 3. The plasma and the Chua signals originate from different systems but have similar broad power spectra with a peak indicating a dominant frequency [13]. Figure 4 depicts both spectra displaying a mismatch of about 50 Hz between the peaks when the two oscillators are uncoupled. A real time visualization of these two signals can be obtained by plotting one versus the other on the oscilloscope. An ellipse-like

303

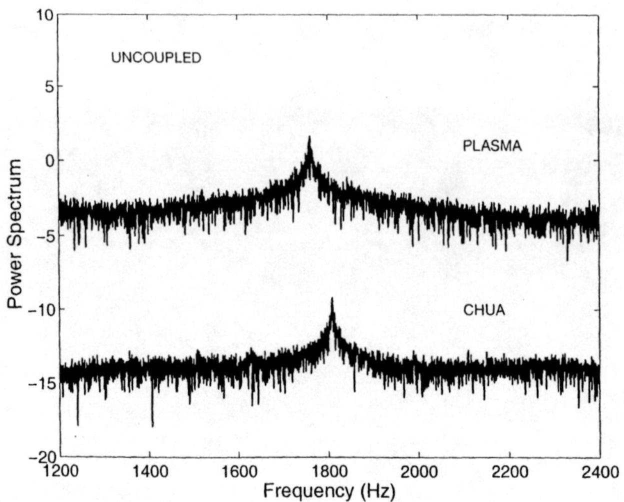

FIGURE 4.   Spectra of uncoupled plasma and Chua.

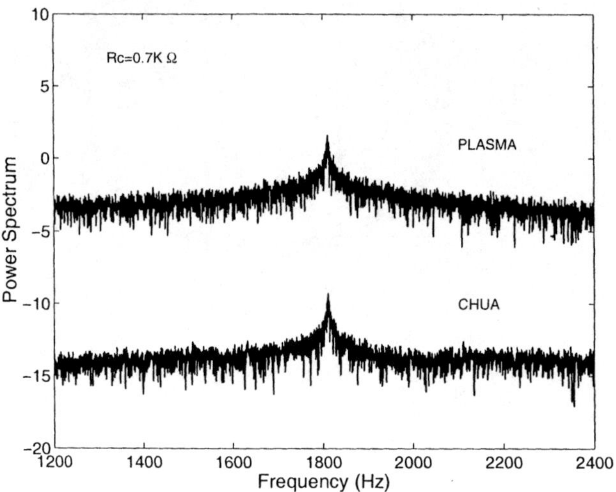

FIGURE 5.   Spectra of coupled plasma and Chua.

figure is obtained oscillating back-and-forth between the angles of 45° and 135°. This back-and-forth motion is a clear indication that the plasma and the Chua are not phase synchronized. However, by coupling the two systems and tuning the circuit to reduce the mismatch between the dominant frequencies on the spectra, we observe that the ellipse on the oscilloscope becomes more stable and ceases oscillating when the frequency peaks get very close to each other. Figure 5 shows the spectra of both signals with Chua and plasma coupled and very small mismatch between the dominant frequencies of the

304

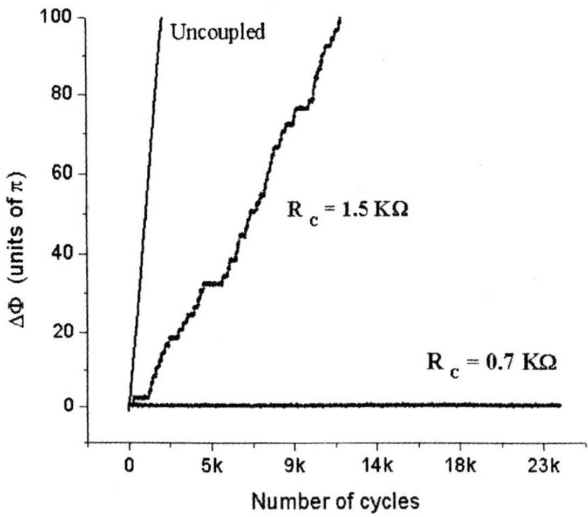

FIGURE 6. Phase synchronization ($R_c = 0.7$kΩ), phase slips of 2π ($R_c = 1.5$ kΩ), and phase unsynchronization (uncoupled).

two oscillators. Notice how the dominant frequency of the plasma moves toward the dominant frequency of the Chua when the coupling is turned on. In this case the ellipse-like plot of one signal versus the other is stable and does not oscillate, typical indication that the two oscillators are phase synchronized.

The coupling resistor R_c plays an important roll in the process when the two oscillators go from phase synchronized to phase unsynchronized chaos. This can be observed in Fig. 6 where the phase difference $\Delta\Phi$ between plasma and Chua is plotted versus the number of cycles. If the oscillators are not coupled, notice that $\Delta\Phi$ grows very fast. However, if the coupling is on with $R_c = 1.5$ kΩ, then $\Delta\Phi$ still grows, but shows plateaus where the two oscillators have their phases synchronized during short time intervals. Finally, when $R_c = 0.7$ kΩ, the coupling is stronger and in this case the phases of the two oscillators remain synchronized for very long time intervals.

CONCLUSION

In conclusion, we have developed a technique for experimental real-time observation of synchronized states between a plasma discharge and a Chua circuit. The power spectra of the two systems associated with the oscilloscope observation of the plasma versus Chua signals provides a concrete and reliable manner for checking the state of the two oscillators. Depending on the strength of the coupling, we observe slips of 2π between the phases of the two oscillators.

REFERENCES

1. I. Blekhman, *Synchronization in Science and Technology* ASME Press, New York (1988).
2. C. Hugenii, *Horoloquim Oscilatorium* (Parisiis, France, 1673).
3. Fujisaka, H., and Yamada, T., *Prog. Theor. Phys.*, **69**, 32–40; L. M. Pecora and T. M. Carroll, Phys. Rev. Lett. **64**, 821 (1990).
4. M. G. Rosenblum, A. S. Pikovsky, J. Kurths, Phys. Rev. Lett. **76**, 1804 (1996); A. S. Pikovsky, M. G. Rosenblum, and J. Kurths, *Synchronization: A Universal Concept in Nonlinear Science*, Cambridge University Press, Cambridge (2001).
5. U. Parlitz, J. Junge, and W. Lauterborn, Phys. Rev. E **54**, 2115 (1996); P. M. Varangis, A. Gavrielides, T. Erneus, V. Kovanis, and L. F. Lester, Phys. Rev. Lett. **78**, 2353 (1997); J. R. Terry, K. S. Thornburg, Jr., D. J. DeShazer, G. D. VanWiggeren, S. Zhu, P. Ashwin, and R. Roy, Phys. Rev. E **59**, 4036 (1999); A. Neiman, X. Pei, D. Russel, W. Wojtenek, L. Wilkens, F. Moss, H. A. Braun, M. T. Huber, and K. Voigt, Phys. Rev. Lett. **82**, 660 (1999); G. M. Hall, S. Bahar, and D. J. Gauthier, Phys. Rev. Lett. **82**, 2995 (1999); A. R. Yehia, D. Jeandupeux, F. Alonso, and M. R. Guevara, Chaos **9**, 916 (1999); J.-W. Shuai and D. M. Durand, Phys. Lett. A **264**, 289 (1999); A. Dabrowski, Z. Gallias, M. Ogorzalek, Int. J. Bifurcation and Chaos **10**, 2391 (2000); D. Mazza, A. Vallone, H. Mancini, and S. Boccaletti, Phys. Rev. Lett. **85**, 5567 (2000); D. J. DeShazer, R. Breban, E. Ott, and R. Roy, Phys. Rev. Lett. **87**, 044101-1 (2001); K. Josic and D. J. Mar, Phys. Rev. E **64**, 056234-1 (2001).
6. E. Rosa, Jr., E. Ott, and M. H. Hess, Phys. Rev. Lett. **80**, 1642 (1998); R. Breban and E. Ott, Phys. Rev. E **65**, 056219 (2002).
7. C. M. Ticos, E. Rosa, Jr., W. B. Pardo, J. A. Walkenstein, and M. Monti, Phys. Rev. Lett. **85**, 2929 (2000).
8. W. B. Pardo, E. Rosa, Jr., C. M. Ticos, J. A. Walkenstein, and M. Monti, Phys. Lett. A **284**, 259 (2001).
9. T. Yalçinkaya and Y.-C. Lai, Phys. Rev. Lett. **79**, 3885 (1997); Z. Zheng, G. Hu, and B. Hu, Phys. Rev. E **62**, 7501 (2000); M. A. Zaks, E.-Y. Park, and J. Kurths, Phys. Rev. E **65**, 026212 (2002); D. L. Valladares, S. Boccaletti, F. Feudel, and J. Kurths, Phys. Rev. E **65**, 055208 (2002); I. Tokuda, J. Kurths, and E. Rosa, Jr., Phys. Rev. Lett. **88**, 014101 (2002); G. V. Osipov, A. S. Pikovsky, and J. Kurths, Phys. Rev. Lett. **88**, 054102 (2002).
10. O. E. Rössler, Phys. Lett. A **57**, 397 (1976).
11. D. Gabor, J. IEE London **93**, 429 (1946).
12. M. P. Kennedy, Frequenz **46**, 66 (1992).
13. E. F. Stone, Phys. Lett. A **163**, 367 (1992).

Chaotic Optical Communication over Turbulent Channel

L. Illling*, N. F. Rulkov* and M. A. Vorontsov[†]

*Institute for Nonlinear Science, University of California, San Diego, La Jolla, CA 93093-0402
†Army Research Laboratory, Adelphi, Maryland 20783

Abstract. We experimentally demonstrate robust self-synchronizing chaos based communication over an approximately 5km free-space laser link operating in a turbulent environment. A binary information sequence is transmitted using a chaotic sequence of short-term pulses and Chaotic Pulse Position Modulation (CPPM), where each pulse is slightly shifted from the original chaotic time position depending on the transmitted binary bit. We report the results of an experimental analysis of the atmospheric turbulence in the channel and the Bit-Error-Rate (BER) performance of this chaos based communication system. The dynamics of error bursts in the CPPM communication system caused by the atmospheric turbulence is discussed.

INTRODUCTION

Thanks to the chaotic origin of chaos, two coupled chaotic oscillators can be synchronized (self-synchronized) to reproduce at the receiver end the chaotic signals generated at the transmitter [1]. Exploiting self-synchronization information encoded in the received chaotic signal can be recovered [2], thereby allowing the use of chaotic carriers for information transmission. Chaos communications is of broad interest because it both enlarges the design possibilities of communication systems and it provides a challenge due to the complexity of nonlinear dynamical issues involved [3].

All practical communication channels will distort the chaotic waveform shape so that the received signal deviates from the transmitter oscillations. Channel noise, filtering, attenuation variability and other distortions in the channel corrupt the carrier and information signal. The presence of distortions can severely degrade the quality and significantly hamper the onset of identical synchronization [4]. Above a critical level of signal distortion self-synchronization fails completely resulting in a loss of the communication link.

A number of special methods have been proposed to overcome the problem of enhanced sensitivity to signal distortions found in communication schemes based on chaos synchronization, thereby removing what is still considered one of the major obstacles for practical implementation of chaos-based communication schemes [5]. At least in theory and numerical simulations, it appears that the regime of identical synchronization in these specially designed systems is significantly less sensitive to channel noise and waveform distortions caused by limited bandwidth of the channel [6]. However, to the best of our knowledge, robust self-synchronizing chaos communication over a channel with significant non-stationary signal distortions has not been demonstrated so far.

CP676, *Experimental Chaos: 7ᵗʰ Experimental Chaos Conference,*
edited by V. In, L. Kocarev, T. L. Carroll, B. J. Gluckman, S. Boccaletti, and J. Kurths
© 2003 American Institute of Physics 0-7354-0145-4/03/$20.00

"

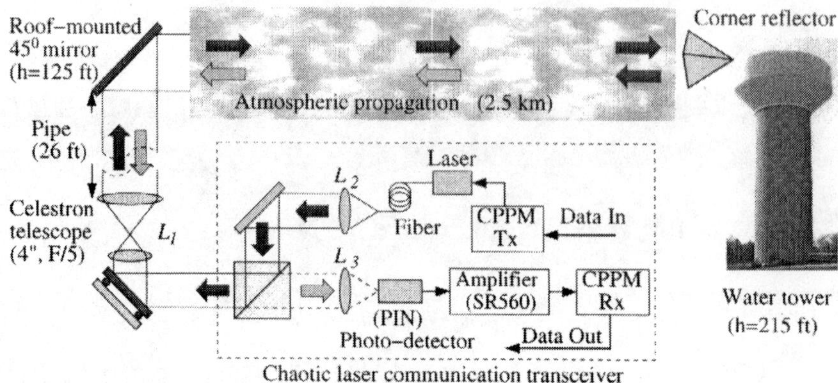

FIGURE 1. Schematic experimental setup for free-space laser communication system with chaotic pulse position modulation transceiver

Here we report the first experimental study of a chaotic self-synchronizing free-space laser communication system operating in the presence of severe communication signal distortions caused by atmospheric turbulence.

The experimental setup is schematically shown in Fig. 1. CPPM Tx and CPPM Rx represent the analog circuit implementation of the Chaotic Pulse Position Modulation transceiver. The optical part of the experiment consisted of an intensity-modulated semiconductor laser beam (10 mW, $\lambda = 690$ nm) that propagated by means of a single-mode fiber, a lens relay system (lenses L_1 and L_2) and telescope (Celestrone), which expanded the beam to a 4" diameter, and a 26 ft long vertical air-locked pipe to a 45° roof-mounted mirror. The light subsequently traveled over an atmospheric path of length $L = 2.5$km to a 4" corner cube mounted on top of a water tower, where it was reflected and propagated back to the roof-mounted mirror. The receiver system used the same telescope and lens relay system as the transmitter. The total double-pass atmospheric laser beam propagation distance was approximately $L \approx 5$km.

TURBULENT CHANNEL

Turbulence in the atmospheric communication channel leads to severe laser beam intensity scintillations that result in deep fluctuations of the received communication signal (received laser beam power). Double-pass wave propagation as in our experiment amplifies these fluctuations due to backscatter enhancement effects [7]. The most deleterious effects from receiver plane scintillations are the loss of signal-to-noise ratio and dropouts (information loss).

To evaluate the level of intensity scintillations in the atmospheric channel we measured the received signal from a continuously running laser with constant output intensity. Figure 2a shows the fluctuations of the normalized received power, measured by the PIN photodetector placed in the lens L_3 focal plane (Fig. 1), amplified by the low noise preamplifier (SR500 with gain of 20), and then acquired with sampling rate 1000 sam-

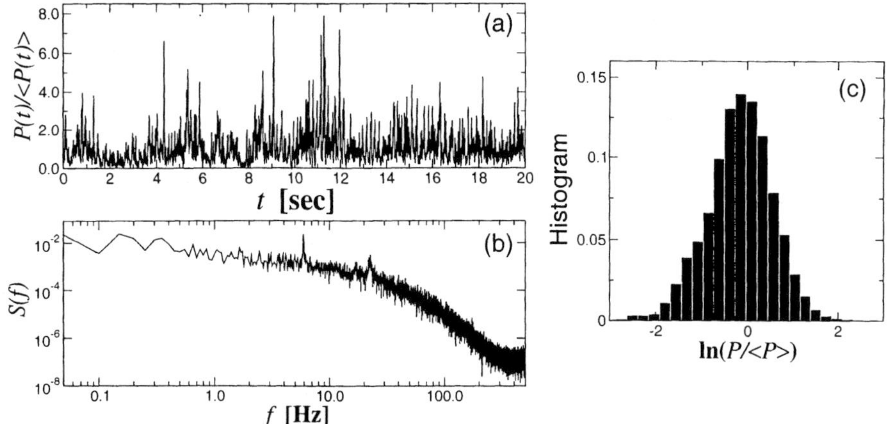

FIGURE 2. Fluctuations of the received power $P(t)$ in the experiment with non-modulated laser generating constant output intensity (10mW). Normalized received power $P(t)/\langle P(t)\rangle$ measured at the photodetector output - (a), and corresponding averaged power spectrum of - (b) illustrate the presence of strong laser beam intensity scintillations. c) Histogram of the probability distribution for the random variable $\ln(P/\langle P\rangle)$.

ples/sec. The received signal standard deviation is as high as 0.8-0.9, which is indicative of a strong scintillation regime. The corresponding ensemble-averaged received signal power spectrum S is shown in Fig. 2b.

In atmospheric optics the laser beam scintillations are traditionally described in terms of the logarithm of the received power (for finite receiver telescope): $\ln(P/\langle P\rangle)$, where $\langle P\rangle$ is the ensemble (time) averaged value [8]. A histogram that represents the distribution of the values of the random variable $\ln(P/\langle P\rangle)$ normalized by the total number of samples is shown in Fig. 2c. The histogram is computed using $N = 10^5$ consecutive samples of the data $P(t)$, a 20 sec segment of which is shown in Fig. 2a. Representing an approximation to the probability distribution of the received power the histogram closely matches the log-normal distribution expected from theory [8].

FREE-SPACE COMMUNICATION

In the communication experiment the laser generated a chaotic sequence of short term pulses ($\sim 1.0\mu s$) that were triggered by a TTL pulse signal from the chaotic transceiver controller CPPM Tx (see Fig. 1). The chaotic sequence of time intervals $\{T_n\}$, where T_n is the time interval between the nth and the $(n-1)$th pulse, corresponds to iterations of a chaotic process with the binary information signal added to the chaotic signal. This method of chaos communication is referred to as Chaotic Pulse Position Modulation (CPPM) [10]. Since both chaos and information are in the timing of the pulses, the particular intensity waveform of the generated light pulses is of little consequence.

In the implementation of CPPM discussed here the chaos is produced by iterations of a one-dimensional tent map (see [10] for details on the design). Denoting with $F(\cdot)$ the

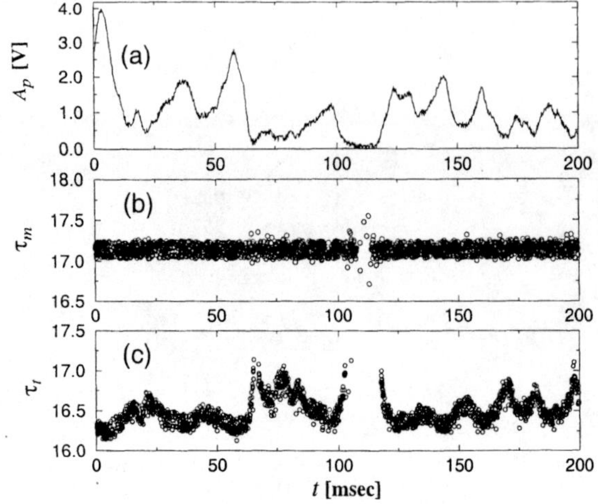

FIGURE 3. Fluctuations of the CPPM light pulses after traveling through atmospheric turbulence. Pulse amplitude A_p measured in volts at the output of amplifier - (a). Propagation times τ_m - (b) and τ_t - (c) in μsec. The pulse propagation times are computed from data acquired simultaneously at the output of CPPM Tx and output of Amplifier (SR560) at a sampling rate of 5×10^6 samples per sec.

nonlinear function of the tent map and with S_n the nth binary information signal (zero or one) we can write the iterative map that generates the chaotic sequence of time intervals $\{T_n\}$ as

$$T_n = F(T_{n-1}) + d + mS_n. \tag{1}$$

The parameter m signifies the modulation amplitude, whereas d is a constant time delay needed for the practical implementation of the communication method. The transceiver circuits were tuned to a regime of robust chaotic behavior. The generated chaotic inter-pulse time intervals $\{T_n\}$ ranged from 10μsec to 25μsec and supported a $\sim$60 kbit/sec bit rate.

The distorted pulses received at the PIN photo detector, placed at the end of the optical propagation path, are applied to the receiver CPPM Rx (see Fig. 1). When the receiver input signal exceeded a chosen threshold level a timer circuit in CPPM Rx was triggered. The receiver acquired two consecutive inter-pulse time intervals T_{n-1} and T_n, which using

$$mS_n = \left[T_n - F(T_{n-1}) - d\right] \tag{2}$$

allows for the decoding of the information signal [11]. Due to the match of the chaotic map $F(\cdot)$ of transmitter and receiver they self-synchronize and the arrival moment of the next pulse can be predicted at the receiver. To improve the system performance by reducing the probability of channel noise falsely triggering the decoder the input of the receiver was blocked until the moment in time when the next pulse was expected.

The turbulent atmospheric channel leads to severe amplitude fluctuations of the received signal, as illustrated by the 200msec timetrace of the received pulse amplitudes shown in Fig. 3a. Despite the significant amplitude fluctuations, the propagation time τ_m, which is the timespan between the leading edge of the TTL pulse applied to the laser and the maximum of the received pulse, only varied within 0.2μsec (see Fig. 3b). Slow and small channel induced pulse propagation time variations are the key to a good performance of the CPPM communication method. However, the detection of the incoming

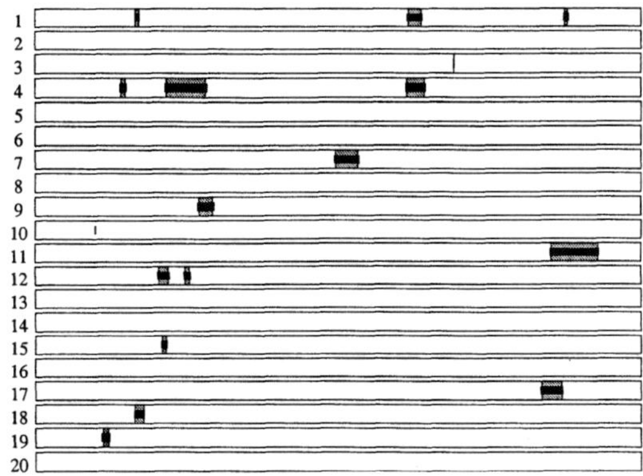

FIGURE 4. Typical structure of errors shown in 20 consecutive measured data streams each of length ~170 msec transmitted at ~2 min intervals. Each strip presents 10000 bits which are transmitted with the CPPM method. White intervals of the strips mark blocks of data received without errors. Narrow black ribbons in the middle of strips mark the blocks of the data received with errors. The gray background shows the blocks of the dropped out data caused by the loss of CPPM pulses due to fading instances.

pulses was implemented using thresholding. The pulse amplitude received by the CPPM Rx circuit had to exceed a certain threshold level ($\sim$200mV), which was chosen so as to minimize the probability of CPPM Rx triggering caused by background noise and reflections of optical pulses off nearby surfaces. Consequently, the actual delay time τ_t, measured between the leading edge of the TTL pulses generated by CPPM Tx and the moment of triggering of the CPPM Rx circuit, depends on the amplitude of the received pulses and fluctuates. Due to thresholding τ_t varies (see Fig.3c). The variations however remain smaller than the modulation amplitude $m \approx 1.5\mu sec$ and therefore self-synchronization and and communication can be maintained most of the time. Exceptions are the instances of pulse fading (drop-outs), where the received laser beam power falls to the receiver noise level and below the threshold level, respectively. These instances appear as gaps in τ_m and τ_t in Fig. 3.

Voice communication using this free-space communication setup was implemented using delta modulators to digitize an analog microphone signal. Except for the short term (< 50 msec) drop-outs, noticeable as occasional clicks, the voice communication was clear and stable.

Figure 4 shows the structure of lost data in the transmission of binary pseudo-random codes. The total Bit-Error-Rate (BER) measured in the experiment is 1.92×10^{-2}. From Fig. 4 it is apparent that data is lost in blocks and a detailed analysis of the error structure reveals the following main contributions to the BER. First, fading in the channel contributes $\sim 1.78 \times 10^{-2}$ or $\sim 92.7\%$ to the BER. Fading moments, that is instances when pulses do not trigger the CPPM receiver, occur at random times during the communication and are associated with the loss of blocks of data of up to 1000 bits. Second, right before and right after the fading instances the amplitude of the received pulses is still close to the threshold and as consequence even small noise in the channel can result in significant fluctuations of the interpulse intervals (see Fig. 3c). This effect contributes $\sim 1.4 \times 10^{-3}$ to the BER ($\sim 7.3\%$). The rest of the errors, those not related to the complete failure of the communication link due to fading, contribute to the BER only $\sim 5.5 \times 10^{-5}$.

CONCLUSION

We have shown experimentally that the CPPM communication method supports robust data transmission over a turbulent atmospheric channel except for instances when the channel fails due to fading. Due to the self-synchronizing feature of the CPPM method and the fact that the CPPM receiver needs to obtain just two consecutive correct pulses to re-establish the regime of chaos synchronization, the communication after drop-out events is re-established almost immediately.

The authors are grateful to L.S. Tsimring, H.D.I. Abarbanel, L. Larson, and A.R. Volkovskii for helpful discussions. This work was supported in part by U.S. Department of Energy (grant DE-FG03-95ER14516), the U.S. Army Research Office (MURI grant DAAG55-98-1-0269). The authors also thank J. Gowens and J. Carrano for support in the development of the Atmospheric Laser Optics Testbed (A_LOT) at Adelphi, Maryland used in the experiments.

REFERENCES

1. H. Fujisaka and T. Yamada, Prog. Theor. Phys. **69**, 32 (1984); L. M. Pecora and T. L. Carroll, Phys. Rev. Lett. **64**, 821 (1990);
2. D. R. Frey, IEEE Trans. Circuits Syst. **40**, 646 (1993); A. R. Volkovskii and N. F. Rulkov, Tech. Phys. Lett. **19**, 97 (1993); U. Feldman, M. Hasler, and W. Schwarz, Int. J. Ciruit Theory Appl. **24**, 551 (1996); L. Kocarev and U. Parlitz, Phys. Rev. Lett. **74**, 5028 (1995).
3. See, for example, special focus issues: IEEE Trans. Circuits Syst. **48** No. 12 (2001); Chaos **6** No. 3 (1996); Chaos **7** No. 4 (1997); Int. J. Bif. Chaos **10**, No. 11&12 (1993); Int. J. Circuit Theory Appl. **27** (1999);.
4. G. Kolumban, M. P. Kennedy and L. O. Chua, IEEE Trans. Circuits Syst. **45**, 1129 (1998); C. Williams, IEEE Trans. Circuits Syst. **48**, 1394 (2001).
5. T. L. Carroll, Phys. Rev. E **53** 3117 (1996); T. L. Carroll and G. A. Johnsón, Phys. Rev. E **57** 1555 (1998); E. Rosa, S. Hayes, and C. Grebogi, Phys. Rev. Lett. **78** 1247 (1997); H. Torikai, T. Saito, and W. Schwartz, IEEE Trans. Circuits Syst. **46**, 1072 (1999).
6. T. L. Carroll, IEEE Trans. Circuits Syst. **42**, 105 (1995); IEEE Trans. Circuits Syst. **48**, 1519 (2001); N. F. Rulkov and L. Tsimring, Int. J. Circuit Theory Appl. **27**, 555 (1999);
7. Backscatter enhancement effects result from correlations in the wavefront phase aberrations between the outgoing and the returned waves which propagate through the same refractive index inhomogeneities [8]. The variance characterizing the received wave phase and intensity fluctuations can exceed the corresponding value for a unidirectional wave that propagates a distance $z = 2L$ by more than a factor of two under strong scintillation conditions [8, 9].
8. L. C. Andrews, R.L. Phillips, and C.Y. Hopen,*Laser Beam Scintillation with Applications*, (SPIE Press, Bellingham, 2001).
9. Yu. A. Kravtsov and A.I. Saichev,*Sov. Phys. Usp.*, **25**, 494 (1982).
10. M. M. Sushchik, *et al. IEEE Communication Letters*, **4**, 128 (2000); N. F. Rulkov, *et al. IEEE Trans. Circ. Syst.* **48**, 1436 (2001).
11. In the CPPM Tx and CPPM Rx devices the consecutive values of T_n and T_{n-1} are generated and stored in the form of voltage signals. Eqs. (1) and (2) are implemented using an analog electrical circuit of the nonlinear function $F(.)$ and a subtracting circuit, see Ref. [10] for details.

DATA ANALYSIS

Development and Application of Chaotic Attractor Property Analysis for Vibration-Based Structural Damage Assessment

Michael D. Todd*, Jonathan Nichols*, Mark Seaver*, Stephen Trickey*, Lawrence Virgin†, Lou Pecora** and Tom Carroll**

*U.S. Naval Research Laboratory, Code 5673, Washington, DC 20375 USA
†Dept. Mechanical Engineering, Duke University, Durham, NC 27708 USA
**U.S. Naval Research Laboratory, Code 6340, Washington, DC 20375 USA

Abstract. Long-term, high-level performance demands on a variety of structures and equipment have stimulated significant research in the field of structural health monitoring. The primary goals of this field are to provide information regarding structural performance capability, damage assessment, and even structural prognosis, all of which may potentially reduce the ownership costs associated with the maintenance and operation of the structure. Much of the research has largely taken a vibration-based approach, whereby the structural dynamic response to either ambient or applied loading is analyzed for changes in certain characteristic "features" that serve as appropriate damage indicators. Many of the features proposed have involved parameters derived from a modal analysis of the structure, e.g., resonant frequencies, mode shapes, damping, strain energy, flexibility, etc.

In this work, we present features taken from measured attractors of the structure. System characterization (including damage detection) by means of geometric invariants, such as attractor(s), is potentially a powerful generic approach which does not rely on implicit assumptions in an underlying model, e.g., linearity. The structure is excited with a chaotic oscillator, and the combined chaotic excitation dynamics and structural response may be thought of as the "filtering" of chaotic data: the structure acts as a "filter" through which the chaotic signal is processed so that small changes to the structure (ostensibly due to damage) will serve to alter the degree to which the signal is filtered. We develop the attractor property trajectory prediction error as a candidate feature and evaluate its utility in detecting clamping force degradation on a continuous metal beam.

INTRODUCTION

Economic constraints constantly drive structural owners towards reducing the costs associated with maintaining their structures. Traditionally, many such owners, both civilian and military, have relied upon human inspection or rote component/system replacement schedules for maintenance, and little recourse was made to any automated methodologies for either assessing current structural health or predicting remaining useful service life. However, with advances in structural materials, sensing technologies, and data processing and management capabilities, structural owners have increasingly embraced research and development in advancing the state of this art in this regard. For example, the United States Navy has recently begun the transition of all of its maintenance of structures (civil, ship, and aerospace) towards a condition basis, requiring that appropriate approaches be developed for augmenting and possibly replacing rote inspection or replacement schedules. This transition has been spawned in a significant way by reduced

CP676, *Experimental Chaos: 7ᵗʰ Experimental Chaos Conference*,
edited by V. In, L. Kocarev, T. L. Carroll, B. J. Gluckman, S. Boccaletti, and J. Kurths
2003 American Institute of Physics 0-7354-0145-4

manning initiatives, such as on DD-X where an almost 70% reduction in total manpower is expected.

Much of the research in this field of *structural health monitoring* has largely taken a vibration-based approach. In this approach, the structural dynamic response to either applied or ambient loading is analyzed for appropriate "signatures" or "features" which serve as indicators of structural degradation. The majority of features investigated in the literature have involved tracking changes in quantities derived from a modal analysis of the structure, such as resonant frequencies, mode shapes, modal damping, modal curvature, flexibility, stiffness, etc. Some studies have presented non-modal approaches such as autoregression, neural networks, and other spatial pattern recognition techniques. A thorough survey of the literature is impossible due to space constraints, but collections or reviews of many techniques and their application may be found in references [1, 2, 3, 4].

More recently, researchers at Los Alamos National Laboratory have shown the importance of statistical assessment of the extracted features for proper structural health identification [5, 6, 7]. This identification or classification is particularly important with regard to making informed engineering judgments about a course of action. For example, many features fluctuate significantly under varying environmental conditions (temperature, humidity, etc.) but a constant structural health condition; Peeters et al. found that the resonant frequencies on a bridge in Switzerland changed by as much as 18% with changes in temperature along [8].

Vibration-based structural health monitoring may thus be implemented as shown in Figure 1: structural vibration excitation, measurement of the structural response, feature extraction from the measured response, and statistical assessment of the feature set for determination of a course of action. Most of the literature has considered the feature extraction step, and, as mentioned, proposed features based on modal analysis. One common problem observed in modal-based features is the significant variability in their values, particularly under an evolving environment (e.g., the bridge study referenced above). Another problem is often their lack of sensitivity to damage itself, sometimes with unchanging values until failure occurs.

This paper seeks to identify a new class of features drawn from the time series analysis community; in particular, the concept of *state space geometry* may yield such features. It is well understood in the dynamics field that any dynamic system (such as a vibrating structure) may be described by a state space formalism. Within this state space, an ensemble of trajectories drawn from initial conditions will evolve in time, and, given inevitable dissipation, migrate towards an invariant steady-state, or *attractor*. System characterization, of which structural damage detection may considered a subset, by means of state space geometric changes–attractors and properties derived from them–appeals as a powerful generic approach which does not rely upon implicit model assumptions, linearity, etc.

This work will present a features based on attractor geometry: *trajectory prediction error*. Specifically, this work will present how a chaotic excitation of a structure and its reconstructed response attractor geometry may be analyzed for a sensitive damage indicator. The combined chaotic input and chaotic structural response may be thought of as the "filtering" of chaotic data: the structure filters the chaotic input, and small changes to the structure (resulting from damage) will alter the nature of the filtering. This approach will be applied to a cantilevered beam subject to clamping force release

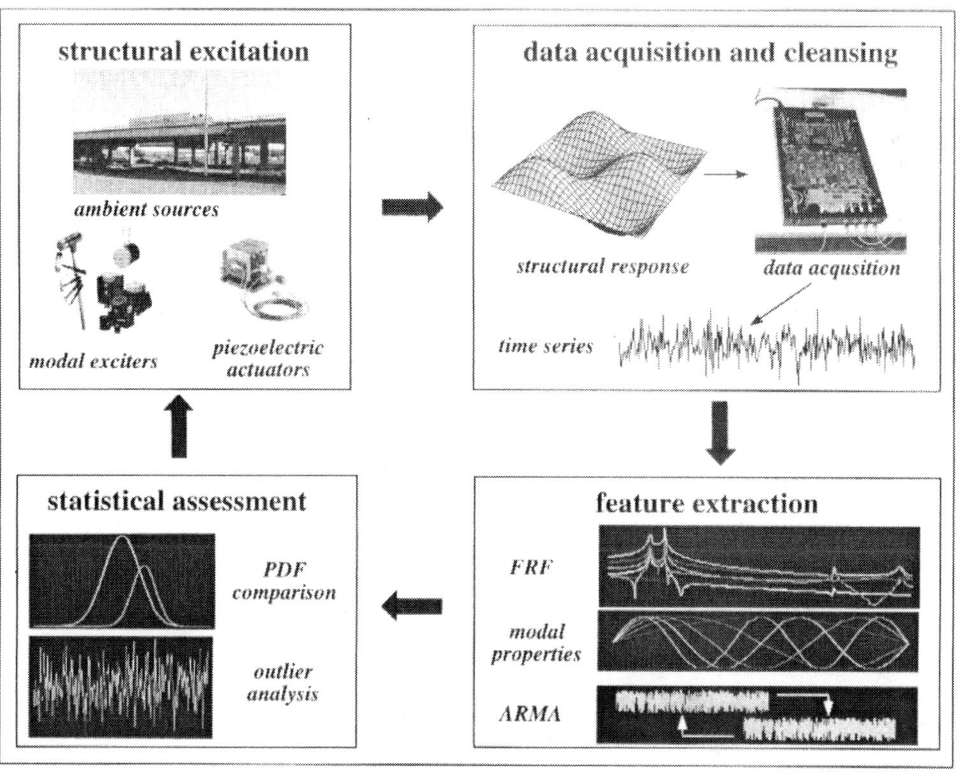

FIGURE 1.   Vibration-based structural health monitoring algorithm.

at one boundary.

STATE SPACE APPROACH

Motivation and Background

One motivating concept for taking a state space approach is the relationship between an attractor's stability and its "dimension." Lyapunov exponents (LEs) are the usual measures of attractor stability, with a full set describing the average exponential rates at which perturbations to the attractor diverge or converge in each of the principal geometric directions in state space. The number of principal directions is determined by the minimum number of dimensions (or coordinates) required to completely "unfold" the attractor in state space [9]. One important property of a chaotic process is the existence of one positive LE, indicating local dynamic instability. Nevertheless, because global dissipation necessarily exists in any real system, the sum of all LEs, representing the stability of the attractor as a whole, must be negative. These apparent contrary observations for a chaotic process–global stability but local instability–suggest that there exists some subset of the full LE spectrum where the net stability is very nearly zero. Kaplan and Yorke [10] used this idea to conjecture a Lyapunov dimension d_L

$$d_L = M + \frac{\sum_{n=1}^{M} \lambda_n}{-\lambda_{M+1}} \tag{1}$$

where M is the maximum number of exponents which may be added before the sum of the exponents becomes negative, and the exponents are assumed ordered from biggest to smallest. This value d_L, identified as the *fractal* dimension of the attractor, essentially gives the smallest-dimensioned subspace in which an initial set of evolving points will be re-captured after a small time passage, and this will serve as the definition of dimension in this work.

Therefore, with this relationship between dimension and stability, it follows that changes in the LEs will change the way points are distributed on the attractor, and vice versa. The implications of this relationship were examined previously by applying a first-order linear filter to a chaotic signal [11]. By changing the filter parameter, the LEs (and hence the dimension) of the system were altered in accordance with Equation 1. These results may be easily extended to higher order linear filters

$$\dot{\mathbf{z}} = \mathbf{A}\mathbf{z}(t) + \mathbf{B}\mathbf{x}(t), \tag{2}$$

where the chaotic excitation signal $\mathbf{x}(t)$ is "filtered" by the matrix $\mathbf{A}$, which is just the state space matrix of the structure, and $\mathbf{B}$ is a matrix which just selects desired components of $\mathbf{x}(t)$ to act on the structure. Both the linear filter output $\mathbf{z}(t)$ and the chaotic input $\mathbf{x}(t)$ will have associated with them a set of LEs. For a linear system, the LEs are simply the real parts of the eigenvalues of $\mathbf{A}$, and if it may be assumed that structural damage will affect the eigenstate of the structure, even subtly, it is thus expected that the structure's LEs will also be affected. The complete Lyapunov spectrum of the structural response λ_i^S to the chaotic excitation is thus given by

$$\lambda_i^S = (\lambda_j^C, j = 1 \ldots d_1 \,;\, \lambda_k^L, k = 1 \ldots d_2), \lambda_1^S > \ldots > \lambda_{d_1+d_2}^S, \tag{3}$$

where the λ_j^C are the exponents associated with the d_1-dimensional chaotic signal, and the λ_k^L are the exponents of the d_2-dimensional structure ("filter"). Although the structure possesses, in theory, an infinite dimension d_2, Equation 1 implies that only the most weakly contracting state space directions (least negative exponents) will determine the geometry of the attractor. This is of great importance to any algorithm seeking to quantify the geometric properties of an attractor. The number of points required to establish reliable computation scales as a power of dimension. Therefore, for a finite amount of data, the smaller the dimensionality of the attractor, the more robust the properties computed on the attractor.

Prediction Error as a Damage-Sensitive Feature

Given the Kaplan-Yorke conjecture, it appears that dimension itself may be the most useful feature to extract from observed time series. However, algorithms calculating Lyapunov dimension from measured time series experience difficulty in trying to detect subtle changes in geometry for two main reasons: first, the algorithms seek *global* geometry properties for calculation bases, which erodes the effects of subtle local attractor geometry changes that may occur due to damage; and second, these calculations are notoriously "noisy," especially with experimental data. Thus, the geometry of the attractor must be analyzed on a local level, in a state space sense. A recent further discussion of the use of dimension as a damage-sensitive feature may be found in [12].

Given this concept, Todd et al. proposed a measure of *local attractor variance* [13], where reconstructed state vectors were locally analyzed over attractors from both damaged and undamaged structural vibration time histories. Details on that approach will be omitted here, but application of the feature to damage scenarios in beam and plate structures may be found in references [14, 15, 16]. The feature developed in this paper is adopted from Shreiber [17], where it was used to detect non-stationarity in time series data. The method is rooted in a supervised learning mode, meaning that access to both "undamaged" (baseline) and and subsequent ("damaged") data sets is possible. In summary, a simple prediction scheme is used on the undamaged data set to forecast the values of the damaged data some number of time-steps s into the future. Each of the baseline time records may be used to reconstruct a set of B baseline attractors

$$\mathbf{x}_i(n) = (x_i(n), x_i(n+T), \cdots, x_i(n+(m-1)T)) \quad i = 1 \cdots B \qquad (4)$$

using the usual embedding approach with embedding dimension m and delay T [18]. Similarly, subsequent data sets will yield B "damaged" attractors, denoted $\mathbf{y}_j(n)$. The method proceeds by selecting a specific baseline attractor i and the attractor whose dynamics are being compared, j. Given a trajectory with time index f on the "damaged" attractor, $\mathbf{y}_j(f)$, the algorithm selects the set of points $U_\varepsilon^X(\mathbf{y}_j(f))$ on the baseline attractor $\mathbf{x}_i(p)$ with time index p that are within some radius ε of that trajectory in a purely geometric sense, or

$$U_\varepsilon^X(\mathbf{y}_j(f)) = \mathbf{x}_i(p) : ||\mathbf{x}_i(p) - \mathbf{y}_j(f)|| < \varepsilon. \qquad (5)$$

Once the set of points $U_\varepsilon^X(\mathbf{y}_j(f))$ is found on the baseline attractor, the time evolution of this set is used in some predictor capacity. Several possible predictors could be used, but one of the simplest is the mean (average) value of the time-evolved set. This point, denoted $\hat{\mathbf{y}}_j(f+s)$, becomes the predicted value for how the damaged trajectory $\mathbf{y}_j(f)$ will evolve. In this sense the baseline attractors are used as "look-up" tables for the various patterns present in the data. The operational hypothesis is that these tables will lose their ability to serve as an accurate predictor as the dynamics are altered by structural damage. Variations of this algorithm have been used for both prediction and data cleansing [19, 20].

Once the prediction database has been constructed, the *average normalized cross-prediction error* may be defined by

$$\hat{\gamma} = \frac{1}{N\sigma^2} \sum_{f=1}^{N} |\hat{\mathbf{y}}_\mathbf{i}(f+s) - \mathbf{y}_j(f+s)|^2 \tag{6}$$

where the average is taken over all N points on the attractor, and σ^2 is the variance of the baseline data. In practice, such a large summation is often prohibitively time consuming, and a good estimate of the average may be obtained by evaluating Equation (6) over some randomly selected subset of N. This computation is repeated for each pair of i, j time-histories between the baseline and candidate attractors to give N_d different sets of prediction error $\hat{\gamma}_{ij}^k$ $i < j$ consisting of $N_m = B(B-1)/2$ members. In addition the *auto-prediction error* is computed for the baseline data by replacing $\mathbf{y}(n)$ by $\mathbf{x}(n)$ in Equation (5). The resulting set $\hat{\gamma}_{ij}^A$ gives some idea of the prediction error one would expect to find in the event that the dynamics are not changing. Excluding duplicate pairings, i.e. $\gamma_{1,2}$ $\gamma_{2,1}$, was done so that each of the resulting N_m members may be treated as independent random variables. Identical indices, i.e. $\gamma_{1,1}$ $\gamma_{2,2}$, are also excluded from consideration. The variation the method seeks to capture is occurring from data set to data set so that adding like pairs serves no purpose other then adding a constant bias to the data.

Statistical Modeling

Although the prediction error feature presented is computed in a deterministic way, there exist several sources of variability which add elements of stochastic behavior to any real data. Measurement noise, unaccounted variables, environmental fluctuation (nonstationarity), and other sources cause the data to populate *distributions* of values rather than reflect unique repeatability. Thus, the damage diagnosis problem is transformed into distinguishing among various distributions of data regarding their proper classification as coming from a damaged structure or not. The practitioner seeks to make a confidence-based, quantitative engineering judgment in this capacity so that proper corrective action may be taken within the paradigm of Figure (1).

A useful procedure for comparing data sets is the analysis of variance (ANOVA). In an ANOVA procedure, the parameters of interest are the *treatments*, which refer to the quantity (or quantities) which are assumed to be changing among data sets, and the *responses*, which are the measured or computed data to be analyzed. In this paper, the driving influence of difference among data sets is assumed to be progression of damage in the structure, and thus only a single-factor treatment is considered (1-D ANOVA). Different amounts of damage are known as treatment *levels*. If the data population mean of the j-th damage level is denoted μ_j, the following null hypothesis may be established

$$H_0 : \mu_j = \mu_{j+1}, \tag{7}$$

where $j = 1 \ldots N_d$, and N_d is the number of different damage levels being considered. The alternative hypothesis is thus that in at least one pairwise comparison of data sets, the means differ. Acceptance of H_0 indicates that no distinguishable difference among data sets may be inferred, and thus no damage to the structure has occurred. Specific details regarding how a 1-D ANOVA procedure is completed may be found in any basic statistical methods text, e.g., [21].

Traditional ANOVA considers single inferences for each μ_j by establishing confidence intervals. If the parent data populations are assumed to be normally distributed, it may be shown that the $100(1-\alpha)\%$ confidence interval about the difference between two means μ_j and μ_k may be given by

$$\mu_j - \mu_k = \left(\overline{X}_j - \overline{X}_k\right) \pm t_{\alpha/2}\sqrt{\mathrm{MSE}\left(\frac{1}{n_j} + \frac{1}{n_k}\right)} \tag{8}$$

where $\overline{X}_j$ and $\overline{X}_k$ the means of the two measured data sets, MSE is the global mean-squared error as obtained from the ANOVA procedure, n_j and n_k are the number of data observations in each set, and $t_{\alpha/2}$ is the Student-t test statistic at confidence level $\alpha/2$.

The interval constructed in Equation 8 is deficient in one key way: each pairwise inference applies individually, and the method cannot be used to draw a family of inferences among several data sets. The deficiency may be alleviated by either constructing wider intervals, meaning less precise estimates, or reducing the confidence level. One popular method for properly estimating pairwise differences is the Bonferroni method. The method requires that the number of pairwise comparisons to be made be specified a priori. The Bonferroni interval estimates are still given by Equation 8, but the Student-t test statistic is modified from $t_{\alpha/2} \rightarrow t_{\alpha/2s}$, where s is the number of pairwise estimates to be considered. For N_d treatments (damage levels), there are $C_2^{N_d} = N_d!/2!/\left(N_d - 2\right)!$ total pairs which may be considered. One advantage of the Bonferroni method, however, is that not all these pairs need to be considered. The final decision whether to accept or reject H_0 depends on whether the intervals constructed by the modified Equation 8 contain zero or not; intervals that contain zero indicate an acceptance of H_0, and intervals which do not contain zero indicate rejection of H_0, i.e., the two means are significantly different.

Two important assumptions are inherent in using the ANOVA procedure as detailed above. First, it must be known in some sense how many levels of data (i.e., damage levels) are contained within all the data sets. This is a realization of the supervised learning mode discussed in the first Section of this paper: a controlled damage study is being performed such that known discrete levels of damage–even if they are not quantifiable as such–are observed. Second, the confidence intervals constructed by the modified Equation (8) rely upon the assumption that the parent data distributions from which the measured data are sampled are normal (Gaussian). This second assumption is usually acceptable provided that the data are not significantly skewed or multi-modal. Even if the parent distributions are skewed or multi-modal, recourse may be made to the central limit theorem by applying some sort of sub-sampling, re-sampling, or bootstrapping technique [22]. The general procedure in these techniques is to take large numbers of randomly sampled subsets of the collected data, take the mean value of each subset, and use the collection of re-sampled mean values as the "new" data set. This

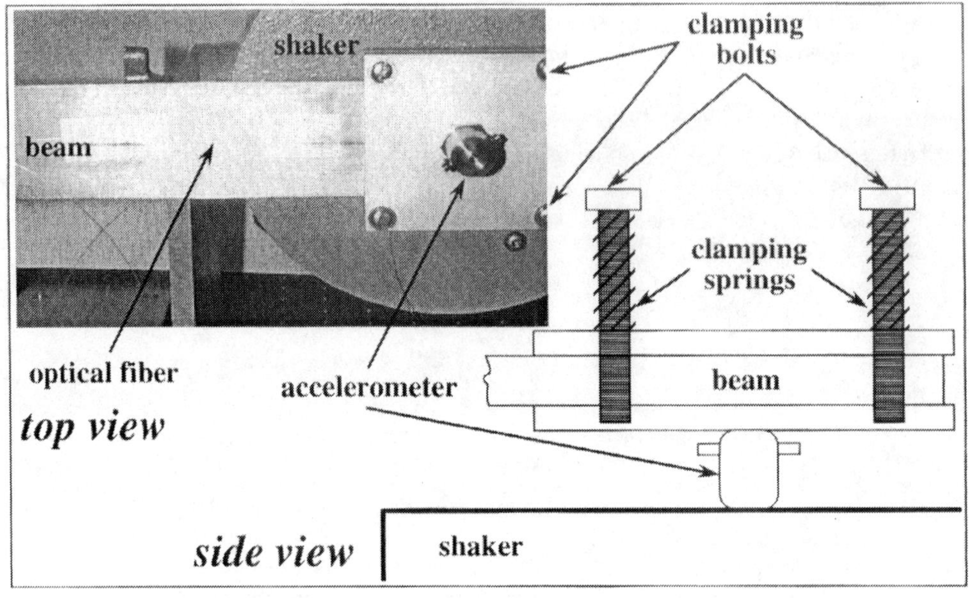

FIGURE 2. The clamped beam experiment.

procedure was implemented on the data taken in this work, and it will be shown later that very Gaussian distributions of data resulted.

RESULTS AND DISCUSSION

The Experiment

This approach was applied to an aluminum beam (0.5 m long by 0.05 m wide by 0.003175 m thick), clamped at one end and free on another. The clamp was specially made with linear springs such that a tunable clamping force could be applied. Clamping force (usually called "pre-load" in mechanical engineering) relaxation is a common failure mode in bolted connections. Seven fiber-optic strain sensors (based on fiber Bragg gratings) were evenly spaced along the central axis of the beam and monitored using a measurement system developed by the Naval Research Laboratory [23]. An accelerometer was attached to the shaker, which was attached to the clamp such that an inertial excitation could be applied. A photograph of the experiment is shown in Figure 2.

The excitation was chosen as the first state variable of the Lorenz oscillator in a chaotic regime, where "tuning parameters" are used to speed up or slow down the oscillator, depending on the inherent LEs of the beam. The matrix $\mathbf{B}$ of Equation (2) incorporates both $\dot{X}_1, X_1$ due to the fact that the excitation is applied inertially as base motion. One obvious difference between the model Equation (2) and the experiment is the number of coordinates required to describe the structure. Numerically such problems are discretized to some finite, usually small, number of degrees of freedom, while the actual beam contains an infinite number of degrees of freedom. In a practical sense this doesn't pose a problem as the time-scales (or alternatively, LEs) become very small (very negative) for only a few natural vibration modes. The Kaplan-Yorke conjecture is essentially providing a means by which the higher modes (dimensions) are filtered out of the dynamics.

For this beam, two different excitation cases were considered. For the first case, the tuning parameters were chosen to yield a complete LE spectrum of $\lambda_i^1 = (1.10, 0.00, -17.90)$, so that only one LE of the structure was replaced by those of the driving signal. The dimension of the response in this case is theorized to be $D_L \approx 2.2$. In the second case, the degree of replacement was maintained, but the oscillator was sped up to yield a spectrum of $\lambda_i^2 = (2.47, 0.00, -40.27)$ and a dimension of $D_L \approx 2.4$. For each of these cases the excitation time-histories were generated using a fourth order Runge-Kutta algorithm. LabVIEW data acquisition software was then used to convert the data files to a voltage for the shaker controller.

Damage was simulated by specifying various tensions on the clamping springs, thus providing various amounts of total clamping force at the base. Under a fully-clamped ("undamaged") condition, all four springs were compressed to 1.18 cm, and three damage levels were produced by relaxing the springs to 1.40 cm (damage level 1), 1.95 cm (damage level 2), and 2.50 cm (damage level 3). These prescribed displacements were precision-controlled with a micrometer.

Results

The experimental test plan involved $N_b = 12$ runs for each the $N_d = 3$ damaged and one undamaged cases, with each run consisting of $N = 45,000$ points, sampled at 2kHz (the first bending mode of the beam was about 10 Hz). Final values for the average prediction error were obtained by evaluating Equation (6) over 5,000 randomly selected trajectories. The sets of prediction error therefore consist of $N_m = 66$ values which were then resampled using the mean values of 20-element random subsets to generate a 10,000-sample data set. Probability density functions of the resampled data using a kernel density estimation technique [24] with a Gaussian kernel are shown on the left side of Figure (3) for both the slow (a) and fast (b) forcing cases.

It is clear from the figure that increased levels of damage lead to a different set of dynamics and hence a higher prediction error. The distributions at all damage levels in both excitation time scales appear very Gaussian. To the naked eye, all distributions appear distinguishable, with the possible exception of damage levels 1 and 2 at the slow excitation time scale. It is interesting to note that the prediction error is larger for this case. The reason for this concerns the time-scales of the two different excitations. The slower time scale will yield, for a given number of N points, fewer oscillations, leaving a more sparsely or "less populated" attractor. For the faster time scale, the same N points will cover more oscillations and give a more complete geometric portrait of the attractor, resulting in a lower prediction error.

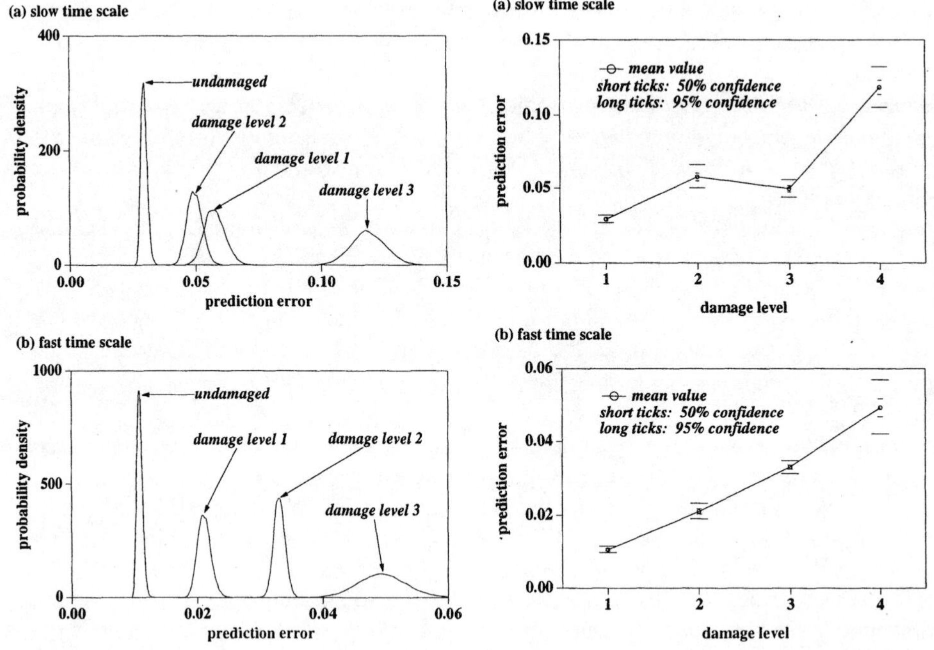

FIGURE 3. Mean values of the prediction error (left) and full distributions (right) at both excitation time scales.

In order to quantify the differences between the distributions, a 1-D ANOVA was performed considering all six possible pairwise differences ($s = C_2^6 = 6$). In all six cases, the intervals constructed by the modified Equation (8) did not contain zero, meaning that all damage levels are distinct from each other. This result was true for both excitation time scales, although the interval at the slow time scale when comparing damage levels 1 and 2 very nearly contained zero.

The right side of Figure (3) shows the progression with damage level of the distribution mean prediction error values for both excitation time scales. Confidence limits are placed on the data at both a 50% and 95% level. These confidence limits are based on the appropriate quantiles for each distribution; this implies, in the 95% confidence case, for example, that 95% of the data in that distribution fall between the confidence limits. Regardless of time scale, it appears that at a 50% confidence level, all damage levels are distinguishable. At 95% confidence, the second and third damage levels have overlapping confidence bands for the slower excitation. Moreover, the trend is not monotonic, rendering damage magnitude classification more difficult. As mentioned, this is because the response attractor was "underfilled" in a geometric sense.

SUMMARY

A framework for vibration-based structural health monitoring using attractor-based methods was presented and demonstrated effectively in an experimental context. By using chaos to interrogate the structure the dimension of the response is kept low, allowing for attractor-based analysis. The random excitation commonly favored for vibration-based structural health monitoring, on the other hand, will produce high-dimensional attractors, rendering attractor-based computations cumbersome and error-prone. Specifically, this work has shown how cross-prediction error may be an effective feature for quantifying the level of degradation in a given structure. The method quantifies the likelihood that a particular empirical model, taken from baseline, or "undamaged", data, describes the damaged data. A poor model is an indicator that the data have begun to deviate from those produced by the baseline dynamics, hence damage is identified. Any method which is to be used in practice must take into account the distributions of features rather than a single realization, as various nonstationary effects are necessarily present in real, noisy data. The cross-prediction error was shown to be robust under variation, as a formal 1-D ANOVA showed almost distinction between damaged and undamaged data sets. In fact, damage magnitude classification also was possible, especially at the faster time scale. It is expected that the slower time scale could be equally as effective if the attractor had been populated more thoroughly; the lower dimension at this time scale would provides computational advantages and ' robustness, as well. Finally, because this method relies on steady-state dynamics at relatively slow time scales (when compared to structural resonances), energy input to the system is minimal, in contrast to many of the sustained resonant driving methods used in common structural health monitoring approaches.

ACKNOWLEDGMENTS

This work has been funded primarily by a 6.1 NRL Advanced Research Initiative in structural health monitoring.

REFERENCES

1. Doebling, S. W., Farrar, C. R., and Prime, M. B., *Shock and Vibration Digest*, **205**, 631–645 (1998).
2. Guemes, J. A., *European COST F3 Conference on System Identification and Structural Health Monitoring*, Graficas, 2000.
3. Chang, F. K., *Structural Health Monitoring 2000*, Technomic, 1999.
4. Chang, F. K., *Structural Health Monitoring: The Demands and Challenges*, CRC Press, 2001.
5. Farrar, C. R., Doebling, S. W., and Nix, D. A., "Damage Identification with Linear Discriminant Operators," in *Proc. 17th International Modal Analysis Conference*, 1999.
6. Worden, K., Manson, G., and Fieller, N., *Journal of Sound and Vibration*, **229**, 647–667 (1999).
7. Farrar, C. R., Cornwell, P. J., Doebling, S. W., and Prime, M. B., Structural health monitoring studies of the alamosa canyon and i-40 bridges, Tech. rep., Los Alamos National Laboratory Report LA-13635-MS (2000).
8. Peeters, B., Maeck, J., and Roeck, G. D., *Smart Materials and Structures*, **10**, 518–527 (2001).
9. Abarbanel, H., *Analysis of Observed Chaotic Data*, Springer-Verlag, 1996.

10. Kaplan, J. L., and Yorke, J. A., "Chaotic behavior of multidimensional difference equations," in *Functional Difference Equations and Approximations of Fixed Points*, edited by H.-O. Peitgen and H.-O. Walther, Springer-Verlag, Berlin, 1979, vol. 730 of *Lecture Notes in Mathematics*.

11. Pecora, L., and Carroll, T., *CHAOS*, **6**, 432–439 (1996).

12. Nichols, J. M., Todd, M. D., Virgin, L. N., and Nichols, J. D., *Mechanical Systems and Signal Processing* (2002 (in print)).

13. Todd, M. D., Nichols, J. M., Pecora, L. M., and Virgin, L. N., "Novel Nonlinear Feature Identification in Vibration-Based Damage Detection Using Local Attractor Variance," in *IMAC XIX: A Conference on Structural Dynamics*, Orlando, Florida, 2001.

14. Todd, M. D., Nichols, J. M., Pecora, L. M., and Virgin, L. N., *Smart Materials and Structures*, **10**, 1000–1008 (2001).

15. Todd, M. D., Seaver, M., Nichols, J. M., Trickey, S. T., and Virgin, L. N., "The Use of a High-Performance Fiber Optic Measurement System in Structural Damage Assessment," in *IMAC XIX: A Conference on Structural Dynamics*, Orlando, Florida, 2001.

16. Todd, M. D., Nichols, J. M., Bement, M., Farrar, C., and Baker, B., "Experimental Demonstration of Local Attractor Variance as a Damage Indication Feature," in *IMAC XX: A Conference on Structural Dynamics*, Los Angeles, California, 2002.

17. Schreiber, T., *Physical Review Letters*, **78**, 843–846 (1997).

18. Takens, F., "Detecting strange attractors in turbulence," in *Lecture notes in Mathematics*, Springer-Verlag, 1981, vol. 898, p. 366.

19. Nichols, J. M., and Nichols, J. D., *Mathematical Biosciences*, **171**, 21–32 (2001).

20. Kantz, H., and Schreiber, T., *Nonlinear Time Series Analysis*, Cambridge University Press, 1999.

21. Quirin, W. L., *Probability and Statistics*, Harper & Row, 1978.

22. Efron, B., and Tibshirani, R. J., *An introduction to the bootstrap*, Chapman and Hall, New York, 1993.

23. Todd, M. D., Johnson, G. A., and Althouse, B. L., *Measurement Science and Technology*, **12**, 771–777 (2001).

24. Silverman, B. W., *Density Estimation for Statistics and Data Analysis*, Chapman and Hall, 1986.

Principal Curves and Chaos

Sandeep Rajput[1] and Duane D. Bruns[2]

Department of Chemical Engineering, The University of Tennessee, Knoxville
419 Dougherty Engineering Building, Knoxville, TN 3799

Abstract. A *Principal Curve* is a hypercurve that passes through the center of the data cloud. We adapt and expand the principal curve algorithm to develop a non-parametric approach called Cluster-linked Principal Curves (CLPC) that locally approximates the structure and scatter in a distribution of data points. The iterative algorithm is based on Expectation-Maximization (E-M) principle. The projections of data points on the principal curve or arc lengths are capable of characterizing the data in fewer dimensions and with greater accuracy than PCA. The distribution of arc lengths is used for gauging stationarity and reversibility, and monitoring. For illustration we use the embeddings formed from chaotic gas pressure time series measurements collected before an electrified capillary nozzle that injects bubbles into a liquid-filled column.

INTRODUCTION

Delay embedding is the standard procedure for analyzing almost all nonlinear and chaotic time series. The upper limit for the embedding dimension required to faithfully reproduce the attractor geometry is established in [1-2], but it is an upper bound and a smaller embedding dimension may be sufficient [3]. It is very profitable to reduce the dimensionality of embedding vector representation while keeping the necessary information. The tasks of gauging stationarity and reversibility, process monitoring and control become much more amenable if the effective dimension of the data is smaller.

In this paper we introduce the concept of Cluster-linked Principal Curves (CLPC) that approximate the distribution of data points by a locally linear hypercurve that is obtained by iterative Expectation-Maximization Principle. The approach is an extension of the Principal Curves [4]. The projections on the hypercurve or *arc lengths* are examined for their potential in reducing the dimension, testing for stationarity and reversibility, and for monitoring and control. The experimental data used in this paper were collected on a liquid-filled column with an electrified capillary through which gas is bubbled. The column is referred to as *bubble column* in this document. The bubble column is a low-dimensional system that exhibits period-doubling route to chaos, and its return maps are quite similar to that of the dripping faucet experiment [5]. Experimental setup and relevant references can be found in [6-8].

Email: rajput@engr.utk.edu[1] , DBruns@utk.edu[2]

CP676, *Experimental Chaos: 7th Experimental Chaos Conference,*
edited by V. In, L. Kocarev, T. L. Carroll, B. J. Gluckman, S. Boccaletti, and J. Kurths
© 2003 American Institute of Physics 0-7354-0145-4/03/$20.00

2. CLUSTER-LINKED PRINCIPAL CURVES

A *Principal Curve*, defined by Hastie and Stuetzle (henceforth referred to as HSPC) is a hyper-curve that locally approximates the data density [4]. The curve is a non-parametric polygonal line. Projecting the point on the polygonal line reduces a point $\mathbf{x}$ to an *arc length* $\lambda(\mathbf{x})$ along it. The arc length is the line integral along the polygonal line from its origin to the point where $\mathbf{x}$ projects on it. The curve is self-consistent, i.e., any point on the curve is the expected value of the distribution at that point. If $\lambda(\mathbf{x})$ is the arc length of the point $\mathbf{x}$, then $\lambda(\mathbf{x})=E(\lambda(\mathbf{x_j})|\lambda(\mathbf{x_j})=\lambda(\mathbf{x}))$. The HSPC algorithm is based on Expectation-Maximization (E-M) principle which successively refines the curve. The iterations are stopped when the fractional change in the sum of squared distances from the curve falls below a predefined threshold.

When dealing with a finite data set, $\lambda(\mathbf{x})$ cannot be found for every possible vector $\mathbf{x}$, because there may not be any other points having the same arc length. In that case, smoothing or kernel estimation techniques can be used to estimate $\lambda(\mathbf{x})$ for a given $\mathbf{x}$. To contend with very small conditional probabilities, HSPC algorithm involves a scatterplot smoothing step, thus leading one to designate one variable as dependent and another as independent. Faced with multivariate data, that choice is not simple to make and is rather restrictive –thus the algorithm is not readily applicable to data with dimension greater than two.

Cluster-linked Principal Curve (CLPC) is an adaptation and expansion of HSPC, and is also a non-intersecting curve through the data space so that it minimizes the orthogonal distances of the data points from it [9]. It treats all dimensions symmetrically and together, and performs smoothing by clustering instead of kernel smoothing. CLPC algorithm forms clusters of data based on their arc lengths. The curve is then redefined by connecting the cluster means with straight lines. The iterations are stopped when the curve stabilizes [9]. Every cluster formed in the CLPC algorithm has an equal number of points. This helps to improve the approximation where the data density is high or where we have more information. Figure 1 shows how CLPC approximates a bubble column return map and a noisy circle. Fifteen clusters were used for both approximations. Note that the return map (figure 1b) is characteristic of many chaotic systems.

CLPC is a continuous polygonal line and can be defined so that it closes on itself, i.e. forms a closed loop. This particular instance can be seen in figure 1a. Our algorithm has only one *hyper-parameter* –viz. the number of cluster centers employed to approximate the structure of data. The optimum number of cluster centers depends on the data density and the complexity of the distribution. Our trials indicate that using a cluster for 10 to 20 points usually yields satisfactory results. Cross-validation methods can be used to check the 'goodness of fit' and to prevent overfitting.

Standard statistical techniques can be used for residual analysis. The residuals are desired to have zero mean, small variance compared to the original data and be independent. The generalized variance of the distribution of points $\{\mathbf{x}\}$ is defined as the trace of $Cov(\mathbf{x})$ or the sum of variances for each dimension. The ratio of the generalized variance of the residuals to that of the original data provides a good estimate of the remaining variability. In the examples shown in fig. 1, the generalized variance of the residuals was less that 0.1% of that of the data sets.

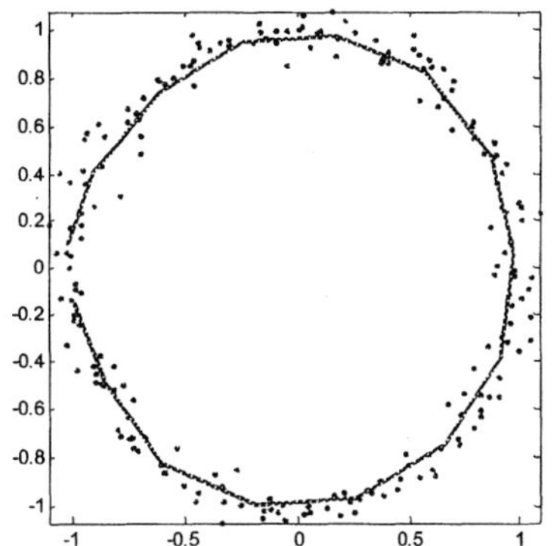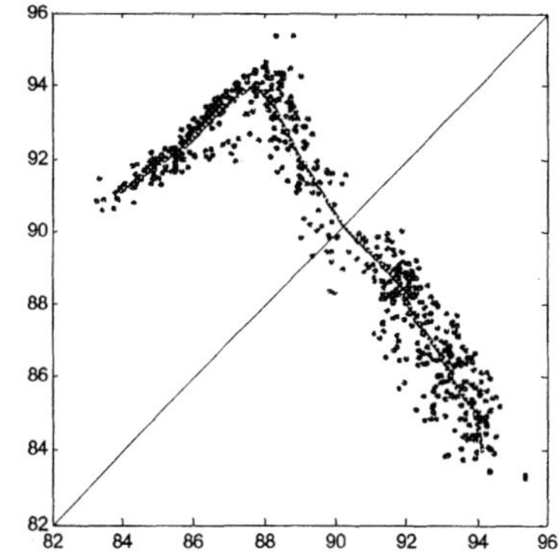

FIGURE 1. Principal Curve fit to a noisy circle (left) and a return map (right)

Beyond these simple examples, CLPC algorithm can be used to approximate the distribution in the embedding space. Figure 2 shows how the CLPC can approximate the attractor of a bubble column, with an embedding dimension of 3. The embedding delay was so chosen to unfold the geometry of the attractor. Figures 2a and 2b show overlaid attractors reconstructed from the first and last one-thirds of a series for Cases A and B respectively. The solid line shows the principal curve. All time series were chaotic. Note that for case A, the electrostatic potential was slightly changed halfway into the experiment and the experiment was allowed to stabilize before resuming measurements. For Case B, the operating conditions were not changed.

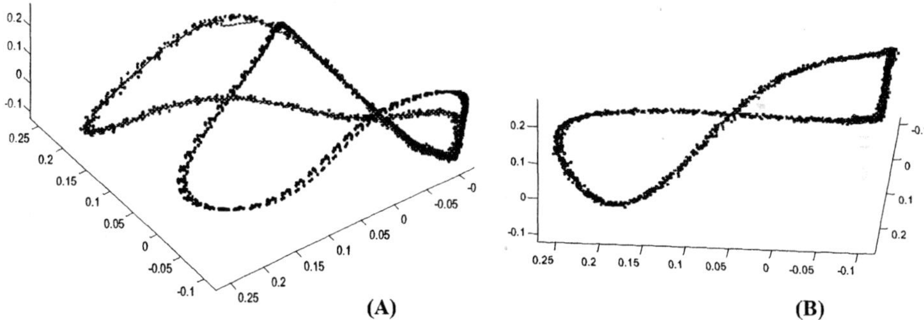

(A)   (B)

FIGURE 2. Principal curves for bubble column embeddings. See text for detail.

It can be seen that the curve approximates the geometry excellently. Some regions of the attractor are much denser than the rest but the curve approximates the trajectories throughout the embedding space quite well. Note that we do not expect the principal curve to capture the intricate fractal nature of the attractor since that aspect cannot be captured by any curve. However, principal curves reduce the effective dimension of the data from three to one. For lower dimensional systems, our investigations indicate that usually one principal curve is adequate and the arc lengths along it characterize the system reasonably well. For the attractors shown in figure 2, the generalized variance of the residuals was less than 0.02% of that of the data. Adding random noise to the embedding data does not alter the principal curve appreciably. We note that the magnitude of residuals can be used as an indicator of the noise level.

3. APPLICATIONS OF PRINCIPAL CURVES: TESTING FOR STATIONARITY AND REVERSIBILITY

Non-stationarity implies the inconstancy of the probability density over time. To explore stationarity, one can compare the first and last one-thirds of the data set. The middle one-third is not used in order to temporally isolate the two segments. Discarding the middle one-third of the data helps detect even slow dynamical changes. A principal curve can be fitted to the first one-third of the data and the arc length distribution **R** can be obtained for the corresponding data points. Data points from the last one-third of the time series can be projected on the principal curve to yield another arc length distribution **S**. The χ^2 statistic provides the easiest way to compare these distributions. The chi-square statistic is defined as $\sum[(R_i\text{-}S_i)^2/(R_i\text{+}S_i)]$ where R_i and S_i denote the elements in the ith bin of **R** and **S**. The corresponding degrees of freedom are N_B-1 where N_B is the number of bins.

If there is no reason to suspect pockets containing very few points in the data space, and some bins are empty, no correction should be made to the degrees of freedom. Otherwise, we suggest reducing the degrees of freedom to account for the empty bins. Note that reducing degrees of freedom increases the Type I error (erroneous rejection of null hypothesis). Figure 3 below shows the overlaid PDFs of the first and last one-thirds of the time series for Cases A and B.

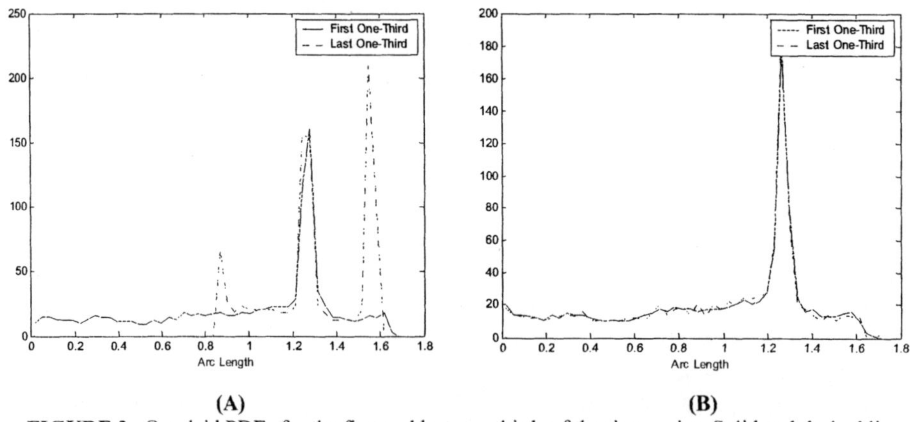

FIGURE 3. Overlaid PDFs for the first and last one-thirds of the time series. Solid and dashed lines represent the first and last one-thirds respectively. See text for detail.

From figure 3a it is obvious that the distributions are very different and the time series was not stationary. On the other hand, from figure 3b it is clear that the distributions are not different. A chi-square test did not reject the null hypothesis of stationarity for case B but did so for case A. Thus the procedure does not reject arbitrarily but captures the shift quite well. Note that one does not have to apply the chi-square test to compare the distributions. The Monte Carlo methods can be used on the surrogate data. Surrogate data [10] can be prepared from the raw time series, and then embedded to form *surrogate embedding vectors*. The arc length distributions of the surrogate embedding vectors then provide confidence limits for the distribution of arc lengths.

This framework can be extended to cover monitoring in a straightforward fashion. The principal curve obtained on the nominal data set can be used to find the arc length distributions for a running window and nominal data. These distributions can then be compared online using the χ^2 statistic. The confidence limits can be found from a probability table. When the χ^2 statistic for the difference between the nominal and the current data exceeds a limit, it indicates a significant change. With à priori information about the classes or régimes and their representative embeddings, a library of arc length distributions can be constructed that can be compared online with the current arc length distribution. Detection of a shift in régimes or states allows better control as well. In addition to the χ^2 statistic, a L^1, L^2 or other measure of distance can also be defined on the distributions and then used for monitoring.

Knowledge of the temporal reversibility of the time series is important since it rules out a random mechanism or its static transformations [11], and limits the modeling approaches suitable for the time series. To gauge for reversibility, one has to explore the differences between the distribution of a time series and that of its time-reversed counterpart. Note that only stationary time series need be tested for irreversibility since non-stationary time series are irreversible.

The procedure detailed above can be utilized to test for reversibility. Time-forward embedding vectors can be formed as $\{x(t)\ x(t-\tau)\ x(t-2\tau)\}$ and time-reversed embedding vectors as $\{x(t-2\tau)\ x(t-\tau)\ x(t)\}$. The distributions of these vectors can then be reduced to that of the arc lengths along the principal curve, and the latter can then be compared under the null hypothesis of reversibility. We use the time series (Case B), which was found to be stationary (see figures 2b and 3b). Figure 4 shows the overlaid PDFs for the time-forward and time-reversed data sets. The distributions are widely different, and a χ^2 test rejected the null hypothesis of reversibility very strongly. The data was chaotic and consequently irreversible. The procedure upheld our belief and thus proved useful. Once again, Monte Carlo simulations can be used for more rigorous hypothesis testing. The definition of principal curves implies reduction of noise, which can be implemented through kernel smoothing applied to a neighborhood of a point **x**.

4. CONCLUSIONS

In this document we showed how the CLPC algorithm can be used to reduce the dimensionality of an embedding, to test for stationarity and reversibility, and extended it to cover process monitoring. The procedures outlined in this document can be applied to the return maps as well. Further research is underway on this subject, especially about using the CLPC framework for prediction.

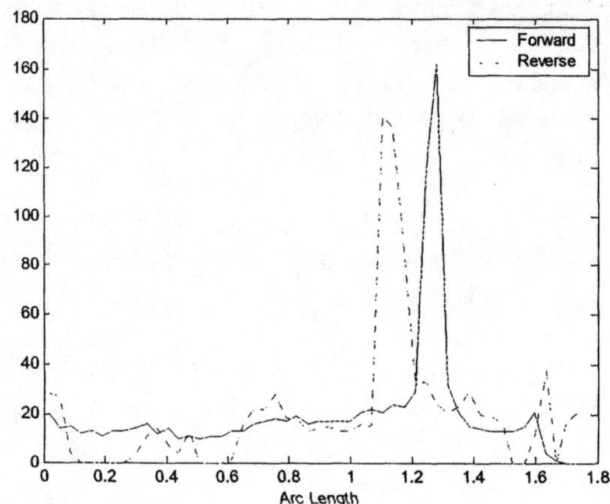

FIGURE 4. Overlaid PDFs for the time-forward (solid line) and time-reversed (dashed line) series. See text for detail.

ACKNOWLEDGMENTS

This work was supported by a grant from the Measurement and Control Engineering Center at The University of Tennessee which is an NSF Industry/University Cooperative Research Center (award number 9908040). The authors thank Dr. C. Stuart Daw for helpful discussions.

REFERENCES

1. Takens, F., in "Dynamic Systems and Turbulence" in *Lecture Notes in Mathematics 898,* edited by D. Rand and L.- S. Young, Berlin: Springer, 1981, p. 366.
2. Sauer, T., Yorke, J. A. and Casdagli, M., *J. Stat. Phys.* **65**, 579-616 (1991).
3. Kennel, M., Brown, R., and Abarbanel, H. D. I., *Phys. Rev. A* **45**, 3043 (1992).
4. Hastie, T. and Stuetzle, W., *J. Am. Stat. Soc.* **84**(406), 502-516 (1989).
5. Bruns, D. D., DePaoli, D. W., Rajput, S. and Menako, R., AIChE Annual Meeting 2002, Indianapolis, IN
6. Menako, C. R., *M. S. Thesis* (2001), The University of Tennessee, Knoxville.
7. Kang, Y., Cho, Y. J., Woo, K. J., Kim, K. I., and Kim, S. D., *Chem. Eng. Sci.,* **55**, 411-419 (2000).
8. Bruns, D. D., Cheng, M., Nguyen, K., Finney, C.E.A., Daw, C. S. and Kennel, M., *J. Chem. Eng.,* **64**(1), 191-197 (1996).
9. Rajput, S., *Ph.D. Dissertation* (2002), The University of Tennessee, Knoxville.
10. Theiler, J., Eubank, S., Longtin, A., Galdrikian, B., and Farmer, J. D., *Physica D* **58**, 77 (1992).
11. Tong, H., *Non-linear Time Series: A Dynamical System Approach*, Oxford: Clarendon Press, 1990, pp. 193-198.

Analysis of Ocean Electromagnetic Data Using a Hilbert Spectrum Approach

Jeffrey Ridgway[*], Michael L. Larsen[*], Cye H. Waldman[*],
Michael Gabbay[*], Rodney R. Buntzen[*] and C. David Rees[†]

[*]*Information Systems Labs., Inc., 10070 Barnes Canyon Road, San Diego, CA 92121*
[†]*Code D857, SPAWAR Systems Center, San Diego CA*

Abstract. We apply a newly developed time series analysis technique, the Hilbert-Huang Transform (HHT), to naturally occurring ocean electromagnetic data obtained from bottom-mounted sensors. The HHT was originally developed as an alternative to the Fourier power spectral density for the analysis of nonlinear phenomena in water waves. The HHT is applied to the data in two steps. In the first step, an empirical mode decomposition is used to extract individual oscillatory modes possessing different characteristic time scales. Unlike Fourier modes, however, these modes can vary in amplitude and frequency. In the second step, the Hilbert transform is used to determine physically meaningful instantaneous frequencies from these modes. We present results showing that the HHT provides a more compact representation of the ocean electromagnetic environment than the Fourier spectrum. In particular, the HHT is able to capture nonlinear wave phenomena associated with ocean swell in a single mode without the need for higher order harmonics.

INTRODUCTION - THE HILBERT-HUANG TRANSFORM

Over the past several years, the Hilbert transform has found growing application in the analysis of nonlinear time series, particularly for nonlinear oscillators and phase synchronization [1]. In this paper, we use the Hilbert transform to analyze underwater electromagnetic data. Such data can manifest rich nonlinear behavior found in the ocean such as nonlinear wave phenomena and turbulence. In the specific approach we employ, application of the Hilbert transform is preceded by an empirically determined modal decomposition technique that facilitates calculation of physically meaningful instantaneous frequencies [2, 3]. This approach was originally applied to the analysis of nonlinear water waves and has since been referred to as the Hilbert-Huang Transform (HHT).

The Hilbert transform permits the formal definition of an instantaneous amplitude and phase of an arbitrary time series $x(t)$. The Hilbert transform of $x(t)$ yields a time series $y(t)$ given by

$$y(t) = \frac{1}{\pi}\mathrm{PV}\int_{-\infty}^{\infty}\frac{x(\tau)}{t-\tau}d\tau, \tag{1}$$

where PV denotes the Cauchy principal value. Hence, the Hilbert transform can be viewed as the convolution of $x(t)$ with $1/\pi t$, which consequently stresses the local nature of the signal. A complex analytic signal can be formed via $z(t) = x(t) + iy(t)$,

CP676, *Experimental Chaos: 7th Experimental Chaos Conference,*
edited by V. In, L. Kocarev, T. L. Carroll, B. J. Gluckman, S. Boccaletti, and J. Kurths
© 2003 American Institute of Physics 0-7354-0145-4/03/$20.00

thereby defining an instantaneous amplitude and phase for the original signal $x(t)$. The instantaneous frequency then corresponds to the time derivative of the phase of $z(t)$.

As the phase evolution should typically reflect progress towards completion of a cycle, only positive instantaneous frequencies correspond to physically meaningful oscillatory behavior. Direct application of the Hilbert transform to data, however, may yield negative frequencies. In order to avoid non-physical instantaneous frequencies, Huang et al [2] developed an empirical mode decomposition (EMD), which separates the data into intrinsic mode functions (IMFs). The Hilbert transform applied to each IMF yields positive instantaneous frequencies. An IMF is a function satisfying two conditions: i) the number of extrema and the number of zero crossings must either be equal or differ at most by one, and ii) at any instant in time, the mean value of the envelopes defined by its local maxima and minima is zero. The first condition is a narrow band requirement, while the second condition reduces undesirable fluctuations induced by asymmetric wave forms. Each IMF has a characteristic oscillatory time scale.

The IMFs are found using a sifting procedure which generates the highest frequency IMF first. This IMF is subtracted from the time series, and the process is iteratively applied to the result until only a non-oscillatory residual, r, which represents the trend in the data, remains. Each IMF $x_n(t)$ has a variable instantaneous amplitude, $a_n(t)$, and frequency, $\omega_n(t)$ calculated via the Hilbert transform. The time-frequency distribution $H(\omega,t)$ of the amplitude is known as the Hilbert spectrum. The original time series can be written as a sum of a finite number of IMFs:

$$x(t) = \mathrm{Re} \sum_{n=1}^{N} a_n(t) e^{i \int \omega_n(t)dt} + r. \tag{2}$$

In contrast, the Fourier expansion of $x(t)$ consists of an infinite number of modes of constant amplitude and frequency. Completeness of the EMD is assured in principle and depends only on the numerical accuracy of the sifting process. Orthogonality of the IMFs, while not guaranteed theoretically, is satisfied in practice [2].

The Hilbert spectrum can be integrated over time to yield the Hilbert marginal spectrum, $\mathrm{HMS}(\omega) = \int_0^T H(\omega,t)dt$. The marginal spectrum represents the cumulated energy at each frequency over the entire data span and is related to the fraction of time that a given frequency can be observed in the system. The marginal spectrum can be more easily compared to the Fourier power spectral density than can the time-dependent Hilbert spectrum.

Although Fourier decomposition can always be used to mathematically represent a given time series, its physical representation of nonlinear and nonstationary systems is often problematic: higher-order harmonics with no independent physical existence apart from the fundamental mode arise from nonlinearity, and a proliferation of global modes are needed to account for rapid transients. The EMD and the Hilbert transform, however, can be sensibly applied to nonlinear or nonstationary systems, and it is often observed that individual IMFs often correspond to identifiable physical processes in the data[2, 3].

APPLICATION TO OCEAN ELECTROMAGNETIC DATA

Electromagnetic fields are generated by ocean flow due to electrical currents induced by the Earth's magnetic field in seawater, a conducting fluid (σ =3-4 mho/m)[4]. Gravity waves (wind waves and far-field swell), internal waves, turbulence, tides and currents all generate flows that produce electromagnetic field fluctuations. Swells have characteristic frequencies in the 50 to 100 mHz range and surface waves driven by local winds are found above the swell frequency to about 0.5 Hz . Although the amplitudes of gravity waves decay exponentially with depth, they can still be significant components of the signal in littoral waters. Turbulence is often present at the bottom boundary layer due to nonlinear interactions between shear flows arising from tidal and wind-driven currents, wind waves, swell, and internal waves [5].

A major component of the measured underwater electromagnetic signal is actually due to sources of non-oceanic origin — geomagnetic, atmospheric and ionospheric electromagnetic radiation. However, for low frequencies (< 1 Hz), these signals are typically coherent over large distances and can be mitigated using a remote reference station [6]. A transfer function is calculated which accounts for the difference in the conductivity in the underlying geology and is used to cancel out magnetospheric sources from the local sensor data.

We employed ocean-bottom magnetotelluric instruments from the Scripps Institution of Oceanography for our underwater electromagnetic measurements [7]. The instrument package consisted of two horizontal, orthogonally-aligned induction coil magnetometers and two horizontal, orthogonally-aligned electric field sensors. These sensors were specially designed to record low frequency signals down to 0.1 mHz. The signals were sampled at 31.25 Hz with 24 bit resolution onto an onboard hard disk. The direction of the instrument was measured with a recording compass. Measurements were taken in a water depth of 150 m at a location 8.5 km southwest of Point Loma in San Diego, California, over several days in June 2001. A land-based magnetic system, stationed 100 km inland in the Anza-Borrego Desert, recorded data concurrently, which was used to remove coherent ionospheric signals from the in-water magnetic data.

Analysis Results

Figure 1 shows the results of the application of the EMD to one component of the underwater magnetic data, from which the coherent geomagnetic background was removed using the land-based sensor. The full time series is shown in the top panel, followed by the eight IMFs generated ($x_1 - x_8$) and the residual (r) on the bottom. The EMD provides a compact representation of the data, requiring only eight modes plus the residual. The amplitudes and frequencies of the IMFs are indeed variable and the fundamental oscillatory time scale becomes longer with each subsequent IMF. Some IMFs appear to be more nonstationary than others.

We now focus on the second IMF and its physical interpretation. The results of applying the Hilbert transform to it are shown in Fig. 2a. The top panel shows a plot of the phase (with the linear trend due to the mean frequency removed) and the

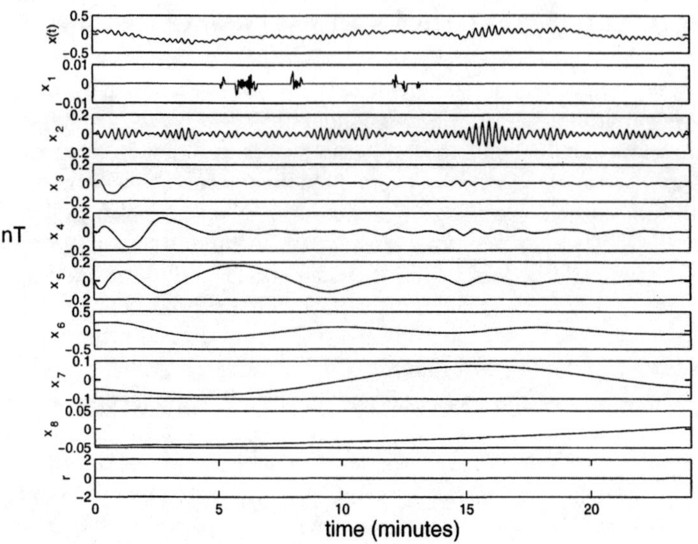

FIGURE 1. Application of the EMD to a segment of magnetic time-series data recorded at 1400 PST on June 21,2001. The top panel is the time series. Note that the vertical axes have different scales.

instantaneous frequency. The bottom panel shows the instantaneous amplitude overlaid on the time series of the second IMF. We see that the frequency is indeed positive (the phase jumps occur at times of zero amplitude, where the oscillations are essentially reset with another phase). Although the data is magnetic, the time series has the characteristic variation in sea surface height due to ocean swell. The Fourier power spectral density (PSD) of the second IMF is plotted in Fig. 2b (obtained using a longer time series than shown in Fig. 1), where a strong peak is seen at 62 mHz. This agrees with the peak frequency in the wave height spectrum as measured by a wave monitoring buoy located a few kilometers to the west in water of similar depth (CDIP Station 91 near Point Loma, CA). The underwater magnetic field produced by a gravity wave can be calculated [8], and so, given the horizontal components of the magnetic field, the wave height and direction can be obtained. This yields a significant wave height in the range of 35–45 cm (depending on the assumed seawater conductivity) which is in accord with the buoy value of 34 cm. The calculated swell direction is from the south, which also agrees with the buoy data. The large size of the magnetic field in the swell peak is due to the relatively shallow water depth of 150 m as compared with the swell wavelength of 400 m, so that the fluid motion is still significant despite its exponential falloff with depth from the surface.

The above results confirm that the second IMF can be identified with ocean swell. An interesting feature in Fig. 2b that appears in the PSD of the second IMF as well as that of the full time series, is the distinct peak at 125 mHz, which is twice the frequency of the primary swell peak. This second harmonic peak, however, is absent in the Hilbert marginal spectrum of the data, also shown in Fig. 2b. In the Fourier view, this second harmonic represents the sharpening or rounding of the crests and troughs of the waves, whereas in the Hilbert view, this is represented as a change in the instantaneous frequency within an oscillation cycle and, therefore, is accommodated by the primary swell peak. A possible physical interpretation of this is that the second harmonic is really a bound wave that is slaved to the fundamental swell frequency by a nonlinear process as discussed in [3]. We now address what that process might be.

As the magnetic field at the primary peak is reasonably attributed to the fluid motion caused by the swell itself, one might initially look to nonlinearity in the swell waves as the source of the second harmonic, such as is present in a Stokes wave train. However, given the small wave height and long wavelength, the wave slope parameter is about 10^{-3} and so such an effect would be many orders of magnitude smaller than the swell peak power. A more probable explanation is the so-called double frequency effect which has been identified as a source of microseismic motions in the seabed as first proposed by Longuet-Higgins [9]. In this effect, two oppositely-directed wave trains with the same frequency, interact nonlinearly to produce a pressure oscillation at twice the frequency. The magnitude of this pressure oscillation does not decay exponentially with depth and would be constant for an infinite area ocean. The double frequency effect has been observed seismically [10], acoustically [11], and with electric field sensors [12]. In the deep ocean, double frequency spectral peaks are far more prominent than those observed at the primary swell frequency. In coastal waters, oppositely-directed wave trains arise from reflections of the incident swell off the coastline, islands, or underwater topography. The amplitude of the pressure oscillations at the seabed is proportional to the product of the incident and reflected waves.

Given the region around the sensor location and the direction of the swell, it is reasonable to assume that reflected swell waves were present during our experiment. The existence of the double frequency effect in the magnetic data would likely be produced as follows: the pressure oscillations would cause a cyclic motion of the seabed which, in turn, would cause a cyclic tilting of the induction coils. Even very small angles of tilt, in the Earth's large magnetic field, would produce detectable changes in the data. As our instruments had neither tiltmeters or pressure sensors, we were not able to measure this effect directly. However, we can estimate the order of magnitude ground motion that would be required to produce the magnetic field amplitude observed at the second harmonic. The oscillation amplitude of the total horizontal field at 125 mHz is 0.015 nT, which given the vertical component of the the Earth's magnetic field of 40,400 nT, corresponds to a tilt angle of 4×10^{-7} rad. Assuming that the tilt angle is on the order of the instrument vertical displacement over the induction coil length (1 m), the estimated vertical displacement is roughly 0.4 μm, which is in line with theoretical estimates [9] and observed seabed displacements [10]. The second harmonic peak can therefore be plausibly attributed to the double frequency effect, and the HHT has properly associated it with the swell fundamental.

In conclusion, we see that the HHT provides a compact and physically meaningful representation of underwater electromagnetic data. The EMD naturally captures the swell, including nonlinear effects, in one mode, and the Hilbert transform allows for the nonlinearity to be represented as changes in the instantaneous frequency. The same behavior is observed in underwater electric field data (to a weaker extent) where the swell is also isolated by the HHT . The phenomenon is robust across time segments (the swell is not always associated with the second IMF), although intermittency problems can arise [3]. We have found the HHT to be a valuable analysis tool that complements Fourier methods. It should find broad applicability to the analysis of geophysical data.

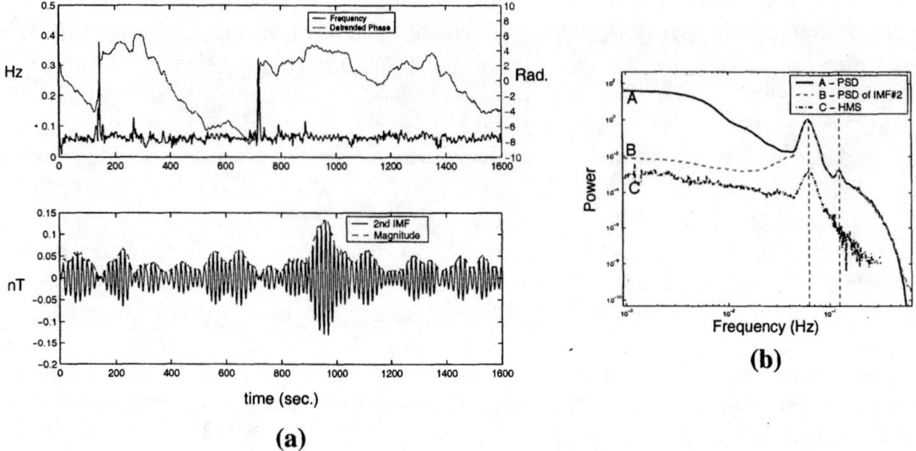

FIGURE 2. (a) Instantaneous phase, frequency, and amplitude of the 2nd IMF. (b) Comparison of the power spectral density of the full time-series (A), power spectral density of the 2nd IMF (B), and the Hilbert marginal spectrum of the full time series (C) (generated from a two hour time-series using 208 second windows with an overlap of 104 seconds).

ACKNOWLEDGMENTS

This work was funded by the US Navy SPAWAR Systems Center San Diego. We thank Charles Cox for enlightening discussions on the double frequency effect, and Steven Constable and Kerry Key for assistance with electromagnetic measurements and analysis.

REFERENCES

1. Pikovsky, A., Rosenblum, M., and Kurths, J., *Sychronization: A Universal Concept in Nonlinear Science*, University Press, Cambridge, 2001.
2. Huang, N., Shen, Z., Long, S., Wu, M., Shih, H., Zheng, Q., Yen, N., Tung, C., and Liu, H., *Proc. Royal Soc. of London A*, **454**, 903–995 (1998).
3. Huang, N., Shen, Z., and Long, S., *Annu. Rev. Fluid Mech.*, **31**, 417–57 (1999).
4. Sanford, T., *J. Geophys. Res.*, **76**, 3476–3492 (1971).
5. Cox, C., Zhang, X., Webb, S., and Jacobs, D., "Electro-magnetic Fluctuations Induced by Geomagnetic Induction in Turbulent Flow of Seawater in a Tidal Channel," in *MARELEC, 3rd International Conference on Marine Electromagnetics*, 2001.
6. Nichols, E., Clarke, J., and Morrison, H., *J. of Geophys. R.*, **93**, 13743–13754 (1988).
7. Constable, S., Orange, A., Hoversten, G., and Morrison, H., *Geophysics*, **63**, 816–825 (1998).
8. Podney, W., *J. of Geophys. R.*, **80**, 2977–90 (1975).
9. Longuet-Higgins, M., *Philos. Trans. R. Soc. London Ser. A*, **243**, 1–35 (1950).
10. Crawford, W., and Webb, S., *Bull. of the Seismological Soc. of Am.*, **90**, 952–963 (2000).
11. Webb, S., *J. Acoust. Soc. Am.*, **92**, 2141–2157 (1992).
12. Webb, S., and Cox, C., *Geophys. Res. Lett.*, **11**, 967–970 (1984).

Multiple Time Series and Attractor Reconstructions

L. Pecora[a], S. Boccaletti[b], D.L. Valladares[c], L. Moniz[a], H.P. Geffert[a], and T. Carroll[a]

[a]Code 6340, Naval Research Laboratory, Washington, DC, USA
[b]Istituto Nazionale di Ottica Applicata, Largo E. Fermi, 6, I50125 Florence, Ital
[c]Departamento de Fisica y Matematica Aplicada, Universidad de Navarra, Pamplona, Spain

Abstract. Many experiments have the ability to record more than one time series of data simultaneously. We explore two issues that are present when multiple time series are used to reconstruct attractors which are not present in the case of one time series. First, we show that there is an algorithmic approach to false nearest neighbors that naturally extends the established, one-series method. Second, the question of redundant information in two or more time series is a new issue and we show that the typical approach of mutual information can lead to erroneous results. The correct approach is a statistic that tests for the existence of a function which leads to the correct results.

INTRODUCTION: ISSUES WITH MULTIPLE TIME SERIES

In many experiments several time series can be simultaneously recorded. For example, for N series we write these as $x_1(t_j)$, $x_2(t_j)$,..., $x_N(t_j)$ which represent a simultaneous recording of N measurements at time t_j. It is understood that we can use the approach of Takens Theorem [1], time delay construction of dynamical vectors to reconstruct an attractor. A general form of this approach leads to the delay reconstruction vector,

$$\mathbf{z}(t) = \left(x_1(t), x_1(t+\tau_1),...,x_1(t+m_1\tau_1), x_2(t), x_2(t+\tau_2),...,x_2(t+m_2\tau_2),..., \right.$$
$$\left. x_N(t), x_N(t+\tau_N),...,x_N(t+m_N\tau_N)\right). \tag{1}$$

where the time delays τ_i and the number of delays m_i are generally different for each time series x_i. We assume the time delays have been obtained for each time series and we now have to determine each m_i. The question is, how do we do this for several time series in a way that is "democratic", i.e. we do not just rely on one or some subset of time series to unfold the attractor in the reconstruction. This is the first issue we address. It is the question of determining the embedding dimension since the embedding dimension m is just the sum of the individual m_i's.

Another issue is the following. If we simply use all the time series from an experiment, we may end up with a large embedding dimension. However, it may be that the number of time series is more than we need. Or to put it another way, some of the time series may not be independent. The question of determining when a time

CP676, *Experimental Chaos: 7th Experimental Chaos Conference,*
edited by V. In, L. Kocarev, T. L. Carroll, B. J. Gluckman, S. Boccaletti, and J. Kurths
2003 American Institute of Physics 0-7354-0145-4

series x_j, say, does or does not add information to the reconstruction is issue 2. We show, in addition that the usual tool of mutual information can give incorrect results. We introduce a new statistic, the function or continuity statistic that always gives the correct result.

ISSUE 1. FINDING THE EMBEDDING DIMENSION

Our goal is to reach Eq. (1) where we know the values of each m_i. To do this we extend the algorithm of False Nearest Neighbors (FNN) in a logical fashion to multiple time series. Recall the in the original FNN tests [2] we begin with a single time series, say $x(t)$ and start by forming a low-dimensional, first reconstruction vector, viz., $\mathbf{z}(t)=(x(t), x(t+\tau))$, where τ, the delay has been previously determined. Then we add another delay to get a new, second reconstruction vector, $\mathbf{z}'(t)=(x(t),x(t+\tau), x(t+2\tau))$. We search the first reconstruction for points that are close (nearest neighbors - NNs). Given two NNs, say $\mathbf{z}_1$ and $\mathbf{z}_2$, we find their partners, $\mathbf{z}'_1$ and $\mathbf{z}'_2$, in the second reconstruction. Note we can find partners because all the vectors are stamped with a unique time t. Now we test to see if $\mathbf{z}'_1$ and $\mathbf{z}'_2$ are close neighbors (not necessarily NNs) in the second reconstruction. This requires choosing some distance σ that characterizes closeness in the second reconstruction. If the distance, $|\mathbf{z}'_1 - \mathbf{z}'_2| >$ σ, then we say the original NNs $\mathbf{z}_1$ and $\mathbf{z}_2$ in the first reconstruction were false (thus FNN). We count the number or percentage of FNN between two reconstructions as a function of increasing dimension. When this percentage falls below an acceptable level, we say we have found the embedding dimension (the right number of delays) and the attractor is suitably unfolded in that dimension. We later showed [3,4] that the FNN statistic is equivalent to testing whether there is a continuous mapping between an embedding at one dimension, say m, and one at a higher dimension $m+1$. This follows from the Takens theorem since if m is large enough to give a faithful embedding of the attractor, then so is $m+1$ and there must to be a diffeomorphism between them. If we have unfolded the attractor in $m+1$, but not m, then there will be points that are separate in $m+1$ dimensions, but projected on top of each other in m dimensions, leading to a discontinuity.

We now proceed with a logical extension of the FNN approach for multiple time series. We approach the problem in a "democratic" way: we test each time series in turn for production of FNN as one delay is added. Then we cycle through the time series set again, testing only those time series that produced FNN previously. We do this until all time series have produced no FNN. The algorithm looks like this:

• Let m_i be the number of dimensions (time delays +1) for the ith component.
• set up the starting embedding vector using just the time series (no delays):
$\mathbf{z}(t) = (x_1(t), x_2(t), ..., x_N(t))$, i.e. start with $m_i=1$ $\forall$ i.

• Let $N_i(m_i)$ be the number of FNN generated when $m_i \rightarrow m_i+1$

• cycle through the time series, one at a time, and for each one check for FNN as one delay is added. Do this until all $N_i(m_i) \leq \delta$, where δ is some minimal, acceptable number of FNN.

```
loop until all N_i(m_i)≤ δ:
    set N_i(m_i)=0
    loop on i:
        N_i(m_i)= number of FNN as m_i→m_i+1
        reset m_i to original value
    end loop on i
    loop on i:
        if N_i(m_i)>δ , set m_i→m_i+1
    end loop on i
end loop until
```

After this process we have, generally, a different number of components m_i for each time series and the total embedding dimensions $m = \sum_{i=1}^{N} m_i$.

An added advantage is that since we do not put the burden on any one time series to carry out the unfolding, when systems are weakly coupled we can detect their separate embedding dimensions with some confidence. We exhibit such an analysis using two, coupled, Rössler-like circuits [5] which each have 3 components (x,y,z) like the usual Rössler system. The attractor for each isolated circuit is shown in FIGURE. 1 below.

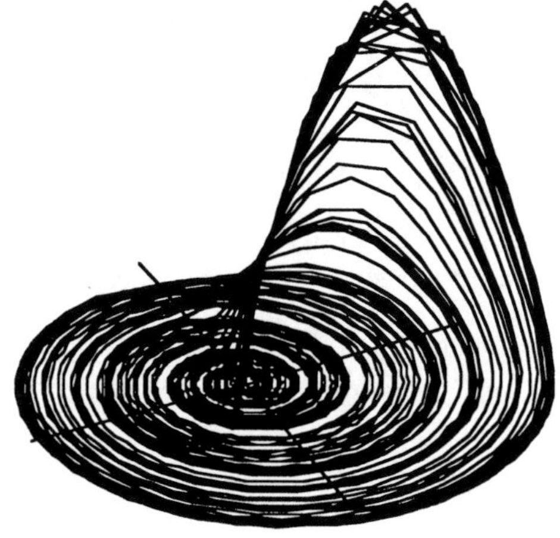

FIGURE 1. Attractor from one Rössler-like circuit.

Two of these circuits were coupled using diffusive coupling through one component (x) of the circuits with a variable coupling strength ε. When we used the multiple-time series FNN method for various coupling strengths we got the results in TABLE 1 below.

TABLE 1.	Embedding Dimension			
Coupling Strength (units of 10^{-4})	x_1-only embedding, $m(x_1)$	x_2-only embedding, $m(x_2)$	*Individual Dimensions* m_1, m_2	Total dimension $m(x_1, x_1)$
5.0	6	6	3, 3	6
5.5	7	6	3, 3	6
6.25	7	6	3, 3	6
7.14	7	6	3, 3	6
8.33	7	6	2, 4	6

We did the embeddings three supposedly equivalent ways. One was using just the x_1 time series (column 2), another using the x_2 time series (column 3), and another using both time series with our FNN multiple time series algorithm (columns 4 and 5). We see that the single time series approaches apparently unfolded the attractor, but give no further information. The multiple time series approach not only unfolds the attractor, but gets the dimension of the subsystems correct until the coupling becomes so large (8.33) that synchronization and subsequent manifold changes begin to appear – although the overall embedding (column 5) remains correct.

We tested this approach on other coupled systems and four-dimensional driven systems with similar results [5]. We expect the multiple time series approach to work increasingly well as the dimension of the system increases or, equivalently, as the number of subsystems increase. In this case there are practical limits to how high in delay dimension one can push single time series approach, but the "democratic" multiple time series approach should distribute the time delays more evenly among the time series and mitigate against the problem of having reconstruction vectors with uncorrelated components [6,7].

ISSUE 2. DETERMINING TIME SERIES INDEPENDENCE

An issue that comes up with multiple time series that is not present for single time series is the following. Suppose we have two time series $x(t)$ and $y(t)$. A sensible question is, if we add the y series to the reconstruction along with the x series, do we gain anything? What do we mean by gain? We would mean here that y would contribute to the reconstruction by helping to reduce the number of time delays we would need with the x component. That is, y adds information that helps unfold the attractor.

A standard way to test for a (possibly nonlinear) relationship between two data time series is to use Mutual Information (MI). MI has become a well-accepted tool for nonlinear time series analysis [6,8]. We introduce below a situation, a counter-example, in which MI actually fails to give the correct answer. We subsequently reveal the flaw that exists in MI and exhibit a statistic that tests for the correct x-y relationship.

For our counter-example we produced a time series from a circuit with limit cycle behavior by recording both voltage components, x and y, which are the two dynamical variables. Then we also generated another series which is the tent map iterated 4 times on the x data. We used the following time series as a test to see which one MI would determine would be most independent from x and thus a good candidate, presumably, to be used with x in a multiple time series attractor reconstruction:

Series A: x itself, used as a sanity check
Series B: 4 iterations of the tent map on x (see FIGURE 2.)
Series C: y time series

When plotted against x in a two-dimensional plane these look like FIGURE 2.

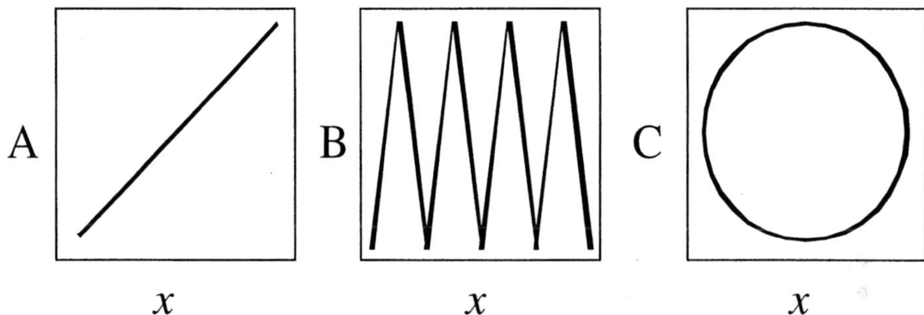

FIGURE 2. Plots of various time series created from the limit cycle circuit vs. x, one of the two voltages describing the dynamics, A: using x itself as the second time series, B: using 4 iterations of the tent map, C: using y the other dynamical variable of the circuit.

In all three cases (A, B, and C) we divide the x-(A,B, or C) planes into square cells (K in each direction, where $K=50$), count points in each cell from the plots, and assign a probability p_{ij} to each cell (cells are labeled by the two "coordinates" i and j). Then we use the MI formula to calculate the MI between x and A, B, or C,

$$\text{MI} = -\sum_{i,j=1}^{K} p_{ij} \ln p_{ij} + \sum_{i=1}^{K} p_i \ln p_i + \sum_{j=1}^{K} p_j \ln p_j, \qquad (2)$$

where p_i and p_j are the marginal probabilities of p_{ij} along the x or A, B, or C axes, respectively. Doing this for the present case shown above yields the following MI values,

Series	MI
A	3.910
B	1.885
C	2.34

The maximum MI for the x series is 3.910 and obviously series A gives that–it's the least independent. Notice that series B has the lowest MI. Thus MI would tell us to use series B, but we know that series B cannot give us more information since it is just a function of x. The correct choice for the right time series to use to reconstruct the attractor would be series C. Series C is the other independent variable in the circuit and even without delays x and series C reconstruct the limit cycle. However, series C does not show the lowest MI as we might expect. Conclusion: MI can give erroneous answers. If we measure time series that, inadvertently, are just functions of another time series being simultaneously measured, then using MI we might conclude that the pair of time series are good partners to use together in a multiple time series reconstruction. If we wanted to minimize the number of time series used (as in the above example), we would choose the wrong time series.

The question is, what is the correct criterion for finding the most "independent" time series? To answer this question we should look not to information theory, but to Taken's theorem. That theorem gives the properties of the coordinates (time-delay coordinates) that give faithful reconstructions. If $\mathbf{s}(t)$ is the actual, physical state space point, then Takens theorem states that if the time-delay reconstruction $\mathbf{z}(t)$ is a good reconstruction, there is a diffeomorphism from $\mathbf{s}(t)$ to $\mathbf{z}(t)$. In the physical state space the attractor is embedded in a manifold of, say, k dimensions. All this means that the attractor described by $\mathbf{z}(t)$ should also be embedded in a manifold of k dimensions or that the Jacobian of the diffeomorphism from $\mathbf{s}(t)$ to $\mathbf{z}(t)$, $\partial s_i/\partial z_j$, should have rank k. In fact, the approach of adding time-delay or other coordinates is precisely to increase the rank of the Jacobian so we get a proper embedding. If we add coordinates that do not increase the rank, then we add nothing to the reconstruction vector. That last sentence is the proper interpretation of our question, about whether an added time series is "independent" of the others in the reconstruction.

Now we can understand why series B is not a good series to add to the reconstruction vector, despite its having the lowest MI with x. If we add another component to $\mathbf{z}(t)$ (like series B) and it is a function of an already existing component (like x), then we end up with two columns in the Jacobian that are related. One has derivatives $\partial s_i/\partial x$, the other has derivatives $\partial s_i/\partial B$. However, B is a function of x and we can use the chain rule to write $\partial s_i/\partial B = \partial s_i/\partial x\ \partial B/\partial x$. Hence, the added column in the Jacobian due to the new B coordinate is just proportional to the existing x column. But as is well known from the theory of Jacobians or determinants, that adding columns that are proportional does not change the rank. Thus, B cannot help unfold the attractor and it adds nothing to the reconstruction.

The problem with MI is that it does not measure functional relationships, but estimates symmetric relationships within a probability distribution (p_{ij}). The more the points are spread around the two-dimensional plane the lower the MI which, in the example is done by the tent map structure which spreads the points around more than the circle from series C (y). We also say that MI is symmetric because the marginals (p_i and p_j) enter the formula (Eq.(2)) symmetrically and this cannot detect a one-way relationship. Functions need not be two ways (invertible) as is evidenced by series B.

What then is the real statistical test to use? We see that functionally **independent** coordinates are necessary in a good reconstruction. Thus, a test for functional dependence between pairs of coordinates would be a good start on this problem. Below we introduce a simplified test for functional dependence and show that it works well in the above A, B, C test. For now we concentrate simply on detecting a functional relation between reconstruction coordinates and neglect the property of smoothness that is needed for a diffeomorphism as required by Takens theorem and the existence of a Jacobian. Other somewhat more sophisticated function statistics have been proposed in the past [3,4] and a more robust statistic is presented in these proceedings [9]. Here we merely want to impress the need for such a statistic.

Let's consider the minimal requirement for properties of a function between two, one-dimensional data sets. We would expect that isolated, discontinuous points which are of zero measure would not be seen in an experiment. Hence, we assume that the minimal function property would be a "piece-wise" continuous function with a finite number of discontinuities on a compact set like an attractor. This means that, generically, we expect nearby points along the x axis to be mapped to nearby points on the y axis at generic points. FIGURE 3 shows this situation using the usual δ-ε approach. Note we do not expect that nearby points in y will be mapped to nearby points in x. That obviously will fail in FIGURE 3. Properly functionality must be tested in both directions.

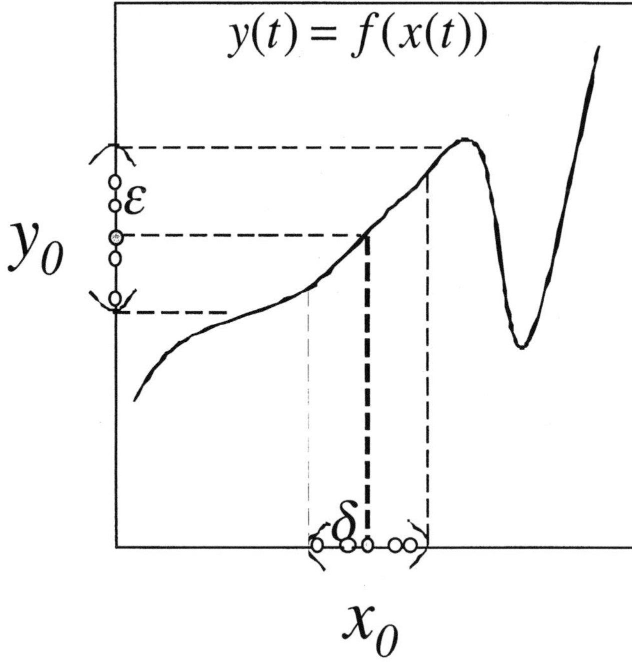

FIGURE 3. Showing that for a continuous function f points close in x (within distance δ) are close as desired in y (within any ε distance).

Of course, unlike mathematical continuity where we can go to arbitrarily small ε and δ points, for statistical continuity we are limited by any noise floor and a finite number of points. Nonetheless, we can use a scale of "smallness" similar to the MI bin size which gives $K \times K$ bins in the two-dimensional space. We simply examine whether a set of "δ" points where δ is the bin size maps into a set of ε points where ε is the bin size. If so we assign that set of points in that δ bin a 1, if not, we assign it a 0. We do this for all x bins and take the average. This average we call the "simple function statistic" SFS. The similarity in scale makes the SFS (simple function statistic) here a fair test against the MI which has the same resolution scale.

Below is a table in which we present the results for the SFS both ways, $x{\to}y$ and $y{\to}x$. In the first row the relationship is that x and y are independent variables and so fill up the

x-y relationship	MI	SFS $x{\to}y$	SFS $y{\to}x$
y (independent)	0.0	0.0	0.0
y (identity)	3.910	1.0	1.0
y (function)	1.885	1.0	0.05
y x (attractor)	2.34	0.04	0.04

plane. Both MI and SFS give correct answers. In the second row the relationship is the identity function. Again both MI and SFS give correct answers. In the third row the relationship is a function and here, as we have seen MI gives a misleading minimum, but the SFS shows that there is a function from x to y , i.e. y is a function of x which is correct. In the fourth row the attractor is mapped out by the choice of y (series C) and here MI suggests that this choice is worse than row three (series B), but the function statistic gives evidence that there is **no** functionality in either direction (values near zero) and hence suggests this is the best choice which is correct.

CONCLUSIONS

We see that because of its symmetric nature and purely probabilistic nature MI can given erroneous answers about time series independence. This probably means that MI is not the best choice for determining proper time delay (τ), but in fact a good function statistic would be a better choice. In that light the use of linear correlation is probably a better choice than MI since linear correlation at least tests for the existence of a special functional relation which is a step in the right direction.

A good function statistic has yet to be developed, but past work [3,4,10] and, especially, present candidates [9] are moving in the right direction. However, this work also suggests that much more needs to be done in time series analysis, especially multiple time series analysis. The recent literature seems to suggest that the problems are all solved and the statistics used are appropriate. But evidence here and contemplation of other issues shows this is simply not true.

We can be more general and ask the question, how do we handle the case when two time series are "partly independent?" The linear situation would be that we have a time series x and a second one $y=x+z$, where x and z are truly independent. Singular value decomposition can handle the linear case, but what about the case where $y=f(x,z)$. Then we are looking at manifolds of data, in a sense. We need some sort of nonlinear singular value decomposition. We also need a test for this situation since the function test might not capture the independent part z if it is small.

Another problem, which we believe can be addressed, is how to find a good scale in a reconstruction. Many current approaches like FNN use a scale (usually a heuristic one) to determine whether distances are above or below some threshold. But this is arbitrary. What scale would make sense from an embedding point of view, but still be justifiable statistically?

What is the best way to look for smoothness in data sets, reconstructions, and mappings between them? Smoothness is a property in addition to continuity that embeddings and their relationships may have. Ref. [4] suggested an approach to this problem, but this has not been explored sufficiently and generally has been ignored in the literature.

As we see there are several outstanding issues that lurk in the area of time series attractor reconstruction. We hope this article will stimulate further research in this area.

ACKNOWLEDGMENTS

We would like to acknowledge conversations about the above topics over the years with the following people: Nikolai Rulkov, Eric Kostelich, Lev Tsimring, Henry Abarbanel, Steven Schiff, Tay Netoff, Paul So, Ernie Baretto, Arkady Pikovskii, Joe Francis, Robert Dodier, James Theiler, Kristin Jerger, Reggie Brown, and Ljupco Kocarev.

REFERENCES

[1] F. Takens, "Detecting Strange Attractors in Turbulence," in *Dynamical Systems and Turbulence, Warwick 1980*, edited by D. Rand and L.-S. Young (Springer, Berlin, 1981), pp. p. 366.

[2] M.B. Kennel, R. Brown, and H.D.I. Abarbanel, "Determining Embedding Dimension for Phase Space reconstruction Using the Method of False Nearest Neighbors," Physical Review **A 45**, 3403 (1992).

[3] Louis M. Pecora, Thomas L. Carroll, and James F. Heagy, "Statistics for Continuity and Differentiability: An Application to Attractor Reconstruction from Time Series," in *Nonlinear Dynamics and Time Series: Building a Bridge Between the Natural and Statistical Sciences, Fields*

Institute Communications, edited by C.D. Cutler and D.T. Kaplan (American Mathematical Society, Providence, Rhode Island, 1996), Vol. 11, pp. 49-62.

[4] L. Pecora, T. Carroll, and J. Heagy, "Statistics for Mathematical Properties of Maps between Time-Series Embeddings," Physical Review E **52** (4), 3420-39 (1995).

[5] S. Boccaletti, Valladares, L. M. Pecora, H. P. Geffert, and T. Carroll 4, "Reconstructing embedding spaces of coupled dynamical systems from multivariate data," Physical Review **E65**, 035204 (2002).

[6] H. Kantz and T. Schreiber, *Nonlinear Time Series Analysis* (Cambridge University. Press, Cambridge, UK, 1997).

[7] H.D.I. Abarbanel, *Analysis of Observed Chaotic Data* (Springer, New York, 1996).

[8] A.M. Fraser and H.L. Swinney, "Independent coordinates for strange attractors from mutual information," Physical Review **A 33**, 1134-1140 (1986).

[9] Linda Moniz, Louis Pecora, Thomas Carroll, Steven Trickey, and Michael Todd, "Vibration-Based Damage Assessment Using Novel Function Statistics with Multiple Time Series," in Proceedings of the The 7th Experimental Chaos Conference, San Diego, (to be published)

[10] G.J. ortega and E. Louis, "Smoothness Implies Determinisim in Time Series: A Measure Based Approach," Physical Review Letters **81** (20), 4345 (1998).

Vibration-Based Damage Assessment Using Novel Function Statistics with Multiple Time Series

Linda Moniz, Louis Pecora, Thomas Carroll, Steven Trickey, Michael Todd

Naval Research Laboratory

Jonathan Nichols

Duke University

Abstract. Analysis of data from experiments on dynamical systems often centers on the embedding of time series data to reconstruct an atttractor. In our system, we consider output expressed as multiple time series from a circuit designed to simulate a spring-mass system in both an undamaged and a damaged state. In order to analyze differences in the reconstructed attractors between the damaged and undamaged states we employ a new version of the Continuity Statistic. This statistic was first introduced by Pecora, Carroll and Heagy [1]. Here we use the statistic in the new setting of an embedding of multiple time series and in order to address noisy data, formulate a new null hypothesis. We show that this new continuity statistic is an appropriate tool for showing differences between reconstructed attractors in the specific case of our simulated spring-mass system in "damaged" and "undamaged" conditions.

INTRODUCTION

Given time-delay embeddings of two different geometric objects from time series data, it is often important to find a functional relationship between the two objects. For instance, in the presence of noisy data from the reconstruction of one object, can we say if it is essentially identical to another object? Proving or disproving the existence of a continuous function between two such objects can be a powerful tool for analysis of nonlinear behavior. However, translating the

CP676, *Experimental Chaos: 7th Experimental Chaos Conference,*
edited by V. In, L. Kocarev, T. L. Carroll, B. J. Gluckman, S. Boccaletti, and J. Kurths

mathematical ε-δ definition of continuity to a time-series reconstruction setting raises two important questions:

1. How can potentially noisy, finite data yield a reasonable definition of continuity either at a point or on an entire attractor?

2. How can such a definition be translated to a meaningful and reliable statistic regarding the absence or presence of a continuous function?

Given a proposed function F: X $\rightarrow$ Y the mathematical definition of continuity at a point x (t) $\in$ X is stated as follows: For all $\varepsilon > 0$, $\exists\ \delta > 0$ such that if d (x (t), x (t_0)) < δ, then d (F (x (t)), F (x (t_0))) < ε.

Three unrelenting facts, therefore, seem to create an impasse with finite data. The first is that ε cannot be made to go to zero. Thus, some finite ε will have to be used. Moreover, for some x, there may be ε for which δ can be found even if there is no continuous F. Secondly, only a finite number of points x $\in$ X can be checked for continuity. Lastly, in the presence of noise, even for an obviously continuous F (e.g. an identity function), all points from a δ-ball may not map to the corresponding ε-ball.

These issues cannot be ignored, but we can remove the impasse by creating a statistical criterion for continuity that is consistent with the ε-δ definition but is difficult to fulfill if there is no continuous functional relationship.

The Continuity Statistic

We set the standard deviation of the Y data, σ_y, as the maximum ε that will be tested. In practice, we normalize and de-mean the data, so that $\sigma_x = \sigma_y = 1$. The statistical definition begins with a null hypothesis that addresses the issue of noise:

Points $x(t_j)$ from any δ-ball will have probability .5 of $y(t_j)$'s being in the ε-ball, regardless of the size of the ε-ball. To reject the null hypothesis, we require a 95% confidence interval. If *n* points are in any δ-ball, the probability of *m* or more of these points' images in the ε-ball must be $\leq$ .05 to reject the null hypothesis.

This differs from the null hypothesis described in [1]. To account for noise, our null hypothesis allows some points from the δ-ball to map outside the ε-ball but requires 95% confidence; the probability of rejecting the null hypothesis lies in the tail of the binomial distribution. We formulate the statistic to be based not only on the acceptance or rejection of the null hypothesis, but also on the *minimum* ε that can be used to reject the null hypothesis. Thus, the statistic can give some idea as to the scaling of any possible functional relationship by comparing minimum and average ε to the minimum δ given by the data.

To compute the continuity statistic, N test points $x(t_i)$ are chosen at random times. This serves to also distribute the points randomly in space. For each test point, initially $\varepsilon = \delta = \sigma$. The number of points in the δ-ball around the representative point $x(t_i)$ is n. Image points in the ball centered around the point $y(t_i)$ are counted; this number is m. Then the binomial distribution with parameters $(n, .5)$ is computed to find the cumulative probability of finding m or more image points in the ε-ball. If this probability is $\leq .05$, the null hypothesis is rejected for this point and ε is recorded as ε^*. Then ε is reduced with the same δ. If the null hypothesis is *not* rejected, δ is reduced. To maintain the 95% confidence interval, there must be at least 5 temporally non-correlated points in the δ-ball. If no ε can be found with any acceptable δ, this point's null hypothesis is not rejected. Note that ε^* for each point represents the *smallest* ε for which the null hypothesis is rejected. The average and distribution of ε^* is recorded, along with the maximum δ for each ε^*, and the number of test points which reject the null hypothesis.

If a continuous function exists between source and target attractors ε^* will exhibit an average close to 0 and a tight distribution clustered around the average and N*/N will be close to 1. Figure 1 compares the ε^* distributions of the continuity test between a Lorenz and itself embedded in 3 space and the continuity test between a Lorenz x signal and a random sequence of numbers. For the Lorenz-to-Lorenz distribution, N*/N = 1, the average $\varepsilon^* = .06$, and *maximum* $\varepsilon^* = .2$. In contrast, the Lorenz-to-random yielded continuity statistics N*/N = .38, *minimum* $\varepsilon^* = .5$, and average $\varepsilon^* = .8947$. We see that with finite data it is possible, even with no functional relationship, to find ε and δ for some points. However, ε^* in these cases was quite far from the smallest ε allowed by the data.

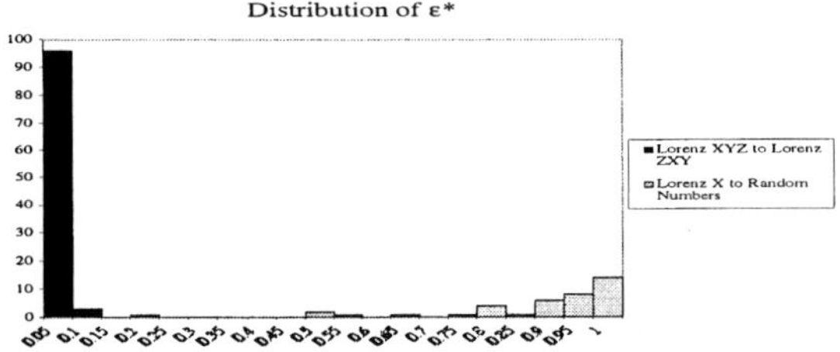

FIGURE 1. Distribution of e* for Lorenz-to-Lorenz(black distribution) and Lorenz-to-random numbers (grey distribution). There were 20K points in each data set.

Now we investigate the continuity statistic as a tool to compare two attractor reconstructions from multiple time series obtained from experiments.

The Spring-Mass Simulation Experiments

An 8-degree of freedom spring-mass damper structure was simulated with a circuit experiment. The intent of these experiments is to determine if the continuity test can detect a small level of damage to the system. The system is driven on the free end (oscillator 8) by a prerecorded Lorenz signal with parameters tuned to the modes of the spring-mass system. The purpose of the prerecorded signal is to obtain output time series for a damaged and an undamaged system with the same input signal. In experiments by Todd, et al [2], changes in the attractor due to small levels of damage to the system were detected using the Local Attractor Variance Ratio (LAVR). The method used here repeats the use of the prerecorded input signal, but system (position) response time series are recorded at each oscillator. The system response is recorded as 8 separate time series at synchronized times.

Time series are recorded from the circuit in both a damaged and undamaged condition. Damage was a 10% loss of position coupling between oscillators 1 and 2 on the fixed end. Output data (position only) was collected from each oscillator in the circuit.

The structure acts as a linear filter of the chaotic time series: $\dot{z} = Az(t) + x(t)$, where $x(t)$ is the Lorenz input signal, $z(t)$ is the position output signal, and A is the filter described by the structure. A has eigenvalues $\lambda^S_i = (\lambda^C_j; \lambda^L_k)$ where λ^C_j are associated with the signal and λ^L_k are associated with the structure and depend on mass, stiffness and damping in the structure.

The Kaplan-Yorke conjecture [5] relates the Lyapunov dimension D_L of an attractor to the Lyapunov Spectrum by the following: $D_L = k + \Sigma^k_{m=1} \lambda_m / -\lambda_{k+1}$, where k is the maximum number of exponents which may be added before the sum becomes negative. If the parameters of the system are varied (here, the stiffness of the system is altered with damage) D_L changes. If the fractal dimension (which is close to D_L) is altered significantly by damage it may be possible to detect the change, and thus the damage, by seeing the loss of a continuous functional relationship between the undamaged and damaged attractor reconstructions.

From Fig.1 we see that it is advantageous to have continuity statistics from a known functional relationship to compare statistics from one that we conjecture is lost or does not exist. Thus, we also performed the continuity test between one undamaged data set and itself as well as between two different experiments with the same undamaged circuit to compare to test results from undamaged to damaged circuits. As stated above, we used the same prerecorded Lorenz signal on all experiments so that a mapping could be made between two different sets of time series.

The original time series (200K points) from the experiments were truncated (in both front, to account for transients, and back) and we obtained time series of 80K. To find an appropriate embedding dimension, a false nearest-neighbor test was performed on each set of 8 time series, using the method suggested in Boccaletti [4]. It was determined that dimension 16 (2 delay coordinates per time series) was sufficient to "unfold" the attractor.

The continuity tests were performed with N=100 randomly (in time) chosen points on each attractor. The Theiler Window was set to 30 time steps. (See, e.g. [3]). This assured that different points counted in the δ-ball were not correlated in time. Delay times for the time series were constant over all 8 time series in the reconstructions and set to the minimum delay window of each *set* of time series. The delay windows were determined by performing a linear auto-correlation on each oscillator's output signal and determining the approximate number of time steps until the auto-correlation decayed to ~1/3 of its original value. The driven oscillator, 8, exhibited the minimum auto-correlation (15 time steps). The results were examined for detectable changes in the mean and distribution of ε^*, the mean of δ^* (the largest δ for each ε^*), and for N^*/N.

Results

FIGURE 2. N^* and ε^* for Continuity tests.

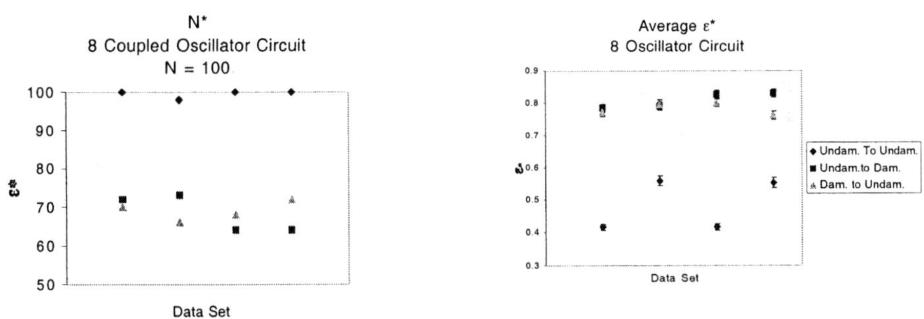

The continuity test from undamaged to damaged and again from damaged to undamaged show an increase in ε^* to .8 σ from the undamaged-to-undamaged ε^* of ~.4σ. Moreover, N^*/N (N=100) was close to 1 in all undamaged-to-undamaged cases; nearly all of the points tested rejected the null hypothesis. In the tests from undamaged-to-damaged and again from damaged-to-undamaged, we see N^*/N on the order of .65-.70. In figure 3, looking at the distribution of ε^* from one undamaged-to-undamaged data set we see a noticeable difference when compared to the distribution for ε* from the undamaged-to- damaged, qualitatively similar to what we saw in fig.1. Thus, it is apparent that there is some loss of a continuous function from the attractor reconstructed from the data from the undamaged circuit to that reconstructed from the data from the damaged circuit.

We note that the continuity statistic on our circuit experiment did not show a total lack of evidence of a continuous function as was clearer with the Lorenz-to-random number example. It did show that there were places on the attractors for which a continuous functional relationship is unlikely. Since the theory points to local changes in the attractor geometry with damage, this is consistent with expected results. However, more theory is needed to explain the relationship between changes in the filter parameters and changes in the geometry of the reconstructed attractor.

FIGURE 3. Distribution of ε* in undamaged to undamaged and undamaged to damaged continuity statistics. For Undamaged to Undamaged, N* = 100 and average ε* = .4183. For Undamaged to Damaged, N* = 64 and average ε* = .8255.

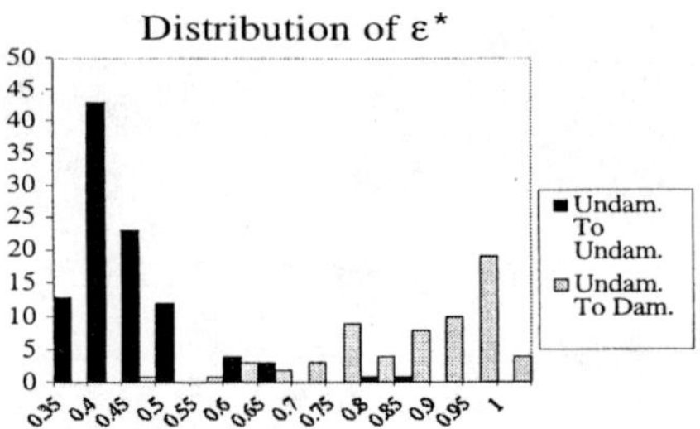

ACKNOWLEDGMENTS

This research was partially supported by an ASEE Postdoctoral Fellowship funded by the Office of Naval Research.

REFERENCES

1.Pecora,L. M., Carroll, T.L. and Heagy,J.F., [1995] *Statistics for mathematical properties between time-series embeddings*, Physical Review E **52**(4), 3420.

2.Todd, M., Nichols, J., Pecora,L.M., Virgin, L. [2001], *Novel Nonlinear Feature Identification in Vibration-Based Damaged Detection Using Local Attractor Variance* IMAC XIX Proceedings Feb. 2001.

3.. Theiler, J. [1986] *Spurious dimension from correlation algorithms applied to limited time-series data*, Physical Review A **34**, 2427.

4. Boccaletti, Valladares, Pecora, Geffert and Carroll [2002], *Reconstructing embedding spaces of coupled dynamical systems from multivariate data* Physical Review E **65**, Preprint.

5. Kaplan, J.L. and Yorke, J.A., [1979] , in *Functional Differential Equations and Approximation of Fixed Points,* edited by H.O. Peitgen and H.O. Walther, Springer Lecture Notes and Mathematics Vol. 730 , Springer-Verlag Berlin.

Experimental Evaluation Of The d_∞ Parameter to Characterize Chaotic Dynamics

A. Bonasera[*], M. Bucolo[§], L. Fortuna[§], M. Frasca[§], A. Rizzo[+]

[§] *Dipartimento Elettrico Elettronico e Sistemistico*
Università degli Studi di Catania
Viale A. Doria, 6 - 95125 Catania Italy
Email lfortuna@dees.unict.it

[*] *Laboratorio Nazionale del Sud*
Istituto Nazionale di Fisica Nucleare
Viale A. Doria, 6 - 95125 Catania Italy

[+] *Dipartimento di Elettrotecnica ed Elettronica*
Politecnico di Bari
Via Re David, 200 – 70125 Bari Italy
Email rizzo@deemail.poliba.it

Abstract. The d_∞ parameter is an asymptotic measure introduced in order to characterise chaotic dynamics. This is computed as the asymptotic distance between nearby trajectories, and is able to take into account both the stretching and the folding phenomena. In this work, after a brief overview on theoretical aspects of the measure, a suitable approach for the experimental evaluation of the d_∞ parameter is described. This is applied to chaotic circuits through the use of an analog circuitry able to perform in real-time and in a totally analog fashion the computation of the d_∞ parameter. Experimental results referring to Chua's and Lorenz circuits are reported in order to validate the approach.

INTRODUCTION

The d_∞ parameter is an asymptotic measure recently introduced to characterize chaotic dynamics [1]. Unlike the parameters adopted so far as indicators revealing the presence of chaos, the computation effort needed for d_∞ is low and, as it will be shown in the following, can be easily performed in real time through a simple analog circuitry. The most used quantities in chaos analysis can be resumed in the following [2]:

- The Lyapunov exponents (LE), indicators of the mean rate of separation between two nearby trajectories in phase space. In particular, the leading LE is often fully representative of the sensitivity to initial condition of a given dynamics, being positive for chaotic behaviours.

- The auto-correlation function, which rapidly decays to zero in the chaotic regimes.

CP676, *Experimental Chaos: 7th Experimental Chaos Conference*,
edited by V. In, L. Kocarev, T. L. Carroll, B. J. Gluckman, S. Boccaletti, and J. Kurths
© 2003 American Institute of Physics 0-7354-0145-4/03/$20.00

-    The power spectrum, which changes from discrete lines to a broad-band continuous spectrum when chaotic dynamics arise.

Several algorithms have been implemented to calculate these quantities either from the nonlinear differential equations describing the system behaviour, or from experimental data recorded from actual experiments [3]. This last case involves a certain care in the implementation of the related algorithms, especially because of the presence of noise in the experimental records.

Focusing attention on the calculation of LEs from field data, some other issues have to be mentioned. First of all, when implementing the algorithm, great care has to be taken on avoiding to include in the calculation the saturation of the trajectories, due to the boundness of the phase space (folding mechanism). Moreover, in some cases the time evolution of a certain variable is not available and only the final space distribution is known. This happens for example in nucleus-nucleus collision experiments, where only the final momentum distribution of almost all particles can be recorded on field. The lack of information about the dynamical evolution of the experiment makes impossible the computation of LEs from field data in these experiments. The d_∞, whose computation is based on asymptotic information, is strictly related to the leading LE, and therefore allows to obtain as much information as that provided from LEs with simpler computations. In this paper, the basic concepts about d_∞ are given. An experimental setup is then described in order to perform the real-time characterization of circuits. Experimental results on the Chua's and Lorenz circuits are reported to confirm the suitability of the approach.

THE D_∞ PARAMETER

The theory of nonlinear, deterministic dynamical systems provides some numerical algorithms to perform quantities characterizing geometrical and dynamical properties of attractors of these systems. Two key aspects of chaos are the stretching of infinitesimal displacements and the phenomenon of folding, which keeps nearby trajectories in a bounded region of the phase space, leading to the existence of complex orbit-like structures, in the form of a vast variety of possible unstable orbits, confined in a region of the phase space called attractor. The stretching property is strictly related to sensitive dependence on initial conditions. It is well known that a quantitative characterization of the stretching property is provided by the leading Lyapunov exponent.

Let us consider a continuous system described by the following differential equation:

$$\dot{x} = f(x, r) \tag{1}$$

where x and r are the state and the parameter vectors, respectively. During chaotic regime, it can be assumed that two nearby trajectories separated by a distance d_0 diverge exponentially.

The trend of the diverging distance (computed in the usual Euclidean sense) can be therefore expressed as:

$$d(t) = d_0 e^{\lambda t} \tag{2}$$

where λ is the leading LE. Eq. (2) leads to the following differential relationship:

$$\frac{\partial d(t)}{\partial t} = \lambda d(t) \tag{3}$$

The folding mechanism can be taken into account by introducing a second-order term, leading to:

$$\frac{\partial d(t)}{\partial t} = \lambda d(t) - \gamma d(t)^2 \tag{4}$$

The asymptotic behavior leads to the following relationship, defining d_∞:

$$\lambda = \gamma \cdot d_\infty \tag{5}$$

Eq. (5) shows a relationship between d_∞ and the leading LE. Due to the formalization of the folding process in which the parameter γ is generally not a constant, eq. (5) has not to be intended as a proportionality relationship. Nevertheless, several numerical studies show that the trend of d_∞ and that of λ are very similar, both exhibiting relevant changes near bifurcations. As an example, let us consider the well known Chua's circuit [4], described in the space state by the following set of dimensionless differential equations:

$$\frac{dx}{d\tau} = \alpha \cdot (y - x - f(x))$$

$$\frac{dy}{d\tau} = (x - y + z) \tag{6}$$

$$\frac{dz}{d\tau} = -\beta \cdot y$$

Fig. 1 shows a comparison between the values of d_∞ and λ versus a change in the parameter β of (6), in correspondence of a transition from double scroll to single scroll attractors [4].

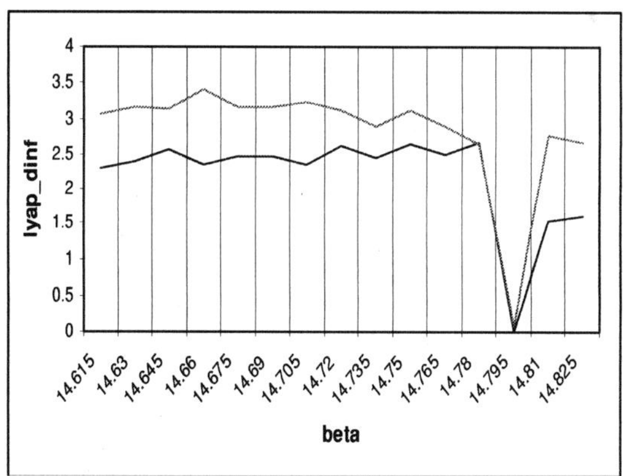

FIGURE 1. Comparison between the trends of both d_∞ and leading L.E. versus β in correspondence of a phase change (double-scroll to single scroll).

It is clear that (5) is a static relationship between the leading LE and the asymptotic distance. Solving the differential equation (4), with $d(0)=d_0$ and $d(\infty)=d_\infty$, a dynamical relationship between those quantities can be easily established:

$$d(t) = \frac{d_\infty d_0}{d_0 + d_\infty \exp(-\lambda t)} \tag{7}$$

Fig. 2 illustrates a comparison between the trend of $d(t)$, and expression (7), computed for a Chua's circuit in a chaotic regime (α=9, β=12.9), revealing good agreement between predicted and simulated data.

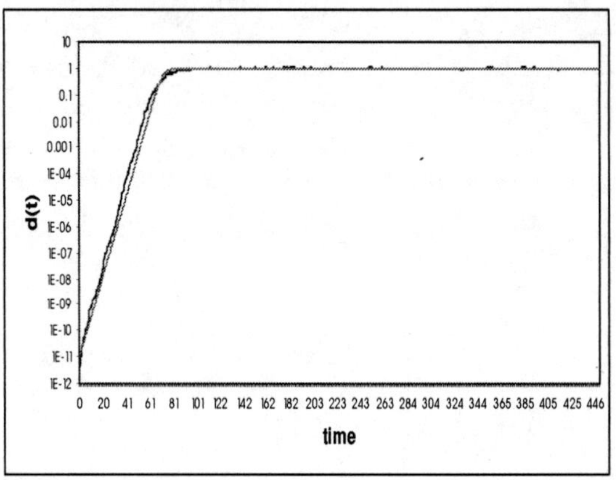

FIGURE 2. Actual versus simulated trend of d_∞ for α=9, and β=12.9.

EXPERIMENTAL EVALUATION OF D_∞

The measure represented by d_∞ parameter can be performed either numerically or on experimental data by computing the asymptotic limit of the following quantity:

$$d(t) = \sum_{j=1}^{n} \left| x_j(t) - x_j'(t) \right| \tag{8}$$

where x_j and x'_j are two nearby (i.e. starting from very close initial conditions) trajectories and n is the dimension of the space state of the chaotic circuit. Distance $d(t)$ can be easily computed by a simple analog circuitry. Displaying the evolution of $d(t)$ on an oscilloscope, d_∞ is revealed after a short transitory phase.

The experimental setup consists of two identical chaotic circuits and the circuit measuring the $d(t)$ distance. The latter circuit is illustrated in Fig. 3. It consists of four stages. The first stage buffers the input, which are the state variables of the chaotic circuits, x,y,z, and $x1$, $y1$, $z1$ (which, for clarity, have been considered of the third order). The second stage performs the difference between corresponding state variables of the two circuits. The third stage is a precision rectifier [6], whereas the last stage is devoted to sum the three contributions due to the three variables of the chaotic circuit.

358

Fig. 5 shows the result of the evaluation of d_∞ for two well-known systems: the Chua's circuit and the Lorenz circuit [5]. The value of d_∞ is experimentally computed as the average value of the output of the circuit of Fig. 3. Fig. 4 illustrates the case in which d_∞ is computed for the two systems operating in chaotic regions, providing an average value which is different than zero in both cases.

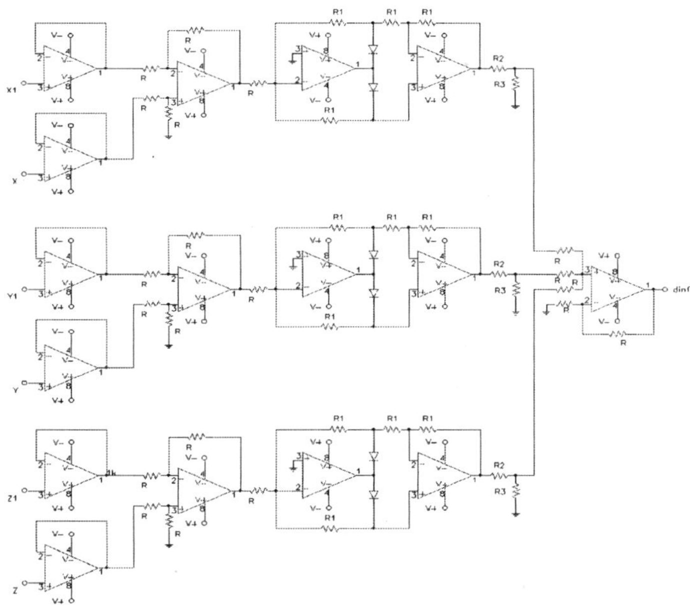

FIGURE 3. Experimental circuit for the evaluation of d_∞.

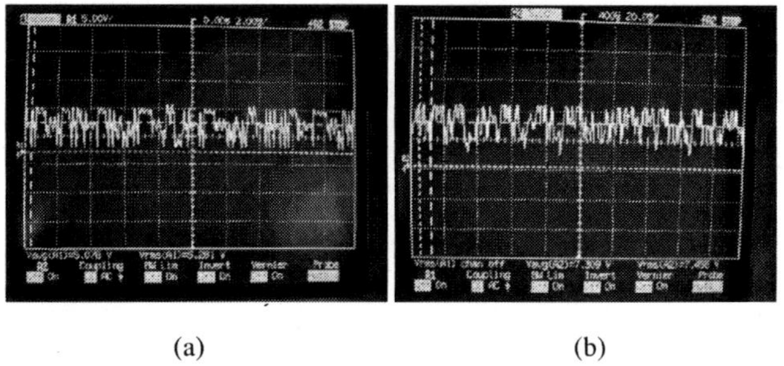

(a)   (b)

FIGURE 4. Experimental evaluation of d_∞ (a) for Chua's circuit; (b) for Lorenz circuit.

CONCLUSIONS

In this paper we introduced a novel approach to characterize chaotic behaviors from asymptotic properties. In particular, the asymptotic distance between trajectories has been defined. This is a complementary quantity to Lyapunov spectrum, often holding as much information as LEs do, and characterizing both stretching and folding processes, whereas LEs refer only to stretching. Moreover, it is a very useful quantity to perform when Lyapunov exponent are hard to work out. Some analytical work has been presented in this paper, especially concerning the static and dynamic relationships between the leading LE and the asymptotic distance. The experimental results, performed on two chaotic systems (Chua's and Lorenz circuits) confirm the suitability of the approach, making the asymptotic distance a very useful tool to analyze and characterize chaos and strange attractors.

REFERENCES

1. A. Bonasera, M. Bucolo, L. Fortuna, A. Rizzo, "The d_∞ Parameter to Characterise Chaotic Dynamics", *IJCNN 2000, International Joint Conference on Neural Networks, Como, Jul 2000.*
2. E. Ott, T. Sauer, J. A. York, *Coping with Chaos*, John Wiley & Sons,Inc.,1994.
3. H. D. I. Abarbanel, *Analysis of Observed Chaotic Data*, Springer, 1996.
4. R. N. Madan, *Chua's Circuit: A Paradigm for Chaos*, World Scientific, 1993.
5. G. Manganaro, P. Arena, L. Fortuna, *Cellular Neural Networks: Chaos, Complexity and VLSI processing*, Springer-Verlag, 1999.
6. J. Millman and A. Grabel, *Microelectronics*, McGraw-Hill, New York, 1989.

POSTER SESSION
ABSTRACTS

P1. EXPERIMENTAL EVIDENCE ON INTERMITTENT LAG SYNCHRONIZATION IN COUPLED CHUA'S OSCILLATORS

P.K.Roy,
Department of Physics, Presidency College, Kolkata 700073, India
S.K.Dana,
Instrument Division, Indian Institute of Chemical Biology, Kolkata 700032, India

Phase synchronization (PS) in coupled chaotic oscillators has been investigated numerically in Lorenz, Rossler models and also experimentally in cardiorespiratory systems by many researchers. In non-identical oscillators, which is a reality, complete synchronization (CS) of amplitude and phase is difficult to arrive at. Lag synchronization (LS) is an intermediate step between complete (CS) and PS. PS shows promises in communication, in the context of pulse position modulation, in homoclinic chaotic systems. As has been observed earlier by others in numerical experiments that there is an intermittent region between PS and LS. This intermediate region is defined as the intermittent lag synchronization (ILS). Experimental evidence on both PS and ILS using two coupled Chua's oscillator (non-identical) is reported here.

P2. CHAOS-BASED DIGITAL COMMUNICATION SYSTEM

Dipendra C. Sengupta
NASA Administrator's Fellow
Elizabeth City State University
Elizabeth City, NC 27909, USA
dcsengupta@mail.ecsu.edu

Monty Andro
Digital Communication Branch
NASA Glenn Research Center
Cleveland, OH 44135, USA
mandro@grc.nasa.gov

A technique for exploiting deterministic chaos through Logistic Map (with DC component) and a Cubic Map (with no DC component) in a new digital communication has been proposed and demonstrated in presence of additive white Gaussian noise (AWGN) by Monte Carlo simulations. Bit-error (BER) graphs were computed in order to compare the performance of the different modulation schemes. The best performance demonstrated, with AWGN noise, was achieved by utilizing chaotic and periodic waveforms as the information bearing components. Results show the BER performance curve implementing a Fourier Transform detection scheme is not far from the theoretical performance for Binary Phase Shift Key (BPSK) modulation.

P3.   SECURE COMMUNICATION IN TS-HVSM SYSTEM BASED ON THE CHAOTIC SYNCHRONIZATION

Ik-Soo Lee and Jin-Kyung Ryeu
Information and Communication, Pohang College,
Hunghae-Eup, Pohang City, 791-940, Korea
Phone : +82-54-245-1253   E-mail : leeis@pohang.ac.kr

In this paper, we proposed a new secure communication scheme using analog TSHVSM (tailed shift hyperchaotic volume-preserving smooth-function maps) model for generating the complicated hyperchaotic signals and implemented the chaotic signal generation circuit on the board level. The proposed TS-HVSM model is consists of three dimensional discrete-time simultaneous difference equations and shows uniquely random chaotic attractors using nonlinear maps and shift modulus function.
In the secure communication system, two hyperchaotic signals are coupled and driven for accomplishing to the chaotic synchronization systems. And the encrypted masking signal which is added speech or digital signal to chaotic signal is transmitted to a subsystem as modulating signal in spread spectrum communications. It is demonstrated that proposed encryption system have shown good performance at the secure chaotic communication by numerical and experimental results.

P4.   ULTRA WIDE BAND COMMUNICATION DEVICE WITH CHAOTIC FREQUENCY MODULATION.

Alexander Volkovskii, Institute for Nonlinear Science, Univ. of California, San Diego
Ian Langmore, Univ. of California, San Diego

The Ultra Wide Band (UWB) communicating systems draw a tremendous attention in the last few years because of many potential advantages the UWB technology can bring to the wireless industry. It can solve the RF spectrum availability problem, improve the security, can provide less expensive, less power consuming equipment for a variety of wireless applications. We study a communication device, in which the Chaotic (Rossler) Oscillator (CO) combined with a linear Voltage Controlled Oscillator (VCO) is used to generate the UWB carrier signal with Chaotic Frequency Modulation (CFM). The information signal can be transmitted via the Frequency Modulation (FM) in analog systems or the Binary Frequency Shift Key-in (BFSK) modulation in digital applications. The receiver has the similar CO and VCO with closed parameters. A simple circuit based on the Phase locked Loop (PLL) is used for both synchronization of the chaotic oscillators and demodulation of the carrier signal. Using the dynamical systems approach we show that the suggested device has a broad parameter region in which the synchronization and demodulation take place. In computer simulations we demonstrate that the system is robust against the noise and parameters mismatch.

P5. DYNAMIC LIMITER CONTROL OF LONG-PERIOD ORBITS AND ARBITRARY TRAJECTORIES IN AN ELECTRONIC CIRCUIT

Ned J. Corron and Shawn D. Pethel
U. S. Army Aviation and Missile Command, Redstone Arsenal, AL 35898

We demonstrate experimental chaos control of long-period orbits and arbitrary trajectories in a piecewise-linear LC oscillator using a new technique called dynamic limiting. Based on limiter control, dynamic limiting uses a predetermined sequence of limited levels to stabilize natural states of the chaotic circuit. The limiter sequence is clocked by the natural return time of the oscillator such that a new limiter level is applied for each peak return. We demonstrate control of period-8 and period-34 unstable periodic orbits and provide evidence that the control perturbations are minimal. We demonstrate control of an arbitrary waveform by replaying a sequence captured from the uncontrolled oscillator, achieving a form of delayed self-synchronization. Finally, we develop a lookup table of limiter levels to control arbitrary symbolic dynamics in the circuit and use it to control all unstable periodic orbits in the oscillator up to period 12.

P6. NONLINEAR FILTERING METHODS IN COMMUNICATION

Jochen Bröcker and Ulrich Parlitz
Drittes Physikalisches Institut
Universität Göttingen
Germany

In recent work on communication, the use of chaotic systems for message transmission has attracted considerable attention. Although potential advantages over classical methods have been pointed out, quantitative analysis on the performance of nonlinear communication systems, especially in noisy environments, seem to be still in its infancy. In our talk we employ nonlinear filtering theory to obtain a representation of the optimal receiver. Using known results on the nonlinear filter we investigate the bit error probability. Finally we discuss problems arising in applications due to the complexity of the nonlinear filter and possible ways to overcome this difficulty.

P7. FERROELECTRICS NONLINEARITY TO REALIZE CHAOTIC CIRCUITS

L. Fortuna, M. Frasca, S. Graziani
Dipartimento Elettrico Elettronico e Sistemistico
Università degli Studi di Catania
Viale A. Doria 6 – 95125 Catania, Italy
E-mail: lfortuna@dees.unict.it
Tel. +390957382307 Fax. +39095339535

In this paper the possibility of observing chaotic behaviour in an electronic circuit including a nonlinear ferroelectrics has been investigated. The ferroelectrics constitutes the medium interposed between the two plates of a capacitor, and is obtained by successive vapour deposition of Strontium, Tantalum and Bismuth on Platinum substrates in small areas. The device is characterized by a nonlinear hysteretic behaviour observed by estimating the output voltage by using a Sawyer-Tower configuration. Different circuits showing chaos and based on hysteretic behaviour of the nonlinearities have been reported in literature. In particular, in the circuit reported in [1] the nonlinearity consists of a piecewise linear resistor, only the two tracts with positive slope are effectively involved in the dynamics and an hysteretic behaviour switching between these two linear segments of the nonlinearity can be addressed as responsible of the emergence of iperchaos in the circuit. Starting from the chaos generator reported in [1], the following dimensionless equations have been developed:

where $f(x)$ is the nonlinearity due to the hysteresis of the ferroelectrics capacitor. The functional $f(x)$ constitutes a simple model of the ferroelectrics device and is defined by two functions (the upper curve and the lower curve based on experimental data on the ferroelectrics). Chaotic solutions of system (1) were searched for, by performing numerical integration with respect to different values of the parameters.

Finally, the circuit implementing equations (1) has been realized. Experimental results show that for a suitable range of parameter a chaotic attractor emerges. Fig. 1 shows the projection of the attractor onto the phase plane $x1$-$x2$. The paper addresses the possibility of exploiting the rich dynamics of ferroelectrics hysteresis to obtain new chaotic attractors with low cost circuits.

References

[1] T. Saito, "An Approach Toward Higher Dimensional Hysteresis Chaos Generator", IEEE Transactions on Circuits and Systems, vol. 37, no. 3, March 1990.

P8.   CHARACTERIZING BIFURCATIONS BY AVERAGES OF CHAOTIC DYNAMIC VARIABLES

José R. Rios Leite and Hugo L. D. de S. Cavalcante
Departamento de Física, Universidade Federal de Pernambuco

Experimental averages for the current and voltage across a nonlinear RLC circuit driven by an external oscilator were obtained and used to characterize the bifurcations to chaos by type-III intermittency. Results are compared to measurements of time duration of laminar phase events through the same bifurcation. Critical exponents for both the average of the variables and the average lenght of laminar phases were extracted. Their values are consistent with predictions made from the normal form map model for type-III intermittency with leading cubic nonlinearity [1,2]. Ongoing experiments in chaotic lasers shall be reported for bifurcations with type-I intermittecy [3].

[1] Averages and critical exponents in type-III intermittent chaos, Hugo L. D. de S. Cavalcante and J. R. Rios Leite, submitted to Physical Review E, 2002.
[2] Characteristic relations of type-III intermittency in an electronic circuit, C.-M. Kim, G.-S. Yim, J.-W. Ryu, Y.-J. Park, Physics Review Letters **80**, 5317 (1998).
[3] Bifurcations and averages in the homoclinic chaos of a laser with a saturable absorber, Physica A **283**, 125 (200).

P9. COMPARISON OF THE NATURE OF CHAOS IN EXPERIMENTAL [EEG] DATA AND THEORETICAL [ANN] DATA

Atin Das, Pritha Das, Department of Mathematics,
Jadavpur University, Jadavpur, Calcutta 700 032, India Email: atin_das@yahoo.com

In this paper, nonlinear dynamical tools like largest Lyapunov exponents (LE), fractal dimension, correlation dimension, pointwise correlation dimension will be employed to analyze electroencephalogram [EEG] data and determine the nature of chaos. Results of similar calculations from some earlier works will be produced for comparison with present results. Also, a brief report on difference of opinion among coworkers regarding tools to characterize chaos will be reported; particularly applicability of LE will be reviewed. The issue of nonlinearity present in experimental time series will be addressed by using surrogate data technique. We have extracted another data set which represented chaotic state of the system considered in our earlier work of mathematical modeling of artificial neural network. By comparing the values of measures employed to the two datasets, it can be concluded that EEG represents high dimensional chaos, whereas the experimental data due to its deterministic nature, is of low dimension. Also results give the evidence that LE exponent is applicable for low dimensional chaotic system while for experimental data, due to their stochasticity and presence of noise- LE is not a reliable tool to characterize chaos.

P11. CODING OF SENSOR INFORMATION BY EXCITED NEURONS

M. M. Sushchik[1], D. G. Zakharov[1], and M. M. Sushchik Jr.[2]
[1] Institute of Applied Physics, Russian Academy of Science
603950 Niznhy Novgorod, Russia, e-mail: sushch@appl.sci-nnov.ru
[2] Therma-Wave Inc., Fremont, CA 94539, US, e-mail: msushchi@thermawave.com

We consider forced behavior of neurons, which without forcing are capable of generating regular and chaotic spike trains and burst sequences. We investigate the effects of sensor inhibitory forcing and study the observed regimes of ordinary, generalized and intermittent synchronizations. We analyze the role of these regimes and of transitions between them in the process of selection of the space for coding of sensor information by a neural system. In our analysis we model a neuron by the four-dimensional generalized system of Hindmarsh-Rose equations [1]. We propose a simplified neuron model, which retains the merits of the Hindmarsh-Rose model and allows for the effects of dynamic unreliability during sensor information coding [2], but is simple and easy to interpret, which is typical of models based on modified radial clocks [3].

1. Pinto R. D. et al., *Phys. Rev. E* **62**, 2644-2656 (2000).
2. Eguia M. C., Rabinovich M. I., and Abarbanel H. D. I., *Phys. Rev. E* **62**, 7111-7122 (2000).
3. Nomura T. et al., *Biol. Cybern.* **72**, 93-101 (1994).

P12. ANTI-PHASE REGULARIZATION OF COUPLED CHAOTIC MAPS MODELLING BURSTING NEURONS

B. Cazelles[1], M. Courbage[2] and M. Rabinovich[3]
• CNRS UMR 7625, Université Paris 6 and UFR de Biologie Université Paris 7
• LPTMC , Université Paris 7, 4, Place Jussieu, 75251 Paris CEDEX 05 FRANCE
• Institute for Nonlinear Science, University of California San Diego, 9500Gilman Drive, La Jolla, 92093-0402 , USA

We introduce a new class of maps that describe the chaotic activity of spiking bursting neurons observed in neurophysical experiments. We show that , depending on the connection (diffusively or reciprocally synaptically) , coupled maps demonstrate several modes of cooperative dynamics : i) weakly correlated chaotic pulstaion, ii) antiphase quasi regular bursting activity independent of the amplitude of the spikes and iii) chaotic synchronisation. Such phenomena have been observed in recent experiments of central patterns generator chaotic neurons.
B. Cazelles, M. Courbage and M. Rabinovich, *Europhys.. Lett.***56,** pp 504-509, (2001)

P13. ROBUST SUPPRESSION OF CHAOS IN SYSTEMS WITH NOVEL TYPE OF ACTIVITY-DEPENDENT COUPLING

Valentin Zhigulin, MC103-33, Department of Physics, California Institute of Technology, Pasadena, CA 91125
Mikhail Rabinovich, Institute for Nonlinear Science, University of California, San Diego, La Jolla, CA 92093-0402

Periodic regimes are known to exist in different weakly coupled chaotic systems. In most cases these deterministic states exist in a very narrow region of parameter space of otherwise chaotic system. In this work we introduce a new type of coupling, namely activity-dependent coupling which strength depends on the time course of activity of connected elements. Properties of this coupling are analogous to those of well known in neurobiology synaptic coupling with Spike-Timing Dependent Plasticity (STDP). Using the example of coupled chaotic Hindmarsh-Rose neurons we show by computer simulations that such activity-dependent coupling is able to self-adjust its strength in such a way that periodic state is achieved independently of initial coupling strength. Hence, STDP-like connections cause the suppression of chaos in a wide range of initial coupling strengths, which we confirm by calculating Lyapunov Exponents. Suppression of chaos is also observed in systems with other types of chaotic oscillators. These results suggest a possible role that STDP plays in the cortex where many types of chaotic neurons are present. Moreover, this novel method of coupling can be applied in physical and other systems where chaos suppression is desired.

P14. THE NATURE OF ESSENTIAL AND PARKINSONIAN TREMORS

J.B. Gao [1,2] and Wen-wen Tung [3]
[1] Department of Electrical and Computer Engineering, University of Florida, Gainesville, FL 32611
[2] Electrical Engineering Department, UCLA
[3] Department of Atmospheric Sciences, University of California, Los Angeles, CA 90095
Email: jbgao@ee.ucla.edu, qc@atmos.ucla.edu

Tremor denotes an involuntary, approximately rhythmic, and roughly sinusoidal movement of parts of the body. Pathological tremors result from disorders of the central nervous system and peripheral nervous system. In this paper we study the two most common pathological tremors, essential and Parkinsonian, using dynamical systems theory. We show that pathological tremors can be characterized as diffusional processes. The time scale range for the diffusional scaling law to be valid starts from about one to several tens of the mean oscillation period. This time scale range contrasts sharply with the predictable time scale for deterministic chaos, which is usually only a small fraction of the mean oscillation period. The mechanism for pathological tremors is also discussed, based on fractal and bifurcation theory.

P15. DERIVATION OF REDUCED ORDER MODELS FOR FLUIDIZED BEDS

Kichol Lee, Antonio Palacios: San Diego State University

Numerical simulations of transport phenomena in fluidized beds are carried out to investigate the complex interaction between gas and solid particles, and to explore the validity of a reduced order model based on the Proper Orthogonal Decomposition. The transport phenomena in the fluidized bed are modeled by conservation of mass, momentum, energy and species equations. In order to get insight, we conduct simulations of gas-solid interactions in a two-dimensional bed filled with sand particles and with a central nozzle to inject the gas. Aided by the Proper Orthogonal Decomposition, spatial dominant features are identified and separated from the spatio-temporal dynamics of the simulations. The numerical results indicate that a significant portion of the gas-solid interaction is confined to a central channel structure. The flow within this structure is successfully captured by a few POD eigenfunctions. This result supports the validity of a reduced order model for fluidized beds, which can then be constructed by projecting the governing equations onto the POD modes, as it is commonly done in the Galerkin method.

P16. ANTI-BUBBLES

Alberto Tufaile and José Carlos Sartorelli
Universidade de São Paulo

An anti-bubble is a striking kind of bubble in liquid that seemingly does not comply the buoyancy, and after few minutes it disappears suddenly inside the liquid. Different from a simple air bubble that rises directly to the liquid surface, an anti-bubble wanders around in the fluid due to its slightly lesser density than the surrounding liquid. In spite of this odd behavior, an anti-bubble can be understood as the opposite of a conventional soap bubble in air, which is a shell of liquid surrounding air, and an anti-bubble is a shell of air surrounding a drop of the liquid inside the liquid.

Two-phase flow has been a subject of interest due to its relevance to process equipment for contacting gases and liquids applied in industry. A chain of bubbles rising in a liquid formed from a nozzle is a two-phase flow, and there are certain conditions in which spherical air shells, called anti-bubbles, are produced. The purpose of this work is mainly to note the existence of anti-bubbling regime as a sequel of a bubbling system. We initially have presented the experimental apparatus. After this we have described the evolution of the bubbling regimes, and emulated the effect of bubbling coalescence with simple maps. Then is shown the inverted dripping as a consequence of the bubble coalescence, and finally the conditions for anti-bubble formation.

P17. NONEXTENSIVITY IN TURBULENCE IN A RAPIDLY ROTATING FLUID*

Sunghwan Jung, Julien Aubert and Harry L. Swinney
Center for Nonlinear Dynamics and the Department of Physics
The University of Texas at Austin, USA.

We study turbulence experiments on a rotating fluid with a zonal flow produced from small-scale eddies, a common feature of many planetary systems. Using enstrophyenergy conservation in the Tsallis nonextensive statistical mechanics, we deduce a potential vorticity profile in accord with the observations. A nonextensive formalism is appropriate because long-range forces between small-scale vorticies give rise to coherent structures in the turbulence. We find that the nonextensive Tsallis statistics leads to predictions that are in good agreement with the observations at large scales.

• This research was supported by the Office of Naval Research.

P18. PLANCK'S NATURAL UNITS ARE CONTAINED IN A TIME SERIES

Claudia Lainscsek and Irina Gorodnitsky
Cognitive Science Department; University of California at San Diego
9500 Gilman Drive, La Jolla, CA 92093-0515, USA

In 1899 Max Planck introduced natural units, also known as Planck quantities [1]. He described these units as intrinsic to the physical world, existing outside our immediate context: 'These necessarily retain their meaning for all times and for all civilizations, even extraterrestrial and non-human ones, and can therefore be designated as "natural units" ...' Equations describing physical processes in natural units take on a form distinct from descriptions that use measuring units associated with a sensor or an instrument. Equations in natural units are known to be more practical and simpler, containing fewer terms than equations expressed in any other units. Well known examples found in the literature include Maxwell's equations, theories of quantum gravity, and string theory. The duality of the mathematical descriptions for a given physical phenomena is relevant to modelers and experimentalists interested in modeling the underlying dynamics from experimentally derived data. All physical measurements are made, by necessity, in some measurement units. For a given vector of observations, the units determine the scale in which the amplitude is measured and the offset from the zero line of this scale. What Planck's theory tells us is that the models obtained from such experimental observations are likely to be unnecessarily more complex than the descriptions that can be obtained in natural units. An important question to entertain here is 'How much more complex the models derived from measured data may be?' More specifically, could our inability to model many dynamical processes from observed time series be related to the measuring scale in which the observations are taken?

In this paper we show that the difference in complexity between two mathematical descriptions, one in natural and the other in arbitrary measurement units, can vary dramatically and, in fact, can impact our ability to recover a model for a given process. We then show that observations of a single coordinate of a dynamical system contain information about Planck's natural units, and we show how this information can be extracted, leading to recovery of simpler models. The points of the paper are illustrated with the help of the Rössler system modeled from its x-component. The differential embedding of this system computed from the x-component has a complex description containing rational polynomials. We show that a simple linear transformation of the xvariable, actually a shift of the time series by a constant, leads to the recovery of a simple differential model containing only monomials. From this differential model we reconstruct the original Rössler system using a procedure demonstrated in the paper. We demonstrate that the simple model is induced only by one particular constant shift of the x-variable and by no other transformation of the x-variable. Thus the model we obtain is a unique simple representation corresponding to the very special, 'natural' units associated with the underlying physical system. Parallel conclusions can be drawn regarding the time scale of the underlying dynamics of a process which we also discuss in the paper.

[1] M. Planck, Sitzungsberichte der Preußischen Akademie der Wissenschaften 5, 479 (1899)

P19.   THE NONLINEAR DYNAMICS OF PACEMAKER DEPENDENCY

Leah Buechley, Department of Computer Science, University of Colorado, Boulder, Colorado, USA

A person is considered pacemaker dependent when most of his or her heartbeats are supplied by a pacemaker. The purpose of this study was to determine whether there are significant differences between the heart dynamics of pacemaker-dependent patients and those of normal patients. Nonlinear dynamics techniques and statistical methods were used to analyze the ECGs of normal patients and pacemaker-dependent patients. Standard embedding of the ECG data yielded inconclusive results, but embedding the beat intervals proved to be much more useful. Lyapunov exponent calculations, recurrence plot analyses, and standard statistical analyses of these data showed significant differences in heart behavior between the two groups of patients. In particular, the beat intervals appear to exhibit chaotic behavior for the normal patients and fixed-point dynamics for pacemaker-dependent patients.

P20.   THE NONLINEAR DYNAMICS OF PROTEIN FOLDING

Elizabeth K. White, University of Colorado at Boulder

Protein folding is one of the most researched, and least tractable, problems in computational biology. Many folding studies search for a protein's lowest-energy conformation. Since these techniques disregard the kinetics of protein folding, they are inherently equilibrium-based. A few researchers are beginning to use nonlinear dynamics to represent a folding protein as a multidimensional pendulum. This approach employs molecular dynamics simulations to model the folding process as a trajectory through the system's state space. At high temperatures, where proteins do not fold, these trajectories present compelling evidence of chaos, including a positive Lyapunov exponent, a broadband power spectrum, and a noninteger Hausdorff dimension for the strange attractor. Decreasing the temperature to a level conducive to folding causes a bifurcation, after which the protein's trajectory exhibits transient chaos, but eventually damps down to a fixed point corresponding to the folded state. The transient chaos model is consistent with the current view of protein folding, in which local parts of the protein fold, in random order, and then interact to propagate formation of the correct structure in the remaining parts. In this research, we explore this bifurcation and the associated basins of attraction by perturbing the conformations of a correctly folded protein, launching molecular dynamics simulations for these conformations at various temperatures, and applying standard nonlinear dynamics techniques to the resulting trajectories.

P21.   A VERY SIMPLE AND FAST MEASURE OF SYNCHRONIZATION AND DELAY BETWEEN SIGNALS.

R. Quian Quiroga [12], T. Kreuz [23] and P. Grassberger [2].
[1] Div. of Biology, Caltech, USA.
[2] Research Centre Juelich, Germany.
[3] Department of Epileptology, University of Bonn, Germany.

We propose a simple method to measure synchronization and time delay patterns between signals. It is based on the relative timings of events in the time series, defined e.g. as local maxima. The degree of synchronization is obtained from the number of quasi-simultaneous appearances of events, and the delay is calculated from the

precedence of events in one signal with respect to the other. Moreover, we can easily visualize the time evolution of the delay and synchronization level with an excellent resolution.
We show the application of the algorithm to intracranial human EEG recordings containing seizure activity and we propose that the method might be useful for the detection of the epileptic foci. It can be easily extended to other types of data and it is very simple and fast, thus being suitable for on-line implementations.

P22. RANDOM NUMBER GENERATORS ON THE BASIS OF SYSTEMS WITH CHAOTIC BEHAVIOR

D. D. Vavriv
Kharkov National University of Radioelectronics, Ukraine

In this work, bit sequences produced by dynamical systems with chaotic behavior are studied from the point of view of their possible applications in cryptographic modules. Lorenz, driven pendulum, and Mackey-Glass equations as well as other classical systems have been used to produce initial chaotic signals. Bit sequences have been obtained by sampling the chaotic signals and generating zero or one depending on the signal value with respect to some threshold. After that, these sequences have been tasted for randomness according to the FIPS 140-1 standard. This standard includes the Monobit, Poker, Runs, and Long Run tests. A sequence is considered as a random one, if these four tests are passed. The regions in the control parameter space of the dynamical systems where random sequences arise have been found and studied. Effects of the sampling rate and the threshold on the randomness onset have been addressed. Possible implementations of such random generators for cryptographic applications have been also considered.

P23. USING CHAOTIC BIDIRECTIONAL TRANSMISSION FOR ASYMMETRIC PRIVATE / PUBLIC KEY ENCRYPTION

Roy Tenny(1,2), Lev S. Tsimring(1), Henry D.I.Abarbanel(1,3), Larry Larson(2)
(1) Institute for Nonlinear Science, University of California, San Diego,
La Jolla, CA 92093-0402
(2) Dept. of Electrical and Computer Engineering, University of California, San Diego,
La Jolla, CA 92093-0354
(3) Department of Physics Department and Marine Physical Laboratory (Scripps Institution of Oceanography),University of California, San Diego,
La Jolla, CA 92093-0402

In the last decade chaotic encryption schemes were developed only for private symmetric key encryption schemes. We propose the first method which based on chaotic dynamics for realizing asymmetric ``public key'' encryption. A high dimensional dissipative chaotic dynamical system is distributed between transmitter and receiver. The transmitter dynamics is public, and the receiver dynamics is hidden. A message is encoded by modulation of parameters of the transmitter and this results in a shift of the overall system attractor. An authorized receiver has full knowledge of the dynamics and therefore knows the state space locations of the attractors corresponding to the different values of the modulation parameters. Unauthorized recipients only know the transmitter part of the full dynamics, and the protocol of communication is chosen in such a way that the attractor can not be reconstructed based on the transmitted signals alone. We present an example using a coupled map lattice. Security can be maintained by changing secret dynamics of receiver frequently, starting each transmitted bit with random initial state, using high dimensional dynamics, and by keeping modulation amplitude as small as possible.

P25.   SEARCHING FOR Non-linearITIES IN NATURAL LANGUAGE

Kiril Ribarov and Otakar Smrz
Center for Computational Linguistics, Charles University, Prague, Czech Republic

Inspired by wide range of applicability of what is commonly referred to as chaos theories, we explore the nature of energy series of a signal of human speech in the light of nonlinear dynamics. Using the TISEAN software package, analyses on various recordings of the language energy series were carried out (single speaker – different speeches; single speech - different speakers; dialogues; talkshows). Also correlated to other tenths of experiments conveyed on different linguistic inputs as written and morphologically analyzed texts, the presented experiment outputs (up to our knowledge, similar experiments have not been performed yet) reveal the complex and tricky nature of the language and are in favor of certain linguistic hypotheses. However, without further research, they do not encourage us to make explicit claims about the language signal such as dimension estimations (although probably possible) or attractor reconstruction.
Our main considerations include: (a) a look into the stochastic nature of the language aiming towards reduction of the currently very large number of parameters present in language models based on Hidden Markov Models on language n-grams; (b) visualization of the behavior of the language and revelation of what could possibly be behind the 'noisy' stream of sounds/letters/word-classes observed in our experiments; and last but not least (c) presentation of a new type of signal to the community exploring natural non-linear phenomena.

P26. MAPPING THE COMPLEX DYNAMICS OF A SEMICONDUCTOR LASER SUBJECT TO OPTICAL INJECTION

T.B. Simpson,[1] S. Wieczorek,[2] B. Krauskopf,[3] and D. Lenstra[2]
[1]Jaycor/Titan, P.O. Box 85154, San Diego, CA 92186-5154
[2]Dept. of Physics and Astronomy , Vrije Universiteit, Amsterdam, De Boelelaan 1081, 1081 HV Amsterdam, The Netherlands
[3]Dept. of Engineering Mathematics, Univ. of Bristol, Bristol BS8 ITR, United Kingdom

Single-mode semiconductor lasers can exhibit stable, bi-stable, periodic, quasi-periodic, and chaotic output characteristics when subjected to monochromatic, near-resonant external optical injection. Experimental measurements of the spectral characteristics of a semiconductor laser under external optical injection identify the various nonlinear output characteristics. Single-mode operation is maintained in all cases. Optical spectra of the laser are augmented by power spectra and spectra of the regenerative amplification of a weak optical probe. The output characteristics are mapped as a function of the strength of the optical injection and the detuning between the injection frequency and the frequency of the unperturbed laser. A comparison of the observed features with the predictions of a rate-equation model shows quantitative agreement over a wide range of injection characteristics [1]. All key model parameters can be determined experimentally. The agreement permits identification of specific routes to chaos and sudden chaotic transitions that are observed in the spectra. This demonstrates the power of dynamical systems modeling for the quantitative prediction of nonlinear dynamics and chaos of semiconductor lasers.

[1] S. Wieczorek, et al., *Phys. Rev. E* **65,** 045207(R) (2002).

P27. NONLINEAR DELAYED DIFFERENTIAL DYNAMICS FOR ENCRYPTION USING CHAOS.

Laurent Larger and Jean-Pierre Goedgebuer and Min Won Lee
Lab. d'Optique P.M. Duffieux, UMR CNRS 6603, Univ. de Franche-Comté
16, route de Gray, 25030 Besançon Cedex – France –

Nonlinear time-delayed differential dynamics are knowing an increasing interest, especially in the area of encryption using chaos. Such dynamics are also met in many other fields, such as mechanics, biology, medicine and optics. In the frame of high dimensional chaotic encryption systems, we have explored several nonlinear oscillators in optics and electronics ruled by nonlinear delayed (or difference) differential equations. After a presentation of the general architecture of such systems, we describe four different experimental set-ups, which are operating respectively with the wavelength of a tunable laser diode, the optical intensity at the output of an integrated electro-optic Mach-Zehnder, the optical path-difference in a coherence modulation scheme, and the electronic frequency at the output of a voltage-controlled oscillator. Numerical bifurcation diagrams are compared with experimental ones, and various dynamical properties are discussed, such as entropy, Lyapunov dimension, time behavior statistics, and spectral properties. Recent developments are also discussed in the view of improving the performances of chaos generators in encryption systems.

P28. SPACE-TIME CHAOS CHARACTERIZATION IN A NON-LINEAR OPTICAL FEEDBACK LOOP

L. Pastur*, U. Bortolozzo, P. Ramazza, F. T. Arecchi
National Institute of Applied Optics, Largo E. Fermi 6, 50125 Florence, Italy

We present an experimental setup that provides the possibility to easily explore different kind of space-time chaos. This allows to extract and compare the relevant features of each of them, so as to further develop efficient adapted control strategies. The system is an optical feedback loop closed through a Kerr-like non-linear medium (liquid crystal light valve). Beyond a critical value of the input light intensity (forcing), transverse instabilities develop in the optical beam. It results in the formation of out-of-equilibrium patterns, whose properties depend on various parameters that can be varied (translated or rotated feedback; focusing or defocusing non-linearity; relative amount of diffraction and interferences, etc). Most of the time, the pattern at threshold is asymptotically stationary, after a diffusive transient. At higher forcing, patterns become non-stationary, and defects are usually created and annihilated continuously, giving rise to a space-time chaos regime. At even higher forcing, many lengthscales can be excited, and the regime becomes turbulent.

We present preliminary results on such a space-time chaos characterisation using different analyzing tools (correlation functions, spectral analysis, Karhunen-Loeve decomposition, wavelet decomposition, mutual information...), as well as the different control strategies that might be implemented in each case.

P29. MULTIFRACTALITY OF THE SATELLITE-DERIVED DEEP CONVECTIVE INDEX IN THE TROPICS

Wen-wen Tung [1] and J.B. Gao [2,3]
[1] Department of Atmospheric Sciences, University of California, Los Angeles, CA 90095
[2] Department of Electrical and Computer Engineering, University of Florida, Gainesville, FL 32611
[3] Electrical Engineering Department, UCLA
Email: qc@atmos.ucla.edu, jbgao@ee.ucla.edu

To study atmospheric convection and its possible roles in the interaction with larger-scale atmosphere, various kinds of data from devices such as radiosonde soundings, satellite imagery, and airborne Doppler radars have been measured. Among them, satellite imagery-deduced data has been wisely used as indexes for convection. In this paper, we carry out multifractal analysis of the "deep convective index" time series based on the equivalent black body temperature (I_{TBB}) derived from hourly geostationary satellite measurements from 1 November 1992 through 28 February 1993. We find that I_{TBB} data is consistent with the random multiplicative model. Since I_{TBB} strongly correlates with the rainfall during the same period, our result also suggests rainfall may also be consistent with the random multiplicative model. Earlier fractal and multifractal analysis of rainfall has been focused on the mature stage of a storm. The I_{TBB} data analyzed here, however, characterizes a collection of convection systems (i.e., storms) in a four-month period. Hence the multifractal variations in I_{TBB} reflect the intermittent occurrence and variation of convection (hence storm) systems in the tropical regions. Various implications of this finding are addressed, in particular, the correlation structure in the I_{TBB} and the implication of this correlation to the interpretation of some concepts and techniques in data analysis, such as signal-to-noise ratio and smoothing.

P30. SYNCHRONIZING THE INFORMATION CONTENT IN DISPARATE SYSTEMS BY CONTROLLING SYMBOLIC DYNAMICS

Ned J. Corron[1], Shawn D. Pethel[1], and Krishna Myneni[2]
[1] U. S. Army Aviation and Missile Command, Redstone Arsenal, AL 35898
[2] Science Applications International Corp., 6725 Odyssey Drive, Huntsville, AL 35806

We demonstrate an extension to the concept of generalized synchronization for coupling disparate chaotic systems, including maps and flows. This broader viewpoint takes multiple systems to be synchronized if their information content is equivalent, and we use symbolic dynamics to quantize and compare the information. We demonstrate generalized drive-response synchronization in an electronic circuit coupled symbolically to a chaotic map. In this experiment, the symbols emitted by a logistics map are encoded in the dynamics of a piecewise-linear electronic oscillator using dynamic limiting chaos control. To assure the response system has the capacity to contain the information generated by the drive, we use kneading theory to show that the topological entropy of the circuit exceeds the Shannon entropy of the logistics map for a relative ordering of the logistics map parameter and oscillator gain. When this condition is met, the symbols exhibited by the map and flow are identical, thereby indicating that the mutual information of the two systems is maximal and generalized synchronization is achieved.

P31. DIFFUSIONAL SCALING LAWS IN OSCILLATORY SYSTEMS WITH STOCHASTIC FORCING

J.B. Gao [1,2] and Wen-wen Tung [3]
[1] Department of Electrical and Computer Engineering, University of Florida, Gainesville, FL 32611
[2] Electrical Engineering Department, UCLA
[3] Department of Atmospheric Sciences, University of California, Los Angeles, CA 90095
Email: jbgao@ee.ucla.edu, qc@atmos.ucla.edu

Rhythmic motions are ubiquitous in nature and in man-made systems, such as in low Reynolds number wake flows, breathing, and pathological tremors including essential and Parkinsonian. Due to the presence of external noise, those sinusoidal movements are typically only approximately rhythmic, or aperiodic, thus may be interpreted as chaotic.

Although existing tests developed for the analysis of chaotic systems may be able to tell some qualitative differences between these stochastically driven oscillatory motions from true chaotic motions, those differences are often not very instructive, because in the study of chaos, one often monitors the motion on fairly short time scales, to be consistent with the one of the key features of chaos---short-term predictability. In this paper, we report a diffusional scaling law for stochastically driven oscillatory motions. By studying a number of measured data such as the fluctuating velocity signals in the near wake of a circular cylinder and pathological tremor data as well as numerically generated data, we shall show that the time scale range for the diffusional scaling law to be valid starts from about one to several tens of the mean oscillation period. Furthermore, we classify the diffusional oscillatory motions into three categories, depending on whether the diffusional exponent less than, equal to, or large than 0.5, and consider the mechanism for each category. It is found that the case with the exponent larger than 0.5 is an anomalous diffusion and is a pre-cursor for noise to induce chaos.

P32. STABILITY BOUNDS FOR SYNCHRONIZED CHAOS

Govindan Rangarajan, Department of Mathematics and Centre for Theoretical Studies, Indian Institute of Science, Bangalore 560 012, India
Mingzhou Ding, Centre for Complex Systems and Brain Sciences, Florida Atlantic University, Boca Raton, FL 33431, USA

Synchronization of coupled chaotic systems has found applications in a variety of fields including communications, optics, neural networks and geophysics. An essential prerequisite for these applications is to know the bounds on the coupling strengths so that the stability of the synchronous state is ensured. We consider the stability of synchronized chaos in coupled map lattices and in coupled ordinary differential equations. Applying the theory of Hermitian and positive semidefinite matrices we prove two results that give simple bounds on coupling strengths which ensure the stability of synchronized chaos. Previous results in this area involving particular coupling schemes (e.g. global coupling and nearest neighbor diffusive coupling) are included as special cases of the present work.

P33. THE PENDULUM WEAVES ALL KNOTS AND LINKS

John Starrett, Departments of Physics and Mathematics
University of Colorado at Denver
3500 Clay Street, Denver, CO 80211 USA

From a topological point of view, periodic orbits of three dimensional dynamical systems are knots, that is, circles (S^1) embedded in the three sphere (S^3) or in R^3. The ensemble of periodic orbits comprising the skeleton of a 3-D strange attractor form a link: a collection of (not necessarily linked) knots. Joan Birman and Robert Williams used a topological device known as the template, a branched two-manifold that results when the stable direction is collapsed out of an attractor, to analyze the knot and link types appearing in the geometric Lorenz attractor. More recently, Robert Ghrist has shown the existence of universal templates: templates that support all knot and link types. I show that the template constructed from the geometric attractor of a forced physical pendulum contains a universal template as a subtemplate, and therefore the orbit set of the pendulum contains every knot and link type.

P34. BRAIN-WAVE DYNAMICS RELATED TO COGNITIVE TASKS AND NEUROFEEDBACK INFORMATION FLOW

Nada Pop-Jordanova*, Jordan Pop-Jordanov, Darko Dimitrovski, Natasa Markovska
*Department of Psychophysiology, Pediatric Clinic, Faculty of Medicine,
Vodnjanska 17, Skopje 1000, Macedonia
Research Center of Energy, Informatics and Materials, Macedonian Academy of Science and Arts, Krste Misirkov 2, P.O.Box 428, Skopje 1000, Macedonia

Synchronization of oscillating neuronal discharges has been recently correlated to the moment of perception and the ensuing motor response, with transition between these two cognitive acts "through cellular mechanisms that remain to be established"[1]. Last year, using genetic strategies, it was found that the switching off persistent electric activity in the brain blocks memory recall [2]. On the other hand, analyzing mental-neural information flow, the nobelist Eccles has formulated a fundamental hypotheses that mental events may change the probability of quantum vesicular emissions of transmitters analogously to probability functions of quantum mechanics [3]. Applying the advanced quantum modeling to molecular rotational states exposed to electric activity in brain cells, we found that the probability of transitions does not depend on the field amplitude, suggesting the electric field frequency as the possible information-bearing physical quantity [4]. In this paper, an attempt is made to inter-correlate the above results on frequency aspects of neural transitions induced by cognitive tasks. Furthermore, considering the consecutive steps of mental-neural information flow during the biofeedback training to normalize EEG frequencies, the rationales for neurofeedback efficiency have been deduced.

References:

[1] E. Rodriguez, N. George, J. Lachaux, J. Martinerie, B. Renault, F. Varela: Perception's shadow: long-distance synchronization of human brain activity, Nature, Vol. 397, 1999.
[2] J. Dubnau, L.Grady, T. Kitamoto, T. Tully: Disruption of neurotransmission in Drosophila mushroom body blocks retrieval but not acquisition of memory, Nature, Vol. 411, 2001.
[3] J. C. Eccles: Do mental events cause neural events analogously to the probability fields in quantum mechanics?, Proc. R. Soc. Lond, B227, 1986.
[4] D. Dimitrovski, J. Pop-Jordanov, N. Pop-Jordanova, E. Solov'ev: Quantum transitions between collective rotational states of water molecules in brain cells as a possible mechanism of memory imprinting, Contributions, Sec. Math. Tech. Sci., MANU, XIX, 1-2, 1998. Also: J. Pop-Jordanov., E. Solov'ev, N. Pop-Jordanova, N. Markovska, D. Dimitrovski: Exploring the quantum concept of memory, International Journal of Psychophysiology, Vol. 30, No 1-2, 1998.

P35. ESTIMATING GOOD DISCRETE PARTITIONS FROM OBSERVED DATA:
SYMBOLIC FALSE NEAREST NEIGHBORS

Matthew B. Kennel, Michael Buhl
Institute For Nonlinear Science
University of California, San Diego
La Jolla, CA 92093-0402

A symbolic analysis of observed time series data typically requires making a discrete partition of a continuous state space containing observations of the dynamics. A particular kind of partition, called "generating", preserves all dynamical information of a deterministic map in the symbolic representation, but such partitions are not obvious beyond one dimension, and existing methods to find them require significant knowledge of the dynamical evolution operator or the spectrum of unstable periodic orbits. We introduce a statistic and algorithm to refine empirical partitions for symbolic state reconstruction. This method optimizes an essential property of a generating partition: avoiding topological degeneracies. It requires only the observed time series and is sensible even in the presence of noise when no truly generating partition is possible. Because of its resemblance to a geometrical statistic frequently used for reconstructing valid time-delay embeddings, we call the algorithm "symbolic false nearest neighbors".

P36. CHAOS-INDUCED RANDOMNESS AND STOCHASTIC RESONANCE IN NONLINEAR CIRCUITS.

Jorge A. González and José J. Suárez.
Centro de Física, Instituto Venezolano de Investigaciones Científicas. Apartado Postal 21827, Caracas 1020-A, Venezuela.

Stochastic Resonance (SR) is a well known process whereby the transmission of a weak periodic signal through some nonlinear systems can be improved by adding a random perturbation (noise). Unlike the conventional SR, we produce this phenomenon by adding a chaotic perturbation to the system. We study chaotic functions as exact solutions to non-linear maps and obtain explicit mathematical expressions to generate the chaotic perturbation. The Lyapunov exponent of these functions can be calculated exactly. The signal to noise ratio (SNR) can be amplified for determined intervals of the Lyapunov exponent. This allows one to control the SR by tuning the level of chaos. We present an experimental evidence in a threshold-type electronic circuit. There exist a common belief that random sequences are produced from very complicated phenomena, making impossible the construction of accurate mathematical models. We show that under appropriated conditions our chaotic functions can be generalized to produce truly random sequences. We build electronic systems to simulate these functions and produce experimentally the so-called Deterministic Randomness (DR). The phenomenon is based in the transmission of deterministic signals through some nonlinear systems to generate stochastic dynamics. In particular we show the DR being produced in the coupling of a chaotic system, the Chua's circuit, with an electronic analogue of the Josephson junction.

[1] V. A. Makarov, E. del Río, W. Ebeling, and M. G. Velarde, Phys. Rev. E 64, 036601 (2001).

P37. EXPERIMENTAL INVESTIGATION ON THE EFFECT OF MESSAGE ENCODING IN CHAOTIC OPTICAL COMMUNICATION

Shuo Tang and Jia-ming Liu
Electrical Engineering Department, University of California, Los Angeles

Chaotic optical communication at high bit rates has attracted great attention because of its potential application in secure communications. In a chaotic optical communication system, a message can participate in the dynamics of the nonlinear system and further affect the characteristics of the chaotic carrier waveform. Message encoding can also change the synchronization quality if an encoding scheme breaks the symmetry between a transmitter and a receiver. In this presentation, we experimentally investigate the influence of message encoding on the characteristics of chaotic dynamics, chaos synchronization, and chaotic communication, using three encoding schemes, namely chaos shift keying, chaos masking, and chaos modulation, with the setup of a chaotic optical communication system with optoelectronic feedback. The chaos shift keying and chaos modulation schemes are found to be able to increase the complexity of the chaotic carrier waveform due to the random nature of the encoded message. Meanwhile, the chaos shift keying and chaos masking schemes are found to deteriorate the synchronization quality when the amplitude of the message is increased. Consequently, only the chaos modulation scheme has the desired feature in increasing the complexity of the chaotic carrier waveform while maintaining the quality of chaos synchronization, which is an important feature in secure communication.

P38. PHASE SYNCHRONIZATION IN POPULATIONS OF CHAOTIC ELECTROCHEMICAL OSCILLATORS

Istvan Z. Kiss, Yumei Zhai, and John L. Hudson
Department of Chemical Engineering, 102 Engineers' Way, University of Virginia, Charlottesville, Virginia 22904-4741

Experiments on the collective behavior and phase synchronization of weakly coupled populations of non-identical chaotic electrochemical oscillators are described. Without added coupling no deviation from the law of large numbers of the mean field is observed with increasing system size. Deviations do arise with very weak global or short-range coupling; large, irregular and periodic mean field oscillations occur along with (partial) phase synchronization of the elements. The transition to phase synchronization with increasing global coupling strength is analyzed using Kuramoto's order parameter and a comparison is made between the synchronization properties of periodic and phase coherent chaotic oscillators. Finally, the constructive effect of noise on the phase synchronization of the chaotic oscillators is demonstrated: at an optimal noise intensity an increased measure of phase synchronization is obtained.

P39. NONLINEAR DYNAMICS OF ADAPTIVE ARRAYS FOR IMPROVED INTERFERENCE SUPPRESSION IN TACTICAL APPLICATIONS

Rachel Goshorn
SPAWAR Systems Center, San Diego
San Diego, California

To date, the use of nonlinear dynamics of adaptive filter weights to provide improved spatial filtering has not been explored. Currently, interference suppression requires adaptation to a wide variety of signals, there may be multiple narrowband and broadband interfering signals from unknown directions. While adaptive arrays are used to control the spatial radiation pattern of antennas (typically by positioning nulls in the directions of interference sources), the methods are often thwarted by non-stationary interference signals. In order to improve the current generation of adaptive arrays, one must now exploit the dynamic behavior of these arrays.

Through the years, adaptive filters have been modeled to attain the classic optimal digital signal processing (DSP) Wiener filter performance, which has recently been shown to be a non-optimal approach. In various applications, the weights of adaptive filters behave nonlinearly; they perform better than the optimal Wiener filter. Understanding this dynamic behavior allows for control of nonlinear effects to improve performance. Recent research shows optimal filter parameters (i.e. time constant, filter order) are significantly different than those predicted from classic adaptive filter theory. For example, classic analysis generally uses a small filter time constant to maximize performance. However, such small time constants suppress the weight dynamics, degrading the performance in important applications. Therefore, a new set of tools is required in the DSP community to understand the dynamics and improve performance. The nonlinear physics community has developed a number of powerful tools (e.g. phase models, bifurcation theory, order parameter equations, and delay-embedding methods) to directly analyze the dynamics of nonlinear systems. These methods, along with digital signal processing analysis, will be applied to the adaptive array problem for performance improvements in spatial filtering.

P40. A CMOS COUPLED NONLINEAR OSCILLATOR ARRAY

Joseph D. Neff
SPAWAR Systems Center, San Diego
San Diego, California

This poster details an experimental nonlinear beam-forming array fabricated in a CMOS process. The unit cell oscillator is a nonlinear second order circuit, which demonstrates self-sustaining oscillation. In this poster experimental results from a test oscillator and a linearly coupled array of oscillators are reported. The circuit equations of motion are shown to be equivalent to the van der Pol oscillator, from which a weakly nonlinear phase-amplitude model is derived. This model forms a basis of understanding for the experimental nonlinear beam-forming array.

P41. INTERCONNECTED RESONANT VIBRATORY GYROSCOPES

David Fogliatti
SPAWAR Systems Center, San Diego
San Diego, California

A brief review of the operation of vibratory gyroscopes and the status of commercial micromachined gyroscopes is presented in the introduction. Following the introduction, an alternative design approach employing multiple, coupled gyroscopes per angular axis is examined as a method to improve performance and redundancy in angular rate sensors. The results are from simulations and explore the effects of non-identical gyroscopes, variations in the driving frequency, and coupling on synchronization in the array. A novel operating approach is presented that requires the array of gyroscopes to be synchronized in phase and frequency to improve the detection of sense axis displacement and to utilize one amplitude demodulator for the entire array of gyroscopes.

P42. APPLIED NONLINEAR DYNAMICS: ADVANCES IN ANTENNA TECHNOLOGY

Ted Heath§, Brian Meadows _, Joseph Neff _, Visarath In_, David Fogliatti_, Paul Hasler*, Stephen Deweerth* and William Ditto*
§ Georgia Tech Research Institute, Atlanta, Georgia
_ SPAWAR Systems Center, San Diego, California
* Georgia Institute of Technology, Atlanta, Georgia

As bandwidth requirements and operating frequencies increase, the fundamental limitations of solid-state devices become significant. These limitations include a considerable decrease in power-combining efficiency due to increased ohmic losses at these higher frequencies and maximum output powers proportional to $_2f-$ owing to size reductions necessary to diminish capacitive and time-delay effects. To address these issues, the coherent addition of the output from arrays of individual solid-state elements has been proposed. Realization of this solution has relied on exploiting the dynamics of coupled, nonlinear oscillators. Viewing these devices as dynamical systems, antenna technology stands to benefit from the leveraging of decades of research in nonlinear dynamics, allowing for novel approaches towards beam forming and signal processing. The focus of this presentation will be a review of the development of the underlying theory of coupled oscillator synchronization and its relevance to antenna design. In addition, current efforts, both theoretical and experimental, in this field by the authors will be presented as well as directions for future research.

P43. MULTI-FREQUENCY OSCILLATIONS IN COUPLED VAN DER POL OSCILLATORS

Patrick Longhini[a], Visarath In[b], Antonio Palacios[a].
[a]San Diego State University, San Diego, California
[b]SPAWAR Systems Center, San Diego, California

Numerical simulations of two symmetrically coupled arrays of N Van Der Pol oscillators are conducted. Under certain conditions on the type of connectivity, ie., which oscillators communicate with each other, and strength of coupling, it is shown that symmetry can force one array to oscillate in a synchronous mode (same waveform and same phase) while the second array exhibits a traveling wave mode, in which the oscillations have the same waveform but they are out-of-phase by a constant amount. More importantly, it is shown that the synchronous mode is forced to oscillate at N times the frequency of the traveling wave mode.

P44. TRANSMISSION OF INFORMATION WITH CHAOTIC SIGNALS

A. Argüelles[1] and H. Estrada[2]
[1]Grupo de Caos y Complejidad
[2]Grupo de Modelación Matemática
Departamento de Física, Universidad Nacional de Colombia, Colombia.

We present a numerical simulation for a transmission of information by applying the synchronization phenomenon[1] of two identical chaotic Lorenz's systems. This system offers the possibility to encrypt information[2]. The numerical method yields to an excellent recovering of voice signals. We also analyze the effect over the recovered signal due to relative variations of the system value of the parameters.

[1] L. M. Pecora and T. L. Carroll, *Phys. Rev. A* **vol 44 N 4,** 2374(1991).
[2] K. M. Cuomo and A. V. Oppenheim, *Phys. Rev. Lett.* **71,** 65 (1993).

P45. SSA, RESONANCES AND CHARACTERIZATION

Ian A. Pawson
Department of Mathematics, Imperial College, London, UK.

Singular Systems Analysis, (SSA), was introduced as a variation on the basic attractor reconstruction technique of delay space embedding, by Broomhead and King, (in 1986). Embedding allows the estimation of a wide range of dynamical invariants, such as the fractal dimensions and the Lyapunov exponents, which characterize attractors. A different application of SSA was introduced by Vautard and Ghil, who used it to extract dominant oscillations in paleoclimatic time series.

An important characteristic of an attractor is the set of eigenvalues of the Perron-Frobenius operator, (PF). This operator describes the mixing properties of the dynamical system. The determination of this set has proved difficult in practice, and has only been achieved for a limited number of systems.

Using the techniques introduced by Vautard and Ghil the relationship between the PF eigenvalues and the singular vectors that result from applying SSA is investigated. We then extend this to show how stack spectra can be used to determine the power associated with the PF eigenvalues. These techniques are applied to both real data (electronic circuits) and synthetic data (1D chaotic map exhibiting intermittency).

P46. UNCOVERING CHARACTERISTIC QUANTITIES FROM CHAOTIC TIME SERIES DISTORTED BY DYNAMICAL NOISE

Achim Kittel[1] and Tobias Letz[1,2]
[1]University of Oldenburg, Faculty of Physics, Department of Energy and Semiconductor Research, Germany
[2]Max-Planck-Institute for the Physics of Complex Systems, Dresden, Germany

Noise in experimental time series plays a crucial role for the analysis of the data. In general one can distinguish between two fundamentally different types of noise in the data: measurement noise, which is passively added to the data during the measurement process, and dynamical noise, which interacts with the system dynamics. In both cases results from time series analysis of the raw data are often questionable especially if one is interested to characterize the underlying deterministic dynamic with chaotic signatures. Here we show the application of a method for extracting the deterministic dynamics from data distorted by dynamical noise. The underlying idea of the method is a Fokker-Planck-Analysis of the data. The investigations are performed from the viewpoint of an experimentalist, i.e., there is only one system variable or a linear combination of a few variables measurable as it is usual for an experimental situation. We show the possibility of uncovering characteristic quantities like dimensions, Lyapunov exponents and power spectrum by the Method. First, the method is demonstrated on numerically integrated data of a non-linear dynamical system perturbed by dynamical noise. In a second step we apply the method to data gained from a laser experiment with an unknown deterministic dynamics.

P47. COMPETITION OF HEXAGONS AND TRAVELING ROLLS IN BINARY FLUID CONVECTION*.

R. Sokolov and C. M. Surko
Department of Physics, University of California, San Diego
La Jolla CA 92093

We describe of experiments to study the patterns in binary fluid convection in ethanolwater mixtures in a large aspect ratio container in the range of parameters in which both the Soret and non-Boussinesq effects are important. When non-Boussinesq effects are negligible, convection in these mixtures close to onset occurs in the form of traveling waves. For pure liquids the dependence of fluid properties on temperature is known to give rise to hexagonal convection patterns. In the range of parameters where both effects are important in the mixtures studied here, we have observed coexistence of traveling waves and hexagons (i.e., at values of separation ratio, Ψ~-0.3). For larger values of |Ψ|, the hexagons become unstable and a transition to chaotic convection is observed.

The patterns and associated bifurcation diagram for these three regimes will be presented, together with the results of other analysis tools.

- This work is supported by the DOE, Division of Basic Energy Sciences.

P48. COMPLETE REPLACEMENT OF CHAOTIC UNCERTAINTY WITH TRANSMITTED INFORMATION

Matthew B. Kennel
Institute for Nonlinear Science,
University of California, San Diego, CA
Shawn D. Pethel
U.S. Army Aviation and Missile Command
Redstone Arsenal, AL

It is now well known that chaotic systems may be controlled with small perturbations to execute orbits yielding a specified symbolic itinerary as long as the grammatical rules of those transitions are allowed by the natural dynamical system. Here, we employ techniques taken from contemporary data compression technology (source modeling and arithmetic coding) but reverse their usual roles to create a channel coder tuned to the observed natural dynamics. With a universal compression technique, we estimate a variable depth Markov-chain model which faithfully approximates the observed symbolic dynamics of the uncontrolled electronic circuit source. Subsequently, we drive the experimental system to the itinerary generated from an arbitrary white binary stream (the message) encoded using the symbolic model and the arithmetic coder. The transmitter's orbits are indistinguishable in grammar and measure from the uncontrolled attractor, demonstrating chaotic stegeanography with no rate loss. All the information naturally generated from chaos has been replaced by message bits at the same rate. Transmission on other chaotic saddles are accessible as well: the measure on the Markov chain may be modified arbitrarily as long as the topology is still respected. One particularly interesting solution, which is derived in explicit form, is that saddle which yields an entropy rate equal to the channel capacity, the upper bound for the given topology. Our algorithms are general, working as stated for arbitrary finite-depth grammars and alphabets, without analytical foreknowledge of the transmitter's equations of motion, though a good symbolic partition is necessary.

P49. PARAMETERS FOR THE ONSET OF PERIOD DOUBLING IN THE DIODE RESONATOR

T. L. Carroll and L. M. Pecora
Naval Research Lab

One of the classic problems in the study of nonlinear dynamics has been the diode resonator. Previous work with the diode resonator sought to explain the causes of period doubling and chaos, and often used simplified models. This paper instead seeks to link the onset of nonlinear dynamical effects to measurable parameters by comparing experiments and numerical models.

P50. LOW-ORDER BUBBLE MODEL FOR BUBBLING FLUIDIZED BEDS – A CASE STUDY OF OZONE DECOMPOSITION

Sreekanth Pannala & Stuart Daw
Oak Ridge National Laboratory
John Halow
National Energy Technology Laboratory
Presented by: Charles Finney, Oak Ridge National Laboratory

A dynamical model for bubble behavior in fluidized beds was developed by Daw and Halow [1] based on experimental observations of bubble interactions. The model is a set of nonlinear, ordinary differential equations, each equation representing the time rate of change of the vertical position for a given bubble. This model has been used for near realtime simulation of 3-D fluidized beds and studied extensively. This agent-based model has been extended for chemical reactions and ozone decomposition in fluidized beds is reported in this paper. The results from these simulations predict the conversion data of ozone decomposition (Fryer and Potter, 1976). This talk will also discuss the next steps and how this model can be extended to design and control fluidized beds.
References:
1. Daw, C. S. and J. S. Halow, "Modeling deterministic chaos in gas-fluidized beds," AIChE Symposium Series, No. 289, 61-69, 1992.

P51. FLAMEDOCTOR™: NONLINEAR BURNER DIAGNOSTIC SYSTEM

Ralph Bailey[1], Stuart Daw[2], Charles Finney[2] (Presenter), Tom Flynn[1] (Presenter), Tim Fuller[2]
[1]McDermott Technology, Inc., Alliance, Ohio
[2]Oak Ridge National Laboratory[2], Oak Ridge, Tennessee

Utility power plants are employing advanced control systems to improve performance over the load range. The performance of the boiler combustion system is critical to the overall performance. Flame Doctor™, which has been developed by McDermott Technology, Inc. and Oak Ridge National Laboratory under sponsorship of Electric Power Research Institute, performs diagnostics on an individual burner basis. The system

consists of analogue-to-digital signal conversion and conditioning hardware, analysis software, and a graphical user interface. Time varying voltage signals from all of the burner flame scanners on a boiler are analyzed simultaneously. Nonlinear techniques such as symbolization and time asymmetry along with linear techniques such as power spectral analysis are used. The nonlinear techniques discriminate stability features in the combustion dynamics not possible with the linear techniques alone. The assessments for a variety of flame conditions are collected in a reference library.
Libraries have been created for a number of flame scanners types. The Flame Doctor™ burner diagnostic system is described. Results from the first utility installation at Ameren UE Meramec power plant are shown. A live hook-up to the power plant is demonstrated. Flame Doctor™ is being offered commercially under alpha and beta demonstrations through the Electric Power Research Institute and Babcock & Wilcox.

P52. ARRAY ENHANCED STOCHASTIC RESONANCE IN A SPATIALLY EXTENDED NONLINEAR OPTICAL SYSTEM.

J. P. Sharpe, K. Carrigan, M. Swaney and N. Sungar
Dept. of Physics, Cal Poly State University, San Luis Obispo, CA 93407
email: jsharpe@calpoly.edu

We report what we believe to be the first experimental observation of array enhanced stochastic resonance in two dimensions. The experimental system comprises an amplitude-modulating liquid crystal light valve in a feedback loop to create a bistable optical system analogous to an array of coupled nonlinear oscillators. The system is driven with an external sub-threshold periodic optical signal and spatially varying additive noise. We find that the response of the optical system (defined as the signal-to-noise ratio of the output at the frequency of the periodic forcing) is improved with the addition of noise and with an increasing array size. This is in qualitative agreement with simple computer models of coupled nonlinear oscillators. We will describe the experimental arrangement and present results.
Work supported by Research Corporation grants CC4461 and CC5315

P53. HYPER-CHAOTIC SETS OF UNCOUPLED 3D OSCILLATORS AND THEIR SYNCHRONIZATION

A. Tamasevicius
Semiconductor Physics Institute, A. Gostauto 11, Vilnius LT-2600, Lithuania

A semi-systematic way of synthesizing higher-dimensional hyper-chaotic oscillators using low-dimensional simple chaotic units has been described in [1]. Hyperchaotic chains are composed of one-way coupled non-identical 3D chaotic oscillators. An alternative method makes use of the mean-field (each-to-each) coupled non-identical oscillators [2]. Both methods require a scalar signal for synchronization of the sets. The present paper suggests to build hyper-chaotic arrays of *uncoupled* 3D chaotic oscillators. Since the building blocks are not coupled the full set is an *apriori*

hyperchaotic one. However synchronization of such two arrays is not a straightforward procedure. To synchronize the arrays the local mean $<x>$ of the individual signals of the chaotic units x_i in the drive system $\mathbf{X}$ is transmitted to the response system $\mathbf{Y}$. In the latter the corresponding local mean $<y>$ is constructed. Then the global difference field $\Delta = <x> - <y>$ is applied to every individual 3D oscillator y_i in the response system. As a result the signal $<y>$ synchronizes to the received signal $<x>$. An interesting feature is that the synchronizable sets in some cases can be composed of *identical* units.
[2] L. Kocarev, U. Parlitz, Phys. Rev. Lett. **74**, 5028-5031 (1996).
A. Cenys, A. Tamasevicius, G. Mykolaitis, Proc. Int. Symp. on Nonlinear Theory and its Applications NOLTA'98, Crans-Montana, Switzerland, 1998, p.519-522.

P54. CONTROL OF CURRENT REVERSAL IN SINGLE AND MULTIPARTICLE INERTIA RATCHETS*

Fereydoon Family, H. A. Larrondo** and C. M. Arizmendi**
Department of Physics, Emory University, Atlanta, GA 30322, USA

We have studied the deterministic dynamics of underdamped single and multiparticle ratchets associated with current reversal, as a function of the amplitude of the external driving force. Two experimentally inspired methods are used. In the first method the same initial condition is used for each new value of the amplitude. In the second method the last position and velocity is used as the new initial condition when the amplitude is changed. The two methods are found to be complementary for control of current reversal, because the first one elucidates the existence of different attractors and gives information about their basins of attraction, while the second method, although history dependent shows the locking process. We show that control of current reversals in deterministic inertia ratchets is possible as a consequence of a locking process associated with different mean velocity attractors. An unlocking effect is produced when a chaos to order transition limits the control range.
* Research supported by a grant from the Office of Naval Research
** Present address: Depto. de Fisica, Facultad de Ingenieria, Universidad Nacional de Mar del Plata, Mar del Plata,, Argentina

P55. BEYOND THE FIRST RECURRENCE IN SCAR PHENOMENA

F. Borondo
Departamento de Química, Universidad Autónoma de Madrid. CANTOBLANCO, 28049 Madrid, SPAIN.
R. M. Benito
Departamento de Física y Mecánica, ETSI Agrónomos. Universidad Politécnica de Madrid, 28040 Madrid, SPAIN

A lot of effort has been devoted to the understanding of scars produced by short unstable periodic orbits in the eigenfunctions of classically chaotic systems, such as the stadium. This scarred wavefunction have been measured in microwave cavities as well as its

influence in magnitudes such as the magnetoconductivity of nanotechnology devices.
Much less is known, however, about what happens past this short--time limit.
In this communication we consider a numerical experiment in which the scarring effect
of longer periodic orbits coexisting in the same region of phase space is studied. It is
shown that the interplay among all of them has a clear manifestation in the intraband
structure of the corresponding spectra (or local density of states).

P56. VORTEX RIPPLES

Markus Abel
Universitat Potsdam, Institut fur Physik Germany

Vortex ripples can be seen in nature on the beach. Their evolution principles are
governed by the fluid-granular equations of motion, which are very complex. We present
a simple model which accumulates most of the physical principles in a nonlinear
interaction function between neighboring ripples. We present an accurate experimental
setup from which the interaction function can be inferred by nonparametric data analysis.